Library Use Only

Günter Kahl

The **Dictionary of Genomics,
Transcriptomics and Proteomics**

Related Titles

Singleton, P.

Dictionary of DNA and Genome Technology

2008
ISBN: 978-1-4051-5608-0

Dunn, M. J., Jorde, L. B., Little, P. F. R., Subramaniam, S. (eds.)

Encyclopedia of Genetics, Genomics, Proteomics and Bioinformatics

8 Volume Set

2006
ISBN: 978-0-470-84974-3

Kahl, G., Meksem, K. (eds.)

The Handbook of Plant Functional Genomics

Concepts and Protocols

2008
ISBN: 978-3-527-31885-8

Meksem, K., Kahl, G. (eds.)

The Handbook of Plant Genome Mapping

Genetic and Physical Mapping

402 pages with 77 figures and 11 tables
2005
Hardcover
ISBN: 978-3-527-31116-3

Günter Kahl

The **Dictionary of Genomics, Transcriptomics and Proteomics**

Fourth, Greatly Enlarged Edition

Volume 2: L–Q

WILEY-VCH Verlag GmbH & Co. KGaA

The Author

Günter Kahl
Johann Wolfgang Goethe-University
Molecular BioSciences, Biocenter
Max-von-Laue-Staße 9
60438 Frankfurt am Main
Germany

and

GenXPro GmbH
Frankfurt Innovation Center Biotechnology (FIZ)
Altenhöferallee 3
60438 Frankfurt am Main
Germany

Chinese Edition published in 2004

Cover Illustration
The Title Page shows a three-dimensional image of a nucleosome, in which the DNA double-helix is depicted as intertwined brownish ribbons (the phosphate backbone), from which the bases protrude (in green-blue colour). About 1.7 turns of DNA are wrapped around the histone core, where histone H2A comes in yellow, H2B in red, H3 in blue, and H4 in green colours. The H_2N-termini of the four histones emerge from the nucleosome as reels in the corresponding colour. The inset portrays the interaction between the phosphate backbone of the DNA with histone H3 K56 mediated by water molecules.

The graph was produced from pdb file 1KX5 with Pymol, and kindly provided by Heinz Neumann and Jason Chin, Division of Protein and Nucleic Acid Chemistry, Evolution and Synthesis of New Function, MRC Laboratory of Molecular Biology, Hills Road, Cambridge CB2 0QH, UK.

■ All books published by Wiley-VCH are carefully produced. Nevertheless, authors, editors, and publisher do not warrant the information contained in these books, including this book, to be free of errors. Readers are advised to keep in mind that statements, data, illustrations, procedural details or other items may inadvertently be inaccurate.

Library of Congress Card No.: applied for

British Library Cataloguing-in-Publication Data
A catalogue record for this book is available from the British Library.

Bibliographic information published by the Deutsche Nationalbibliothek
The Deutsche Nationalbibliothek lists this publication in the Deutsche Nationalbibliografie; detailed bibliographic data are available on the Internet at <http://dnb.d-nb.de>.

© 2009 WILEY-VCH Verlag GmbH & Co. KGaA, Weinheim

All rights reserved (including those of translation into other languages). No part of this book may be reproduced in any form – by photoprinting, microfilm, or any other means – nor transmitted or translated into a machine language without written permission from the publishers. Registered names, trademarks, etc. used in this book, even when not specifically marked as such, are not to be considered unprotected by law.

Composition SNP Best-set Typesetter Ltd., Hong Kong
Printing betz-druck GmbH, Darmstadt
Bookbinding Litges & Dopf GmbH, Heppenheim

Printed in the Federal Republic of Germany
Printed on acid-free paper

ISBN: 978-3-527-32073-8

It is a pleasure to dedicate these three volumes to my friends, colleagues, and foreign visitors of

 GmbH

in the Frankfurt Research Innovation Centre Biotechnology in Frankfurt am Main (Germany).

Preface

> *Η γνώση των λέξεων οδηγεί στην γνώση των πραγμάτων*
> *(The knowledge of words leads to a knowledge of things)*
> *(Πλάτων, Platon, 427–347 B.C.)*

The glamour and excitement of genetic engineering during the past three decades have given way to routine and almost trivial daily work in the laboratory. Many of the young researchers are nevertheless still fascinated by the precision of the various gene technologies and the surprising possibilities they offer, and even seasoned researchers are enchanted by this ever-growing field of molecular genetics. It is fair to say that gene technology has now infiltrated all areas of molecular biology, and massively contributed to the vast information in this field, that is accumulating at a more explosive rate than ever before. This phenomenal development forces to divide the field of gene technology into at least three subsections: **genomics, transcriptomics,** and **proteomics,** and this is exactly done in the present opus.

> *"This book contains a considerable volume of informations.*
> *I deeply regret this, but unfortunately it was inevitable."*
> *(Samuel Langhorne Clemens alias Mark Twain, 1835–1910)*

With the immense growth of the three, and other related areas of molecular biology, the number of novel technologies, procedures and technical terms is soaring. So, the present three volumes contain a total of 12,000 different terms, many of them describing recent developments and brand-new technologies. It is therefore the most comprehensive collection of descriptions of molecular processes and techniques worldwide. Some of the terms and their multiple variants dominate. For example, traditional PCR and its numerous facets comprises some 190 entries, surpassed by the terminology around microarray with more than 250 entries, and well over 100 "omics" neologisms mess up both literature and daily language. The second ("next"), and emerging third ("next-next") generation sequencing technologies brought a burst in novel terminology, not to speak of the many other cutting-edge techniques that appear almost daily.

> *"Make everything as simple as possible, but not simpler!"*
> *Albert Einstein (1879–1955)*

This flood of terms and acronyms sometimes leaves the researcher a bit helpless, especially since a single term might mean different things, and many different terms may mean the same single thing. The present volumes aim at ordering this chaos a bit, and are not restricted to the omics trilogy (or even gene technology), but link to other related disciplines and describe relevant terms, if considered to be necessary or helpful for a better understanding. Obviously, the growing number of proteins cannot be treated with in such a dictionary, especially since the peptides, proteins, and their isoforms will probably be in their millions. Therefore only a limited selection is portrayed. Another problem was, is, and will be, the extent of description. Some entries are described in some depth, some others only defined spartanically.

"Going too far is as bad as not going far enough"

(Chinese proverb)

My prime appreciations go to my son Uwe Kahl, who took the tantalizing task to draw a multitude of figures and schemes from partly fragmented and absolutely insufficient samples. He did a great job! I also thank Achim Wilz for his patient introduction into the various facets and pitfalls of the computer world. Sigrid – as always – gave me all her support and the freedom needed for such a work.

These three books are dedicated to all people of GenXPro GmbH in the Frankfurt Innovation Center Biotechnology (FIZ, Frankfurt am Main, Germany). Since excellent science is nowadays also daily work in companies, I learned a lot and hopefully could give a tiny bit to the energetic people of this dynamic enterprise. I am in fact grateful for all the discussions, scientific turns and innovations, and all the adventures that accompany a young spin-off company: a very rewarding experience.

I appreciate the hospitality of various institutions in different countries, where I have been working on this opus over the last years, as there is The Research Institute for Bioresources (Kurashiki, Japan), the Department of Biology and Molecular Biology (University of California at Los Angeles, USA), the International Center for Agricultural Research in the Dry Areas (Aleppo, Syria), the Centro Agronomico Tropical (Turrialba, Costa Rica), the Iwate Biotechnology Research Institute (Kitakami, Japan), and the Pharma Center (University of Vienna, Austria).

Frankfurt am Main, September 2008   Günter Kahl

Contents

Instructions for Users XI
A to Z 1–1828
Appendix 1 Units and Conversion Factors 1829
Appendix 2 Restriction Endonucleases 1834
Appendix 3 Databases 1841
Appendix 4 Scientific Journals 1851
Appendix 5 Acknowledgements 1862

The Dictionary of Genomics, Transcriptomics and Proteomics, 4th Ed., Günter Kahl
Copyright © 2009 WILEY-VCH Verlag GmbH & Co. KGaA, Weinheim
ISBN: 978-3-527-32073-8

Instructions for Users

- All the entries are arranged in strict alphabetical order, letter by letter. For example, "mismatch**ed** primer" precedes "mismatch **g**ene synthesis", and this is followed by "mismatch **r**epair". Or, "photo-**d**igoxygenin" precedes "photo-**f**ootprinting", which in turn precedes "photo-**r**eactivation". In case an entry starts with, or contains a Roman, Greek or Arabic numeral, it has first to be translated into Latin script. A few examples illustrate the translation:

cI	: c-**o**ne
exonuclease VII	: exonuclease **s**even
exonuclease III	: exonuclease **t**hree
5′	: **f**ive prime
G 418	: G **f**ourhundred and eighteen
λ	: **l**ambda
P1	: p-**o**ne
ΦX 174	: phi **X** one-seven-four
Qβ	: q-**b**eta
RP 4	: RP **f**our

For help, the user may consult the Greek alphabet and the Roman numerals below.

- The main entry title, printed in bold type, is followed by synonyms in parentheses. Italicized letters in titles (and text) of entries indicate use of these letters for abbreviations.

- Cross referencing is either indicated by an arrow, or the words "see", "see also", and "compare".

- By using the cross-references as a road map between definitions, the reader will gain an appreciation of molecular biology as an integrated whole rather than a collection of fragments of isolated information.

- Organismal name: The formal Latin binomial names of organisms are italicized, whereas common names and derivatives of the Latin names are not.

- Etymology of the terms: Most biological terms originate from Greek or Latin language. Only the most common word roots are defined in this dictionary.

The Dictionary of Genomics, Transcriptomics and Proteomics, 4th Ed., Günter Kahl
Copyright © 2009 WILEY-VCH Verlag GmbH & Co. KGaA, Weinheim
ISBN: 978-3-527-32073-8

Greek Alphabet and Roman Numerals

Greek alphabet:

Capital	Lower case	Name	Capital	Lower case	Name
A	α	alpha	N	ν	nu
B	β	beta	Ξ	ξ	xi
Γ	γ	gamma	O	o	omicron
Δ	δ, ∂	delta	Π	π	pi
E	ε	epsilon	P	ρ	rho
Z	ζ	zeta	Σ	σ, ς	sigma
H	η	eta	T	τ	tau
Θ	θ, ϑ	theta	Y	υ	ypsilon
I	ι	iota	Φ	φ	phi
K	κ	kappa	X	χ	chi
Λ	λ	lambda	Ψ	ψ	psi
M	μ	mu	Ω	ω	omega

Roman numerals:

I	II	III	IV	V	VI	VII	VIII	IX	X
1	2	3	4	5	6	7	8	9	10
XX	XXX	XL	L	LX	LXX	LXXX	XC	IC	C
20	30	40	50	60	70	80	90	99	100
CC	CCC	CD	D	DC	DCC	DCCC	CM	XM	M
200	300	400	500	600	700	800	900	990	1000

The Dictionary of Genomics, Transcriptomics and Proteomics, 4th Ed., Günter Kahl
Copyright © 2009 WILEY-VCH Verlag GmbH & Co. KGaA, Weinheim
ISBN: 978-3-527-32073-8

Abbreviations and Symbols

a	– atto (10^{-18})
A	– adenine or adenosine, absorbance
Å	– Ångstrom unit (1 Å = 0.1 nm)
~	– approximately
≅	– approximately equals
A/D	– analog-to-digital
aa	– amino acid
Ab	– antibody
Ag	– antigen
Ap	– ampicillin
ATP	– adenosine triphosphate
B	– any nucleo*base* (A,C,G,or T)
BAC	– bacterial artificial chromosome
Bis	– *N, N*′-methylenebisacrylamide
BLAST	– basic local alignment search tool
bp	– base pair(s)
Bq	– Becquerel
BSA	– bovine serum albumin
c	– centi (10^{-2})
C	– cytosine or cytidine
^{14}C	– radioactive carbon
°C	– centigrade (degrees Celsius)
Ca	– Calcium
CBB	– Coomassie Brilliant Blue
CCD	– charge-coupled device
cDNA	– complementary DNA
CE	– capillary electrophoresis
CGE	– capillary gel electrophoresis
Ci	– Curie
cm	– centimeter(s)
Cm	– chloramphenicol
CO_2	– carbon dioxide
cpm	– counts per minute
CTAB	– cetyltrimethylammonium bromide
Cy	– cyanine
D, Da	– Dalton
DAF	– DNA amplification fingerprinting
dATP	– deoxyadenosine triphosphate
dCTP	– deoxycytosine triphosphate

The Dictionary of Genomics, Transcriptomics and Proteomics, 4th Ed., Günter Kahl
Copyright © 2009 WILEY-VCH Verlag GmbH & Co. KGaA, Weinheim
ISBN: 978-3-527-32073-8

Abbreviations and Symbols

ddNTP	– 2′, 3′-dideoxynucleotide triphosphate
DGGE	– denaturing gradient gel electrophoresis
dGTP	– deoxyguanosine triphosphate
DMF	– N, N′-dimethylformamide
DMSO	– dimethyl sulfoxide
DMT, DMTr	– dimethyloxytrityl
DNA	– deoxyribonucleic acid
DNase	– deoxyribonuclease
dNTP	– deoxynucleotide triphosphate
ds	– double-stranded
dT	– deoxythymidine
DTT	– dithiothreitol, Cleland's reagent
dTTP	– deoxythymidine triphosphate
dUTP	– deoxyuridine triphosphate
EC	– enzyme classification number
ECL	– enhanced chemiluminescence
E. coli	–Escherichia coli
EDTA	– ethylenediaminetetraacetic acid
EGTA	– ethylene glycol-bis (β-aminoethylether) N,N,N′,N′-tetraacetic acid
e.g.	– for example
ELISA	– enzyme-linked immunosorbent assay
ESI	– electrospray ionization
ESI-MS	– electrospray ionization mass spectrometry
EST	– expressed sequence tag
EtBr	– ethidium bromide
EtOH	– ethanol
f	– femto (10^{-15})
Fab	– antigen-binding region of an antibody
FACS	– fluorescence-activated cell sorter
FIGE	– field inversion get electrophoresis
FITC	– fluorescein isothiocyanate
fmol	– femto mol
5′	– carbon atom 5 of deoxyribose
g	– gram(s) or gravity
G	– guanine or guanidine, giga (10^9)
Gb	– gigabase
GC	– gas chromatography
GFP	– green fluorescent protein
Gm	– gentamycin
GMO	– genetically modified organism
>	– greater than
h	– hour(s)
HAC	– human artificial chromosome
^{3}H	– tritium, radioactive hydrogen
HCl	– hydrochloric acid

HEPES	– N-(2–hydroxyethyl) piperazine-N'-(2–ethanesulfonic acid)
HGP	– human genome project
HIV	– human immunodeficiency virus
HPCE	– high-performance capillary electrophoresis
HPLC	– high pressure liquid chromatography
HRP	– horseradish peroxidase
HTE	– high Tris-EDTA buffer
H_2O	– water
H_2O_2	– hydrogen peroxide
HTML	– hypertext mark-up language
HVR	– hypervariable region
i.e.	– that is
IEF	– isoelectric focusing
Ig	– immunoglobulin
IP	– intellectual property
IVS	– intervening sequence, intron
k	– kilo (10^3)
kb	– kilobase(s)
KB	– kilobyte
kbp	– kilobase pairs
kD (kDa)	– kilo Dalton
kg	– kilogram(s)
Km	– kanamycin
l	– liter(s)
<	– less than
LC	– liquid chromatography
LiCl	– lithium chloride
LIF	– laser-induced fluorescence
LTE	– low Tris-EDTA buffer
mAb	– monoclonal antibody
MALDI-MS	– matrix-assisted laser desorption/ionization-mass-spectrometry
m	– meter(s) or milli (10^{-3})
μ	– micro (10^{-6})
μg	– microgram(s)
μl	– microliter(s)
M	– molar or mega (10^6)
Mb (Mbp)	– megabase pairs
MB	– megabyte
MCS	– multiple cloning site
Mg	– magnesium
mg	– milligram(s)
$MgCl_2$	– magnesium chloride
$MgSO_4$	– magnesium sulfate
min	– minute(s)
ml	– milliliter(s)

Abbreviations and Symbols

mm	– millimeter(s)
mM	– millimolar
mmol	– millimole
mol	– mole
M_r	– relative molecular mass (no dimension)
mRNA	– messenger RNA
MS	– mass spectrometry
MS/MS	– tandem mass spectrometry
mtDNA	– mitochondrial DNA
MW	– molecular weight
m/z	– mass-to-charge ratio
n	– number or nano (10^{-9})
NaCl	– sodium chloride
Na_2EDTA	– disodium-EDTA
NC	– nitrocellulose
ng	– nanogram(s)
NH_4Cl	– ammonium chloride
NH_4OAc	– ammonium acetate
nm	– nanometer(s)
NMR	– nuclear magnetic resonance
nt	– nucleotide
OD	– optical density
ODN	– oligodeoxynucleotide
OH	– hydroxy
oligo	– oligonucleotide(s)
ORF	– open reading frame
ORN	– oligoribonucleotide
P	– phosphorus
p	– pico (10^{-12})
P_i	– inorganic phosphorus
^{32}P	– radioactive phosphorus
PAGE	– polycrylamide gel electrophoresis
PBS	– phsphate buffered saline
PCR	– polymerase chain reaction
PEG	– polyethylene glycol
Petabyte (PB)	– 10^{15} bytes
PFGE	– pulsed field gel electrophoresis
pfu	– plaque forming unit
pg	– picogram(s)
pH	– logarithm of reciprocal of hydrogen (H) ion concentration
pI	– isoelectric point
PMS	– phenazine methosulfate
PMSF	– phenylmethylsulfonyl fluoride
PNA	– peptide nucleic acid
pp	– page(s)

ppm	– parts per million
PSD	– post-source decay
PTFE	– polytetrafluoroethylene
PVDF	– polyvinylidene difluoride
PVP	– polyvinyl pyrolidone
RAPD	– random amplified polymorphic DNA
RFL	– restriction fragment length
RFLP(s)	– restriction fragment length polymorphism(s)
RIA	– radioimmunoassay
RNA	– ribonucleic acid
RNase	– ribonuclease
RP	– reversed phase
rpm	– revolutions per minute
rRNA	– ribosomal RNA
RT	– room temperature (also reverse transcriptase)
RT-PCR	– reverse transcriptase PCR
^{35}S	– radioactive sulfur
SAGE	– serial analysis of gene expression
SD	– standard deviation
SDS	– sodium dodecyl sulfate, lauryl sulfate
SE (SEM)	– standard error (standard error of the mean)
sec	– second(s)
Σ	– sum of
Sm	– streptomycin
S/N	– signal-to-noise ratio
SNP	– single nucleotide polymorphism
ss	– single-stranded
SSC	– sodium chloride sodium citrate (saline sodium citrate)
SSCP	– single-strand conformation polymorphism
ssDNA	– single-stranded DNA
SSO	– sequence-specific oligonucleotide
SSP	– sequence-specific probe
SSPE	– sodium chloride-sodium phosphate-EDTA
STR	– short tandem repeat
T	– thymine or thymidine, tera (10^{12})
$\tau_{1/2}$	– half-life
TAE	– Tris-acetate-EDTA
TBE	– Tris-borate-EDTA
TBS	– Tris-buffered saline
Tc	– tetracycline
TCA	– trichloroacetic acid
TE	– Tris-EDTA-buffer
TEMED	– N, N, N′, N′-tetramethylethylene diamine
Terabyte (TB)	– 10^{12} bytes
3′	– carbon atom 3 of deoxyribose

TLC	– thin-layer chromatography
T_m	– melting temperature
TOF	– time of flight
Tp	– trimethoprim
Tris	– tris (hydroxymethyl) aminomethane
tRNA	– transfer RNA
U	– unit(s)
U	– uracil or uridine
URL	– uniform resource locator
UV	– ultraviolet
V	– voltage, volt(s)
VNTR	– variable number of tandem repeats
vol	– volume
v/v	– volume/volume
w/v	– weight/volume
www	– world wide web
$\bar{X}$	– mean
χ^2	– chi squared
YAC	– yeast artificial chromosome
yr	– year(s)

L

L:
a) See → linking number.
b) Abbreviation for the long arm of a → chromosome.

Label (tag): Any atom or chemical group introduced into a molecule for its identification.

Labeled compound
a) Any molecule that contains one or more radioactive atoms of the same or different kinds.
b) A molecule to which a non-radioactive → label has been attached. See → labeling.

Labeled oligonucleotide test (LO): A technique for the detection of non-specific endonuclease, 5′,3′-exonuclease or phosphatase contamination in → restriction enzyme preparations. LO involves the incubation of the restriction endonuclease with ^{32}P-labeled 17-mer single- or double-stranded oligonucleotide substrates that do not contain the corresponding restriction recognition site. The reaction products are then separated by → polyacrylamide gel electrophoresis and degradation products detected by → autoradiography or → phosphorimaging. If the test is negative (i.e. no degradataion products visible), the restriction enzyme is considered to be pure.

Label-free detection: The discovery of molecules or molecular interactions (as e.g. DNA-DNA-, DNA-RNA-, DNA-protein-, or protein-protein interactions) without the use of radioactive or fluorescent labels. For example, such interactions can be detected on e.g. → microcantilever surfaces as a bending of the cantilever due to a change in the surface stress after an interaction of a cantilever-bound target molecule and a ligand. This change in surface stress is then transformed into a change in the integrated piezoresistor, which can be easily monitored by a simple instrumentation. Label-free detection techniques circumvent problems with e.g. hazardous radioactive substances or the light-induced bleaching of → fluorochromes.

Labeling: The introduction of radioactive or non-radioactive markers into DNA, RNA, or protein molecules. A great variety of techniques has been devised for this purpose, a number of which are described in this book in some detail. See → DNA labeling, → conjugated antibody, → gold labeling, → non-radioactive labeling, → psoralen labeling, → radioactive label.

Lab-on-a-chip (*micro total analysis system*, microTAS, mTAS): A microfabricated glass or plastic chip containing a network of interconnected microchannels that allows the handling of extremely small volumes and the analysis of minute amounts of DNA, RNA, or proteins in extremely short time periods. All the necessary experiments as e.g. sample application, reagent dispensing and mixing, incubation and reaction, electrophoretic separation (by electrokinetic forces generated between two pin electrodes), detection of e.g. fluorescence signals and data analysis can be fully automated, require only small amounts of reagents and extremely short time (e.g. separation of DNA fragments is complete in 1.5–2 minutes only). Moreover, the experimentor is only minimally exposed to potentially hazardous compounds and little waste is produced. See → automated lab-on-a-chip.

Lab-on-a-slide (m-slide, micro-slide, µ-slide):

a) A microfabricated glass or plastic chip containing a central chamber accommodating a → microarray with spotted oligonucleotides, DNAs, or → cDNAs and a microchannel, into which the hybridisation solution (including the target sequence) can be pipetted. After inserting the separately spotted microarray, the center chamber is closed tightly and the hybridisation reaction started with the injection of the target. Compare → lab-on-a-chip.

b) A microfabricated plastic slide with one or multiple cell chamber(s) each with a thin bottom for high resolution microscopy and a special coating (e.g. poly-L-lysine (PLL)/ poly-D-lysine (PDL) polymer as an adhesion substrate for cell cultures, collagen, or the glycoprotein fibronectin) that allows optimal growth of a variety of cells. The plastic material has only negligible autofluorescence, and therefore suits inverse fluorescence and confocal microscopic imaging of cells in real-time. This type of lab-on-a-slide comes in several variants, as e.g. the channel micro-slide that allows to expose adherent growing cells to hydrodynamic shearing.

lac I: See → *lac* repressor.

lac **operon (lactose operon, *lac*):** A 6 kb DNA segment of the *E. coli* chromosome containing the → operon for lactose uptake and lactose catabolism. The *lac* operon is organized into an → operator and the structural genes *Z* (coding for a → β-galactosidase, see → *lac Z* gene), *Y* (encoding a β-galactoside permease) and *A* (encoding a β-galactoside transacetylase). These genes are coordinately transcribed into a single → polycistronic mRNA molecule. Transcription is regulated by a → promoter 5′ upstream of the operator. If a → repressor protein, encoded by *lac* I (a gene located 5′ upstream of the promoter), is bound to its recognition sequence in the *lac* operator, it effectively blocks *lac* transcription. The block is released by allolactose (a side-product of the β-galactosidase reaction which yields glucose and galactose from lactose), which functions as an → inducer by binding to the → *lac* repressor and inducing conformational changes in it. The altered repressor has a greatly reduced affinity towards the operator and dissociates from it. This induces *lac* expression. The *lac* operon thus is an → inducible operon, whose expression is regulated by → negative gene control.

lac **repressor (lactose repressor, lactose repressor protein):** Any one of a family of about 20 homotetrameric acidic, allosteric protein (monomer: 37 kDa, tetramer: 152 kDa) of *E. coli*, encoded by gene *lac* I located 5′ upstream of the *lac* promoter. The repressor, in the absence of lactose, binds with high affinity to its recognition sequence in the *lac* → operator region, and effectively blocks *lac* expression by either preventing the binding of → RNA polymerase or the → elongation of the → transcript. If combined with lactose ("inducer") or isopropyl-β-D-thiogalactoside ("gratuitous inducer"), cAMP or CAP, transcription of the *lac* operon starts. See → *lac* operon.

Lactose (4-O-β-D-galactopyranosyl-β-D-glucose): A sugar component of milk (e.g. of humans, bovine).

Lactose operon: See → *lac* operon.

Lactose repressor: See → *lac* repressor.

***lac Z* gene:** A gene of the → *lac* operon of *E. coli* that encodes → β-galactosidase. This enzyme catalyzes the conversion of the disaccharide lactose into the monosaccharides galactose and glucose. The *lac Z* gene is constituent of various → cloning vectors and functions as a → reporter gene in transformation experiments.

Ladder: Any mixture of DNA (DNA ladder) or RNA fragments (RNA ladder) that cover a specific, usually broad molecular weight range, and allow the exact determination of the molecular weight of target fragments ("sizing"). See for example → kilobase ladder, also → marker.

Laemmli gel: A → polyacrylamide gel used for the separation of proteins differing in their molecular mass and charge that is composed of an upper stacking gel, in which the protein sample is concentrated, and a lower separating gel (running gel), in which the different proteins of the sample are separated from each other. This combination of stacking and running gel ensures separation of the molecules into sharp bands corresponding to their electrophoretic properties. Laemmli gels are used for → SDS polyacrylamide gel electrophoresis.

Lagging strand: The DNA strand that is discontinuously synthesized in a 5' to 3' direction away from the → replication fork during DNA → replication. This strand contains the ligated → Okazaki fragments. Compare → leading strand.

Lag phase

a) The time period between a stimulus and the response to this stimulus.
b) An initial phase shortly after the inoculation of a bacterial starter culture into a fresh medium, during which the number of cells remains relatively constant, and after which the number of cells rapidly increases as a consequence of cell division.
c) The first of several growth phases of reactor-grown cell suspension cultures, during which the cells adapt to the new medium and the environment within the bioreactor.

LAM: See → *l*uminescence *a*mplifying *m*aterial.

Lambada phage: A highly infectious human virus originating from Colombia that rapidly spread in South-American countries, where it was mutated into a so-called passionate virus in Brazil. Through hightech communication it was disseminated throughout the Western hemisphere. The immediate symptoms of the infection include rhythmic convulsions of the hip and an almost total lack of contact inhibition. The virus still resists being designed as a cloning vector.

Lambda (λ)

a) Equivalent to microliter (μl).
b) Temperate phage of *E. coli*, see → lambda phage.

Lambda (λ) arms: Two regions of the → lambda phage genome that are generated by the enzymatic removal of the so-called → stuffer fragment, and carry genes for the synthesis of head and tail proteins (left arm), and for regulatory functions, as e.g. for host cell lysis (right arm). In cloning experiments, foreign DNA is ligated between both arms to form a packageable genome.

Lambda (λ) autocloning vector: See → autocloning vector.

Lambda (λ) EMBL vectors: See → EMBL vectors.

Lambda (λ) exonuclease: An enzyme catalyzing the removal of 5′ phosphomononucleotides from the 5′ termini of DNA duplex molecules. Blunt-ended DNA with a 5′ phosphate group is the preferred substrate.

Lambda (λ) gt vectors (generally transducing): A series of → insertion vectors designed for the cloning of cDNA, some of which are described here. They are listed in numerical, not alphabetical order:

a) λ gt 10: An → insertion vector of 43 kb that accomodates → cDNA fragments of up to 7.6 kb in length at a unique → cloning site within the *imm* 434 *cI* repressor gene (see → immunity region). Insertion of cDNA into the *cI* repressor gene inactivates it and causes the appearance of a turbid → plaque if the phage is grown on appropriate bacteria. The plaques generated by λ gt 10 are very large, and thus ideal for screening with nucleic acid probes.

b) λ gt 11: An → expression vector (*lac 5 nin 5 cI* 857 S100) of 43.7 kb that accomodates → cDNA fragments of up to 8.3 kb in length at a unique *Eco* RI site 53 bp upstream of the *E. coli* → β-galactosidase (*lac Z*) → stop codon. Insertion of linkered cDNA into this site leads to → insertional inactivation of the β-galactosidase gene so that recombinants can be easily selected as white plaques on → X-gal and → IPTG. Any cDNA cloned into the *Eco* RI-site is transcribed and translated as a β-galactosidase → fusion protein in *E. coli hfl* A (high frequency of lysogeny) mutant strains. The lysogens produce immunologically detectable amounts of hybrid proteins that allow antibody screening. Derivatives of λ gt 11 are → λ gt 11 *Sfi-Not*, → λ gt 18 and → λ gt 22.

c) λ gt 11 *Sfi-Not*: A derivative of the conventional lambda gt11 vector that additionally carries *Not* I and *Sfi* I sites in close proximity to the *Eco* RI site within the *lac Z* gene. The *Sfi-Not* vector is designed for → forced cloning of cDNA exploiting the unique restriction sites in the cloning region and using a *Not* I primer-adaptor. Since all the cDNA molecules are cloned in the same orientation relative to the *lac Z* gene, the likelihood of in-frame expression of cDNA inserts as β-galactosidase fusion proteins is increased.

d) λ gt 18, λ gt 19: An → expression vector of 43 kb. The vector is a derivative of lambda gt 11, in which the two natural *Sal* I sites have been destroyed by *Sal* I digestion and ligation of the resulting l arms to an oligodeoxynucleotide preventing the reconstruction of the *Sal* I sites. This → insertion mutation causes a red⁻ gam⁻ phenotype. A → polylinker allows the cloning of insert DNA of up to 7.7 kb that can be expressed and detected by → nucleic acid hybridization or → immunological screening procedures. λ gt 19 differs from λ gt 18 in the orientation of the polylinker. A derivative of λ gt 18 is λ gt 20.

e) λ gt 20, λ gt 21: An → expression vector of 42.7 kb, derived from lambda gt 18 by elimination of the *Sac* I and *Xba* I → recognition sites in the → polylinker and the insertion of a synthetic → chi sequence. These manipulations deleted some 500 bp of the vector, so that its cloning capacity could be increased to accomodate 8.2 kb of foreign DNA. Any insert DNA can be expressed and detected by → nucleic acid hybridization or → immunological screening procedures. Lambda gt 21 differs from lambda gt 20 in the orientation of the polylinker.

f) λ gt 22, λ gt 23: An → expression vector of 43 kb for the → forced cloning of cDNA. This vector is derived from λ gt 11 by replacing the unique *Eco* RI site by an in-frame → polylinker with *Not* I, *Xba* I, *Sac* I, *Sal* I, and *Eco* RI recognition sites. The expression of an insert cloned into one of these sites leads to a β-galactosidase → fusion protein. For directional cloning an oligodeoxynucleotide → primer-adaptor with a *Not* I recognition site upstream of an oligo(dT)$_{15}$ stretch is used to prime → first strand synthesis. After → second strand synthesis the double-stranded cDNA is ligated to *Eco* RI → adaptors or → linkers and then digested with the appropriate restriction enzyme. Then the cDNAs can be ligated into λ gt 22

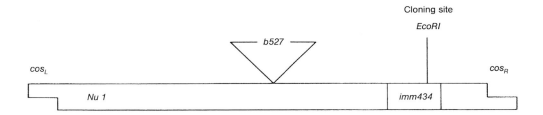

Simplified map of lambda gt 10

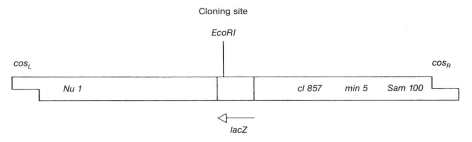

Simplified map of lambda gt 11

Lambda (λ) gt vectors (generally transducing)

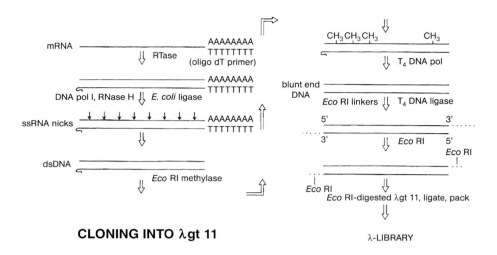

CLONING INTO λgt 11

λ-LIBRARY

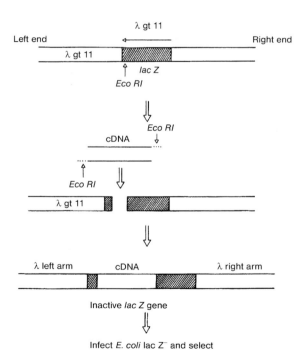

Insertional inactivation of β-galactosidase gene

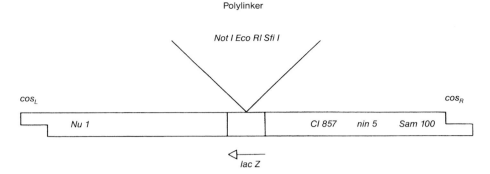

Simplified map of lambda gt 11 Sfi-Not

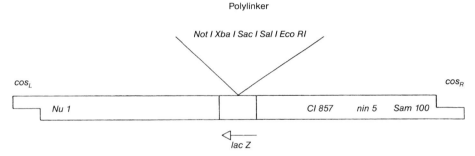

Simplified map of lambda gt 22

arms in the same orientation relative to the → lac Z gene. λ gt 23 differs from λ gt 22 in the orientation of the polylinker. A derivative of λ gt 22 is → λ gt 22A.

f) λ gt 22A: An → expression vector for the → forced cloning of cDNA. This vector is derived from → λ gt 22 by converting a Sac I restriction site within the → polylinker into a unique Spe I site.

h) λ gt WES: A series of replacement vectors that carry → amber mutations in their W, E and S genes and act as generally transducing phages for any foreign gene (generally, DNA) cloned into their central (replaceable) region. Compare → lambda WES. See also → lambda phage derived cloning vector.

Lambda (λ) ORF 8: A modified → lambda phage designed as a → cloning vector for the construction of → cDNA libraries. The 42.8 kb vector contains the → lac operon DNA, an → ampicillin resistance gene, and unique Eco RI, Bam HI and Hind III → recognition sequences in the 5′ lac Z coding region. Foreign DNA fragments of up to 9 kb in length can be directionally cloned (→ forced cloning) into the cloning site. Their expression results in lac Z → fusion proteins, which can be detected by → immunological screening.

Lambda (λ) WES: A modified → lambda phage that carries → amber mutations in its W, E and S genes, see also → lambda gt WES.

Lambda (λ) ZAP: A modified → lambda phage designed as an → insertion vector for the construction of → cDNA libraries. The 40.8 kb vector contains a → polylinker with six unique → recognition sequences (*Sac* I, *Not* I, *Xba* I, *Spe* I, *Eco* RI, *Xho* I). Foreign DNA fragments of up to 10 kb in length can be directionally cloned (→ forced cloning) into the cloning site, which is located in the C-terminal region of the *lac Z* gene, allowing easy selection with → IPTG and → X-gal ("blue-white selection"). Any insert DNA can be expressed as a β-galactosidase → fusion protein, so that lambda ZAP libraries can be screened with both → nucleic acid hybridization and → immunological screening procedures. In addition, phage → T3 and → T7 RNA polymerase promoters flank the polylinker. Thus strand-specific RNA transcripts can be generated in either direction (sense and anti-sense RNA, see → *in vitro* transcription).

The lambda ZAP vector combines the advantages of lambda phage cloning systems with the versatility of a → plasmid cloning vector, because the insert DNA can be excised from the phage vector with the aid of → f1 or → M13 → helper phages. Excision occurs at specific sequences (I: initiator; T: terminator) and leads to a → phagemid vector (→ Bluescript) that can be recircularized. Bluescript vectors allow the sequencing of insert DNA (or cDNA), the synthesis of RNA probes, the expression of the insert as a fusion protein and → site-directed mutagenesis. See also → lambda phage-derived cloning vector.

Lambda-mediated recombination: See → ET recombination.

Lambda phage (phage lambda, λ): A temperate → bacteriophage that infects *E. coli* (→ coliphage). Its linear double-stranded DNA genome of about 49 kb is packaged into an icosahedral head and contains two 12 bp complementary 5′ protrusions (→ cohesive ends, → *cos* sites) that allow the → circularization of the phage DNA after its injection into the host cell. Once inside the host cell, the l genome can enter either of two pathways of replication:

a) Lysogenic cycle. Lambda DNA replication is repressed. The circularized DNA can, however, integrate into the host chromosome at specific sites (l *att* site, see → attachment site). After its insertion the phage DNA (→ prophage) is

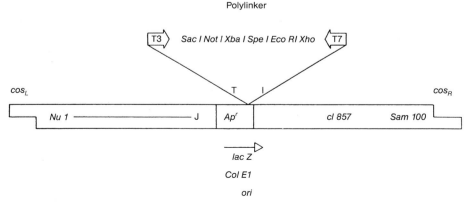

Simplified map of lambda ZAP

transcriptionally silent, but replicates in concert with the host chromosome. No phage progeny is produced. The prophage may, however, be activated (see → lysogeny) and then also enters the second pathway:

b) Lytic cycle. After the injection of the phage DNA into the host cell, it is replicated by a rolling-circle mechanism (→ rolling-circle replication) which generates long multimeric → concatemers whose monomers are linked by the annealing of their *cos* sites. This multimer is then cut down into its monomers by a terminase (ter protein, product of gene A) that recognizes the

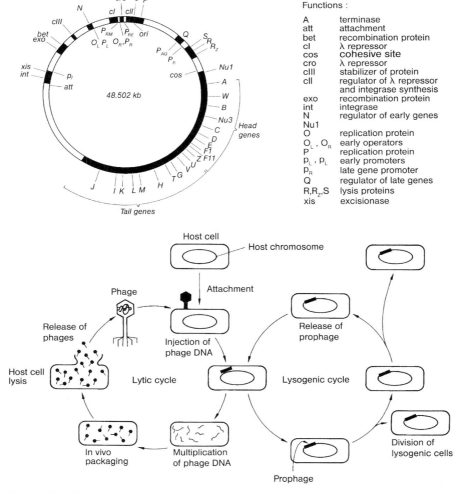

Lambda phage

cos sites, producing 12 bp long, sticky ends. The resulting monomers can only be packaged into the viral head if they are of correct size (44–52 kb, see → cloning capacity). The attachment of a tail to the head completes the production of infective phage particles.

See also → cI, → cro protein, → lambda arms, → lambda exonuclease, → lambda phage-derived cloning vector.

Lambda phage-derived cloning vector: Any one of a series of modified → lambda (l) phages functioning as → insertion or → replacement vectors in recombinant DNA experiments. Lambda-derived vectors described in some detail are → charon phages, → EMBL vector, → lambda gt vectors, → lambda ORF 8, → lambda ZAP.

Lambda terminase: An endonuclease from → lambda phage that catalyzes the formation of the → cohesive termini of the lambda DNA molecule and the → packaging of the molecule into the prohead. The 117,000 Da enzyme consists of two non-identical subunits (hetero-oligomer) encoded by gene gpA and gpNu1, respectively, and is used to cleave → cosmid DNA at the cohesive (cos) sites to generate 5′ ends with 12 bases single-stranded extensions. These sequences provide targets for the hybridization of synthetic oligodeoxynucleotides that allow partial restriction endonuclease digestion and specific end-labeling strategies to be used for the → restriction mapping of genomic DNA cloned in cosmids. See → cosmid insert restriction mapping.

Lambdoid phage: Any one of a group of temperate bacteriophages whose genomes can recombine with each other (e.g. phage l, F 80, P22).

Lamin: Any one of a family of *i*ntermediary *f*ilament *p*roteins (IFPs). The IFPs are either cytoplasmic (subclasses I–IV, VI) or nuclear (subclass V), and share common features, e.g. a central conserved α-helical domain (composed of four α-helical coils that are separated by short linker sequences). The class V lamins carry a C-terminal → *n*uclear *l*ocalization *s*equence (NLS), which together with the phosphorylation of some internal serine residues directs the nuclear transport of the molecule. The association of lamins with the inner nuclear membrane is mediated through isoprenylation, proteolytic trimming and methylation at H_2N-CaaX-COOH motifs (*c*: cysteine; a: *a*liphatic amino acid; X: any amino acid). The variable N-("head") and C-terminal ("tail") regions harbor serine residues for cdc2-kinase catalyzed phosphorylation that regulates the mitotic degradation and postmitotic synthesis of the nuclear envelope.

Vertebrates possess two types of lamina genes, the A and C families on one, and the B family (with B_1 and B_2) on the other hand. The lamin A gene type derived from lamin B type genes by → exon shuffling during vertebrate evolution, which also experienced an increase in the number of lamin genes generally, and in the number of splice variants specifically. Distinct lamin splice variants are expressed in meiotic and post-meiotic germ cells that are involved in the lamin assembly during chromatin remodelling in spermatogenesis and spermiogenesis. Expression of lamin genes A and C is developmentally regulated, the lamin B genes are constitutively active (therefore B type lamins are present in the nuclear envelope of any eukaryotic cell any time). *Xenopus laevis* has five, *Drosophila melanogaster* two lamin genes.

Lamins are principal components of the nuclear lamina that forms a fibrillar meshwork of intermediary filaments of 10–20 nm in diameter. This lamina attaches to the inner nuclear membrane, surrounding the karyoplasm, and anchors the → nuclear pore complexes, functions in the nuclear-cytoplasmic exchange of macromolecules, and stabilizes the interphase nucleus by coordinating the three-dimensional organization of the interphase chromatin. Hyperphosphorylation of lamins induces the decay of the nuclear lamina.

Lamin A gene (LMNA gene, «progeria» gene): A nuclear gene, located on human chromosome 1 and encoding the protein lamin A that is part of multi-protein complexes in the nuclear envelope controlling nuclear architecture and molecular trafficking from cytoplasm into the nucleus. A single base exchange in the LMNA gene activates a → cryptic splice site, leading to the skipping of a 150 bp sequence out of → pre-messenger RNA (pre-mRNA). This truncated mRNA is translated into a shorter protein ("progerin"). Progerin still attaches to the inner nuclear membrane, but cannot be detached, because a farnesyl group on the protein cannot be removed. Therefore progerin no longer functions correctly and leads to instability of the → nuclear envelope with dramatic consequences. The mutated protein causes the Hutchinson-Gilford Progeria Syndrome (HGPS), an extremely rare (1/4–8 millions of people), but fatal disease with many symptoms of severe premature aging (e.g. limited growth and weight [maximum: 16 kg], strokes, heart attacks). The disease equally affects both sexes and all races. Life expectancy of the afflicted individuals is about 13–14 years on average. Parents of HGPS patients do not carry the → mutation in their somatic cells, so that it probably occurred in a sperm cell. The mutation causes the disease in the → heterozygous state (one normal and one mutated LMNA gene). Farnesyl transferase inhibitors (FTIs) reverse the effects of the progeria lamin A mutation *in vitro*. Other mutations in the lamin A gene cause muscular dystrophy, dilatative cardiomyopathy, and familial partial lipodystrophy.

Laminar flow: A slow flow in which the momentum of transfer may be considered as taking place in (infinitely thin) plates (lamina) sliding relative to one another, creating a velocity gradient from center to edge. Under conditions of laminar flow, there is no radial convective mixing to counteract peak broadening due to velocity differences. Laminar flow is a major feature of most flow benches used for sterile work in gene technology laboratories.

Laminar *f*luid *d*iffusion *i*nterface (LFDI): The border line between two (or more) parallelly flowing individual streams of liquids in a single microfluidic channel (of e.g. a chip). Proteins or other biological molecules flowing in the LFDIs form a diffusion interaction zone, in which their local interactions can be studied. LFDIs are produced in the channels of a → lab-on-a-chip.

Laminopathy: Any one of (usually human) diseases caused by mutation(s) in either the → lamin A gene (phenotype: Hutchinson-Gilford progeria syndrome, HGPS), the gene coding for emerin (phenotype: Emery-Dreifuss type of muscular dystrophy), or genes encoding other *i*nner *n*uclear *m*embrane (INM) proteins such as the *l*amin *B* *r*eceptor (LBR; phenotype: Pelger-Huet anomaly). The molecular causes of the laminopathies are not clear, but a non-functional lamin protein A leads to a disturbance in the binding of constitutive and facultative → heterochromatin to the → nuclear lamina.

Lampbrush chromosome: A chromosome in the nucleus of primary oocytes from invertebrates and vertebrates, characterized by paired loops which extend laterally from the main axis of its chromomeres, and give it a lampbrush-like appearance. These loops may vary in size (from 1 to over 100 µm, corresponding to 3 to more than 300 kb) and shape from chromomere to chromomere. They are unwound and actively transcribed regions of the lampbrush chromosome.

Figure see page 835

LAM-PCR: See → linear amplification-mediated polymerase chain reaction.

Landmark map: Any → physical map of a → genome that contains → genetic markers dispersed at regular intervals. Such landmark maps are crucial for refined mapping, i.e. creating higher marker density throughout the map. Compare → comparative mapping, → gene map, → genome mapping, → physical map. See → map.

Lane: The part of an agarose or polyacrylamide gel, in which one single sample (protein, RNA, DNA) is running during electrophoresis. The samples are pipetted into small pockets ("slots") in the upper part of a gel and electrophoretically forced into the gel and towards the opposite electrode (in case of nucleic acids, towards the positive electrode).

Language gene ("speech gene"): A laboratory slang term for any gene that encodes a protein necessary for the development of a capacity for mouth and facial movements and the maintenance of neural circuitries related to speech. For example, the forkhead-domain gene *FOXP$_2$* is such a language gene, whose mutation leads to severe speech and language disorders. This gene encodes a 715 amino acids protein, which suffered only three amino acid exchanges since the diversion of human and mice some 70 million years ago. Only two mutations occurred in the human lineage since humans and chimpanzees diverged roughly 6 million years ago.

Lantibiotic (*lan*thionine-containing anti*biotic* peptide): Any one of a group of amphiphilic polycyclic anti-microbial peptides produced by Gram-positive bacteria (e.g. *Staphylococcus epidermidis*) that contains the unusual dehydroamino acids dehydroalanine and dehydrobutyrine and the thioether amino acids lanthionine and 3-methyl-lanthionine. Lantibiotics are derived from ribosomally synthesized precursor peptides ("prepeptides") by extensive post-translational enzymatic modifications (e.g. dehydration by a series of enzymes collectively designated LanB, LanC, and LanM; addition of cystein-SH groups to form thioethers) and proteolytic processing (e.g. the removal of the N-terminal → leader sequence by serine proteases). The genes involved in lantibiotic synthesis are arranged in → gene clusters, which comprise genes encoding modification enzymes, proteases, transporters, regulatory proteins and peptides for self-protection ("immunity") of the host (e.g. the membrane-associated LanI, or ABC transporters as e.g. LanEFG). On the basis of structural and functional features, lantibiotics fall into two categories. Class A peptides are elongated and mostly act by the depolarization of cytoplasmic membranes and the transient formation of pores (example: nisin). To this class belong

Lampbrush chromosome

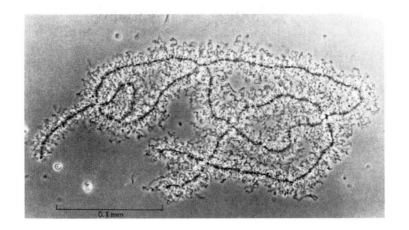

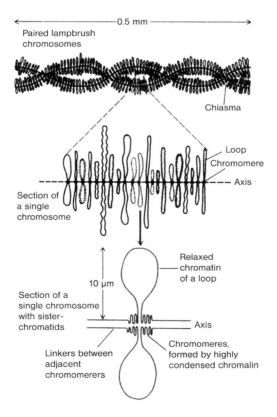

Lampbrush chromosome

carnocin, cytolysin L1 and L2, epidermin, epilancin K7, galldermin, lacticin 481, nisin A, nisin Z, pep5, salivaricin, streptococcin, subtilin, variacin. Class B peptides adopt a globular structure and inhibit the transglycosylation step in the peptidoglycan biosynthesis pathway (examples: actagardin, ancovenin, cinnamycin, duramycin, mersacidin). Genetic engineering of lantibiotic-encoding genes lead to optimized peptides (e.g. peptides with increased solubility and stability) that can be used as food preservatives. Compare → antibiotic, → bacteriocin.

LA-PCR

a) See → *l*igation-*a*nchored *p*olymerase *c*hain *r*eaction.
b) See → *l*inker-*a*daptor *p*olymerase *c*hain *r*eaction.
c) See → long and accurate polymerase chain reaction.

LAR: See → *l*igation *a*mplification *r*eaction.

Large fragment: See → Klenow fragment.

Large-gel *two-d*imensional electrophoresis (large-gel 2-D electrophoresis): A variant of the conventional → two-dimensional gel electrophoresis, which uses large gels (i.e. a 40 cm long capillary tube for the → isoelectric focusing in the first dimension, and 40×30 cm gels for the → denaturing gel electrophoresis in the second dimension. Large gels own higher resolution power and allow to separate up to 30,000 proteins.

Large-scale *c*opy variation (LSC, large-scale *c*opy *n*umber variation, LCV, large-scale *c*opy *n*umber *p*olymorphism, lsCNP): Any DNA → polymorphism between two (or more) individuals that comprises hundreds of thousands of base pairs (in humans >100 kb). Originally, the term sequence polymorphism was reserved for smaller → insertions or → deletions (INDELs), or → transition-transversion-type → single nucleotide polymorphisms (SNPs). LSCs, on the contrary, represent large polymorphisms that represent genetic variations in populations, and may be diagnostic for a specific disease or sensitivity towards a drug in human beings. LSCs can be detected by e.g. → representational oligonucleotide microarray analysis. See → segmental aneuploidy. Compare → copy number polymorphism, → gene copy number polymorphism.

Large-scale duplication: The duplication of a whole → genome, a chromosome or a large chromosomal fragment in evolutionary times. Whole-genome *d*uplication (WGD) is a consequence of either → autopolyploidy (i.e. the doubling of every set of homologous chromosomes in a genome), or → allopolyploidy (the creation of a genome with doubled chromosome number through interspecific hybridization). Duplication of individual chromosomes (→ aneuploidy) leads to an abnormal chromosome number in a → karyotype (e.g. trisomy). The duplication of a chromosomal fragment occur through DNA → transposition or → translocation followed by meiosis. In → comparative mapping, such regional duplications manifest themselves as segmentss enriched for → paralogous pairs in genome self-comparisons. See → polyploidy.

Large single copy sequence: See → long single copy sequence.

Large-step chromosome walking: See → cosmid walking.

Large T (T): As opposed to "small t" (t), see → T antigen.

Large unilamellar vesicle (LUV): A lipid bilayer vesicle containing self-assembled supramolecular pores (SSPs) spanning its membrane that allows the diffusion of low molecular compounds as e.g. a → fluorochrome from inside to the surrounding medium exclusively through the SSPs. If, however, a second compound binds inside the pore, it suppresses the diffusion of the former substance (here: a fluorophore). SSP-LUVs can therefore be used to monitor enzymatic reactions. For example, LUVs containing self-quenching concentrations of a fluorochrome can be constructed. The fluorophore can only diffuse through the SSP channels. If, however, the SSPs are blocked by either a substrate or the product of an enzymatic reaction, the dye is not at all released or its rate of release is decreased. The enzyme activity is then measured as the ratio between fluorophore release before and after blockage. The SSP structure can be modified by e.g. p-octaphenyl group inncorporation. The eight phenyls in each octaphenyl chain will not form a planar structure, but every second one is oriented perpendicular to the plane of the others. If short peptides are attached to each phenyl group, then every second peptide is also oriented perpendicular to the other ones. The peptide chains interdigitate with chains from another stave, and this configuration leads to the formation of β-sheet structures and a barrel-shaped pore. The inner surface of the SSPs can be functionalized by different amino acids, which trap molecules that bind to these residues and thereby block the pore cavity. To monitor an enzymatic reaction, substrate and enzyme are incubated, and a fraction of the substrate will be converted to the product. Now, if the substrate binds to the interior of the SSPs of an LUV filled with a fluorochrome, the enzymatic reaction takes place within the pore. Since the fluorochromes within the LUV are highly concentrated, any fluorescent light is reabsorbed and not emitted. The dye can only escape from the inside of the LUV, where it is completely quenched, if the SSPs allow its diffusion. The conversion of the blocking substrate to a nonblocking product by the enzyme relaxes the pore, and fluorochrome diffuses through the pores. Generally, the number of SSPs per LUV determines the maximal diffusion rate, which is diminished prportional to the concentration and affinity of pore-blocking molecules in the analyte solution.

Lariat: The looped structure arising during the → splicing of PRe-mRNA, and consisting of → intron RNA. In a first step, pre-mRNA is cut at the junction of exon 1 and the intron. Simultaneously the lariat RNA containing the intron and exon 2 is shaped in consequence of the formation of a 2′, 5′-phosphodiester bond between the 2′-OH group of the last A in the conserved → TACTAAC box upstream of the intron's 3′ end and the pG at the intron's 5′ end. In a second step, the intron-exon 2 junction is cut, and exon 1 and exon 2 are ligated to yield the mature mRNA. This endonucleolytic cut releases intron RNA in a lariat form.

Figure see page 838

Laser (light amplification by stimulated emission of radiation): An intense, monochromatic and collimated light beam that is used to excite → fluorochromes in various techniques of molecular biology (e.g. in → cytogenetics, → DNA chip technology, → flow cytometry, → labelling of e.g. → probes).

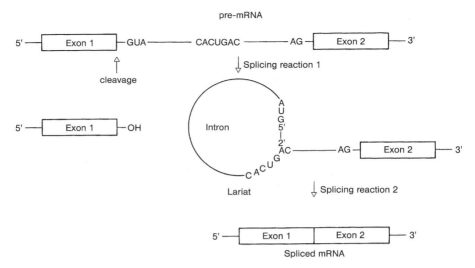

Lariat

Laser ablation: A technique for the controlled removal of material from a sheet of thermoplastic by a high-powered pulsed laser that allows to fabricate µm channels for disposable polymer microfluidic chips. Since most plastics absorb UV light, pulsed excimer UV lasers with pulse rates of 10–10^4 Hz are most frequently employed for laser ablation. More specifically, polystyrene, polycarbonate, cellulose acetate and polyethyleneterepthalate (PET) materials are ablated with ArF lasers (193 nm), polymethylmethacrylate (PMMA), polycarbonate, polyethyleneterepthalate glycol (PETG), polyvinylchloride (PVC) and polyimide with KrF excimer lasers (248 nm), and PMMA also with CO_2 lasers (10.6 µm). A lithographic metal mask allows to protect specific areas of the chip and to expose others for laser ablation. A pulse energy of some hundred mJ ablates ~0.5 µm of thermoplastic material. See → embossing, → soft lithography, → TPE molding.

Laser-capture microdissection (LCM): A technique for the isolation of specific intact cells from complex tissues under direct microscopic visualization. In short, a 5 µm tissue section is first placed on a microscope slide, and a transparent ethylene vinyl acetate (EVA) thermoplastic film placed onto the specimen. Then an infrared laser beam focused onto the target cell melts the film directly above it, such that the polymer expands and impregnates (embeds) the cells beneath it. After the cooled polymer solidified, it is lifted off the tissue section (leaving the unimpregnated tissue still attached to the microscope slide), taking the target cells with it. These cells can then be analyzed for their → proteome or → transcriptome, or other parameters.

Laser-induced decomposition (LID): The fragmentation of a target molecule (e.g. a protein) by a laser beam. LID is part of the

→ matrix-assisted laser desorption-ionization mass spectrometry. See → collision-induced decomposition.

Laser-induced fluorescence (LIF): The excitation of a fluorochrome by laser light at the wavelength (in nm) of its maximal absorbance.

Laser microbeam irradiation ("laser micropuncture", laserporation): A technique to produce submicrometer holes in the membranes of human, animal or plant cells, using a highly focused laser microbeam in order to facilitate → direct gene transfer. Proposed to be particularly suited for plant cells which could be transformed without removal of the rigid cellulose cell wall.

Laser microdissection (LMD): A technique for the isolation of specific chromosomes or cells from tissue sections or also culture dishes. The mounted tissue is moved either manually or robotically around a stationary laser that cuts the target cell from the surrounding neighbors. The excised cell is then trapped in various ways (see → laser-capture microdissection, → laser pressure catapulting). LMD can also be used to destroy a particular cell amidst the surrounding tissue ("negative selection").

Laserporation: See → laser microbeam irradiation.

Laser pressure catapulting (LPC): A technique for the isolation of a microdissected cell from the surrounding cells that uses the energy pulse of e.g. a UV-A laser beam focussed below the specimen by the objective of an appropriate microscope to catapult the excised cell (or cells) into a microfuge tube cap against gravity. The cells usually survive this drastic treatment, and can then be processed further (for e.g. chromosome isolation). For this purpose, the cells are centrifuged and can be recultivated. Sources for the cells could either be old, archived material, cytospins, chromosomal probes or cells from a cell culture. See → laser microdissection, → laser capture microdissection.

Laser scanning: The screening for fluorescently labeled molecules and their detection in an → agarose or → polyacrylamide gel, in which they were separated by → gel electrophoresis. Laser scanners are composed of a photomultiplier tube, a laser source and a *c*harge-*c*oupled *d*evice (CCD) camera. Commercially available light sources are diode lasers (635 nm), argon-ion lasers (488 and 514 nm), helium-neon lasers (633 nm), neodymium-yttrium-aluminium garet lasers (532 nm) and so-called *s*econd *h*armonic *g*eneration (SHG) lasers (473 and 532 nm). Two (or more) such sources can be combined such that different fluorophores can simultaneously be detected (i.e. different fluorochromes can be used to label biomolecules, e.g. proteins). The scanners serially pass a beam of coherent light over each point of the gel in a raster pattern, driven by mechanical or optical devices. The laser excites the fluorophore in the gel, the emitted fluorescence signal is collected by two optical fiber bundles, and distinct fluorescent signals are separated by interference filters and then converted to electrical signals by dual photomultiplier tubes. These electrical signals are in turn transformed into images for data analysis by analog-to-digital converters.

Last update: The most recent date, at which new, complementary or corrective information(s) for a given sequence or genetic locus were introduced into a database.

Last common ancestor: A single genome (or organism), from which all contemporary genomes (or organisms) in a particular group are descended (derived).

Late embryogenesis abundant (LEA): A family of proteins in prokaryotes, invertebrates and plants that is part of the → heat-stable proteome. For example, LEA proteins are major components of plant seeds, and are associated with *d*esiccation *t*olerance (DT) by disintegration of the hydration shell and the fast removal of bulk water from other proteins, thereby preserving the functional state of the dehydrated proteins. Genes encoding LEAs are expressed to high levels during dehydration of e.g. plant tissues, and are inactivated during re-hydration.

Late gene: Any gene that is expressed only late in the life cycle of a virus or cell. If, for example, a → simian virus 40 particle infects permissive cells, the viral DNA is uncoated and transferred to the host cell nucleus. During the subsequent 4 hours (early phase), the activation of → early genes leads to the synthesis of early proteins (e.g. large T and small t proteins). Then the expression of late genes starts and extends over a period of 36 hours (late phase) during which the viral proteins VP1, VP2, and VP3 are produced.

Latent splice site: Any nucleic acid sequence conforming to either a canonical 5′ or 3′ → splice site, but normally not used for → pre-messenger RNA → splicing. Such latent splice sites are abundant in the human genome, particularly in → introns of protein-coding genes.

Latent virus: Any viral genome that is integrated into a host's genome, but not expressed. It may be activated by certain stress factors (e.g. ultraviolet irradiation). Activation leads to the synthesis of infective virus particles.

LATE-PCR: See → linear-after-the-exponential polymerase chain reaction.

Lateral gene transfer: See → horizontal transmission.

Lateral genomics: The whole repertoire of techniques for the detection of → lateral gene transfer. This discipline will doubtless gain interest in future, since the various genome sequencing projects detect substantial amounts of DNA exchanged between different organisms. For example, about 20% of the genes of the *E. coli* K12 genome were introduced by lateral gene transfer in the past 100 million years, and about 25% of the genome of the hyperthermophilic bacterium *Thermotoga maritima* originates from archaeal hyperthermophilic organisms.

Latex agglutination: A procedure to detect specific antigen-antibody reactions in which an → antibody is either covalently bound or adsorbed to spherical polysterene (latex) beads. The antibody-containing particles are then mixed with → antigen. The agglutination can be visualized using a dark field microscope or quantified by turbidimetric or nephelometric measurement.

Figure see page 841

Laurell rocket technique: See → electroimmunoassay.

Lawn: See → bacterial lawn.

Layered expression scanning (LES): A technique for the molecular analysis of cells,

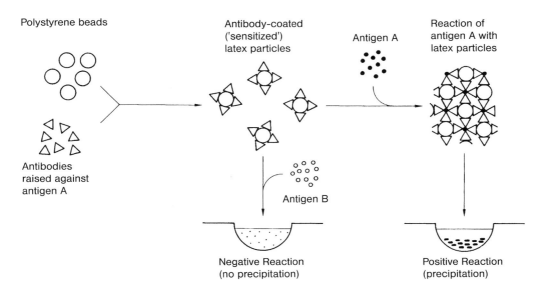

Latex agglutination

cell lysates, microdissected cells, or tissue sections that is based on the transfer of target molecules (e.g. specific proteins or RNAs) from these specimens through a series of membranes with immobilized antibodies (protein detection) or cDNAs (nucleic acid detection), the capture of the traversing target molecules, and their subsequent detection by either immunoblotting or hybridisation detection methods. In short, the biological sample (e.g. cell lysates, dissected cell populations, or tissue sections) is covered with stacks of nitrocellulose membranes or agarose gel layers, and the proteins or nucleic acids from the sample transferred through the membranes by capillary movement (compare → Southern blotting). Each of the membranes carries a specific antibody or other capture molecule. On their passage through the membranes (whose numbers can exceed 100, which means that at least 100 different target molecules can be tested), the various target molecules are captured. Then the membranes are separated from each other, and the captured molecule detected membrane by membrane. During the transfer process, the overall two-dimensional architecture of the samples is preserved, such that the different detected molecules can be localized to specific cell types in e.g. a tissue section, thereby producing a molecular profile of each cell type present in the specimen.

LB medium (Luria-Bertani medium): A rich growth medium for bacterial cultures, containing bacto-tryptone, bacto-yeast extract and NaCl.

LCM: See → laser-capture microdissection.

LCN: See → low copy number DNA sample.

LCR

a) See → ligase chain reaction.
b) See → locus control region.
c) See → low-copy repeats.

LCV: See → large-scale copy variation.

LD: See → linkage disequilibrium.

LD block: See → haplotype block.

L-DNA: See → locked nucleic acid.

LD-PCR: See → *long-d*istance *p*olymerase *c*hain *r*eaction.

LDR: See → ligation detection reaction.

LEA: See → late embryogenesis abundant.

Lead compound: Any (usually synthetic) compound that binds to and/or inhibits and/or activates a → validated target from a high-throughput screen. Lead compounds usually underwent a process of modification and re-testing called optimization, before a lead candidate is identified, which is selected on the basis of its toxicology and efficacy.

Leader peptide: See → signal peptide.

Leader sequence (leader):

a) The transcribed part of a eukaryotic gene that follows the → cap site and precedes the → start codon with the consensus sequence 5'-AGNN-3' (animals) or 5'-CGAANN-3' (higher plants).

b) The untranslated part of an mRNA molecule (also 5'-untranslated region, 5'-UTR, untranslated sequence) that extends from its 5'-terminus (cap site) to the translational start codon ATG (excluded). The leader may contain an → attenuator sequence, or → Shine-Dalgarno sequences. Leader sequence length is remarkably consistent from fungi and plants to invertebrates and vertebrates (including humans). It

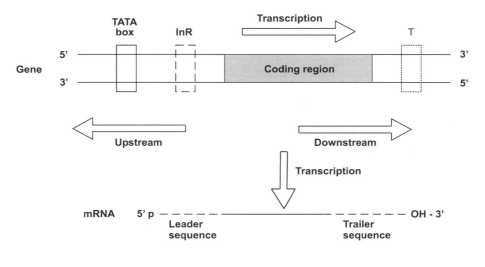

Leader sequence

spans 80 (plants), 100 (cold-blooded vertebrates), 120 (fungi, rodents) and 150 nucleotides (invertebrates, humans). Leader sequences comprise a short → open reading frame encoding a socalled leader peptide and a downstream region capable of forming → stem-loop secondary structures. The formation of a distinct stem-loop either affects the continuation of transcription (see → transcriptional attenuation) or the initiation of translation (see → translational attenuation). For example, the socalled *iron-responsive elements* (IREs) are such hairpin structures in the 5'- (also 3') UTR of various messenger RNAs coding for proteins involved in cellular iron metabolism. The IREs are recognized by trans-acting proteins ("*iron regulatory proteins*", IRPs) that control mRNA translation rate and stability. Two IRPs (IRP-1 andf IRP-2) bind to IREs and become inactivated (IRP-1) or degraded (IRP-2), when the iron content within the cell increases. See → trailer, → messenger RNA circularization.

c) The N-terminal amino acid sequence of secretory proteins (leader sequence peptide, leader peptide) that is cleaved off during or after the secretion process. See → signal peptide.

Leader sequence peptide: See → signal peptide.

Leading strand: The DNA strand that is continuously synthesized in a 5' to 3' direction towards the → replication fork during DNA → replication. Compare → lagging strand.

Leaf disk transformation: The introduction of foreign DNA into plant cells of a leaf disk via → *Agrobacterium*-mediated gene transfer. Such disks are aseptically cut from sterile leaves, and incubated with virulent → *Agrobacterium tumefaciens* cells that are induced to transfer part of their → Ti-plasmid together with cloned foreign DNA into wounded cells at the rim of the disk. Then the bacteria are killed by appropriate antibiotics and the transformed cells regenerated into transgenic plants. Regeneration of such cells is only possible with a very limited number of plants (e.g. tobacco, petunia, potato). Compare also → agroinfection.

Leaky mutant: A mutant carrying a → leaky mutation.

Leaky mutation: Any gene mutation that does not completely abolish gene function and allows the synthesis of a protein which still partly functions. See → leaky protein.

Leaky protein: Any protein, encoded by a mutated gene, but still possessing residual function(s). See → leaky mutation.

Leaky scanning: The movement of a → ribosome along a → messenger RNA molecule in search of an AUG → initiation codon such that it skips → upstream AUG codons and initiates → translation further downstream.

LEAPS: See → light-controlled electrokinetic assembly of particles near surfaces.

LED: See → localized expression domain.

Lederberg technique: See → replica-plating.

LEE: See → *linear expression element*.

Left-handed DNA: See → Z-DNA.

Left splice junction: See → donor splice junction.

Left splicing junction: See → donor splice junction.

Legitimate recombination: See → homologous recombination.

***Leishmania* gene expression vector:** A eukaryotic gene expression vector that combines the advantages of a bacterial cloning system (easy handling, generation of expression constructs) with the protein synthesis-folding-modification machinery of the eukaryotic trypanosomatid protozoon host *Leishmania tarentolae* (into which the construct is transferred by → electroporation). In short, a plasmid vector for cloning of an interesting insert contains → selectable marker gene(s) as e.g. → hygromycin B, → bleomycin, → neomycin or nourseothricin resistance genes, and a cloning site in addition to the normal vector sequences (as e.g. the → origin of replication from *E.coli*). The target gene is inserted into such a vector, the vector transfected into a *Leishmania* host strain, the transformants selected with the appropriate antibiotic(s), and the *Leishmania* cultivated. The expressed proteins are supposed to be folded correctly, and undergo mammalian-type → post-translational modifications such as e.g. → glycosylation, → prenylation, or phosphorylation, to name few. The protein of interest can then be extracted from *Leishmania*, purified, and characterized.

LEM domain: A 40 amino acids long domain of inner nuclear membrane and nucleoplasmic proteins. For example, *Caenorhabditis elegans* owns three LEM domain genes (*emr-1, lem-2* and *lem-3*, encoding Ce-emerin, Ce-MAN1 and LEM-3, respectively. Ce-emerin and Ce-MAN1 are the only integral membrane proteins containing a LEM domain, and both depend on Ce-lamin for their localization to the → nuclear envelope. Ce-emerin partly replaces Ce-MAN1, since downregulation of both proteins, or downregulation of Ce-MAN1 in emerin-null worms causes 100% lethality by the 100-cell stage.

Lentivirus-*i*nfected *c*ell *m*icroarray (LICM): Any → microarray (here: a coated glass slide), onto which nanoliter volumes of highly concentrated lentiviruses (1×10^9 IFU/ml) pseudotyped as vesicular stomatitis virus glycoprotein (that infect a wide variety of mammalian cells, also non-dividing cells with high efficiency) are deposited. Onto such slides, adherent mammalian cells (e.g. 2×10^6 HeLa cells) are added and cultured for 3–4 days. Cells landing on the lentivirus spots become infected, and each cluster of cells are transduced with a single type of lentivirus. Thereby a living array of stably transduced cell clusters within a monolayer of uninfected cells is formed. These clusters can then be fixed and processed for e.g. → immunofluorescence. The lentivirus can be engineered to encode → short hairpin RNA or → cDNA expression cassettes, so that the mammalian cells can be tested for specific functions. LICMs therefore serve the high-throughput screening of gene function in diverse mammalian cells.

Lesion-specific DNA repair protein: Any one of many nuclear proteins that specifically recognizes a particular primary lesion in DNA, binds there and initiates repair processes. For example, MutS proteins bind to mismatched bases, the Ku heterodimer to → double-strand breaks (DSBs), and the *Xeroderma pigmentosum* (XP) group C protein (XPC) involved in →

*n*ucleotide *e*xcision *r*epair (NER) is among several proteins selectively recognizing UV-induced DNA photoproducts.

Lethal allele (lethal gene): Any, usually heavily mutated gene, whose expression inevitably leads to the death of the carrier organism. See → lethal mutation.

Lethal mutation: Any → mutation that changes a normal gene to a gene encoding a faulty protein, which does not function and leads to the death of the carrier organism. See → lethal allele.

Letsinger-Caruthers solid phase oligonucleotide synthesis: A technique to synthesize oligonucleotides (e.g. → antisense oligonucleotides, or → ribozymes) that starts with the binding of a → nucleoside onto a solid phase (e.g. glass, plastic) and a series of activation, coupling, oxidation, and detritylation steps to generate oligonucleotides of up to 200 bases in length. See → chemical DNA synthesis, → solid phase cDNA synthesis.

LF: See → long form.

Letsinger-Caruthers solid phase oligonucleotide synthesis

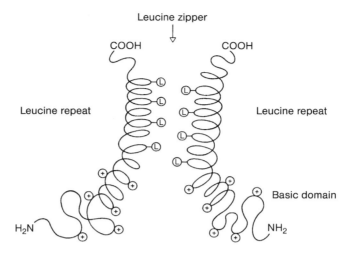

Leucine zipper: A region in a specific class of → DNA-binding proteins (e.g. transcriptional regulators such as yeast GCN4 and transforming proteins such as Myc and Fos) that spans about 30 amino acids with a periodic repeat of hydrophobic leucines every seven residues (heptad repeat). This region is thought to adopt an → α-helical structure. These α-helices from two such proteins interact through interdigitation of the leucine zippers so that protein dimers are formed. Dimerization results in the appropriate juxtaposition of basic amino acid regions of the protein dimer and the DNA-binding domain. At least some of the leucine zipper proteins bind DNA as a dimer. Compare → helix-loop-helix, → helix-turn-helix, → zinc finger protein.

Leucinopine: An amino acid derivative that is synthesized in plant cells transformed by the soil bacterium → *Agrobacterium tumefaciens*. Leucinopine belongs to the so-called → opines. See also → crown gall.

$$\begin{array}{c} \text{CH}_3 \\ | \\ \text{CH}-\text{CH}_2-\text{CH}-\text{COOH} \\ | \qquad\qquad\quad | \\ \text{CH}_3 \qquad\qquad \text{NH} \\ \qquad\qquad\qquad | \\ \text{HOOC}-(\text{CH}_2)_2-\text{CH}-\text{COOH} \end{array}$$

Leucinopine

Lex A two-hybrid system (Lex A interaction trap): A variant of the conventional → two-hybrid system, designed for the detection of protein-protein interaction(s) in *Saccharomyces cerevisiae* (especially of cytotoxic or cytostatic proteins). It uses the prokaryotic B42 acidic ("acid blob") transcription → *a*ctivation *d*omain (AD) and an *E. coli* Lex A → DNA-binding domain (DBD), Lex A operator sites upstream of the → reporter gene and the inducible *GAL1* promoter for tight control of → bait and prey expression. In essence, AD-fusion proteins are expressed only during the actual experiment such that they exert but limited cytotoxic effect(s) on the yeast host cell. Usually, only one or a few Lex A operator sites are needed (more such sites upstream of the reporter gene actually weaken the interaction(s) between the proteins and lead to increased numbers of false positives, i.e. activation of the reporter gene without concomitant protein-protein interaction). Various modifications of this system exist (e.g. the selectable marker could be *LEU2* or *HIS3*, the reporter gene either *lacZ* or the → green fluorescent protein). See → dual bait yeast two-hybrid system, → interaction mating, → interaction trap, → mammalian two-hybrid system, → one-hybrid system, → repressed transactivator (RTA) yeast two-hybrid screen, → reverse two-hybrid system, → RNA-protein hybrid system, → split-hybrid system, → split-ubiquitin membrane two-hybrid system, → three-hybrid system, → two-hybrid system.

Lexosome: A → nucleosome whose → histone core has been partly relaxed so that the whole structure becomes extended. This relaxation may be brought about by → histone acetylation (especially of histone H3 and H4) which introduces repulsion forces into the → core particle (negatively charged acetyl residues of histones – negatively charged phosphate groups of DNA). The nucleosome – lexosome transition is thought to be necessary for the movement of RNA polymerase molecules along the primary sequence of a gene.

L fragment: See → chromosome *l*inking clone library.

L gene: An operational term for any DNA sequence that show only *l*ow → homology to → genic sequences deposited in the data banks, and therefore does not resemble a gene. L genes own confidence scores of <50%. See → H gene, → M gene.

LGT agarose: See → low melting point agarose.

Library (bank): See → gene library.

Library amplification: The growth of host cells containing the recombinant → plasmids belong ing to a → gene library, or the development of the corresponding recombinant → phages within host cells with the concomitant multiplication of the foreign DNA inserted into these plasmids or phages, respectively.

LICM: See → lentivirus-infected cell microarray.

LIC-PCR: See → *l*igation-*i*ndependent *c*loning of PCR products.

LID

a) See → laser-induced decomposition.
b) See → lysis-inducing domain.

LIF: See → laser-induced fluorescence.

Life cycle: The development of an organism from fertilization to death.

Life vector (LV): Any recombinant living bacteria (also viruses) that specifically localize to, and colonize tumors, locally replicate and express specific genes (e.g. → reporter genes). LVs may be used as therapeutic agents. For example, specific recombinant vaccinia viruses colonized tumors in animals and led to their regression.

Lifting: See → plaque hybridization

Ligand: A molecule that binds to a specific complementary site of another molecule. For example, a substrate molecule whose three-dimensional configuration allows its binding to the catalytically active site of an enzyme molecule, is a ligand.

Ligandomics: A variant of → metabolomics that aims at isolating, characterizing and modifying the whole set of small molecular weight cellular compounds (ligands) interacting with macromolecules such as DNA, RNA, proteins, or peptides. See → kinomics.

Ligase: Synonym for → DNA ligase.

Ligase chain reaction (LCR): See → ligation amplification reaction.

Ligase-independent cloning: A technique for the → cloning of DNA fragments previously amplified by conventional → polymerase chain reaction protocols. In principle, large (12 or more nucleotides long) complementary single-stranded 5' → overhangs between a → vector and the fragment to be inserted are generated that hybridize to form → nicked circular molecules. These in turn are competent for → transformation of a host (e.g. *E. coli*), in which the remaining → gaps in the circular DNAs are closed by → DNA repair enzymes.

Ligated oligonucleotide probe (LOP): An → oligonucleotide from 16 to more than 500 bp in length that is used for the detection of sequence polymorphisms in → variable numbers of tandem repeats (VNTR) loci of complex genomes (e.g. the human genome). This probe is enzymatically generated *in vitro*. In short, a number of short, partially complementary oligos are synthesized. When annealed, they form a long double-stranded DNA molecule with nicks that can be ligated using → T4 DNA ligase. Usually one of the oligos is 5' endlabeled with T4 polynucleotide kinase before annealing. Thus a radioactive probe is generated.

Ligated *p*hospho*p*rotein (LPP): Any protein generated by the fusion of a specific phosphopeptide to a carrier protein by → ?intein-mediated protein ligation.

Ligated protein: See → mature protein.

Ligation: See → DNA ligation, also → blunt end and → cohesive end ligation.

Ligation *a*mplification *r*eaction (LAR; *l*igase *c*hain *r*eaction, LCR): An *in vitro* DNA amplification procedure that uses → DNA ligase to amplify a template. In short, a pair of synthetic oligo-deoxynucleotides is allowed to anneal to adjacent complementary regions of one (upper) strand of the target dsDNA, and two other oligos to adjacent complementary regions of the other (lower) strand. Then each pair of oligos is ligated by DNA ligase, and the ligation product used as template for subsequent cycles of ligation of complementary oligos. Therefore the product(s) accumulate exponentially. The ligation process requires perfect base pairing at the ligation site. Therefore the reaction will proceed for one allele but not for others, if there is sequence variation at the ligation site. LAR can therefore be used to reveal polymorphisms in genomic DNA or → polymerase chain reaction amplified DNA. See → gapped ligase chain reaction.

Figure see page 849

Ligation-anchored PCR: See → ligation-anchored *p*olymerase *c*hain *r*eaction.

Ligation-*a*nchored *p*olymerase *c*hain *r*eaction (LA-PCR; ligation anchored PCR): A technique for the isolation of the 5' termini of → messenger RNAs. In short, messenger RNAs are first isolated and the first strands of → cDNA synthesized with → reverse transcriptase. Then → T4 RNA ligase is employed to add an anchor

Ligation detection reaction (LDR)

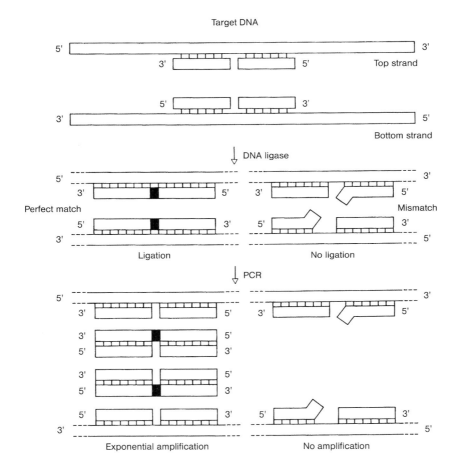

Ligation amplification reaction

oligoribonucleotide to the 3′ end of this first-strand cDNA. Subsequently an anchor-specific → primer and a primer complementary to a known sequence within the cDNA are both used to amplify the 5′ end of the transcript via conventional → polymerase chain reaction.

Ligation *detection* reaction (LDR): A technique for the detection of small → deletions, → insertions or → single *n*ucleotide *p*olymorphisms (SNPs) in target DNA that relies on the potential of → DNA ligase to ligate adjacent oligonucleotides hybridised to the target DNA only if a perfect complementarity exists at the junction site. In short, the → template is first amplified by conventional → polymerase chain reaction, and the socalled common → probe annealed to the template immediately downstream of the nucleotide in question. Then two allelic probes are added, one with a 3′-terminal nucleotide complementary to the → wild-type allele, and another one with a 3′-terminal nucleotide corresponding to the mutant allele. Both allelic probes compete for the template (i.e. anneal to the template adjacent to the common probe). A double-stranded region containing a nick (i.e. a missing phosphodiester bond) is formed at the target nucleotide position. Now DNA ligase can only ligate the allelic probe (with perfect complementarity) to the common probe, but not the mismatched probe. The allelic probes can be designed such that each has a unique length, so that the two ligation products can easily be separated by size. Or, the allelic probes can be labeled with different → fluorochromes that can be discriminated on the basis of their different emission wave-lengths. LDR can also be multiplexed, i.e. several SNPs can be typed simultaneously. Multiplexed LDR can therefore be employed for the diagnosis of diseases that result from multiple mutations in a gene. For example, *cystic fibrosis* (CF) is such a disease, where more than 400 mutant alleles exist worldwide.

Ligation-independent cloning of PCR products (LIC-PCR): The cloning of complex mixtures of DNA fragments amplified in the conventional → polymerase chain reaction, by creating single stranded complementary tails at the termini of both the amplified DNA and the vector. These tails allow the joining of vector and insert with subsequent circularization of the resulting molecule. These hybrid molecules are then transformed into *E. coli* host cells and ligated *in vivo*, so that neither → restriction endonucleases nor → DNA ligases are necessary for the cloning process.

In short, the linearized plasmid vector DNA is amplified by PCR, using primers with an overhang of 12 identical nucleotides at their 5′-termini (oligo(dA), oligo(dT) or oligo(dC), but no dGMP residues) that do not hybridize to the vector. PCR amplification results in accumulation of products that lack dCMP residues at their 3′ ends. The 3′ → 5′ exonuclease activity of → T4 DNA polymerase is now used to degrade the DNA from the 3′ ends. The reaction is carried out in the presence of dCTP and therefore stops at the first dCMP residue. Thus a 5′ overhang of 12 identical nucleotides at each end of the vector molecule is generated which therefore cannot circularize. Similarly, the target DNA is amplified, but using primers whose 5′ ends are complementary to the 5′ ends of the vector primers (oligo(dA), oligo(dT), oligo(dG), but no dCMP residues) in the 5′-terminal 12 nucleotide sequence. Again, single-stranded tails can be generated with T4-DNA polymerase, this time in the presence of dGTP. The complementary termini of both vector and target DNA are allowed to anneal and form a circular molecule. This is transformed into competent *E. coli* cells and repaired (i.e. covalently closed) in the host cell.

Ligation-mediated amplification of RNA: A modification of the conventional → 5′-RACE technique for the detection and analysis of the termini of RNA molecules that starts with the → T4 RNA ligase-mediated ligation of a defined ribooligonucleotide to isolated cellular RNA. This step preserves the termini of the RNAs. Then cDNA is synthesized, amplified and analyzed (e.g. sequenced). In short, cellular RNA is treated first with → calf intestinal phosphatase to remove the 5′ phosphates from truncated transcripts, then with → *to*bacco *a*cid *p*yrophosphatase (TAP) to remove the cap (exposing the 5′ phosphate). The 3′ OH group is removed by sodium periodate (preventing self-ligation of the RNAs). Then a defined RNA fragment (pretreated with phosphatase to remove the 5′ phosphate, thus preventing its self-ligation) is ligated to the TAP-treated RNAs using → T4 RNA ligase. Subsequently → cDNA is synthesized using an oligonucleotide primer complementary to the gene of interest, an oligo (dT) primer (or primers of arbitrary sequence), and → reverse transcriptase. After removal of the RNA by alkali treatment, the cDNA is amplified with a primer complementary to the RNA ribooligonucleotide and a gene-specific primer. Only cDNA containing the ribooligonucleotide sequence will be amplified. The resulting

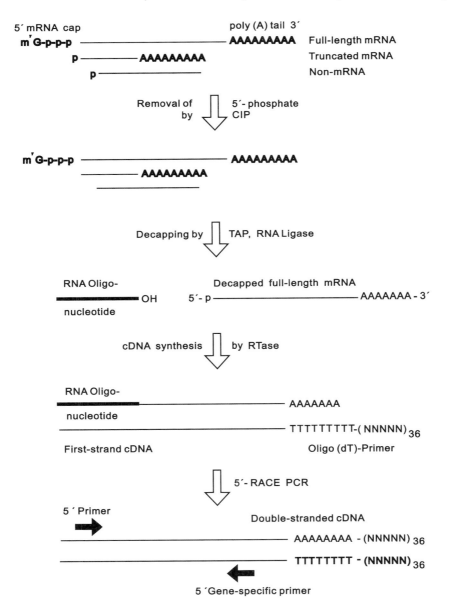

Ligation-mediated amplification of RNA

double-stranded product is then cloned and sequenced. Ligation-mediated amplification of RNA is used to isolate RNA molecules with their 5′ terminus ending at the → cap site (i.e. full-length → messenger RNAs), truncated molecules, and molecules that possess additional nucleotides 5′ to the regular cap site.

Ligation-mediated mutation screening (ligation-dependent mutation screening): A method for the detection of mutations in DNA that employs a thermostable DNA ligase with extreme hybridization stringency and ligation specificity such that only perfectly matching probes are ligated to each other. A frequently used DNA ligase catalyzes the NAD$^+$-dependent ligation of adjacent 3′-hydroxy and 5′-phosphorylated termini in duplex DNA. Since the enzyme is derived from a thermophilic bacterium, it is highly heat-stable and more active at higher temperatures than conventional DNA ligases, so that strin-

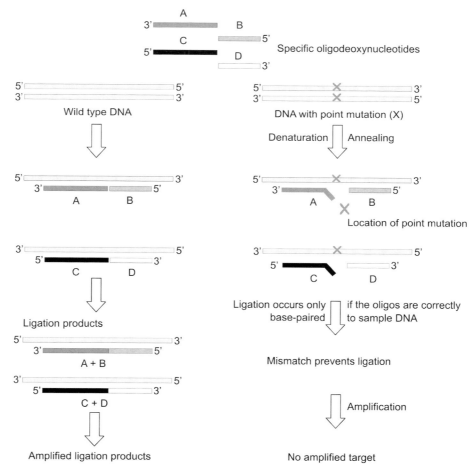

Ligation-mediated mutation screening

gency of hybridization can be extraordinarily rigid and specificity of ligation be absolute.

Ligation recut assay: A test system to determine the quality of → restriction endonuclease preparation by testing, whether the DNA fragments generated by the endonuclease in question can be ligated and whether the restored sites can be cleaved by the same restriction endonuclease (recut).

Light box: See → light-inducible promoter.

Light chain: Constituent polypeptide chain of → antibody molecules.

Light-controlled electrokinetic assembly of particles near surfaces (LEAPS): A technique for the manipulation of biomolecules (e.g. DNA) in AC or DC electric fields. LEAPSs allow to form planar arrays of particles (e.g. beads, cells) in designated areas of a silicon/silicon oxide or a conductive indium/tin oxide electrode. These areas may be interactively manipulated (i.e. merged, placed, shaped, split, or translocated) by illumination. LEAPS is part of → DNA chip technology.

Light-inducible promoter: Any → promoter, originating from a light-inducible plant gene (e.g. the nuclear gene for the ribulose-1,5-bisphosphate carboxylase small subunit, *rbs* gene; the chlorophyll a/b-binding protein gene, *cab* gene; the chalcone synthase gene, *CHS* gene), and containing one or more so-called light boxes (i.e. consensus sequences that function as address sites for → transcription factors).

Light nuclear RNA: See → lnRNA.

Light repair: A → DNA repair system that involves the binding of a photolyase, in *E. coli* encoded by gene *phr*, to the mutagenic site in the DNA (e.g. a → thymine dimer), and the subsequent cleavage of the dimer, using light energy.

Light strand: See → H-strand.

Light upon extension primer **(LUX primer):** A variant of a → sunrise primer, in which the function of a quencher → fluorophore is replaced by the secondary structure of the 3'-end of the → primer oligonucleotide that reduces the initial fluorescence to a → minimum. Therefore a quencher dye is not necessary. The primed → *p*olymerase *c*hain *r*eaction (PCR) amplification product is visualized after → agarose gel electrophoresis. See → duplex scorpion primer, → scorpion primer.

LIM domain (LIM motif): A 50–55 amino acid long, cysteine-rich motif, consisting of $xxCxxCx_{17-19}HxxCxxCxxCx_{16-20}Cxx[D/H/C]x$ residues that is characteristic for at least 50 different eukaryotic proteins (LIM proteins). The LIM domains are structurally similar to → zinc fingers, bind zinc ions and are necessary for protein-protein interactions. The number of LIM domains within proteins varies from 1 to 5. The LIM domain proteins fall into three broad groups. Group 1 proteins always contain LIM domains near the N-terminus that belong to the sequence class A and B and are frequently associated with a → homeodomain (socalled LHX proteins). For example, LHX proteins Lmx-1 and LMO interact functionally and physically with the → *h*elix-*l*oop-*h*elix (HLH) proteins Pan1 and TAL1, respectively. Group 2 proteins contain one or two copies of a single type of LIM domain, and group 3 proteins harbor more heterogeneous LIM domains that are located in the C-terminal region. LIM group 3 proteins interact with the cytoskeleton or organelles, function in adhesiveness, defining cell shape, and act in intracellular trafficking. LIM motifs are also part of developmental regulators (as e.g. the rat insulin gene enhancer binding protein ISL-1, the LIN-11 and MEC-3 proteins of *Caenorhabditis elegans*, the APTEROUS protein of *Drosophila* and the XLIM-1 protein of *Xenopus*, but also the mammalian oncoproteins TTG-1 and TTG-2 of the rhombotin family).

Limiting dilution *polymerase chain reaction* (limiting dilution PCR): A variant of the conventional → polymerase chain reaction that allows to determine the number of target template molecules initially present in an amplification mixture. The template DNAs are serially diluted and the individual dilutions separately used for template amplification, unless a particular dilution step is defined, at which amplification does no longer occur.

LIM motif: See → LIM domain.

***LIM*-only protein (LMO, LIM domain protein):** Any one of a series of nuclear proteins that harbors a conserved → LIM domain ("LIM motif"). LMOs form heterodimeric LIM complexes with CLIM-2 (LNI, LBD1) that recruites LIM proteins together with other → transcription factors, and enhances the nuclear retention of the LIM proteins. LMO1 and 2 are expressed in specific cells of the adult brain, where they regulate the differentiation of the cellular phenotype of neurons and are in turn regulated by neuronal activity.

Lineage-specific *position* effect (LSPE): The differential expression of two different genes at the same location in the genome, based on their unique interactions with the surrounding chromatin ("genomic milieu"). See → generalized position effect.

Linear acrylamide: A → synthetic non-polymerized acrylamide that is free from biological contaminants (e.g. traces of DNA or RNA) and used as → coprecipitant, aiding the ethanol precipitation of picogram quantities of DNA fragments larger than 20 bp. Linear acrylamide is not effective for shorter fragments or free → nucleotides, and can therefore be used to separate nucleotides and → oligonucleotide primers from amplification products after a → polymerase chain reaction.

Linear-*after*-the-exponential (LATE) *polymerase chain reaction* (LATE-PCR): A variant of the conventional → polymerase *chain reaction* (PCR), or more precisely → asymmetric *polymerase chain reaction* (asymmetric PCR) that uses low → T_m probes whose loop T_m is 5–10 °C below Tm_L of the *limiting* → primer, and a low-temperature detection step either before or after the → extension temperature. LATE-PCR produces efficient linear kinetics for more than 80 PCR cycles, i.e. almost 90% higher than that of → symmetric PCR reactions, and generates → single-stranded amplicons. The technique permits uncoupling of → primer annealing from product detection. As a result, the T_m of the probe does not need to be higher than the T_m of either primer. Low-T_m probes own more → allele-discriminating power, generate lower background, and can be used at saturating concentrations without interfering with the efficiency of amplification, which is comparable to symmetric PCR, and allows the use of primers over a wide range of concentration ratios.

Linear amplification DNA sequencing (linear polymerase chain reaction sequencing; double-stranded DNA cycle sequencing; cycle sequencing): A technique to sequence native double-stranded DNA after its amplification in the conventional → polymerase chain reaction. In short, purified dsDNA and a 5′ endlabeled sequencing primer are mixed in four conventional → Sanger sequencing reactions with → *Thermus aquaticus* DNA polymerase. It is then submitted to repeated programmed temperature cycles in a → thermocycler, which lead to repeated denaturing and reannealing of template DNA and primers. The Taq polymerase extends the primer, until incorporation of a → dideoxynucleoside triphosphate (ddNTP) stops the reaction. In this way a series of fragments are generated and amplified that can be electrophoretically separated in → sequencing gels. This allows the determination of the base sequence of the original DNA (compare also → Sanger sequencing). Linear amplification DNA sequencing allows to read more than 500 bases in one step, and due to the signal amplification appreciably reduces both background problems and the amount of DNA needed for sequencing. Compare → genomic amplification with transcript sequencing. See also → single colony sequencing.

Linear amplification-mediated polymerase chain reaction (LAM-PCR): A variant of the

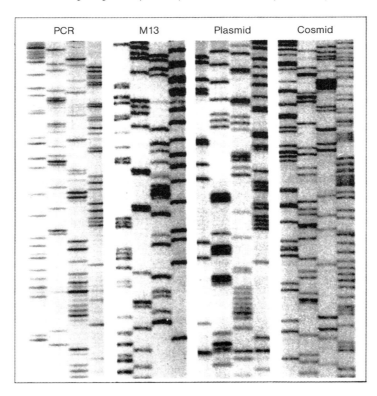

Linear amplification DNA sequencing

conventional → polymerase chain reaction (PCR) that allows the amplification, isolation and sequencing of an unknown region of a bacterial, fungal, plant, animal or human → genome flanking a known segment. In short, the junction between the known DNA segment and the adjacent flank is first linearly amplified by repeated → primer extension with a 5'→ biotinylated primer complementary to part of the known DNA, using e.g. → Taq DNA polymerase (see → linear amplification). Usually 100 amplification cycles are employed, and fresh Taq DNA polymerase added after 50 cycles. Then the amplified fragments of the target DNA are captured on → streptavidin-coated magnetic beads. Subsequently, a second strand of each enriched target sequence is synthesized via → random hexanucleotide priming catalyzed by a → Klenow fragment of DNA polymerase. The resulting → double-stranded DNA is specifically digested with a suitable → restriction endonuclease (e.g. Sse91, cutting → genomic DNA approximately every 256 bp). Restriction serves to reduce fragment size. The length of each fragment is dependent on the distance between the end of the fragment and the next Sse91 recognition site. An asymmetric double-stranded oligonucleotide ligation cassette (LC) is then ligated to the over-

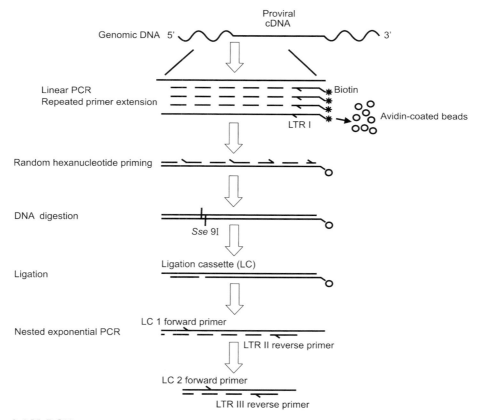

hanging ends of the *Sse*91 fragments by T4 DNA ligase to provide for a primer-binding site of known sequence, and the construct denatured (e.g. with 0.1 NaOH). Finally, part of the amplified fragments serve as templates for a nested exponential PCR amplification that is performed with LC-specific → forward primers (first LC1, then LC2) and target DNA-specific → reverse primers (nested in the known DNA). Sequencing of the whole fragment directly (e.g. by → cycle sequencing) or after → cloning (e.g. into a → TOPO TA-cloning vector) then discloses the base sequence of the unknown region.

Linear bacterial chromosome: A bacterial chromosome that does not represent a → *c*ovalently *c*losed *c*ircular DNA (cccDNA) molecule, but has two ends containing → *t*erminal *i*nverted *r*epeats (TIRs) ranging from 24–210 kb in e.g. *Streptomyces*, and covalently bound *t*erminal *p*roteins (Tps). Its replication usually starts at an internal → origin of replication. Linear chromosomes are constituents of many bacteria (e.g. *Borrelia burgdorferi* [960 kb], whose chromosome carries → hairpin structures at both ends, i.e. the 3'OH group of one strand is covalently linked to the 5'OH terminus of the complementary strand via a phosphodiester bond, *Agrobacterium tumefaciens* C58 [2.07 Mb], *Streptomyces coelicolor* A3 [8.66 Mb], *Rhodococcus fascians*, *Actinoplanes philippinensis*, *Micromonospora chalcea*, *Nocardia asteroides*, *Saccharopolyspora erythrea*, *Streptoverticillium abikoense*), and challenge the view of an apparently universal circularity of bacterial chromosomes. Frequently, circular and linear chromosomes exist sidy by side in the same bacterial cell (e.g. chromosome I [circular, 2.84 Mb] and chromosome II [linear, 2.07 Mb] in *Agrobacterium tumefaciens*, strain C58.

Linear cloning vector: An artificial → plasmid cloning vector that is linear (as opposed to most other plasmid vectors, which are circular). Linear vectors accommodate up to 50 kb of → insert DNA that is highly stable inspite of containing e.g. usually unstable AT-rich regions. Since such vectors remain linear also during → replication, their ends rotate freely, thereby avoiding → supercoiling that may induce instability in conventional circular plasmid vectors. The termini of some of these linear vectors are protected by telomere-like sequences that are supported by e.g the product of a protelomerase gene (e.g. *tel*N). Linear vectors are used for the cloning of large insert genomic libraries or difficult-to-clone DNA such as e.g. DNA with extremely high GC or AT content, toxic genes, multiple repeats, or → strong promoters.

Linear expression element (LEE): Any artificial construct consisting of an → *o*pen *r*eading *f*rame (ORF), a → promoter and a terminator sequence, in which the different modules are noncovalently linked by complementary → overhangs. Each module is separately amplified by conventional → polymerase chain reaction techniques, then the whole construct assembled from the parts, precipitated onto gold particles, and transferred into a target tissue by biolistic bombardment. LEEs are expressed in the target tissue, and represent elements for the functional testing of genic sequences or promoters.

Linearization: The introduction of a single double-strand → cut into a → *c*ovalently *c*losed *c*ircular DNA (cccDNA) molecule which converts it into a linear DNA duplex. Contrary to → circularization.

Linearized vector: Any circular vector that has been cleaved once so as to interrupt both DNA strands and to break the circle. See → linearization.

Linear plasmid: A somewhat incorrect term for an extrachromosomal low-molecular weight linear double-stranded DNA element of yeasts, filamentous fungi and higher plants that is flanked by → terminal inverted repeats (TIR) of considerable length (e.g. 1 kb) with proteins covalently bound to their 5′ termini. Linear plasmids encode DNA and RNA polymerases, and are either localized in mitochondria or, exceptionally, in the cytoplasm (e.g. in yeast). They resemble viral genomes, though they are not associated with → capsids and do not lead to an infection cycle. The *Kluyveromyces lactis* linear plasmid encodes a toxin (killer toxin) that kills other yeast cells. Other linear plasmids may be involved in fungal senescence.

Linear polymerase chain reaction sequencing: See → linear amplification DNA sequencing.

Linear reverse transcriptase primer: Any → oligonucleotide → primer for → reverse transcriptase that is not folded into a secondary structure (i.e. does not contain a → stem-loop structure at its 3′-end) and serves to prime → reverse transcription of a → messenger RNA into a → cDNA. See → stem-loop reverse transcriptase primer.

Linear RNA amplification: See → Eberwine procedure.

Linear *time-of-flight mass* spectrometer (linear TOF-MS): A → time-of-flight mass spectrometer, in which the ionised peptide fragments are accelerated on a straight trajectory between source and detector. Compare → reflector time-of-flight mass spectrometer.

LINE1-mediated transduction (L1-mediated transduction): The recruitment of sequences flanking a → long *in*terspersed *e*lement (LINE) of the LINE1 (L1) family at either 3′ down- or 5′-upstream; by overriding → transcription termination site(s) or transcription initiation in a → promoter upstream of L1 elements. The recruited extra DNA can then be moved around in the → genome: L1-mediated transduction may be one mechanism for → exon shuffling. In humans, transduced sequences range from 30–970 bp (3′end) and 145–215 bp (5′end).

LINES (*long in*terspersed *e*lements, LINE elements; long period interspersion; long interspersed nucleotide element, long interspersed repeat elements): A fraction of → repetitive DNA that is widely distributed in eukaryotic genomes and able to change its position within a chromosome or between chromosomes (similar to → retrotransposons). LINES are usually longer than 1 kb (L1 LINE: 1.5 kb) and accumulate up to 20–40,000 copies per genome, which in turn alternate with → single-copy DNA. Some LINES contain → open reading frames that encode integrase or → reverse transcriptase proteins, and are flanked by short → direct repeats at the site of integration. As a consequence of → retrotransposition, LINES carry a poly(A) tract at the 3′ end. LINES frequently suffer from → deletions of variable extent at their 5′ end, so that socalled incomplete LINES (truncated versions of the complete LINE prototype) exist. In primates, the most prominent LINES belong to the socalled *Kpn* I family (human genome: 50,000 copies), the members of which are characterized by a cluster of → recognition sites

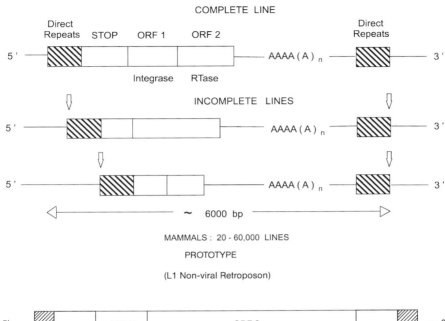
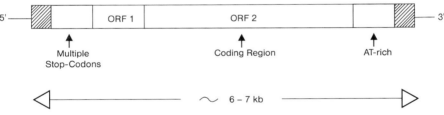

LINES

for the → restriction endonuclease *Kpn* I. Compare → *Alu* I sequence, → SINES.

Linkage: The close physical association of two or more genes on the same chromosome which allows a high frequency of concomitant inheritance (genetic linkage). Generally, linked genes do not show independent assortment, although linkage may be disturbed by crossing-over during meiosis. This leads to recombination, the formation of new linkage groups in the gametes, and a new genotype in the progeny. Thus the frequency of recombination between two (or more) different loci is a measure for their linkage. The closer the loci are spaced, the rarer they are separated by recombination. See → linkage analysis, → linkage drag, → linkage group, → linkage group homology, → linkage map.

Linkage analysis: The estimation of the frequency of → cross-overs or → recombination events between DNA sequences (→ recombination frequency). For example, in plant breeding individuals with many differing traits are crossed, producing a genetically uniform F1 generation. In the second generation (F2), the traits (i.e. the

→ alleles of the genes encoding these → traits) segregate either independently from each other (if the corresponding genes are located on different chromosomes) or linked with each other (if the genes are closely located on the same chromosome). Linkage can be disrupted by → crossing-over (i.e. recombination between → homologous chromosomes during meiosis). This type of recombination occurs more frequently, if the two linked gene loci are more distantly located on the same chromosome, and less frequently, if the two genes are located close to each other. The frequency of recombination allows to estimate the distance (measured in → centiMorgans, cM) between the two linked genes: 1 cM = 1% recombination frequency = 99% linkage. If the alleles of two genes are linked to each other in 99 of 100 F2 individuals (relatively tight linkage), they have a distance of 1 cM. This genetic distance translates to large distances on the → physical map (e.g. from 1000 to 10,000 base pairs, or more). Therefore, linkage analysis is used to establish the position of a particular sequence (e.g. a gene) on a chromosome. See → linkage, → linkage drag, → linkage group, → linkage map.

Linkage *d*isequilibrium (LD, "allelic association", gametic disequilibrium): The occurrence of two (or more) linked → alleles (or loci) at a higher frequency than should be expected from their frequency in a particular population. The tighter the genetic

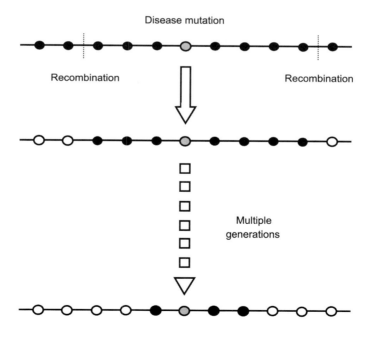

Linkage disequilibrium (LD)

linkage between such loci, the higher the degree of linkage disequilibrium.

Linkage disequilibrium block: See → haplotype block.

Linkage disequilibrium map (LD map, LD unit map): A → physical map of a → genome that does not only depict the distance between markers (or genes, generally loci) in → base pairs, but additionally indicate the number of times two adjacent markers (genes, loci) segregate together in a given number of meioses. See → haplotype block, → linkage disequilibrium mapping.

Linkage disequilibrium mapping (LD mapping, association mapping): The determination of the → linkage disequilibrium of two (or more) linked loci (e.g. a gene and a marker locus) in a genome. For example, LD mapping localizes genes affecting disease susceptibility or other phenotypes through association with closely linked sequence polymorphisms (e.g. → single nucleotide polymorphisms).

Linkage drag: The association between a desirable (positive) and undesirable (negative) trait or gene.

Linkage group: Any group of genes that are contiguous on a linear → chromosome map (i.e. located on the same chromosome), and show a high degree of → linkage. Usually, a linkage group is equivalent to all genes on the same chromosome. Therefore, the number of linkage groups is identical to the haploid number of chromosomes. See → linkage analysis, → linkage group homology, → linkage map.

Linkage group homology: The presence of identical markers on → linkage groups from two different crosses, so that a → homology can be inferred.

Linkage map: See → genetic map.

Linked genes: Two or more genes that are located on the same chromosome in a nucleus, cell, or organism. See → linkage group.

Linked marker: Any → molecular marker(s) located closely to a target gene on the same chromosome such that the → recombination frequency between them approaches zero. Linked markers are exploited for the isolation of the linked gene(s) via → positional cloning.

Linker:

a) A synthetic oligodeoxyribonucleotide of defined sequence containing one or several → restriction endonuclease → recognition site(s). A DNA linker can be ligated to any blunt ended DNA duplex using T4 DNA ligase to prepare this DNA for cloning. See → linker tailing. See also → linker mutagenesis, → linker scanning, → TAB linker.

b) A DNA segment of variable length (linker DNA) connecting two adjacent → nucleosomes in eukaryotic → chromatin. Linker DNA sequences, especially at the entry and exit site of DNA in a nucleosome, are recognized by → histone H1 which binds there.

c) Any aliphatic chain of variable length that can be covalently attached to specific positions of nucleic acid bases in → oligonucleotides and also to solid supports (as e.g. glass slides). Several linker categories can be discriminated. The socalled amino linkers of various lengths (e.g. C2 or C6 amino linker) can be covalently bound to e.g. the C5 of a base (mostly → thymidine) and carry an amino group at the terminus. The 5′carboxyl linkers, carrying a carboxy group at one end, can be used for the conjugation of oligonucleotides to amine-bearing supports (e.g. glass slides for → microarrays), and thiol linkers with an SH (thiol) group at one end can link an oligonucleotide via stable thioether links with maleimides.

Linker-adaptor polymerase chain reaction (LA-PCR): A variant of the conventional → polymerase chain reaction for the amplification of genomic → restriction fragments of unknown sequence. Genomic DNA is first isolated, restricted with an appropriate → restriction endonuclease (e.g. *Rsa*I), and the resulting restriction fragments ligated to double-stranded oligonucleotides (linker adaptor). Then → primers complementary to the adaptors are used to amplify the sequences between the adaptors.

Linker capture subtraction (LCS): A variant of the conventional → subtractive hybridisation technique for the identification of genes differentially expressed in two (or more) cell types, tissues or organs. In short, total RNA is first isolated from two different samples (the tester and driver, respectively), → poly(A)$^+$ → messenger RNA purified, and double-stranded → cDNA synthesized by → reverse transcriptase and an oligo(dT) → primer. The resulting cDNA pools are then restricted by the → four-base cutter → restriction endonuclease *Alu* I and *Rsa*I, producing blunt-ended fragments. These fragments are ligated to a synthetic → linker that carries the → recognition sites for the restriction enzymes *Alu*I and *Sac*I. The linkered cDNA fragments are purified and amplified by conventional → polymerase chain reaction (PCR). Then the driver cDNA is successively restricted with *Alu*I and *Sac*I to remove the linker sequences. After → denaturation of both pools, hybridisation starts with an excess driver cDNA (to remove sequences from the tester pool that represent genes with similar or identical expression levels). cDNA fragments present in both driver and tester are eliminated by single strand-specific → mung bean nuclease, and sequences highly represented in the tester pool are amplified and enriched by PCR. Several round of subtraction further enrich target sequences. The enriched tester cDNA finally is cloned into an appropriate → plasmid vector ("capture"), the → insert sequenced or used as → probe for a → Northern analysis to verify their differential expression. The subtraction product can also be labeled by PCR in the presence of fluorescent nucleotide analogues and then used as probe on expression → microarrays. See → adapter-tagged competitive PCR, → enzymatic degrading subtraction, → gene expression fingerprinting, → gene expression screen, → module-shuffling primer PCR, → preferential amplification of coding sequences, → quantitative PCR, → targeted display, → two-dimensional gene expression fingerprinting. Compare → cDNA expression microarray, → massively parallel signature sequencing, → microarray, → serial analysis of gene expression.

Linker mutagenesis (linker scanning mutagenesis): The introduction of → mutations into a DNA molecule by the insertion of → linkers. First, a circular DNA molecule is treated with → DNase I under conditions that allow random cutting of the duplex. Such treatment leads to the generation of a set of linear molecules with different termini. Then linkers are ligated to these ends and cut with the → restriction endonuclease whose → recognition site is specified by the linker, which in turn generates single-stranded → overhangs that are used to recircularize the molecules. This procedure then leads to the accumulation of DNA molecules with → insertion mutations at random positions that can easily be localized by → restriction mapping, since the specific restriction site of the linker is known. Compare → linker scanning. See also → TAB linker mutagenesis.

Linker scanning: A technique to estimate the optimal spacing between two adjacent regulatory sequences (boxes) of a → promoter, using synthetic homopolymeric → linkers of variable length such as oligo(dA), oligo(dG), oligo(dC) or oligo(dT). They are cloned between the two boxes (for a description of the cloning procedure see → linker mutagenesis) and the effect of linker length on e.g. the expression of the linked gene(s) is determined. Linker scanning allows the determination of the distance between regulatory boxes which is optimal for transcription, and the mapping of additional sites within the spacer that function in the binding of transacting proteins. In this way the sequence between the -35 box and -10 box in specific, strong *E. coli* promoters has been found to be 16 or 17 base pairs in length. Any variation (e.g. 14, 15, 18 or 19 bp) leads to a decrease in promoter efficiency.

Linker tailing: The ligation of short synthetic oligodeoxyribonucleotides containing one or more → restriction endonuclease → recognition sites (→ linkers) to termini of a DNA duplex molecule using → DNA ligase. Subsequent restriction enzyme cleavage generates → cohesive ends suitable for cloning into an appropriate restriction site of a → vector. Internal recognition sequences for the restriction enzyme have to be methylated in order to be protected from cleavage (→ methylation-protection). Linker tailing can also be performed by ligating one strand of an unphosphorylated linker duplex to a normal 5' phosphorylated terminus of a target DNA duplex molecule. This results in termini carrying covalently linked single-stranded self-complementary tails that can be annealed to produce a hybrid molecule. See also → non-palindromic cloning, where non-complementary linkers are used.

Figure see page 864

Linking clone (L fragment, L-junction): Any specific clone that contains an internal restriction recognition site of the rare-cutting → restriction endonuclease originally used to produce the macrorestriction fragments from the target DNA (example: *Not*I). The clone therefore contains overlap sequences from two adjacent large DNA fragments. When used as → hybridization probe, a linking clone will identify two DNA fragments originally adjacent in the genome.

Linking number (L; topological winding number, α): The number of times with which two strands of a closed circular double-helical DNA molecule are wound around each other. A positive linking number signals overwinding.

Linum **insertion sequence:** See → LIS-1.

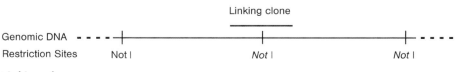

Linking clone

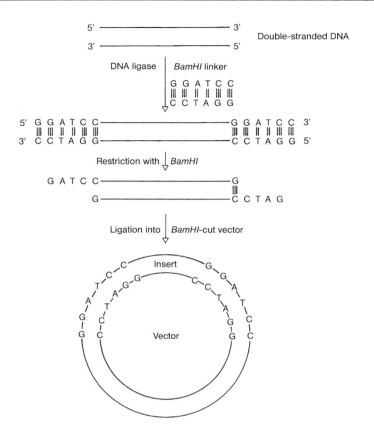

Linker tailing

Lipidome: Another term of the → ome era, describing the complete set of lipids of a cell, a tissue, an organ or an organism, their intracellular and intrabody distribution, their trafficking over time, their half-life times, their interactions with peptides and proteins, and their postsynthetic modifications. Subsets of the lipidome are described by sphingolipidome or phospholipidome. See → lipidomics, → lipoproteomics.

Lipidomics: The whole repertoire of techniques for the detection, characterization, localization, movement, degradation, synthesis and function of the complete set of lipids (the → lipidome) of a cell, a tissue, an organ or an organism. It also encircles techniques to identify genes involved in various aspects of lipid metabolism and its disorders (as e.g. obesity, diabetes, or atherosclerosis) and the characterization of lipid-binding proteins in normal life and disease. See → lipoproteomics. Consult the European Lipidomics Initiative http://www.lipidomics.net/

Lipofection (liposome-mediated gene transfer, vesicle-mediated gene transfer): A simple and effective technique to introduce up to 120 kb of DNA into eukaryotic cells by its entrapping into small unilamellar → liposomes consisting of synthetic cationic lipids (e.g. DOTMA, N-[1-(2,3-

dioleyloxy)-propyl]-N,N,N-trimethyl-ammonium chloride). A method for → direct gene transfer. DNA can be entrapped in such liposomes by simply mixing phospholipids (e.g. phosphatidylserine) and buffer containing DNA by brief sonication. Loaded vesicles are then fused with membranes of recipients (in case of plant cells the plasma membranes are only accessible after enzymatic removal of the cellulose cell wall) and deliver the DNA into the cell. The sonication step in this procedure is most critical because high molecular weight DNA molecules may be broken down by this treatment. Moreover, the efficiency of loading is poor (only one in 100–10,000 liposomes contains DNA), and most of the liposomes do not fuse with the cell membrane, but are taken up through endocytosis (which leads to excessive breakdown of DNA). See also → lipofection-mediated RNA transfection.

Lipofection-mediated RNA transfection (RNA transfection): A method for the introduction of foreign RNA into eukaryotic cells, using inclusion of the RNA in liposomes which consist of a synthetic cationic lipid, N-[1-(2,3-dioleyloxy)-propyl]-N,N,N-trimethyl-ammonium chloride (DOTMA). This method provides an efficient and reproducible way to express exogenous proteins in a wide range of cultured cells. Compare also → lipofection.

Lipophylic silencing RNA (lipophylic siRNA): Any → small interfering RNA that is chemically modified by the introduction of a lipophylic group. Such amodification facilitates the uptake of the corresponding siRNA by target cells and expands its life-time within the cell.

Lipoproteomics: The whole repertoire of techniques for the detection and characterization of protein → domains that interact (and bind) lipids, and the mutations occurring in the exons of the underlying genes (i.e. the amino acid substitutions in the domains). Lipoproteomics aims at developing → biomarkers for the diagnosis and treatment of e.g. cardiovascular diseases.

Liposome (Greek lipos = fat; soma = body):
a) A cytoplasmic lipid globule of specialized cells (e.g. gland cells, secretory cells).
b) An artificially generated lipid or phospholipid vesicle of about 25 nm to 1 µm in diameter, consisting of a lipid bilayer enclosing a single aqueous compartment (unilamellar liposome) or several concentric bilayers entrapping a number of aqueous spaces (multilamellar liposome). Liposomes can be used for → direct gene transfer experiments (see → lipofection) or for the transfection of RNA, see → lipofection-mediated RNA transfection. See → liposome entrapment.

Liposome entrapment ("entrapment"): The process of encapsulating macromolecules (e.g. DNA, RNA, proteins) within a → liposome. See also → lipofection.

Liposome-mediated gene transfer: See → lipofection.

Liquichip: See → bead array.

Liquid hybridization (solution hybridization): The annealing of complementary single strands of nucleic acids to double-stranded helical molecules (DNA-DNA, DNA-RNA, RNA-RNA) in solution. Compare → hybridization. See for example → reverse Southern hybridization.

Liquid scintillation counter (LSC): A spectrophotometer whose photomultiplier tube allows the detection of light flashes emitted by a fluorescent chemical (scintillator, e.g. POPOP, POP) after its reaction with an ionizing particle or photon. Serves to measure radioisotopes or compounds containing them dissolved in a scintillation medium ("cocktail"). See for example → fluor diffusion assay.

LIS-1 (*Linum* insertion sequence): A 5.8 kb sequence element of flax (*Linum usitatissimum* L.) that contains a complex set of short repeats in both orientations, and appears in the genome of various flax lines at exactly the same position after an environmental stress, though it was absent in the original line. A 3 bp duplication at the → insertion site suggests a → transposition, but the exact mechanism of action is unknown. See → *hth* gene, → template-directed correction.

Live cell microarray: See → cell chip.

Living antibiotic: A somewhat misleading term for a bacterium species that kills another bacterial species, but does not attack the host, in which the second bacterium develops a disease. For example, the bacterium *Bdellovibrio bacteriovorus* is such a living antibiotic. *Bdellovibrio* first identifies any prey with its chemical receptors, moves chemotactically towards it and reversibly captures it with the help of a pilus. The pilus allows the predator to come into close contact with its prey. *Bdellovibrio* then secrets an enzyme cocktail that decomposes the wall structures of the attacked cell locally, such that a hole is generated. After penetration of the prey cell, the pore is closed by *Bdellovibrio* that itself remains in the periplasmic space of the host cell and immediately starts growing at the expense of the host's metabolism. *Bdellovibrio* uses amino acids, sugars, nucleotides, and other nutrients to increase its mass and length, unless all the metabolites are depleted. The invader then divides into a series of 15 daughter cells that destroy the wall of the host cell and start a new cycle of invasion. *Bdellovibrio* does not recognize eukaryotic cells and possesses an only weakly immunogenic surface. Moreover, distinct *Bdellovibrio* strains are specific for certain bacterial species and do not attack others.

Living array (biological array): A membrane on which pools of proteins, cell fractions or also living cells are systematically arranged for a high-throughput screening of DNA-protein, RNA-protein, protein-protein, or protein-ligand interactions. For example, a living array with about 6000 yeast colonies, where each colony expresses a different Gal4 activation domain-ORF-fusion protein, and the → ORF sequences are derived from the *Saccharomyces cerevisiae* genome sequencing project, is such a biological platform. Any interaction(s) of protein candidates with the ORF-derived proteins can be monitored by → reporter gene activation. Likewise, living arrays with about 8000 yeast strains of identical genetic background, but different → transposon insertions, can be screened for mutant phenotypes by simply monitoring growth under various culture conditions. See → nonliving arrays. Compare → living chip.

Living chip: A misleading term for a → microtiter or nanotiter plate with bottomless wells, in which a series of biochemical reactions (up to 100) can simultaneously be initiated and monitored in 10–100 ml cell suspensions. Compare → living array.

Living microarray: See → cell chip.

L-junction: See → chromosome linking clone library.

LMNA gene: See → lamin A gene.

LMO: See → LIM-only protein.

LMP agarose: See → low melting point agarose.

LNA: See → locked nucleic acid.

LNA gapmer: A nucleic acid construct composed of a core of 5–15 deoxynucleotides ("DNA core") and two flanking sequences ("arms") of variable length consisting of → locked nucleic acid monomers. LNA gapmers bind their target DNA with high affinity and are extremely stable *in vitro* and *in vivo*. LNA gapmers e.g. block → gene expression by specifically binding to target → messenger RNAs or also DNAs.

LNA inhibitor: Any → locked nucleic acid sequence that is partly complementary to, and specifically binds to an → intron, thereby blocking the amplification of the adjacent exons in a conventional → polymerase chain reaction (PCR). LNA inhibitors are employed for the exclusive amplification of → cDNA, and avoid the simultaneous amplification of the corresponding → genomic DNA, a frequent contaminant in → reverse transcriptase PCR experiments. The LNA inhibitor usually comprises 14 LNA bases, which must be fully complementary to the intronic target sequence to achieve complete inhibition. Only one single → mismatch or a single LNA ⇆ DNA base exchange is tolerated with no significant reduction in inhibitory activity.

LNA microarray (LNA array): Any → microarray, onto which T_m-normalized → locked nucleic acid oligonucleotides (usually 12–50 nucleotides long) are immobilized (e.g. by photo-coupling procedures) with a spot-to-spot distance of 100–200 nm. Such LNA microarrays are used for the detection of → single nucleotide polymorphisms in target DNA or also for transcription profiling. Specific LNA microarrays (see → miChip) are loaded with >1200 capture oligonucleotides complementary to microRNAs covering all human, mouse and rat microRNA sequences annotated in miRBase 10.0, and used to profile the expression of the miRNAs during e.g. developmental processes in a target organism.

LNAzyme: A variant of the conventional → DNAzyme, in which the DNA of the two binding arms flanking the catalytic core is replaced by → locked nucleic acid (LNA) monomers. The LNA parts of the DNAzyme bind strongly and specifically to an RNA substrate adjacent to the cleavage site and cleave the corresponding → phosphodiester linkage more efficiently than an unmodified DNAzyme.

lnRNA (light nuclear RNA, low molecular weight nuclear RNA): A nuclear, 80–260 nucleotides long → RNA that has been found in animal cells only. It comprises about 0.4–11% of the total nuclear RNA and is relatively stable. Its function is obscure.

LO: See → labeled oligonucleotide test.

Loading buffer (sample loading buffer): A mixture of → tracking dyes and a viscous solution of glycerol, sucrose or Ficoll, used to increase the density of DNA, RNA or protein samples before their loading into the slots of a gel, and to monitor the electrophoretic run (compare → gel electrophoresis). For example, in 0.5 × TBE buffer (Tris-HCl, pH 7.5; boric acid, EDTA), → bromophenol blue as tracking dye migrates through agarose gels as fast as linearized dsDNA of 0.3 kb, whereas → xylene cyanol migrates with linearized dsDNA of about 4–5 kb.

Local alignment: The computer-assisted comparison of any region of similarity between two (or more) DNA or RNA sequences that spans only part of their lengths. See → global alignment, → sequence alignment.

Local genome duplication: Any duplication of part of a → genome that is restricted to within 1 Mb. See → global gene duplication.

Local hopping: A laboratory slang term for the repeated → insertion of different "daughter" → retrotransposons at a site in the host → genome, where an original "parental" integration of a retrotransposon occurred. Local hopping leads to the accumulation of retrotransposons in particular regions or at specific sites in the host genome.

Localisome (localizome): The subcellular localization of (preferentially) all peptides and proteins of a cell at a given time, including the organellar proteins (of e.g. the nucleus, the mitochondrium, and in green plants the chloroplast). For example, about 47% of yeast cell proteins are cytoplasmic, 13% are mitochondrial, 13% are exocytic, and 27% are nuclear.

Localisome mapping: The process of establishing a graphical depiction of the localisome of a cell at a given time ("localisome map").

Localized expression domain (LED): Any group of genes (here called "domain"), whose expression is enriched in a defined morphological or anatomical region of an organ (e.g. a root of a plant).

Localized hypermethylation: The occurrence of methylated cytosines in a specific genomic region at a significantly higher frequency than in the rest of the genome. See, for example, → promoter hypermethylation.

Locally multiply damaged site: See → clustered lesion.

Local mutator: Any triplet → microsatellite repeat that itself expands, or induces expansion of other triplet microsatellite repeats.

Location proteomics: A special field of → proteomics that aims at determining the complete protein patterns in various cell compartments and their changes over time during development, in disease, and before and after stress or drug administration. One of the prominent techniques of location proteomics is based on the introduction of a short artificial sequence ("tag") or a sequence encoding an → epitope into the gene(s) of interest, and the detection of the expressed proteins by fluorescently labelled → antibodies, their excitation by laser light and their monitoring by fluorescence microscopy.

Loci-spanning probe (LS probe): Any → oligonucleotide that is designed to span at least two, preferably more non-contiguous regions of genomic DNA maximally 58 nucleotides apart from each other, to hybridize to them, and to be detected by e.g. → fluorescence resonance energy transfer (FRET), if labeled, or melting temperature analysis with appropriate → fluorochromes, and therefore serves to analyze multiple gene (or generally, locus) variants simultaneously. The intervening sequence on the DNA template, omitted in the LS probe, loops out. For example, the wild-type and the three mutations in the β-globin gene responsible for the corresponding hemoglobinopathies (HbS for sickle cell anemia, characterized by an A → T → transversion in nucleotide 62206, haemoglobin variant HbC, characteristic for its G → A → transition at nucleotide 62205, and variant HbE, characteristic for its G → A transition at nucleotide 62265 of the β-globin gene) are simultaneously detectable with both fluorescently labeled and unlabeled LS probes. Variants can be analyzed in a 100-nucleotide range with LS probes spanning three loci and creating two template loops.

Locked *n*ucleic *a*cid (LNA, L-DNA, "bridged nucleic acid"): A nucleic acid derivative that contains one or more 2'-C,4'-oxymethylene-linked bicyclic ribonucleotide monomers (furanose rings locked in a 3'-endo conformation) embedded among DNA nucleotides as constituents of an → antisense oligonucleotide. This restricted conformation allows the formation of extremely stable → Watson-Crick base-pairing between the LNA and complementary DNA or RNA (T_m = +3 to +10 °C per LNA monomer introduced). Therefore, LNA-DNA mixmers (see → gap-mer) are potent duplex stabilizers, but biologically inert (non-toxic). Moreover, they are resistant to 3'-exonucleolytic degradation, soluble in aqueous media, and can be cut by → restriction endonucleases. LNA primers are recognized by various DNA polymerases and → reverse transcriptases, and show excellent → mismatch discrimination in e.g. → SNPing. Moreover, the high binding affinity of LNA oligonucleotides allows to reduce → probe length without hampering its hybridization to the target DNA. See → DNA-LNA mixmer capture probe.

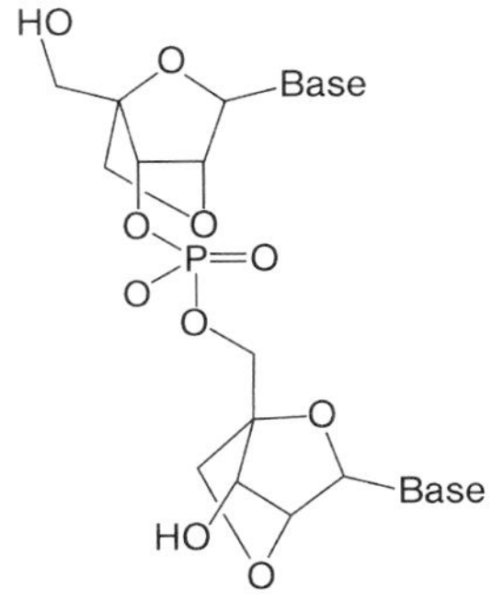

LNA-dimer

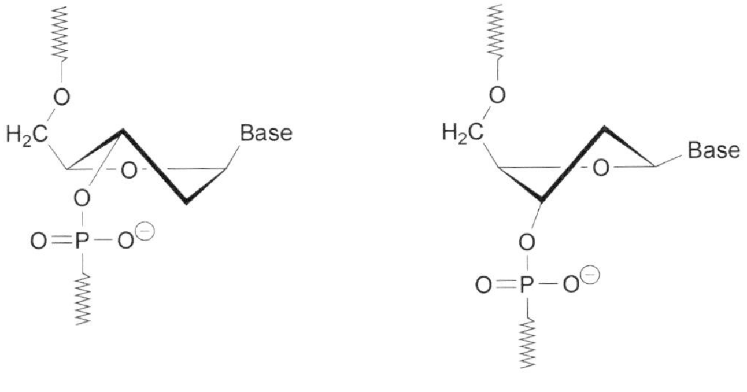

N-type   S-type

LNA nucleotide

LNA (Locked Nucleic Acid)

Locus (plural: loci): A specific position on a chromosome, a → genetic map, or a → physical map, usually identified by a → marker (e.g. → molecular marker) that does not generally represent a gene, but may contain non-coding sequences.

Locus control region (LCR): Any DNA sequence that exerts a dominant, activating effect on the transcription of genes in a large → chromatin domain (10–100 kb). LCRs prevent the influence of e.g. → heterochromatic silencing on neighbouring sequences, and therefore are used in transgenic experiments as insulators (insulator elements) that protect themselves and linked genes against the repressive action of heterochromatin. See → position effect, → position effect variegation, → specialized chromosome structure.

Locus-specific primer technology (LSPT): A comprehensive term for → genome scanning techniques that employ → primers specific for a particular → locus to amplify this locus by conventional → polymerase chain reaction methodology. LSPT enables to detect specific alleles, allele size differences and allele frequencies, and generates locus-specific → molecular markers, which are inherited in a → codominant way. See → sequence-tagged microsatellite sites.

Locus-specific probe: A single nucleic acid fragment or a selected collection of fragments, whose sequences are homologous to one specific region (or → locus) of a → genome. Locus-specific probes usually detect genes, → translocation breakpoints, or also → microsatellite-flanking regions that all occur only once in the target genome. Compare → whole chromosome probe.

LOD score (z value): A mathematical description of genetic → linkage, defined as the decadic logarithm of the relation of probabilities that the observed results are produced by linked or unlinked loci. A LOD score of 3 or more indicates linkage.

LOF: See → loss-of-function mutation.

LOH: See → loss of heterozygosity.

L1 retrotransposon: See → non-LTR retrotransposon.

Long and accurate polymerase chain reaction (LA-PCR): A variant of the conventional → polymerase chain reaction technique that employs *Avian Myeloblastosis Virus* (AMV) → reverse transcriptase in combination with a specific oligo(dT)-adaptor → primer for → first strand cDNA synthesis from polyadenylated RNA (that can take place at 42–60 °C) and a → *Taq* DNA polymerase for the → second strand synthesis. LA-PCR allows to amplify cDNAs of up to 12.2 kb in length and runs in one single tube. Do not confuse with → linker-*a*daptor (LA) polymerase chain reaction, or → ligation-*a*nchored (LA) polymerase chain reaction.

Long-*d*istance *p*olymerase *c*hain *r*eaction (LD-PCR; long PCR; long fragment PCR; long range PCR): A variant of the conventional → polymerase chain reaction that is designed to amplify DNA fragments of up to 40 kb (as opposed to 6–8 kb in traditional PCR using → *Taq* polymerase only) with 10-fold higher fidelity as compared to standard techniques. LD-PCR requires the combination of two thermostable DNA polymerases, e.g. → *Thermus thermophilus* (*Tth*) DNA polymerase (no proofreading activity) and → *Pfu* DNA polymerase (3′-5′ exonuclease activity = proofreading activity). The specificity of *Tth* polymerase can be enhanced by inclusion of a → monoclonal antibody raised against this enzyme that blocks polymerase activity completely. At the start of the thermal cycling process, the enzyme-antibody complex dissociates, and renders the antibody inactive. This step greatly reduces artifacts of inaccurate amplification. The longer product in LD-PCR results from the greater thermodynamic stability of one of the DNA polymerases at elevated temperatures.

Long-*e*xtension *p*olymerase *c*hain *r*eaction (LX-PCR): A variant of the conventional → polymerase chain reaction, which allows to amplify extremely long stretches of DNA. For example, a pair of → primers complementary to closely opposed sequences in → mitochondrial DNA, with their 3′ ends facing away from each other, can be used to amplify virtually the entire mitochondrial genome. In this case, amplification products can be detected that differ in length from the wild-type mtDNA (e.g. shorter fragments are indicative for → deletions, and longer fragments for duplications or → amplifications in genomes of *Chlamydomonas reinhardii* and higher plants).

Long *f*orm (LF): A laboratory slang term for the longer, normally spliced wild-type → messenger RNA (or its → cDNA) transcribed from a particular gene, as compared to the → short form(s) arising from the → transcript → of the same gene undergoing → alternative splicing. A longer form can also arise from alternative splicing with → intron retention. The shorter form then is the wild-type form.

Long fragment PCR: See → long-distance polymerase *c*hain *r*eaction.

Long *in*terspersed *e*lements: See → LINES.

Long *i*nterspersed *n*ucleotide *e*lements: See → LINES.

Long-lasting fluorophore: Any → fluorochrome that remains in an excited state for extended periods of time (e.g. for 500–600 msec; average fluorochrome: 5–6 nsec). For example, chelates of the "rare earth" metal lanthan, socalled lanthanide chelates, are such long-lasting fluorophores. They are very stable organic molecules with a strong Stoke's shift, i.e. the wavelength of the emitted light is up to 200–300 nm longer than the excitation light, exceeding that of conventional fluorochromes (e.g. → fluorescein). Moreover, the excited state may last up to 600 msec, which increases sensitivity and reduces background (e.g. the prompt fluorescence light emitted from cells, cell components, compounds of reaction mixtures, or plastic material). Long-lasting fluorophoreses are the basis for non-radioactive detection assays (as e.g. *time-resolved fluorometry*, TRF).

Longmer: Any → oligonucleotide with a length of more than 50 nucleotides. See → shortmer.

Long patch repair: The excision of more than 1500 nucleotides around and including a site of DNA damage (e.g. a missing base or a thymine dimer) and the repair of the resulting gap by → DNA polymerase that uses the undamaged strand as a template. The genes involved in this type of repair (*uvr* genes) are constitutively expressed in *E. coli*. Compare → short patch repair.

Long period interspersion: See → LINES.

***Long primer random amplified polymorphic DNA* (LP-RAPD):** A variant of the conventional → *random amplified polymorphic DNA* (RAPD) technique that employs comparably long → primers (18–25 nucleotides) for the amplification of anonymous genomic regions using → polymerase chain reaction. These primers are designed from consensus sequences in several families of short interspersed repetitive elements of eubacteria, and allow amplification to be performed at higher stringency than normal RAPD (primer length: 10-mer). LP-RAPD detects sequence → polymorphisms between the genomes of different organisms and is used for identity testing, population studies and phylogenetic relationships.

Long range PCR: See → long-distance polymerase *c*hain *r*eaction.

Long-range repeat (LRR): A special type of repeated DNA sequence with a unit length of about 100 kb. For example, a cluster of LRRs is located in band D of mouse chromosome 1, the unit number ranging from 20–70 copies in different mice. This cluster contains a family of related genes (LRR gene family) that originated from a single-copy ancestor gene by amplification and subsequent diversification. The LRR copy number per haploid genome is highly variable in different mouse strains (40–1800) without any phenotypic consequence. Compare → leucine-rich repeat.

Long-range restriction map: The linear array of → rare-cutter → recognition sites on large DNA fragments isolated by → pulsed-field gel electrophoresis (PFGE). Compare → restriction map.

Long-range restriction mapping: A procedure for the construction of relatively coarse → maps of genomic DNA, using → restriction endonucleases that recognize relatively rare → restriction sites (→ rare cutter). Such maps detail the positions, where the DNA molecule is cut by the particular restriction enzyme(s). See → restriction map, → restriction mapping.

Long reverse transcriptase polymerase chain reaction (long RT-PCR): A variant of the conventional → reverse transcriptase polymerase chain reaction for the generation of → cDNA from → messenger RNA that allows to synthesize full-length ("long") transcripts. Long RT-PCR employs a mutant M-MLV reverse transcriptase (lacking → RNase H), coined Superscript RT, and a mixture of → Taq DNA polymerase and a second DNA polymerase with intrinsic 3′→ 5′ exonuclease activity. AMV RT and M-MLV RT can principally be used, but they produce much less cDNAs than the mutant enzyme in the above combination.

Long RNA (lRNA): Any one of a series of cytoplasmic and nuclear → poly-adenylated RNAs longer than 200 nucleotides. Such lRNAs potentially represent parts of nuclear → primary transcripts that encode conserved functional → short RNAs. See → promoter-associated long RNA.

Long RT-PCR: See → long reverse transcriptase polymerase chain reaction.

Long serial analysis of gene expression (LongSAGE, LS): A variant of the conventional → serial analysis of gene expression (SAGE) technique for the quantification of → transcript abundance in the RNA population of a cell, a tissue, an organ, or an organism that generates 21 bp → tags derived from the 3′-ends of → messenger RNAs (mRNAs) rather than the 14 bp in the original SAGE protocol. In short, RNA is first extracted from the target cells, and mRNA isolated. This mRNA preparation is then treated (e.g. converted to → cDNA) according to the conventional SAGE procedure with the following changes. After digestion of the cDNAs with NlaIII, → linkers containing an MmeI recognition site are ligated to the 3′ends of the cDNAs. Linker-tag molecules are then released from the cDNA using the type IIS restriction enzyme MmeI. The resulting tags are then directly ligated with → DNA ligase. Tag → concatemers are sequenced, and the longer tags analysed and matched to genomic sequence data. Matching of tags to genomic sequences allows precise localization of genes, from which the tags ultimately are derived. See → SuperSAGE.

Long single copy sequence (LSC; large single copy sequence): An 80–90 kb region of the → chloroplast DNA that is flanked by the two → inverted repeat regions and carries unique chloroplast genes as single copies (e.g. the gene encoding the large subunit of ribulose-1,5-bisphosphate carboxylase/oxygenase [Rubisco], rbcL). Together with the → short single copy sequence it represents the unique part of the → plastome.

Long terminal repeat (LTR): The repeat sequences, several hundred base pairs in length, at the ends of a retroviral nucleic acid. In → proviruses the upstream LTR functions as → promoter/→ enhancer, the downstream LTR as a → poly(A) addition signal. LTR sequences can be used as elements of → integration vectors.

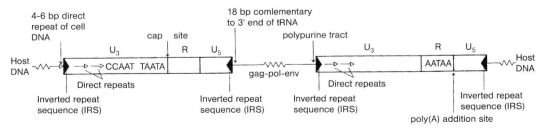

The structure of MLV long terminal repeats

Long term repression: The inhibition of the expression of a gene or a group of genes in differentiated eukaryotic cells that persists under normal physiological conditions. See also → repression.

Loop:

a) A single-stranded region at the end of a hairpin in DNA or RNA, see → foldback DNA.
b) See → looped domain.

Looped domain (loop domain, loop): A special form of packaging of eukaryotic → chromatin that generates discrete and topologically constrained loops (see → supercoil). Such loops contain in between 10 (yeast) to 230 kb (higher eukaryotes) of nucleosomally arranged and solenoidally packed DNA that is anchored at the → nuclear matrix at sequences close to active genes. At these anchorage sites → DNA topoisomerase II is located. The loops probably are individual → replicons, and also transcriptional domains where only those genes close to the anchorage point (attachment site) are active. The loop architecture is very dynamic and varies in different cell types. Synonym: →chromosomal loop.

Looping model: The hypothetical description of the interaction of → transcription factors bound to → enhancers and → promoters, with the transcription machinery.

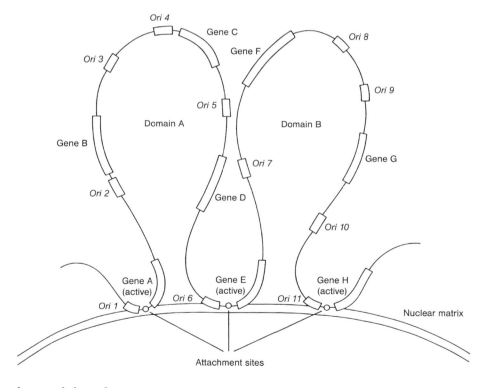

Looped domain

According to this model, the initiation of transcription is stimulated by the interaction of proteins attached to enhancer and/or → upstream regulatory sequences (URS) and proteins attached to proximal promoter elements (e.g. → CAAT box, → TATA box). Thereby sequences that are remote from each other on the linear DNA are brought into close proximity, whereas the DNA between the distal and proximal promoter elements is looped out.

Loop-sheet-helix protein (LSH protein): Any one of a family of → transcription factors that contains a loop protruding out of the main body of the protein, followed by a small β-sheet, an → α-helix, and another loop leading back into the protein. A Zn ion is coordinated by three cysteines and a histidine residue in the two loop regions. LSH proteins bind with their → α-helix in the → major groove of the DNA, and with the loops in the → minor goove. For example, the tumor suppressor protein p53 is a loop-sheet-helix protein, with each of its four subunits contacting a separate 5 bp recognition sequence positioned one after the other.

LOP: See → ligated oligonucleotide probe.

Lorist vector: A → cosmid cloning vector that allows the isolation of large regions of eukaryotic genomes and the ordering of overlapping recombinant clones. Lorist vectors carry an → origin of replication of bacteriophage l (contributing to a more uniform copy number and yield of recombinant cosmid vector), two strong promoters (e.g. → T7 and → SP6 promoter) facing each other and flanking a unique cloning site (→ polylinker) and a → selectable marker (e.g. a → neomycin resistance gene). *In vitro* transcription of the insert DNA, catalyzed by T7 or SP6 RNA polymerase, yields sense or antisense RNA probes that can be used to screen for overlapping genomic clones.

Loss-of-expression mutation: Any mutation in a gene that silences the gene (i.e. leads to the disappearance of its transcript). A loss-of-expression mutation represents a → loss-of-function mutation. See → gain-of function mutation, → reduction-of-function mutation.

Loss-of-function mutation (lf): Any → mutation that completely abolishes the function of the encoded protein. See → gain-of-function mutation, → loss-of-expression mutation, → reduction-of-function mutation.

Loss of heterozygosity (LOH; allele imbalance): The disappearance of one of two heterozygous loci in specific cell types, e.g. tumor cells. For example, a → microsatellite → marker closely linked to a putative colorectal tumor suppressor gene is represented as two equivalent, heterozygous loci, i.e. a microsatellite site of shorter and one of longer size, but both at the same concentration. In contrast, in colorectal cancer cells the shorter microsatellite allele is either reduced in concentration (i.e. is underrepresented) or lost, probably a consequence of → mutation(s) in the microsatellite flanking regions. Compare → chromosomal instability, → microsatellite instability.

Loss of imprinting (LOI): The reversal of the methylation of → cytosine residues at strategic sites in a gene (i.e. in → exons and also → introns) or its → promoter, leading to the cessation of epigenetic silencing and the activation of transcription of the gene. For example, LOI in the gene encoding the insulin-like growth factor II (*IGF2*), an important tumor growth factor, leads to the activation of the normally silenced gene. Therefore, LOI of the *IGF2* gene is associated with a family history of *c*olorectal *c*ancer (CRC) and a personal history of colon adenomas and CRC. LOI is inherited or acquired early in life, and LOI at the *IGF2* locus serves as → biomarker for a distinct risk for CRC.

LoTE: A modified → TE buffer with a *lo*w (Lo) *T*ris-HCl (3 mM; pH 7.5) and *E*DTA concentration (0.2 mM).

Low abundance proteome: A → subproteome that contains preferentially low-abundance proteins (i.e. proteins present in low copy numbers or even traces). The low abundance proteome is isolated with e.g. → protein equalizer technologies.

Low abundancy messenger RNA (low abundance mRNA, low abundancy message): A subfamily of eukaryotic → messenger RNAs, comprising messages encoded by unique genes and present in some 5–10 copies per cell. Compare → high and → intermediate abundancy messenger RNA.

"Low cop" mutation: Any chromosomal mutation which leads to a decrease in the → copy number of plasmids per cell. Not desired in recombinant DNA experiments. The "low cop" mutants can be counterselected by high → antibiotic concentrations. Under certain conditions, however, → low copy number plasmid vectors are favored, see there.

Low copy number DNA sample (LCN): Any sample from living organisms (e.g. human blood, buccal swaps, semen, hairs, skin, fingerprints or also material contaminated with such remains) that contain extremely low amounts of DNA as a consequence of deterioration by e.g. fire, chemicals, age, or other harsh environmental factors. Conventionally, LCNs are examined by → microsatellite fingerprinting, amplification of remaining → mitochondrial DNA by → polymerase chain reaction, or → single *n*ucleotide *p*olymorphisms (SNPs) in nuclear DNA (e.g. in specific genes).

Low copy number plasmid (single-copy plasmid, stringent plasmid): A → plasmid that is present in one or only a few copies per bacterial cell (e.g. → pSC 101). Derivatives of pSC 101 carrying three → antibiotic resistance markers and unique → restriction sites are favored vectors for the cloning of genes which disturb the cell's normal metabolism if present in high copy number (e.g. genes encoding surface membrane proteins). Low copy number plasmids are replicated under → stringent control. Compare → multicopy plasmid, → runaway plasmid.

Low copy polymerase chain reaction (LC-PCR): A variant of the → polymerase chain reaction that uses very low concentrations of the template DNA for amplification, so that a specific DNA sequence (e.g. a gene) is represented in only one to ten copies.

Low-copy repeat (LCR): Any one of a series of highly conserved, duplicated chromosomal regions, operationally defined by their length (1–20 kb) and degree of sequence conservation (90–99,5%). LCRs contain fragments of coding sequences and are likely the sites of new gene formation by → domain shuffling. LCRs probably origin from → translocation followed by transmission of unbalanced chromosomal complements in human sub-telomeric regions, → *Alu*-mediated transposition in peri-centromeric regions, copy number expansion through → non-allelic homologous recombination (NAHR) mediated by DNA repeats, and chromosomal instability owing to variations in → DNA supercoiling. The LCR content substantially varies between different lineages, with more LCRs in human and chimpanzee than in other species.

Low-coverage sequencing: The sequencing of a → genome to an extent that all potential coding regions and → promoters, also the content of repeats are covered. Low-coverage sequencing leaves an appreciable part of the genome unknown, but is more cost-effective than whole genome sequencing.

Low density array: See → low density chip.

Low density chip (low density array): A laboratory slang term for a → DNA chip, onto which about 10–100 → probes (e.g. → single nucleotide polymorphisms) are spotted. Compare → high density chip, → medium density chip.

Low-density screening: The identification of specific DNA sequences in a → genomic or → cDNA library that is plated out at low density (i.e. as several hundred colonies per plate) before hybridization to a suitable → probe.

Low-fidelity DNA polymerase: Any → DNA- or RNA-dependent DNA polymerase that erroneously incorporates wrong bases into the polymerization product. For example, → avian myeloblastis virus → reverse transcriptase introduces 1 error per 17,000 bases, → Moloney murine leukemia virus reverse transcriptase 1 error per 30,000 bases. The low-fidelity reaction is undesired, because it yields mutated products in e.g. → cDNA synthesis.

Low level promoter: See → weak promoter.

Low melting point agarose (LMP agarose; low gelling temperature agarose, LGT agarose): A specific → agarose that melts at 65 °C, remains fluid at 37 °C, and solidifies below 25 °C. Since the resolving characteristics of both low melting point agarose and standard agarose are similar, the former is ideally applicable for the recovery of DNA fragments after → gel electrophoresis.

Low molecular weight nuclear RNA: See → lnRNA.

Lowry technique: A method to determine protein concentrations in the range from 1–25 µg that exploits the interaction of a phosphorotungsten-molybdate complex (Folin-Ciocalteu reagent) with tyrosine residues of the proteins. The reaction leads to the development of a blue color that can be quantified spectrophotometrically.

Low stringency: Any set of reaction conditions during nucleic acid → hybridization that allow the formation of duplexes from single-stranded molecules with a certain degree of base mismatches. Compare → high stringency.

Low-stringency single specific primer polymerase chain reaction (LSSP-PCR): A technique for the generation of a sequence-specific gene fingerprint ("gene signature"). In short, the target DNA fragment (preferable more than 1 kb in length) is purified, then amplified with very high concentrations of → *Taq* DNA polymerase and a single oligodeoxynucleotide → primer complementary to a sequence close to one of the termini of the template. Under specific conditions (low → stringency), the primer does not only anneal to this complementary region, but also to multiple sites within the target fragment, producing a complex set of reaction products after → polymerase chain reaction amplification. The PCR products are then separated by either → agarose or → polyacrylamide gel electrophoresis. LLSP-PCR allows to detect differences of only one nucleotide between two target DNAs ("gene variant signatures"). The technique is mainly used for identity testing and mutation screening.

lox P (*locus of crossing-over*): A 34 bp DNA sequence element of bacteriophage P1, composed of two → inverted repeats of 13 bp each, separated by a central asymmetric 8 bp core sequence that functions as address site for → *cre* recombinase (*cre* = *c*auses *re*combination). First, four *cre* recombinase molecules bind to two adjacent lox P sites and form a DNA-protein complex. Within this complex, the two core regions are cut. Subsequent reciprocal strand exchange leads to → Holliday junctions (see → crossover). This recombination process results in different products, depending on the position and relative orientation of the two lox P sequences. The polarity of the element is determined by the core sequence, and the

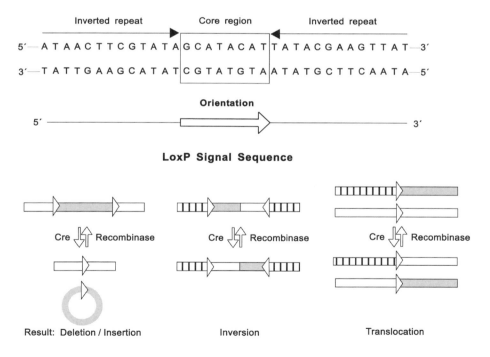

LoxP Signal Sequence

Cre - mediated Recombination

lox P

polarities of two adjacent elements determine the type of recombination. Tandemly repeated lox P sites (identical polarity) on one DNA strand cause *cre* recombinase-catalyzed excision of the intervening sequence and its circularization, leaving one single lox P site behind in the genome. If two adjacent lox P sites possess opposing polarity, the enzyme will introduce an → inversion of the intervening sequence. And if the two lox P sequences with identical polarity are localized on two different DNA strands (e.g. on different chromosomes, or a plasmid and a chromosome), a translocation of one lox P-flanking region onto the other results. The integrated DNA resides between two directly repeated lox P sites.

LPP: See → ligated phosphoprotein.

LP-RAPD: See → *l*ong *p*rimer *r*andom *a*mplified *p*olymorphic *D*NA.

L RNA (*l*arge RNA): The largest linear single-stranded 8.9 kb RNA molecule of the tripartite genome of Tospoviruses (family: Bunyaviridae) that is associated with the nucleocapsid proteins, forms a circle (pseudocircle) due to the complementarity of its 5′ and 3′ ends, and encodes the viral polymerase. See → M RNA, → S RNA.

LRR:
a) See → *l*eucine-*r*ich *r*epeat.
b) See → *l*ong-*r*ange *r*epeat.

LR reaction: The exclusive recombination between → lambda phage *att*L sites and *att*R sites. See → bp sites.

LSC:
a) See → large-scale copy variation.
b) See → *l*iquid *s*cintillation *c*ounter.
c) See → *l*ong *s*ingle *c*opy sequence.

LSD: See → lysine-specific histone demethylase.

L-shuffling: A variant of the → directed molecular evolution technique that randomly ligates fragments from related genes without the use of any → DNA polymerase and therefore avoids introducing unwanted mutations. In short, double-stranded fragments of the parental genes are first denatured, then hybridized ("assembled") onto one or several → matrices ("templates", assembling matrices), and their ends ligated to each other by a suitable → DNA ligase. Then ligated and non-ligated fragments are denatured and cycles of → hybridization, → denaturation and ligation repeated, until full-length novel genes are created. The assembling template is then removed, and the resulting recombined genes cloned, expressed, and the corresponding recombinant proteins screened for desirable characteristics. L-shuffling then enables randomized *in vitro* recombination of gene fragments of high or low sequence homology, and at the same time maintains and accumulates the sequence information of the parental genes (no → polymerase chain reaction involved), which generates a high number of novel functional variants (>10,000). See → DNA shuffling, → incremental truncation for the creation of hybrid enzymes, → pool-wise directed evolution, → protein complementation assay, → ribosome display, → staggered extension process.

L-shuffling

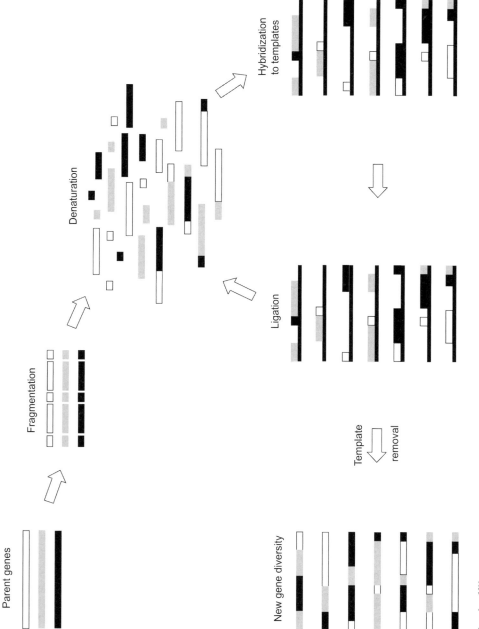

LSP: See → random amplified polymorphic DNA.

LSPT: See → locus-specific primer technology.

LSSP-PCR: See → low-stringency single specific primer polymerase chain reaction.

L strand (light strand): The strand of the double-stranded → mitochondrial DNA that floats in a → buoyant density gradient (e.g. of CsCl) at a position of lower density as compared to the → H strand.

LTR: See → long terminal repeat.

LTR retrotransposon: Any → retrotransposon that is flanked by → long terminal repeats (e.g. → copia-like or gypsy-like retrotransposons of *Drosophila melanogaster*).

Luc: The gene for → luciferase.

luc gene (Luc): A gene from the eukaryotic North American firefly, *Photinus pyralis*, encoding the 63 kDa enzyme → luciferase, which is used as → reporter gene in genetic engineering experiments. Compare → lux gene.

Luciferase (luciferin: oxygen-4-oxidoreductase, EC 1.13.12.7): An enzyme catalyzing the oxidative decarboxylation of D-luciferin (→ luciferin) to oxyluciferin in the presence of ATP, Mg^{2+} and O_2. The formation of an enzyme-bound luciferin-adenylate induces a conformational change in the protein providing a hydrophobic active site for deprotonation and addition of hydroperoxide to the C4 of luciferin. Subsequent decarboxylation and splitting of the linear peroxide produces CO_2, H_2O, AMP and an excited dianionic form of oxyluciferin. The reaction generates a flash of light that can be easily monitored. The gene encoding luciferase has been isolated from fireflies (→ *luc* gene), bacteria (*lux* gene) and coelenterata (*R luc* gene), and used as → reporter gene for the analysis of → promoters, transcription termination signals, translational enhancer elements, → transient expression, and generally the detection of the transformed state of transgenic cells, tissues or organisms. See also → bioluminescence. Compare → chemiluminescence.

Luciferin (D-luciferin, D-(-)-luciferin): Any one of structurally unrelated compounds of various organisms that are substrates for the enzyme → luciferase (luciferin: oxygen-4-oxidoreductase; decarboxylating, ATP-hydrolyzing: EC 1.13.12.7). For example, the natural substrate for luciferase from the American firefly, *Photinus pyralis*, is chemically a D-(-)-2-(6'-hydroxy-2'-benzothiazolyl)-D^2thiazoline-4-carbonic acid. Another luciferin has been isolated from the ostracode *Cypridina* and called *Cypridina* luciferin.

Formula see page 882

D-(−)-luciferin   **Oxyluciferin**

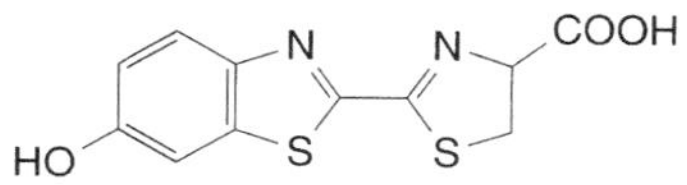

Luciferin from *Photinus pyralis*

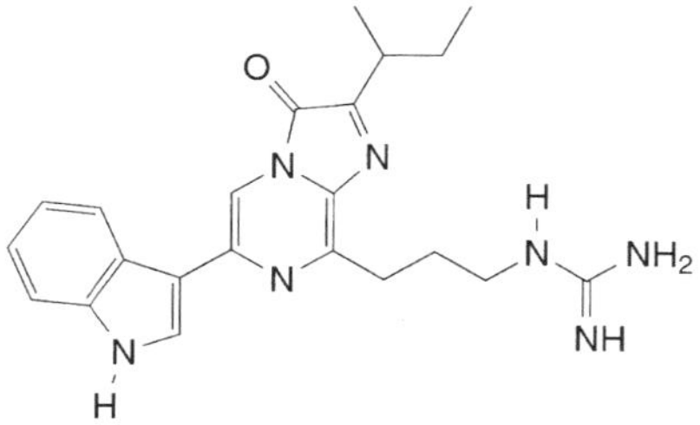

Luciferin from *Cypridina*

Luciferin

Luminescence: The emission of photons (light) by molecules excited by a chemical reaction. This may be enzymatically catalyzed, for example the naturally occurring conversion of → luciferin into oxyluciferin by the enzyme → luciferase (→ bioluminescence). In → chemiluminescence reactions, on the other hand, synthetic substrates are used for the detection of enzymes like → horseradish peroxidase or → alkaline phosphatase. See → chemiluminescence detection, → fluorescence.

Luminescence amplifying material (LAM): Any one of a series of chemicals that provide chemiluminescent substrates (e.g. → AMPPD or → CSPD) with a hydrophobic environment preventing quenching of → chemiluminescence detection. LAMs therefore serve to enhance light emission from chemiluminescent substrates ("enhancer"). Compare →enhanced chemiluminescence detection.

Luminescent reporter: Any protein that can be detected by its autofluorescence or luminescence. To this category of proteins belong aequorin (from *Aequorea victoria*), β-galactosidase (from *E. coli*), β-glucuronidase (from *E. coli*), → green fluorescent protein and its variants, bacterial → luciferase (from *Vibrio fischeri*), firefly luciferase (from *Photinus pyralis*), *Renilla* luciferase (from *Renilla reniformis*), *Vargula* luciferase (from *Vargula hilgendorfii*) and → secreted alkaline phosphatase.

Luminol: A substrate for horseradish peroxidase (HRP), employed for → chemiluminescent detections. Luminol reacts with hydrogen peroxide in the presence of horseradish peroxydase, yielding products and light that can be measured and quantified. The chemical is used for the detection of peptides and proteins in → Western blots and microplate-based immunoassays involving HRP-labeled antibodies, for nucleic acid identification in → Southern and → Northern blots, and for → in situ hybridisation techniques.

LUV: See → large unilamellar vesicle.

***lux* gene:** A gene from the prokaryotic *Vibrio harveyi*, *Vibrio fischeri* or *Photobacterium phosphoreum*, encoding the enzyme → luciferase which is used as → reporter gene in genetic engineering experiments. Compare → *luc* gene.

Lux primer: See → light upon extension primer.

Luxury gene (tissue-specific gene, cell-specific gene): A somewhat misleading term for a gene that is expressed in one or a few cell types of an organism only (e.g. the gene for hemoglobin is expressed in erythrocytes but not in other cell types).

Luxury protein: A misleading term for any protein specifically expressed in, and unique for a certain cell type. See → luxury gene.

LX-PCR: See → long-extension polymerase chain reaction.

Lymphochip: Any → microarray, onto which genes, → cDNAs or → oligonucleotides are spotted in an ordered pattern that originate from, and are selectively expressed in lymphocytes or lymphomas. Lymphochips are used to monitor changes in expression of lymphocyte-specific genes during developmental or pathological processes, and to characterize the different lymphomas.

Lyophilization: See → freeze-drying.

Lysate: A solution enriched in mature bacteriophage particles which are released from bacterial host cells during the process of host cell → lysis.

Lysate RNase protection assay (lysate RPA): A technique for the detection, quantitation and characterization of specific → messenger RNA molecules directly in crude cell or tissue lysates without prior RNA isolation. In short, the cells or tissues are lysed in the highly chaotropic → guanidinium thiocyanate buffer that rapidly inactivates cellular → ribonucleases and effectively solubilizes the cells or tissues. Then a labeled single-stranded anti-sense RNA → probe is directly hybridized to the cellular RNA in the lysate, where all the RNAs are available (as opposed to RNAs after their isolation, which inevitably leads to losses). Then → RNase A and → RNase T1 are added that digest excess RNA probe and single-stranded RNAs. The RNases are then inactivated with → proteinase K and the RNA:RNA hybrids precipitated with isopropanol. These hybrids (= protected fragments) are subsequently separated on denaturing → polyacrylamide gels. These gels are then dried, exposed to X-ray films, and the quantity and length of the hybrids determined by → autoradiography or → phosphorimaging. See → nuclease protection assay, → RNase protection assay, → S1 nuclease protection assay.

Lysine-specific histone demethylase (LSD 1): A highly conserved nuclear amine oxidase homolog that contains a C-terminal amine oxidase → domain and a central SWIRM domain mediating protein-protein interaction(s), and catalyzes → histone H3- lysine 4 (K4)-specific demethylation as a step towards → repression of the adjacent gene, or methylation of histone H3- lysine 9 (K9) as a prelude for gene activation. The oxidation reaction catalyzed by LSD1 is dependent on flavin adenine dinucleotide (FAD) and generates an unmodified lysine and formaldehyde.

Lysine tagging: The derivatisation of lysine-rich peptides with e.g. 2-methoxy-4,5-dihydro-1H-imidazole (MDHI) to increase the signal-to-noise rate of mass spectrometer signals. Tagging starts with the excision of protein spots from two-dimensional polyacrylamide gels, the in-gel tryptic digestion of the protein and subsequent reaction with MDHI. Lysine tagging reduces the problem of low mass spectrometer sensitivity towards lysine-rich peptides (as compared to e.g. arginine-rich peptides). Non-lysine amino acids are unaffected by the tagging reagent.

Lysis: The destruction of a cell, in particular by infecting virus particles, with the concomitant release of (infective) virus progeny. Cell lysis may, however, also be brought about by enzymes, see for example → lysozyme.

Lysis-inducing domain (LID): One of three → domains of a bacterial protein that is involved in the self-assembly of filaments for the bacterial cytoskeleton. Normally, all three domains are necessary for a successful formation of filaments. If two domains are deleted, the remaining domain cannot substitute the deficiency, which leads to a destabilization of the cytoskeleton. This effect is exploited for the induced lysis of bacteria. First, the one-domain protein-encoding gene is engineered into the target bacterium and expressed. Then the resulting protein competes favourably with the endogenous three-domain proteins for binding sites on the cytoskeleton. This in turn leads to the incorporation of the "faulty" protein, interruptions in the filaments, destabilization of the cytoskeleton and lysis of the cell. This procedure avoids the traditional use of either mechanical devices or chemical agents for the lysis of bacteria. See → Birnboim-Doly method.

Lysogen (lysogenic bacterium, lysogenized bacterium, lysogenic host): Any bacterium that contains DNA from a → temperate bacteriophage (a → prophage). This DNA may be integrated into the bacterial genome at specific sites (e.g. *E. coli att* site in case of the → lambda phage), or may be maintained as an independent entity (e.g. as a plasmid). Presence of a prophage confers immunity against a secondary infection by the same phage, see → phage exclusion. See also → lysogenic bacteriophage, → lysogeny.

Lysogenic bacteriophage (lysogenic virus, lysogenic phage): A bacteriophage that does not lyse the host cell but either integrates into its genome (→ prophage) or stays as a separate entity (e.g. in a plasmid) within the cell. In both cases, the DNA of the lysogenic phages is replicated coordinately with the host chromosome. See also → lysogeny.

Lysogenic bacterium: See → lysogen.

Lysogenic conversion: See → phage conversion.

Lysogenic immunity: See → phage exclusion.

Lysogenic phage: See → lysogenic bacteriophage.

Lysogenic repressor: A protein that prevents a → prophage to enter the → lytic cycle.

Lysogenic response: The response of a bacterial host cell that is infected with a → temperate bacteriophage whose DNA does not enter the → lytic cycle but instead becomes lysogenic. See → lysogen, → lysogenic bacteriophage, → lysogenic repressor; compare → lytic response.

Lysogenic virus: See → lysogenic bacteriophage.

Lysogenization: The establishment of a strain of lysogenic bacteria (→ lysogen) by mixing permissive host cells with either the DNA of a → temperate phage, or the temperate phage itself.

Lysogenized bacterium: See → lysogen.

Lysogeny: The integration of → temperate bacteriophage DNA into the host cell chromosome (e.g. → lambda phage integration at *E. coli att* sites, leading to a → prophage) or its maintenance as a physically independent → replicon ("episome", e.g. bacteriophage P22), in which the lytic functions are repressed. The lysogenic phage DNA is replicated coordinately with the host DNA, but may escape host control e.g. after UV irradiation, and enter the so-called → lytic cycle.

Lysopine (N2-[1-D-carboxyethyl]-L-lysine): An amino acid derivative that is synthesized in plant cells transformed by the soil bacterium → *Agrobacterium tumefaciens*. Lysopine belongs to the so-called → opines. See also → crown gall.

Lysozyme (muramidase, mucopeptide N-acetylmuramoyl hydrolase, EC 3.2.1.17): An enzyme (e.g. from chicken egg white, but also found in bacteria and plants) that catalyzes the hydrolysis of the cell wall peptidoglycans of many bacteria by cleaving the β-1,4 linkage between N-acetylmuramic acid and N-acetyl-glucosamine. This leads to the removal of the cell wall, and eventually to → lysis of the cell. Integral component of antibacterial defense systems (e.g. tears, saliva, mucosa).

lysY: An *E. coli* mutant that carries a mutant gene for lysozyme from the bacteriophage T7. The mutation K128Y encodes a protein still binding to, and inhibiting T7 RNA polymerase, but is defective in lysozyme activity.

Lyticase: An enzyme preparation from culture media of *Oerskovia xanthinelytica*, containing β-(1-3)-glucanase and protease activities. It is used for the preparation of fungal → spheroplasts and yeast chromosomes in agarose plugs as a preparative step for → pulsed field gel electrophoresis.

Lytic cycle (lytic pathway): A viral or bacteriophage life cycle, in which progeny viruses are formed at the expense of the host which will finally be lysed. During this phage multiplication, the host DNA is degraded and its nucleotides are used for phage DNA synthesis. Furthermore the host ribosomal system is misused for the production of viral proteins. See also → lambda phage.

Lytic infection (productive infection): The infection of host cells by → bacteriophages, which enter the → lytic cycle. Generally, lytic infection is any viral infection of a host cell leading to the production of viral progeny.

Lytic pathway: See → lytic cycle.

Lytic response: The → lysis of a bacterial host cell in response to its infection by a → virulent bacteriophage or a → temperate bacteriophage that enters the → lytic cycle.

Lytic virus: Any virus whose multiplication leads to the → lysis of the host cell, also a → virulent bacteriophage. Compare → lysogenic bacteriophage.

M

M:

a) Abbreviation for either → adenine or → cytosine (a*M*ino in large groove), used in sequence data banks.
b) See → mismatch, → perfect match.

^{m}A (mA): Abbreviation for adenine carrying a methyl group (e.g. at N^6).

MAAP: See → arbitrarily amplified DNA.

mAB: See → *m*onoclonal *a*nti*b*ody.

MAB: See → marker-assisted breeding.

MAC:

a) See → *m*ammalian *a*rtificial *c*hromosome
b) See → *m*ap-*a*ssisted *c*loning.
c) See → mutagenesis in aging colonies.

Macroarray (nylon macroarray): Any nylon or nitrocellulose membrane (also plastic support), onto which several hundreds to a few thousand target molecules (e.g. cDNAs, oligonucleotides) are regularly spotted ("gridded"). In contrast, → microarrays contain thousands, hundred thousands, or millions of spots. Macroarrays are typically hybridised to radioactively labelled → probes, and the hybridization and washing procedures are identical to the development of a → Southern blot. Also, detection of the hybridisation event occurs via → autoradiography. Macroarrays are optimal for the expression analysis of a limited set of genes, as e.g. genes encoding enzymes of a particular metabolic pathway.

Macroautoradiography: See → autoradiography.

Macrochromosome: Any one of several → chromosomes of most (if not all) avian orders and some primitive vertebrates that has an average size of 50–200 Mb. For example, the domestic chicken (*Gallus gallus*) genome consists of three chromosome size classes: five macrochromosomes (GGA 1–5), measuring from 50–200 Mb in size, five intermediate chromosomes (GGA 6–10) with sizes ranging from 20–40 Mb, and 28 → microchromosomes (GGA 11–38) spanning from 3–12 Mb.

***Macro*H2A (mH2A):** Any one of a family of variants of the canonical → histone H2A, whose amino-terminal third is almost identical to full-length H2A, but additionally carries a unique nonhistone carboxy-terminal tail ("macrodomain"), part of which resembles a → leucine zipper that makes it nearly three times bigger than the conventional H2A histone. This large non-histone region distinguishes mH2A from all other known core histones. mH2A binds to → nucleosomes in various target genes (e.g. *Fox* genes, Hedgehog and Wnt signalling protein-encoding genes, genes coding for T-box transcription factors, hormone receptors, paired-box proteins, DLX, POUF and MEIS homeobox proteins, and retinoic acid signalling proteins, to name few), and is

The Dictionary of Genomics, Transcriptomics and Proteomics, 4th Ed., Günter Kahl
Copyright © 2009 WILEY-VCH Verlag GmbH & Co. KGaA, Weinheim
ISBN: 978-3-527-32073-8

preferentially concentrated on the inactive X chromosome of female mammals, and may be involved in the transcriptional silencing of this chromosome. The enrichment of macroH2A at the Xi forms a characteristic structure in the female nucleus, the *macrochromatin body* (MCB). About 10% of all human genes contain mH2A in the nucleosomes of their promoters. Two distinct mH2A proteins exist in mammalian tissues, are called mH2A1.1 and mH2A1.2 mH2A2, differ from one another in only one region, and are both enriched in Xi chromatin. These subtypes are encoded by the same gene, arise by alternative splicing, and have distinct expression patterns during development and in different adult organs. A third mH2A subtype is transcribed from a separate gene.

Macromolecule: Any molecule whose molecular weight exceeds a few thousand daltons (e.g. polysaccharides, proteins, nucleic acids).

Macronucleus: The larger of the two nuclei of certain protozoan species (ciliatae) that actively transcribes its genes during asexual growth, replicates during asexual reproduction, but is destroyed and re-formed during sexual reproduction. Therefore, macronuclei do not transmit genetic information to sexual offspring. Destruction of the macronucleus is followed by the development of a new macronucleus, which starts with multiple rounds of → DNA replication and leads to the formation of → polytene chromosomes. The extent of → polyploidization varies with the species, but it reaches ploidy levels of up to 64. The polytenic chromosomes undergo fragmentation, vesicle-like structures form and enclose the different chromosome fragments: short macronuclear chromosomes appear. The vesicles persist, but large amounts of DNA are eliminated (see → DNA deletion). In some cases, up to 90% of the micronuclear genome vanishes. Finally, the vesicles decay, and multiple rounds of DNA replication produce the ultimate ploidy level of the mature macronucleus. Compare → gene-sized DNA, → micronucleus, → nuclear dimorphism.

Macro-restriction map: A graphical description of the linear arrangement of an ordered set of large DNA fragments generated by → rare-cutting → restriction endonucleases of → genomic DNA. This type of → physical map spans long DNA stretches and can be used to localize interesting sequences (e.g. genes) by hybridization of fluorescent or radiolabeled → probes to the different cloned restriction fragments. See → ordered clone map.

Macrosatellite: A somewhat imprecise term for a → satellite DNA that exceeds a certain size (which is not precisely fixed). See → megasatellite, → microsatellite, → minisatellite.

Macrosynteny: The conserved order of large genomic blocks (in the megabase range) in the genomes of related (but also unrelated) species, as detected by e.g. → chromosomal *in situ* suppression hybridization, → fluorescent *in situ* hybridization. See → homosequential linkage map, → microsynteny, → synteny.

MAD-DNA: See → moderately affected Alzheimer disease DNA.

MADGE: See → *m*icroplate *a*rray *d*iagonal *g*el *e*lectrophoresis.

MADS box: A 56 bp conserved motif of → transcription factors that function in the regulation of various genes (e.g. *MCM1* of yeast, *ag*amous homoeotic gene *AG* of *Arabidopsis thaliana*, *DEF A* [*def*icient flower] gene in *Antirrhinum majus*).

MADS box gene: Anyone of a series of genes encoding → transcription factors containing a highly conserved domain of 56 amino acids (→ MADS box domain) functioning as a DNA-binding site.

MAF: See → minor allele frequency.

Magic spot: See → guanosine tetraphosphate.

Magnet-*a*ssisted *s*ubstraction *t*echnique (MAST): A method for the detection of sequences expressed in only one of two cell types. In short, total → RNA is separately isolated from both cell types, and each RNA separately chromatographed over oligo(dT)$_n$ fixed to → magnetic beads. The → polyadenylated RNA, including most → messenger RNAs, is bound to the oligo(dT) and can easily be separated from the → poly (A)$^-$-RNA by magnetic force and washing. Then the poly(A)$^+$-RNA from cell type A is converted to → cDNA, using → reverse transcriptase, and the resulting mRNA-cDNA complexes denatured such that the cDNA remains attached to the paramagnetic beads ("driver cDNA"). The same procedure produces cDNA from mRNA of cell type B ("tracer cDNA"). Now all cDNAs that are present in equal amounts in both cell types, are removed using a 25-fold excess of driver cDNA. Only those cDNAs will remain that are expressed in cell type B specifically. Compare → subtractive hybridization.

Magnetic bead (*p*ara*m*agnetic *p*article, PMP): Paramagnetic materials (e.g. iron oxide), coated with → polyacrylamide and → agarose and packaged into submicron-sized particles that have no magnetic field but form a magnetic dipole when exposed to a magnetic field. Magnetic beads serve as a solid-phase support for the separation of DNA or RNA molecules from complex mixtures of biomolecules. Specific binding is usually achieved via specifically designed DNA fragments (e.g. → oligonucleotides) coupled to the magnetic beads. See → magnetic crosslinking, → magnetic polyvinyl alcohol, → paramagnetic particle technology.

Magnetic crosslinking: A somewhat incorrect term for a simple technique to separate contaminating → cloning vector sequences from labeled (e.g. by → nick-translation) DNA inserts that uses → magnetic beads to which single-stranded capture DNA sequences are covalently attached. These capture sequences will hybridize to the undesired vector DNA, and can be removed together with the magnetic support by centrifugation or magnetic separation. Compare → paramagnetic particle technology.

Magnetic *p*oly*v*inyl *a*lcohol (M-PVA) microparticle: A somewhat misleading term for → magnetic beads, composed of the hydrogel matrix polyvinyl alcohol with encapsulated magnetite, charged with a functional group (e.g. –COOH, –NH$_2$, –NHR, –CHO, → streptavidin, → oligo d(T), or → protein A) that allows interaction with and binding of the corresponding site on e.g. a protein. Since M-PVA shows only minimal unspecific protein adsorption compared to other carrier media, it is used for immunoassays, affinity separations, → messenger RNA isolation, and detection of → DNA- or RNA-binding proteins.

Magnetic *r*esonance *i*maging (MRI): A technique for the creation of an *in vivo* image of gene expression that uses an MRI contrast agent with the ability to indicate → reporter gene expression. For example, (1-[2-(b-galactopyranosyloxy)propyl]-4,7,10-tris(carboxymethyl)1,4,7,10-tetra-azacyclododecane)gadolinium(III); EgadMe), in which access of water to the first coordination sphere of a chelated paramagnetic gadolinium ion (Gd^{3+}) is blocked by a galactopyranose residue, is such a contrast agent. The galactopyranose cap is released by cleavage catalyzed by → beta-galactosidase, which exposes the Gd^{3+} to water, allowing modulation of water proton relaxation times and increase in magnetic resonance signal intensity. Regions of a cell or an organelle with higher MR image intensity therefore correlate with higher expression of the reporter gene.

Magnetofection: An *in vitro* and *in vivo* technique for the rapid and efficient transfer of any nucleic acid (e.g. a → plasmid vector and its → insert) into target cells by loading it onto the surface of superparamagnetic nanoparticles (SMNs) using salt-induced colloid aggregation. In short, the magnetic particle (usually superparamagnetic iron oxide, SPIO, particle diameter >50 nm, or ultrasmall SPIO, uSPIO, particle diameter <50 nm) is first coated with a polyelectrolyte (e.g. polyethyleneimine, PEI), then mixed with naked DNA in a salt-containing buffer. The DNA binds to or co-aggregates with the particles. Target cells are then incubated with the particle-DNA cocktail and exposed to a magnetic gradient field that attracts the particles toward the cells and arrests them on the cell's surface. Magnetofection increases the number of transfected cells in comparison to other techniques of → artificial gene transfer, because it expands the time of exposure. The technique is also employed for magnetic drug targeting (MDT), in which the magnetic nanoparticles (MNs) are loaded with e.g. chemotherapeutics that finally are enriched in targeted tumor tissue. A suspension of superparamagnetic nanoparticles can also be injected into a tumor and heated by an alternating magnetic field, which destroys the tumor directly.

Magnetophage: Any → bacteriophage that is loaded with → paramagnetic particles. In short, the selected phage is first treated with iron oxide particles pre-treated with epichlorhydrin. The epichorhydrinated particles can then react with amino groups of the phage's coat proteins, which conjugates an average of 80 iron oxide particles to each phage. Magnetophages can be used as magnetic reporter and contrast agent for magnetic resonance imaging (MRI).

Magnification: The increase in 18S and 28S → ribosomal RNA genes that occurs in germ-line cells of rDNA-deficient *Drosophila* flies. The → amplification process probably occurs via unequal sister chromatid exchange (see → unequal crossing-over).

Main band: A broad band of genomic DNA that appears after → isopycnic centrifugation in cesium chloride density gradients in the presence of ethidium bromide. It contains most of the cellular DNA, including → cryptic satellites.

Maintenance gene: Any one of a set of genes that are turned on early in fetal development and remain active throughout the lifetime of the organism. For example, ATP synthase of mitochondria, elongation factor EF-1 a, histone deacetylase, RNA polymerase II, and ubiquitin-conjugating enzymes are encoded by such maintenance genes. See → housekeeping gene.

Maintenance methyltransferase: Any one of several nuclear enzymes that catalyze the transfer of methyl groups onto pairs of cytosines on complementary DNA strands at CpG dinucleotides after DNA replication.

Major gene: Any → gene, whose contribution to the expression of a particular polygenic trait is superior to the contribution(s) of other → minor gene(s).

Major groove: The indentation on the surface of a DNA → double helix molecule, formed by the sugar phosphate backbones and the edges of the base pairs (linked by → Watson-Crick base pairing forces) that contain the N6, N7, O6 (in → purines) or O4 and N4 atoms (in → pyrimidines). See → double helix, → minor groove.

Major transcript: The most abundant → messenger RNA among two (or more) alternatively spliced transcripts from the same gene. See → alternative splicing, → alternative transcript.

MALDI: See → matrix-assisted laser desorption-ionization mass spectrometer.

MALDI-MS: See → matrix-supported laser desorption-ionization mass spectrometry.

MALDI post source decay mass spectrometer (MALDI-PSD-MS): A specially designed mass spectrometer that allows to determine the masses of peptide fragments, generated by ionization of isolated proteins. The mass spectrometer contains a reflector ("reflectron") that diverges the ions from their normally linear flight, such that their speed is first slowed down and then their direction and speed of flight are changed. After reflection, they reach the detector according to their mass-charge ratio. Since fragmentation occurs at the reflector (i.e. after the acceleration), this type of analysis is called post source decay (PSD) mass spectrometry.

X: helix axis
The oxygen of the guanine deoxyribose ring lies above, the oxygen of the cytosine deoxyribose ring below the level of the base pair.

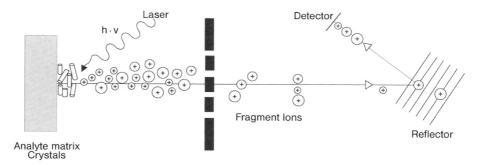

Maldi

As a result of MALDI-PSD analysis, a spectrum of peptide fragment ions becomes available that can be compared to the theoretically expected fragment ions of the known proteins, or peptide sequence accessions in appropriate data banks, so that proteins and their post-translational modifications can be identified, using software packages as e.g. SEQUEST. See → tandem mass spectrometer.

MALDI-PSD-MS: See → MALDI *post source decay mass* spectrometer.

MALDI-TOF: See → matrix-assisted laser desorption ionisation-time of flight.

Maltose *binding protein* (MBP): A protein whose gene is used in gene fusion experiments. See → protein fusion and purification technique.

Mammalian *artificial chromosome* (MAC): A high-capacity → cloning vector for mammalian cells that contains a mammalian → origin of replication, → telomeres, → centromeres, and other sequences necessary for its function in mammals. Since MACs are not integrated into the host cell genome, but nevertheless stably maintained at one copy per cell, they may be used in → gene therapy. See → bacterial artificial chromosome, → human artificial chromosome, → plant artificial chromosome, → P1 cloning vector, → *Schizosaccharomyces pombe* artificial chromosome, → transformation-competent artificial chromosome vector, → yeast artificial chromosome.

Mammalian cell display (mammalian cell surface display): Any one of a series of techniques for the presentation of distinct peptides or proteins close to, or on the surfaces of mammalian cells. For example, genes encoding the display peptides or proteins can be integrated into the single-stranded DNA genome of a mammalian virus, and the corresponding peptides expressed as → fusion proteins with a viral coat protein under the control of the cytomegalovirus immediate early promoter (high transcriptional rates in most cell lines) or the human ubiquitin C promoter (less active). The fusion proteins are then exposed to the surrounding medium. Another system exploits engineered human cells that display functional single-*c*hain Fv (scFv) antibodies. To that end, the anti-CD22 scFv fused to the N-terminal portion of the transmembrane domain of human *p*latelet-*d*erived growth *f*actor *r*eceptor (PDGFR) displayed on human embryonic kidney (HEK) 293T cells. The selected scFv can easily be converted to whole IgG antibodies or immunotoxins for clinical use. See → *Bacillus* spore display, → baculovirus expression system, → CIS-display, → mammalian cell display, → mammalian cell surface display, → messenger RNA display, → phage display, → phagemid display, → ribosome display, → yeast display.

Mammalian Gene Collection (MGC): A project of the National *I*nstitute of *H*ealth (NIH) that provides a complete set of → full-length → open reading frame sequences and → cDNA clones of expressed human and mouse (in future other mammal) genes.

Mammalian two-hybrid system: A variant of the conventional yeast → two-hybrid system (YTH, Y2H) for the detection of even relatively weak and transient protein-protein interactions *in vivo* and the identification of genes encoding interacting proteins in mammalian cells. The system is based on the dual modular composition of many eukaryotic → transcriptional activators (e.g. the *GAL*4 protein of yeast) that contain two discrete, physically separable, functionally independent molecular → domains, a target-specific *D*NA-*b*inding *d*omain (DBD) that binds to a specific → promoter sequence, and a target-independent → *a*ctivation *d*omain (AD). The DBD serves to target the → transcription factor to specific → promoter sequences (e.g. → *u*pstream *a*ctivation *s*equences [UAS] in yeast), whereas the AD directs the → RNA polymerase II(B) transcription complex to transcribe the gene downstream of the DNA-binding site. Both domains are required for transcriptional activation, and neither domain alone can activate transcription. However, a non-covalent interaction of two independent hybrid proteins containing a DBD and an AD, respectively, leads to a reconstituted (i.e. active) transcription factor, triggering the expression of a → reporter gene by the linked promoter. The system therefore exploits the interaction of proteins expressed from two → hybrid genes that are constructed *in vitro* and then independently transferred into, and maintained in mammalian cells on two separate, but compatible → plasmids. In short, for the mammalian two-hybrid system, one protein of interest is expressed as a → fusion to the *Gal*4 DNA-binding domain, and another protein is expressed as a fusion to the activation domain of the VP16 protein of the herpes simplex virus. The plasmid vectors expressing these fusion proteins under the control of the *Si*mian *v*irus 40 (SV40) promoter are co-transfected into a mammalian cell line (e.g. HeLa cells, also *C*hinese *h*amster *o*vary, CHO, cells, or CV1 cells) together with a reporter → *c*hloroamphenicol *a*cetyl*t*ransferase (CAT) vector. The reporter plasmid contains the *cat* gene under the control of five → consensus *Gal*4 binding sites upstream of the E1b → minimal promoter. If the two fusion proteins interact, as strong increase in cat reporter gene expression ensues. Mammalian two-hybrid systems probably mimic the *in vivo* situation in mammalian cells better yeast two-hybrid systems, because protein folding, → post-translational modifications (e.g. phosphorylation, acetylation, or glycosylation) and sub-cellular localization of the hybrid proteins may well be different in the two cellular systems. See → dual-bait two-hybrid system, → interaction mating, → interaction trap, → LexA two-hybrid system, → one-hybrid system, → repressed transactivator (RTA) yeast two-hybrid screen, → reverse two-hybrid system, → RNA-protein hybrid system, → split-hybrid system, → split-ubiquitin two-hybrid system, → three-hybrid system, → two bait system.

Mammalian vector: Any → cloning vector that functions in mammalian cells.

Mammalian-wide interspersed repeat (MIR): Any one of about 120,000 to 300,000 copies of → transfer RNA-derived → short interspersed nuclear elements (SINEs) of the primate genome that can transpose either in → sense or → antisense orientation, also into genic sequences. The original MIR seems to be a 260 bp SINE, fragments of which are found as 70–100 bp elements. A central core region of about 25 bp is conserved in the MIRs of different mammals.

Mannopine: An amino acid derivative that is synthesized in plant cells transformed by the soil bacterium *Agrobacterium tumefaciens*. Mannopine belongs to the so-called → opines. See also → crown gall.

$$\begin{array}{c} CH_2OH \\ | \\ (CHOH)_4 \\ | \\ CH_2 \\ | \\ NH \\ | \\ HOOC-CH-(CH_2)_2-CO-NH_2 \end{array}$$

Mant nucleotide: See → N-Methylanthraniloyl nucleotide.

Map:

a) A graphical description of genetically or physically defined positions on a circular (e.g. → plasmid) or linear DNA molecule (e.g. → chromosome) and their relative locations and distances. A map may show the distribution of specific → restriction sites (→ restriction map), genes (→ gene map), markers (→ marker map, chromosome markers (→ chromosome map), or the distance between two loci (e.g. a marker and a gene) in base pairs (→ physical map) or → centiMorgans (→ genetic map). The term is now also used for the illustration of peptide-peptide-, peptide-protein-, and protein-protein interaction networks in a cell or an organelle, and for the intracellular distribution of low-molecular weight cellular compounds (metabolites). See → BAC map, → base pair map, → biallellic genetic map, → bit map, → cDNA map, → cell map, → chromosome expression map, → chromosome features map, → chromosome map, → circular linkage map, → circular restriction map, → Cleveland map, → clone-based map, → contact map, → content map, → contig map, → cytogenetic map, → deletion map, → denaturation map, → diallelic map, → difference map, → diversity map, → DNA map, → doublet frequency map, → epitope map, → EST map, → expression map, → expression imbalance map, → fine-structure map, → frequency distance map, → functional map, → functional map atlas, → gene expression map, → gene expression terrain map, → gene map, → genetic map, → genome control map, → genome fingerprint map, → genome map, → haplotype map, → high density genetic map, → high-density map, → high resolution genetic map, → high resolution physical map, → homology map, → homosequential linkage map, → *in silico* map, → integrated map, → integrated physical-genetic map, → interactome map, → landmark map, → linkage map, → long-range restriction map, → macro-restriction map, → map, → marker map, → metabolic map, → methylation map, → microsatellite map, → nucleotide diversity map, → ordered clone map, → peptide map, → pharmacophore map, → physical map, → protein expression map, → protein interaction map, → protein linkage map, → protein-protein interaction map, → proteome map, → quantitative chromosome map, → radiation hybrid map, → recombinational map, → recombination frequency map, → response regulation map, → restriction map, → RN map, → SAGE map, → segregation map, → self-organizing map, → sequence map, → sequence-tagged sites map, → SNP map, → telomere map, → transcript map, → transcriptome map, → two-dimensional gel map, → ultra-high density map, → YAC map.

b) See → mutagenesis assistant program.

mAP: See → *m*essenger *a*ffinity *p*aper.

Map-assisted cloning: See → positional cloning.

Map-based cloning: See → positional cloning.

Map-based sequencing: See → clone-by-clone sequencing.

Map distance: The distance between two genes on a linear DNA molecule, expressed as → map units or centiMorgans (see → Morgan unit).

MAPH: See → multiplex amplifiable probe hybridization.

MAP kinase: See → mitogen-activated protein kinase.

Mapmaker: An interactive computer program for the construction of genetic → linkage maps that allows the estimation of the most likely order of specific genetic loci (e.g. → RFLP), and recombination frequencies between them. Calculations presuppose extensive data on meiotic segregation.

Mapped restriction polymorphism (MRP): See → mapped restriction site polymorphism.

Mapped restriction site polymorphism (MRSP; *m*apped *r*estriction *p*olymorphism, **MRP):** A variant of the → restriction *f*ragment *l*ength *p*olymorphism (RFLP) technique for the genetic fingerprinting of individual genomes that is based on the amplification of target sequences (e.g. genes) with → primers complementary to conserved parts of these sequences. In short, genomic DNA is first isolated and primers directed against conserved gene sequences used to amplify these genes in a conventional → polymerase chain reaction in the presence of a ^{32}P-labeled deoxynucleoside triphosphate (e.g. dCTP). Frequently employed primers span conserved regions in e.g. 16S rRNA (*rrs*) and 23S rRNA (*rrl*) genes of eubacteria. After amplification, the products are restricted with → restriction endonucleases that cleave frequently in the target sequence. The restriction fragments are then separated by native → polyacrylamide gel electrophoresis and visualized by → autoradiography. Differences in the electrophoretic mobility of bands represent differences in the distribution of restriction sites along the target gene(s).

Mapping:

a) The plotting of gene positions or other defined sites along a strand of DNA. See also → acetylation mapping, →

admixture mapping, → antigenic mapping, → association mapping, → bottom-up mapping, → cell mapping, → centromere mapping, → chromosome mapping, → clinical mapping, → comparative gene mapping, → comparative mapping, → compositional mapping, → contact mapping, → contig mapping, → cosmid insert restriction mapping, → cross-mapping, → deletion mapping, → denaturation mapping, → domain mapping, → epitope mapping, → exon-intron mapping, → expressed sequence tag mapping, → expressed sequence tag polymorphism mapping, → expression mapping, → fine mapping, → function mapping, → gene mapping, → genetic mapping, → genome mapping, → haplotype mapping, → HAPPY mapping, → heteroduplex mapping, → high density mapping, → high resolution physical mapping, → H-mapping, → homozygosity mapping, → *in silico* mapping, → integrative mapping, → interphase mapping, → intron-exon mapping, → localisome mapping, → long-range restriction mapping, → map, → megabase mapping, → nucleotide analogue interference mapping, → nucleotide mapping, → optical mapping, → pathway mapping, → peptide mapping, → protein expression mapping, → protein-protein interaction mapping, → proteome mapping, → QTL mapping, → radiation hybrid mapping, → receptor mapping, → restriction mapping, → retentate mapping, → saturation mapping, → Smith-Birnstiel mapping, → S1 mapping, → STS content mapping, → telomere mapping, → top-down mapping, → visual mapping. Compare → epitope mapping.

b) MAPPing, see → *m*essage *a*mplification *p*henoty*p*ing.

Mapping population: The group of related organisms used for the construction of a → genetic map.

MAPREC: See → *m*utant *a*nalysis by *P*CR and *r*estriction *e*nzyme *c*leavage.

MAPS: See → *m*inisatellite-primed *a*mplification of *p*olymorphic *s*equences.

Map unit: One centiMorgan (cM). See → Morgan unit.

Map Viewer: A software component of → Entrez Genomes providing special browsing capacities for a subset of organisms that allows to view and search an organism's complete → genome, display → chromosome maps, and zoom into progressively greater levels of detail, down to the sequence data for a region of interest. In case multiple maps of a chromosome exist, Map Viewer aligns and displays them based on shared → marker and → gene names and, for the sequence maps, on a common sequence coordinate system. The organisms currently represented in Map Viewer are listed in the "Entrez Map Viewer help" document, which provides general information for its use. The number and types of available maps vary from organism to organism, and are described in the "data and search tips" file provided for each organism.

MAR: See → scaffold-associated region.

Mariner: Anyone of a family of animal → transposons (originally detected in insects and related arthropods, but also present in the genomes of other animals, including man).

Marker:

a) A → genetic marker.
b) Any protein, RNA or DNA molecules of known size or molecular weight that serve to calibrate the electrophoretic or chromatographic separation of proteins, RNAs and DNAs, respectively. See → binning marker, → ladder, → molecular weight standard.

Marker-assisted breeding (MAB): The use of → MOlecular markers for the development of new animals and plant varieties, e.g. by → marker-assisted selection. See → marker-assisted introgression.

Marker-assisted introgression: A technique to facilitate → introgression of desirable genes into target organisms that is based on the detection of → molecular markers closely linked to the gene encoding the trait of interest, and the monitoring of their fate in the progeny of sexual crosses. Marker-assisted introgression therefore avoids lengthy evaluation processes (e.g. the continuous monitoring of the phenotype of plants in the field over several years). See → marker-assisted breeding, → marker-assisted selection.

Marker-assisted selection (marker-based selection, MAS; marker-mediated selection, MMS): A technique to select individual organisms (bacteria, fungi, plants, animals) carrying a desirable gene with the aid of genetic → markers linked to the gene. For example, marker-based selection allows to screen for pathogen-resistant plants in germplasm collections via markers closely linked to the gene(s) for resistance without exposing them to the pathogen. Additionally, MAS appreciably speeds up the process of conventional animal and plant breeding.

Marker-based patient selection (MBPS): The identification of a specific → genotype in a population of human individuals (e.g. patients) that is diagnostic for a specific disease susceptibility, or sensitivity (insensitivity) to a particular drug (responder → non-responder). Compare → marker-assisted selection.

Marker-based selection: See → marker-assisted selection.

Marker bracket: The location of two or more → genetic or → molecular markers in the vicinity of a → gene, so that it is tagged both → upstream and → downstream ("bracketed").

Marker exchange: See → homogenization.

Marker map: Any → genetic or → physical map that is either based on phenotypic (→ morphological) or → molecular markers. See also → chromosome expression map, → chromosome map, → cytogenetic map, → denaturation map, → diversity map, → expression map, → fine-structure map, → frequency distance map, → gene map, → genetic map, → integrated map, → landmark map, → linkage map, → macrorestriction map, → map, → nucleotide diversity map, → ordered clone map, → quantitative chromosome map, → recombinational map, → recombination frequency map, → response regulation map, → restriction map, → RN map, → sequence map, → SNP map, → ultra-high density map.

Marker-mediated selection (MMS): See → marker-assisted selection.

Marker rescue:

a) The survival of gene(s) from an irradiated, inactive bacteriophage, by recombination with an unirradiated active bacteriophage. If a bacterial host is infected with two genetically marked phages (mixed infection) of which only one type is irradiated (and hence inactivated), rare recombination processes occur between both phage types. Thus recombinants can be found that contain genes from the irradiated parent. These are referred to as "rescued".

b) The re-isolation of a → genetic marker from a transgenic host, into which it has been transferred e.g. by → direct gene transfer techniques. Marker rescue allows the detection of marker alterations (e.g. truncations, → deletions, → inversions, generally → rearrangements) which have occurred during its transfer to the host and/or its integration into the host's genome.

MAS:

a) See → *m*arker-based *s*election.
b) See → *m*askless *a*rray *s*ynthesizer.

MASA: See → *m*utant *a*llele-specific *a*mplification.

Masked *m*essenger *RNA* (masked mRNA): An inactive, stable and longlived → messenger RNA that has to be unmasked before its translation. Such masked messages occur in such diverse systems as unicellular algae (e.g. *Acetabularia*), angiosperm seeds, and echinoderm oocytes. Masking is brought about by RNA-binding proteins ("mRNA masking proteins") that probably need phosphorylation for their activity. Activation of masked messenger RNAs is catalyzed by → polyadenylation.

Masked mRNA: See → masked messenger RNA.

Maskless *a*rray *s*ynthesizer (MAS): A computerized instrument for the light-directed synthesis of high-resolution oligonucleotide microarrays, using a digital micromirror array generated on a computer to form virtual masks, instead of the conventional chrome/glass photolithographic masks. In short, microscope slides are first silanized. Then the photolabile protecting group (R, S)-1-(3,4-(*m*ethylene-dioxy)-6-*n*itro*p*henyl) ethyl chloroformate (MeNPOC) is attached to the nucleotides and *h*exaethyleneglycol (HEG) as a spacer molecule. The photoprotected HEG is converted to a phosphoramidite, which in turn is covalently bound to the silanized slide. This procedure produces a microscope slide covered with a monolayer of spacer molecules containing hydroxyl groups protected by photolabile MeNPOC groups. These protective groups are conventionally removed by UV-light revealing all free hydroxyl groups. This deprotection does not occur at random, if MAS is used. Instead, a high-resolution pattern of UV light directed with the 786,000 individually regulatable aluminum micromirrors of the socalled *d*igital *m*icromirror *d*evice (DMD) of the MAS and projected onto the microscope slide reproduces an identical pattern of free hydroxyl groups on its surface. The DMD then creates digital masks that replace the rigid chromium masks used for the production of conventional high-density microarrays. Coupling of nucleotides then occurs at the free hydroxyl groups. Such MAS-produced microarrays accomodate about 100,000 oligonucleotides on spaces of $16\,\mu m^2$ with the potential to discriminate single-base → mismatches in thousands of genes simultaneously. Do not confuse with → marker-based selection.

Massively parallel picoliter reactor sequencing: See → fiber-optic reactor sequencing.

Massively parallel signature sequencing (MPSS): A high-throughput technique for the sequencing of millions of → cDNAs conjugated to oligonucleotide tags on the surface of 5 µm diameter microbeads that avoids separate cDNA isolation, template processing and robotic procedures. In short, 32-mer capture oligonucleotides are

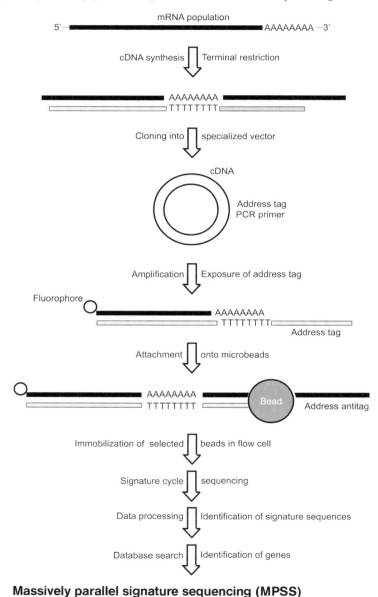

Massively parallel signature sequencing (MPSS)

attached to the surface of separate microbeads (diameter: 5 µm) by combinatorial synthesis, such that each microbead has a unique tag for its complementary cDNA. Then → messenger RNA is reverse transcribed into → cDNA using oligo(dT) primers, restricted at both ends with e.g. *Dpn* I, complements of the capture oligonucleotides are attached to the poly(A) tail of each cDNA molecule and the construct cloned into an appropriate vector containing PCR handles, which serve as primer-binding sites for → polymerase chain reaction based amplification of the tagged cDNA. The cDNA is now amplified with a → fluorochrome-labeled primer, denatured, and the single-stranded address tag-containing fragments annealed ("cloned") to the surface of microbeads containing address tag sequences as hybridization anchors, and then ligated ("*in vitro* cloning"). Each microbead displays about 100,000 identical copies of a particular cDNA ("microbead library"). The fluorescent microbeads (all containing a cDNA) are then separated from non-fluorescent ones (not containing a cDNA) by a *f*luorescence-*a*ctivated *c*ell *s*orter (FACS). Each single microbead in the library harbors multiple copies of a cDNA derived from different mRNA molecules. If a particular mRNA is highly abundant in the original sample, its sequence is represented on a large number of microbeads, and vice versa. In the original version of MPSS, 16–20 bases at the free ends of the cloned templates on each microbead are sequenced ("signature sequences"). First, millions of template-containing microbeads are assembled in a densely packed planar array at the bottom of a flow cell such that they remain fixed as sequencing reagents are pumped through the cell, and their fluorescence can be monitored by imaging. Then the fluorophore at the end of the cDNA is removed, and the sequence at the end of the cDNA determined in repetitive cycles of ligation of a short → adaptor carrying a restriction recognition site for a class IIS → restriction endonuclease (binding within the adaptor and cutting the cDNA remotely, producing a four nucleotide overhang; e.g. *Bbv*I). Next, a collection of 1,024 specially encoded adaptors are ligated to the overhangs, and the coded tails interrogated by the successive hybridization of 16 different fluorescent decoder oligonucleotides. This process is repeated several times to determine the signature of the cDNA on the surface of each bead in the flow cell. The abundance of each mRNA in the original sample is estimated by counting the number of clones with identical signatures. Compare → serial analysis of gene expression.

Massively parallel single molecule sequencing: A technique for the parallel sequencing of hundred thousands of DNA, oligonucleotide, cDNA, messenger RNA or genomic DNA molecules spotted on a → DNA chip. In short, the probe molecules are first immobilized on a chip surface optimised for single molecule detection in a distance of about 400 nm from each other. After → priming, a DNA polymerase starts sequencing reactions at all spotted DNAs simultaneously, using a selected "temporarily terminating" and fluorescently labeled nucleotide (structure not disclosed), which leads to a reversible chain termination. This socalled "pausing" of the polymerase allows to detect all incorporated bases on the complete chip surface with a fluorescence microscope. Then the modification (not disclosed) of the incorporated base is removed with a proprietary technology, which leads to the liberation of the 3'-ends of the DNA molecules and allows continuation of the process with another labeled nucleotide. Once a sequence has been determined over 15–35 nucleotides, it can be compared to entries in the databanks.

Mass spectrometry (MS): A technique for the measurement of the molecular mass of a molecule (e.g. a protein) by determining the mass-to-charge ratio (m/z) of ions generated from the target molecule. A mass spectrometer is basically composed of a source to generate these ions (usually a laser) and to deliver them into the gas phase, an analyzer for separating and sorting the various fragment ions, and a detector to sense the sorted ions. One round of mass spectrometry generates a spectrum of fragments that displays ion intensity as a function of m/z. See → electrospray ionization mass spectrometry, → electrospray ionization time-of-flight, → *m*atrix-*a*ssisted *l*aser *d*esorption *i*onization (MALDI), → parent-ion-scan technique, → tandem mass spectrometer. Compare → peptide fingerprinting.

MAST: See → *m*agnet-*a*ssisted *s*ubstraction *t*echnique.

Master circle: The idealized → restriction map of the → mitochondrial DNA of a plant cell. Since a single cell, and even a single mitochondrium contain mtDNAs of different size, composition and gene order, it is impossible to isolate the mitochondrial genome per se. Instead, the total mtDNA of a plant species is restricted and the restriction fragments arranged in the socalled master circle.

Master gene:

a) Any → gene that controls one or more other genes.
b) See → source gene.

Master mix: A laboratory slang term for a pre-mixed solution consisting of a suitable buffer, optimised Mg^{2+} concentrations, all four dNTPs and a heat-stable DNA polymerase (e.g. → *Taq* DNA polymerase). This master mix is usually made for several (mostly 10, or 100) → polymerase chain reactions, is therefore first aliquoted, and each aliquot pipetted once into a reaction tube. Then → primers and → template DNA are added to start the reaction. Variants of the master mix exist, but in each case such mixes avoid multiple pipetting steps and multiple pipetting errors.

MAT: See → mating type.

Matching gene: Any host gene that possesses a pathogen gene counterpart (and *vice versa*) in a gene-for-gene interaction. For example, socalled virulence gene(s) of a pathogenic fungus encode peptides or proteins that produce usually low molecular weight substances. These elicitors are recognized by a receptor protein anchored in the host cell membrane or located in the cytoplasm, and as a consequence of the interaction, a signalling cascade is incited leading to the activation of host genes and host defense reactions. The particular gene in the fungal genome "matches" the corresponding receptor gene in the host's genome.

Mate pair: See → paired-end sequence.

Maternal inheritance: See → cytoplasmic inheritance.

Maternal messenger RNA (maternal messenger, maternal mRNA): Any mRNA that is transcribed from the maternal genome during the oogenesis of animals. Maternal mRNA may be deposited in the oocyte and is needed for early embryogenesis. See → maternal effect genes.

Maternal mRNA: See → maternal messenger RNA.

Maternal programming: The presence of various → maternal messenger RNAs and proteins in the animal egg cell prior to fertilization that altogether are required for normal development of egg and embryo. Most of the components are synthesized and accumulate during oogenesis, and are stored in an inactive form in the oocytes, until they are activated stage-specifically during egg maturation and subsequent development.

Maternal to embryonic transition in gene expression (MET): The reprogramming of the transcription patterns in egg and sperm nuclei between fertilization and activation of the newly formed, combined embryonic genome. Early embryonic development is largely dependent on maternal RNAs and proteins synthesised during oogenesis, whereas later developmental stages are fully dependent on zygotic transcription that starts at a species-specific time after fertilisation. Without this transition in gene expression the embryo will die. The MET requires changes in chromatin structure induced by acetylation of core histones and commencement of DNA replication.

Maternal X chromosome (X^m): One of the two X chromosomes of female diploid organisms that originates from the female parent. Compare → paternal X chromosome.

Mates: A pair of DNA sequence reads with overlapping end sequences that are randomly sampled from a genomic library and assembled with a special computer program. Mates are critical for → whole genome shotgun sequencing.

Mating: See → conjugation.

Mating-based transformation: The transfer of → plasmid DNA from two separate haploid yeast cells into one single diploid

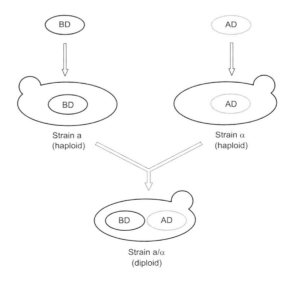

Mating-based transformation

cell by mating. In short, one of the individual haploid yeast cells, belonging to mating type a is transformed with a plasmid harbouring a → cDNA encoding e.g. the → DNA-binding domain of a → transcription factor, using the lithium acetate technique. Then a second haploid yeast cell of the complementary mating type α is also transformed with a second plasmid carrying e.g. the sequence encoding the → activation domain of the transcription factor. Both yeast cells then mate (i.e. fuse with each other), producing one single diploid cell (a/α) carrying both plasmids (double → transformant). Mating-based transformation is more efficient than successive transformation of a target cell, and is used for → yeast two-hybrid screening.

Mating type: Any one of two different cell types of → yeast (*Saccharomyces cerevisiae*) that allows → conjugation with the respective other type. In short, meiosis in yeast leads to the production of four haploid cells from an original diploid mother cell. This tetrad remains in a sac (ascus), formed by the cell wall of the mother cell, and is composed of two a and two *a* cells. Conjugation can only occur between an a and an *a* cell, never between cells of the same mating type, and leads to a diploid cell. A single gene locus (MAT, *mating type*) regulates the formation of the cell type. Its allele a is necessary for the generation of the a cell, its allele *a* for the *a* cell type.

Matrix-assisted laser desorption-ionization mass spectrometer (MALDI; gene balance): An instrument that allows to determine the mass of a gene (generally, a DNA sequence). Basically, the gene balance is a mass spectrometer (*m*atrix-*a*ssisted *l*aser *d*esorption-*i*onization mass spectrometer, MALDI). The DNA sample is first embedded into a matrix, which is evaporated by a short laser pulse. This releases the DNA molecules into the gas phase, where they are ionized by collision with matrix molecules. These ionized molecules are accelerated into a field-free channel. DNA molecules with differing base sequences (i.e. different masses) reach a detector at different times, which allow to calculate their precise masses. The gene balance can be used to e.g. determine the different masses of → alleles.

Matrix-assisted laser desorption ionisation-time of flight (MALDI-TOF): See → matrix-supported laser desorption-ionization mass spectrometry.

Matrix attachment region: See → scaffold-associated region.

Matrix CGH: See → array comparative genomic hybridization.

Matrix comparative genomic hybridisation: See → array comparative genomic hybridization.

Matrix-supported laser desorption-ionization (MALDI) mass spectrometry (MALDI-MS; MALDI-TOF-MS): A technique for the production and mass analysis of intact gas phase ions from a wide variety of biomolecules (e.g. peptides, proteins, oligonucleotides, carbohydrates, or glycolipids, to name few). The various analytes are prepared for MALDI-MS analysis by dissolving them in a solution containing a matrix compound that absorbs at the wave-length of the employed laser light (UV-laser, $l = 337$ nm, for example). The matrix compounds are either cinnamic or benzoic acid derivatives (e.g. α-cyano-4-hydroxycinnamic acid or 2,5-dihydroxybenzoic acid, respectively) that additionally function to individualize the analyte molecules. The solution is then spotted on a metal target plate, and dried. The metal plate may hold up to several hundred samples at once. The solvent is then evaporated and the resulting analyte crystals irradiated with a short pulse of laser light (e.g. a nitrogen laser firing at 337 nm, or a ND:YAG laser at 355 nm) to destroy the crystal structure, to desorb and ionize the analyte molecules, thereby creating a burst of ions. These ions in the particle cloud are then accelerated in the electric field of the mass spectrometer (voltage: 20–30 kV) and directed towards a detector. The *time of flight* (TOF) of the ions from the original location to the detector is measured and transformed into ion masses. Advanced variants of MALDI-MS work with a socalled *d*elayed *e*xtraction (DE): the acceleration tension does not act on the ionized particles at the time of ionization, but a few hundreds of nanoseconds later. This delay allows the ions to move into the acceleration channel, driven by the surplus energy of ionization. Therefore the ions are no more fully accelerated by the separation tension, which altogether leads to an improved resolution. Another improvement is the use of ion reflectors in → MALDI post source decay mass spectrometry.

The raw data are collected, processed and analyzed. Usually the range of mass resulution is not unlimited, because the kinetic energy of the laser-produced ions is too widely distributed. MALDI-MS is increasingly being used in → proteome research, allowing the analysis of e.g. 100–200 kDa proteins and the determination of the molecular weights of the resulting peptide fragments in the fmol range (with an accuracy of few ppm).

For peptide analysis, the solubilized protein is pipetted onto a carrier, whose surface is densely packed with either one (e.g. trypsin) or several immobilized proteases (e.g. trypsin, α-chymotrypsin and V8 protease). The fixation of these proteases prevents their autolysis, but allows the digestion of the protein analytes into a series of peptides. After limited proteolysis, the reaction is terminated by the addition of a socalled acidic matrix solution, dried at room temperature, and laser irradiation started. Basically the same technique can be applied to DNA analysis (with immobilized phosphodiesterases) or oligosaccharides (immobilized exoglycosidases).

Mass spectrometry therefore replaces the whole repertoire of traditional fragmentation of the analyte molecule (by e.g. → restriction) and the gel electrophoretic separation of the fragments. See also → electrospray ionization mass spectrometry, → electrospray ionization time-of-flight, → parent-ion-scan technique, → tandem mass spectrometer.

Maturation:

a) of proteins, see → post-translational modification.

b) of RNA, see → post-transcriptional modification and → RNA editing.

Mature protein (ligated protein, spliced protein): The product of → protein splicing. A mature protein consists of → exteins, combined by peptide bond formation after the → cleavage of extein-intein junctions in the → precursor protein and the joining of the free exteins.

Mature RNA: Any RNA that underwent one or several → post-transcriptional modification(s). For example, → pre-messenger RNA is first synthesized, then trimmed by → splicing (removal of →

introns and joining of → exons), → polyadenylated at its 3'-end and → capped at its 5'-end. Only after these modification is the then mature → messenger RNA transported to the cytoplasm and translated on cytoplasmic ribosomes. See → RNA precursor.

Maxam-Gilbert sequencing: See → chemical sequencing.

Maxi-cells: *E. coli* or *B. subtilis* cells (*recA*, *uvrA*) irradiated with UV light, which leads to an extensive degradation of the chromosomal DNA and to cessation of chromosomal DNA synthesis. Plasmids contained in these cells are not damaged by UV light and therefore continue to replicate and to express their genes. Thus maxi-cells can be used to study cloned genes in a system without appreciable chromosomal background (*in vivo* transcription-translation system).

Maxizyme: A short allosterically regulatable synthetic → ribozyme, consisting of one molecule that binds to the substrate region, and another that cleaves the substrate RNA at the sequence NUX (where N=any nucleotide; X=A,C or U). The name derives from *m*inimized, *a*ctive, *x*-shaped [functions as dimer], intelligent [allosterically regulatable] ribo*zyme*.

MazF-based SPP system: See → single protein production.

Mb: See → *m*ega*b*ase.

MB-PCR: See → methyl-binding polymerase chain reaction.

ᵐC (mC): Abbreviation for cytosine carrying a methyl group (e.g. at C5).

Mc: See → mis-cleavage.

MCAC: See → immobilized metal affinity chromatography.

MCAM: See → methylated CpG island amplification and microarray.

Mcm: See → *m*ini*c*hromosome *m*aintenance.

MCR: See → minimal common region.

Mcr system: See → *m*odified *c*ytosine *r*estriction system.

MCS:

a) See → *m*ultiple *c*loning *s*ite.
b) See → *m*ultispecies *c*onserved *s*equence.

MDA: See → multiple displacement amplification.

MDE: See → *m*utation *d*etection *e*lectrophoresis gel.

M-DNA (metal DNA, metallo-DNA): A complex of double-stranded DNA and divalent ions (e.g. Zn^{2+}, Co^{2+}, Ni^{2+}) formed above pH 8.0, in which the imino proton of each base in the duplex is substituted by a metal ion. Therefore, the DNA is coated with metal ions, and consequently possesses special conductive properties not owned by normal DNA (for example, an electron transfer can proceed along the molecule). M-DNA is also called a molecular wire ("nanowire"). See → A-DNA, → B-DNA, → C-DNA, → D-DNA, → E-DNA, → ε-DNA, → G-DNA, → G4-DNA, → H-DNA, → P-DNA, → V-DNA, → Z-DNA. See → DNA wire.

MD-PAP: See → multiplex dosage pyrophosphorolysis-activated polymerization.

MEA: See → microelectronic array.

Mechano-stimulated gene: See → touch gene.

MeCP: See → *m*ethyl-*C*pG-binding *p*rotein.

Mediator (mediator complex, *co*regulator, CoR, adaptor): A 1,5 MDa multi-protein complex of about 28 largely conserved eukaryotic proteins necessary for the transcriptional activation in a fully reconstituted → RNA polymerase II transcription system. About 22 of these proteins are highly conserved between yeast and man. The mediator proteins of yeast fall into three broad categories: Sin 4/Rgr 1 proteins (function in repression as well as activation), Srb (*s*uppressor of *R*NA polymerase *B*), and Med proteins. The complex itself binds to activators specifically recognizing enhancers and interacts with the non-phosphorylated → *c*arboxy-*t*erminal *d*omain (CTD) of the large subunit of → RNA polymerase II (B) to form a 1.5 Md holoenzyme. During this interaction the mediator unfolds, envelops the globular polymerase molecule, and controls the phosphorylation of the CTD. Mediator complexes therefore act as interface between → activators (definition a) and the RNA polymerase II (i.e. between → enhancer sequences and → promoters).

Medical genomics: The detection, isolation and characterization of genes and the encoded proteins with medical relevance. See → behavioral genomics, → comparative genomics, → environmental genomics, → epigenomics, → functional genomics, → genomics, → horizontal genomics, → integrative genomics, → medical sequencing, → nutritional genomics, → pharmacogenomics, → phylogenomics, → proteomics, → recognomics, → structural genomics, → transcriptomics, → transposomics. Compare → clinical mapping.

Medical sequencing: The repeated → sequencing of the same genomic region from various individuals, the alignment of these sequences, and the detection of sequence polymorphisms (e.g. → *s*ingle *n*ucleotide *p*olymorphisms, SNPs, or insertions-deletions, → Indels) between different individuals that may be associated with, or even cause diseases.

MeDIP: See → methylated DNA immunoprecipitation.

Medium copy plasmid: Any → plasmid that is present in 40 to 60 copies per bacterial host cell. For example, → pBR322 is such a medium copy plasmid.

Medium density chip: A laboratory slang term for a → DNA chip, onto which from 100–10,000 → probes are spotted. Compare → high density chip, → low density chip.

Medium overlap: The number of bases matched between two clones (e.g. → bacterial artificial chromosome clones) that are not matched using the strictest criteria, but are matched using less strict criteria. See → strong overlap, → total overlap, → weak overlap.

medRNA: See → mini-exon-derived RNA.

Megabase (Mb): One million nucleotides or nucleotide pairs; 1000 kb. See → base pair. See also → megabase mapping, → megabase marker.

Megabase cloning: A technique for the → cloning of extremely large fragments of DNA (in the range from one to several megabases) into suitable vectors.

Megabase mapping: The establishment of a linear → gene map using markers that are separated from each other by one million bases (a megabase).

Megabase marker: A series of DNA fragments that range in size from about 50 to more than 1000 kb, used in → pulsed-field gel electrophoresis as size markers for the estimation of the molecular weight of large DNA molecules. For this purpose, → lambda phage → concatemers can be used that cover a molecular weight range of 48.5 kb to 1.2 Mb.

Megadalton **(Md):** Equivalent to 10^6 → daltons.

Megagene: Any unusually large gene whose length exceeds 10–20 kb (e.g. the X-linked Duchenne muscular dystrophy [DMD] gene with about 1000 kb, or the dystrophin gene with a total length of 2300 kb and 100 introns).

Megalinker (megalinker I-*Sce*I): The oligodeoxynucleotide → linker 5′-GATCCGC TAGGGA-TAACAGGGTAATATA-3′ that contains a unique → meganuclease I-*Sce*I site. This linker permits the insertion of the unique I-*Sce*I recognition sequence into any *Bam* HI site of a → cloning vector or, generally, target DNA.

Megalinker I-*Sce* I: See → megalinker.

Meganuclease: Any sequence-specific → endonuclease that recognizes and binds relatively large (>12 bp) recognition sites, and therefore cuts only one site in a → genome. Therefore, meganucleases are perfect tools for genome engineering. Meganucleases are essentially represented by → homing endonucleases. See → "meganuclease".

"Meganuclease" (meganuclease I-*Sce*I; omega nuclease, omega transposase): An → endonuclease encoded by a mobile group I → intron of yeast mitochondrial ribosomal RNA gene sequences that catalyzes the cleavage of the 18 base pair recognition sequence 5′-TAGGGATAA/CAGGGTAAT-3′ to generate 4 bp 3′ → cohesive ends. Since an 18 bp recognition sequence will statistically occur only once in 6.9×10^{10} bp of genomic DNA, the enzyme represents an extreme → rare cutter that can be used for the cloning and mapping of artificially inserted sequences in pro- and eukaryotic genomes, and the mapping of large DNA fragments in genome analysis. Since the enzyme from yeast also introduces → *d*ouble-*s*tranded *b*reaks (DSBs) in genomic DNA, it can be used for the specific fragmentation of whole chromosomes.

Meganuclease I-*Sce*I: See → meganuclease.

Megaplasmid: An imprecise term for any → plasmid whose size exceeds 200 kb.

Megaprimer method: See → megaprimer PCR mutagenesis.

Megaprimer mutagenesis: See → megaprimer PCR mutagenesis.

Megaprimer PCR mutagenesis (megaprimer mutagenesis, megaprimer method): A technique for the introduction of site-specific mutations (i.e. single base exchanges) into a target DNA. In short, the template is first amplified in a conventional → *poly-*

merase chain reaction (PCR) by using a flanking → primer (A or B) and an internal mutagenesis primer (M1 or M2). M1 as well as M2 should be designed such that the mutational mismatch is about 10–15 bases away from the 3′ terminus to allow for normal amplification. Also, a mixture of two thermostable DNA polymerases is employed, → *Taq* DNA polymerase without proofreading, but an → extendase activity, and → *Pfu* DNA polymerase with a 3′ → 5′ exonuclease function. This combination of enzymes reduces the extendase activity of *Taq* polymerase, which would otherwise lead to undesirable additional mutations in the final product. The amplification leads

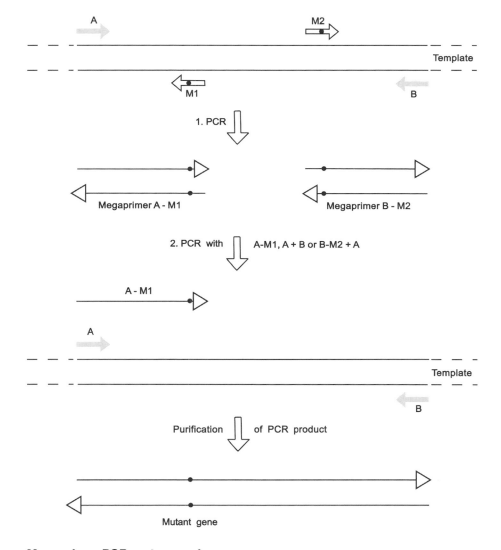

Megaprimer PCR mutagenesis

to a socalled megaprimer that contains the desired mismatch mutation. After electrophoretic purification and extraction of this megaprimer from the gel matrix, a second PCR with the flanking primers A and B and the product of the previous PCR (A-M1, or B-M2) are used to introduce the mutation into the target gene.

Megasatellite (megasatellite repeat): Any → satellite DNA with a repeat size of more than 1kb that is tandemly arranged with other repeats of the same size to form large domains at specific sites in mammalian chromosomes. For example, a 4.7kb human megasatellite repeat, which is arranged with other repeats in head-to-tail tandem clusters of 50–70 copies per haploid genome, even contains a → promoter and an → open reading frame encoding a deubiquitinating enzyme. See → macrosatellite, → microsatellite, → minisatellite, → satellite.

Megasequencing (megabase sequencing): The determination of the primary sequence of DNA fragments (see → DNA sequencing) of at least 1 Mb (1 million bases). Compare → kilosequencing.

Megatranscript: A somewhat misleading term for any one of a series of → messenger RNAs (→ transcripts) originating from genomic regions of several → megabases in size.

Mega-Yac: See → mega-yeast artificial chromosome.

Mega-yeast artificial chromosome: Any → Yac clone that contains an insert of more than one million base pairs.

Meiotic drive: The preferential transmission of a particular → allele (or chromosome, also gene, centromere) of a heterozygous pair to the progeny, occuring in natural populations of fungi, plants, insects, and mammals. For example, female meiosis exerts a drive pressure, since only one of four meiotic cells typically develops into a gamete (the egg cell), whereas the other three cells degenerate. In the developing gamete, whose selection is determined by its position in the female tetrad, meiotic drive of e.g. centromeres ("centromere drive") may account for the rapid evolution of complex centromeres in plants and animals. So an expansion of a satellite array within the centromere may generate a stronger centromere with a larger kinetochore that attracts more microtubules. CENP-A, CENP-C and other centromeric proteins undergo positive selection to suppress this centromere drive (by "adaptive evolution"). Meiotic drive is absent in yeast, and therefore the centromere is stable.Meiotic drive does not conform with classical Mendelian genetics. See → segregation distorter.

MEK kinase (MAP kinase kinase kinase, MKKK, MAP3K, or MEKK): Any one of a family of kinases that phosphorylate and thereby activate MAP kinase kinases or MKKs). The phosphorylated MAP kinase kinases in turn phosphorylate and activate MAP kinases. For example, MEK kinase-1, MEK kinase-3, MEK kinase-4, MEK kinase-5 (also called ASK 1) Raf-1, Raf.B, and Mos are such MEK kinases.

MELK: See → multi-epitope ligand cartography.

Melting (DNA melting, RNA melting): The dissociation of the complementary strands of double stranded DNA or RNA, as well as of DNA-RNA heteroduplex molecules to form single strands. In the laboratory melting is usually achieved by heating, while *in vivo* various nucleic acid binding proteins catalyze strand-separation in e.g. DNA → replication, or RNA → translation. See also → melting curve, → melting temperature, and compare → C₀t analysis. Also → denaturation, → denatured DNA, → G + C content.

Melting curve (DNA melting curve): The graphical display of the dissociation of strands in a DNA duplex molecule to form single strands as a function of temperature. Compare → C₀t curve, see → melting temperature.

Melting point: See → melting temperature.

Melting protein: See → DNA topoisomerase I, also → helix-destabilizing protein.

Melting temperature (T_m, t_m, t_m; melting point): The temperature at which fifty percent of existing DNA duplex molecules are dissociated into single strands. For measurement of T_m, a DNA solution is heated and its absorbance at 260 nm is continuously monitored. Transition from double- to single-stranded DNA occurs over a narrow temperature range and shows a characteristic increase in absorbance at 260 nm, so that a sigmoidal (S-shaped) curve results. T_m is defined as the temperature at the midpoint of the absorbance increase that is, the temperature at which fifty percent of the molecule(s) are dissociated. Melting temperature (T_m) calculation:

a) Simplified calculation:
$T_m = [2°C × (\#A + \#T)] + [4°C ×)\#G + \#C)]$

For example, the melting temperature of the 10-mer oligonucleotide ACG TAC GTA C is: $[2°C × (3 + 2)] + [4°C × (2 + 3)] = 30°C$

b) Alternative calculation:
$T_m = 81.5°C - 16.6 + [41 × (\#G + \#C)]/$oligonucleotide length $- (500/$oligo length$)$

For example, the melting temperature of the 10-mer ACG TAC GTA C is: $81.5°C - 16.6 + [41 × (5)]/10 - (500/(10)) = 35.4°C$

Melting temperature-shift genotyping (T_m-shift genotyping): A single-tube technique for the detection of → single *n*ucleotide *p*olymorphisms (SNPs) in genomic DNA that is based on the discrimination of SNP → alleles by the different → melting temperature profiles of their amplification products. In short, → genomic DNA is first amplified in a conventional → *p*olymerase *c*hain *r*eaction (PCR) with two → allele-specific primers, of which either only one, or both contain a GC-rich tail at the 5′-end. If only one allele-specific primer is tailed, then the tail comprises 26 bp. In case both allele-specific primers are tailed, then the 5′end of one primer extends by 6 bases only that of the other primer by 14 bases:

Allele-specific primer tail 1:
5′-GCGGGC-3′
Allele-specific primer tail 2:
5′-GCGGGCAGGGCGGC-3′

This difference of only 8 base pairs discriminates the melting profiles between the two allelic products, but only marginally influences the priming and amplification procedures. In addition, the primers differ by the 3′-terminal base that corresponds to one of the two allelic variants. Therefore, for each SNP two 15–22 bases long → forward allele-specific primers (optimized T_m: 59–62°C) with the 3′ base of each primer matching one of the SNP allele bases, and a common 22–27 bases long → reverse primer (optimal T_m: 63–70°C) are employed in PCR. The common primer typically binds no more than 20 bp

→ downstream of the SNP, thereby producing relatively short PCR products with a good amplification efficiency. Amplification is catalyzed by the → Stoffel fragment of → DNA polymerase to enhance discrimination of 3' primer-template → mismatches. Samples homozygous for allele 1 are amplified with the short GC-tailed primer (6 bases), and produces one product with lower temperature peak in the melting profile. Samples homozygous for allele 2 will be amplified with the long GC-tailed primer (14 bases) and present only one higher temperature peak. Heterozygous samples are amplified with both GC-tailed primers, and correspondingly the melting curves exhibit two temperature peaks. Depending on the SNP configuration in two (or more) → genotypes, either one or the other, or both allele-specific primer(s) is (are) extended. Since the allele-specific primers differ by their GC-rich tails, the corresponding PCR products also differ by their distinct T_ms that in turn depend on which of the two primers is used for amplification. Genotypes can finally be determined by inspection of melting curves on a real-time PCR instrument.

meltMADGE: See → programmable melting display microplate-array diagonal gel electrophoresis.

Membrane-associated transcription factor (MTF): Any one of a series of membrane-anchored → transcription factor proteins that contains alpha-helical *t*ransmembrane *m*otifs (TMs) in the C-terminal region, and is released from the membrane by either *r*egulated *i*ntra-membrane *p*roteolysis (RIP) or *r*egulated *u*biquitin/proteasome-dependent *p*rocessing (RUP). Many NAC MTFs are upregulated by diverse stresses and DNA-damaging agents. The controlled proteolytic cleavage of MTFs and their concomitant activation ensures rapid transcription-independent responses to external (and internal) stimuli. MTFs regulate many cellular functions in prokaryotes, yeast and animals. In the plant *Arabidopsis thaliana*, a NAC MTF mediates cytokinin signalling during cell division.

Membrane-based two-hybrid system: See → split ubiquitin two hybrid system.

Membrane cutting: An infelicitous term for the excision of stained protein spots from → nitrocellulose or other membranes. Usually the proteins are first separated by → polyacrylamide gel electrophoresis, then blotted onto e.g. → nitrocellulose membranes, interesting spots visualized by e.g. staining (with e.g. fluorescent Sypro-Ruby or Pro-Q-Diamond dyes) and excised by socalled spot-cutters. See → membrane processing.

Membrane microarray:

a) Any → microarray that contains target sequences (e.g. → cDNAs, → oligonucleotides, peptides, proteins) on a membrane support (e.g. nylon or → nitrocellulose). Other microarrays are made of glass or quartz (also plastic) supports.

b) A misleading laboratory slang term for a → microarray, onto which a series of membrane-bound proteins are spotted.

Membrane processing: A somewhat misleading term for the tryptic digestion of protein spots on a membrane (e.g. nitrocellulose, onto which proteins separated by → polyacrylamide gel electrophoresis are transferred by a blotting procedure) as a prelude for the estimation of the masses of the resulting peptide fragments by → mass spectrometry. See → membrane cutting.

Membrane proteome: The complete set of membrane-bound or membrane-associated peptides and proteins of a cell, a tissue, an organ, or an organism. Since the membrane proteins are not readily soluble in aqueous media, their isolation, solubilization, separation and characterization by e.g. mass spectrometry or X-ray crystallography requires special and individually adapted technologies. See → glycoproteome, → phosphoproteome.

Membrane slide: A microscope slide carrying a thin layer of a microporous polymer with high capacity to bind either DNA or oligonucleotides, RNA or proteins (e.g. antibodies). Such membrane slides accomodate up to several thousand spotted probes that are e.g. crosslinked by UV, and serve as → microarrays (which can even be produced manually with the aid of an appropriate → arrayer ["MicroCaster™"]).

Membrane-tethered transcription factor (MTTF): Any membrane-bound → transcription factor that coordinates the → expression of nuclear genes with the metabolic state or the membrane properties of a particular → organelle (e.g. a → mitochondrium, endoplasmic reticulum, ER, or plastid). The transcription factor → domain of such MTTFs is released from the membrane by regulated intramembrane proteolysis, moves into the nucleus, and activates its target gene(s). For example, the → *b*asic *h*elix-*l*oop-*h*elix (bHLH) zip *s*terol *r*egulatory *e*lement *b*inding *p*rotein (SREBP) of mammals is such an MTTF and is involved in feedback control of cholesterol and fatty acid synthesis in the ER. When cholesterol levels are high, the SREBP is retained within the ER by its interaction with the *S*REBP *c*leavage-*a*ctivating *p*rotein (SCAP) and the *in*sulin-*i*nduced *g*ene protein (INSIG). With decreasing cholesterol levels, INSIG dissociates from SCAP, which triggers transfer of the SREBP-SCAP complex to the Golgi apparatus. There it is cleaved by protease S1P within its central loop (site1). Subsequently protease S2P cuts the trans-membrane spanning domain, the site2, resulting in the release of the bHLH-zip transcription factor domain. The latter activates the gene encoding HMG-CoA reductase in the nucleus, and thereby triggers cholesterol biosynthesis.

Membrane-translocating sequence (MTS): A short (e.g. 12 amino acids long) hydrophobic peptide sequence at the C-terminus of proteins that mediates thens into cells for functional tests. For example, an MTS from the h region of the signal sequir translocation across the cellular membrane. Such MTSs are used to deliver cargo proteience of the Kaposi fibroblast growth factor, if fused to the C-terminus of reporter proteins, efficiently imported these proteins into fibroblasts and also other cells.

Memory suppressor gene (long-term memory suppressor gene): Any one of a series of genes encoding proteins that function to inhibit memory formation and long-term memory storage by e.g. decreasing synaptic strength and forcing neurons to learn only salient features. For example, long-term memory is at least partly a consequence of synaptic plasticity (i.e. the ability of neurons to alter the strength of their synaptic connections with prolonged activity and experience), which is controlled by a series of protein kinase signalling cascades and positive regulators of transcription as e.g. *c*yclic *a*denosine *m*ono*p*hosphate (cAMP) *r*esponse *e*lement *b*inding protein 1 (CREB1) and C/EBP.

Activation of these positive regulators is essential for the consolidation of short-term memory into long-term memory. However, the removal of negative, inhibitory elements is equally important. In *Aplysia*, for example, the cAMP-dependent protein kinase A (PKA) pathway, mediated by CREB, stimulates the growth of new synaptic connections between sensory and motor neurons of the gill-withdrawal reflex after repeated exposure to serotonin (or behavioral training). Now, CREB2 is a repressor of these morphological and also functional changes, because an anti-CREB2 injection replaces serotonin functionally. The gene encoding this CREB protein (in *Aplysia*, ApCREB2) belongs to the family of memory suppressor genes. Compare → tumor suppressor gene.

MEMS: See → micro-electromechanical system sequencing machine.

Mendelian trait: See → monogenic trait.

MEPS: See → minimum efficient processing segment.

Mercaptoethanol: See → β-mercaptoethanol.

Mercaptopurine (6-mercaptopurine): A purine analogue that blocks the conversion of → inosine to → adenine (as well as the biosynthesis of 4-aminoimidazol-5-carboxamide ribotide).

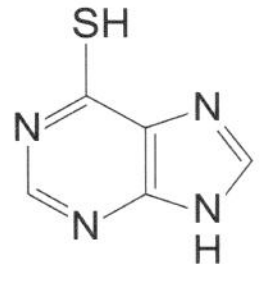

6-Mercaptopurine

Merged open reading frame (mORF): Any → open reading frame (ORF) that is merged from two existing annotated and adjacent ORFs by read-through of a → stop codon in the 5′-ORF of the pair, forming a single complete ORF. In yeast, about 25% of all mORFs are located within 20 kb of subtelomeric DNA. See → stop codon read-through.

Merging genes: A misleading term for the combination of two gene names into one, after experimental evidence (e.g. the isolation of a full-length cDNA) proved that the two genes are representing only one single locus. The new name corresponds to the locus with the majority of sequences, the abandoned name is kept associated to the locus of the merged gene. See → splitting genes.

Message amplification phenotyping (MAP-Ping): A rapid and sensitive technique to analyze multiple mRNAs present in a single cell or a small population of cells simultaneously. In short, mRNA is isolated by a guanidinium thiocyanate/cesium chloride microscale procedure, reverse transcribed into → cDNA, primed with amplification → primers (amplimers) specific for the target messages (which may for example be derived from sequence data), and amplified in the → polymerase chain reaction (PCR). The figure shows MAPPing results obtained with cytokine primers.

Figure see page 914

Messenger affinity paper (mAP): A diazothiophenyl paper to which poly(U) chains of more than 100 nucleotides in length are covalently bound. This paper is used to isolate polyadenylated mRNA that binds to poly(U) via hydrogen bonds (→ base-pairing).

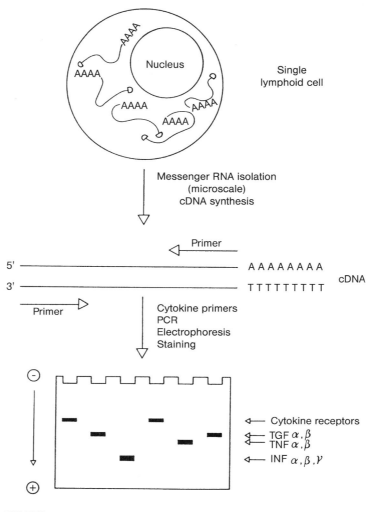

MAPPing

Messenger ribonucleoprotein (mRNP): The fully processed → messenger RNA molecule, complexed with a series of proteins, representing the transport form of mRNA. mRNPs may also be associated with → translational control RNA.

Messenger RNA (mRNA): A single-stranded RNA molecule synthesized by → RNA polymerase (RNA polymerase II or B in eukaryotic organisms) from a protein-encoding gene template (→ structural gene) or several adjacent genes (→ polycistronic mRNA). An mRNA specifies the sequence of amino acids in a protein during the process of → translation.

Messenger RNA-based vaccine: Any, mostly recombinant → messenger RNA (mRNA) encoding a specific protein (e.g. a tumor-specific → antigen) that is transfected into target cells, where it is translated into the antigen. This antigen in turn triggers the activation of tumor-specific lymphocytes and elicits an adaptive response directed against the target tumor. In this case, the messenger RNA serves as as → genetic vaccine. See → DNA vaccine → genetic vaccination, → stabilized messenger RNA vaccine.

Messenger RNA circularisation (mRNA circularisation, transcript circularization): The interaction between the 3'-end (see → trailer) and the 5'-end (see → leader sequence) of a eukaryotic → messenger RNA (mRNA) molecule, mediated by protein-protein interactions that lead to the formation of a loop structure ("closed loop", "circular structure"). For example, the *poly(A)-b*inding *p*rotein (PABP), once bound to the → poly(A)-tail of a particular mRNA, contacts → translation initiation factor eIF4G, which in turn interacts with the → cap-binding protein eIF4E, thereby effectively and physically circularising the mRNA in a head-to-tail loop. Transcript circularisation increases translational efficiency, which can be compromised by the intervention of a protein (or proteins) bound at the trailer.

Messenger RNA decapping (mRNA decapping, "decapping"): The removal of the m7GpppN →cap at the 5'-end of a → messenger RNA. The decapping process starts with the dissociation of → translation initiation factors (e.g. eIF-4E) from the → transcript ("translation exit") and the assembly of the decapping complex on the mRNA. Socalled decapping activators (e.g. Dhh1p and Pat1p) assist the decapping by moving mRNAs from → polysomes and into the translationally inert state that accumulates in → P bodies.

Messenger RNA display (mRNA display, *in vitro* virus, mRNA-protein fusion): A technique for the *in vitro* discovery and → directed molecular evolution of new peptides and proteins from combinatorial libraries, in which the → messenger RNA molecules are covalently attaches to the peptide or protein they encode. In short, a synthetic oligonucleotide containing → puromycin at its 3'-end is first enzymatically ligated or photochemically attached to the *in vitro* transcribed messenger RNA. This mRNA is then *in vitro* translated by e.g. a → rabbit reticulocyte lysate. The ribosome reads the message in the 5' → 3' direction, and puromycin as a chemically stable, small mimic of aminoacyl → transfer RNA binds to the ribosomal A-site and attaches the mRNA to the C-terminus of the nascent peptide. The resulting covalently linked mRNA-peptide complex is isolated, reverse-transcribed and used for *in vitro* selection experiments. After binding to a target molecule (e.g. a drug), the fused molecule complex is eluted and the mRNA recovered by RT-PCR. Therefore, phenotype and genotype are elegantly linked. See → *Bacillus* spore display, → *Baculovirus* expression system, → CIS-display, → mammalian cell display, → messenger RNA display, → phage display, → phagemid display, → ribosome display, → yeast display.

Messenger RNA expression array: See → cDNA expression array.

mRNA initiation site: See → transcription start site.

Messenger RNA-interfering complementary RNA: See → countertranscript.

Messenger RNA isoform: Any one of a series of → messenger RNAs that all originate from one single gene, but differ in the combination of their exons. Isoforms are generated by → alternative splicing.

Messenger RNA loop (mRNA loop, "closed loop", "circular structure"): A specific structure formed by the interaction between the m7GpppG → cap at the 5'-end and the → poly(A)-tail at the 3'-terminus of eukaryotic → messenger RNA (mRNA). In short, the cap recruits the cap-binding protein eIF4E and the scaffold protein eIF4G that together form eIF4F. eIF4G interacts with the poly(A)-binding protein (PABP) and forms a bridge between both ends of the mRNA, creating the mRNA loop, and additionally binds other initiation factors such as eIF4A and eIF3. This protein complex removes the secondary structure from the → 5'-untranslated region (5'-UTR) of the mRNA, using the free energy of hydrolyzed ATP and the → helicase activity of eIF4A. This configuration therefore facilitates the scan for a correct → initiation site on the 40S ribosomal subunit. See → messenger RNA circularization.

Messenger RNA profiling (mRNA profiling): The simultaneous detection of thousands of → messenger RNAs (indicative for the transcription of thousands of genes) upon developmental, physiological, environmentally influenced or pathological processes. Profiling can be achieved by → cDNA expression arrays, → massively parallel signature sequencing, or → serial analysis of gene expression, to name only few techniques.

Messenger RNA scanning (mRNA scanning): The movement of a → ribosome along a → messenger RNA, bound to the ribosome by its methylated → cap, until the → initiation codon 5'-AUG-3' is reached, where → translation starts.

Messenger RNA tagging (mRNA tagging): A technique for the profiling of → gene expression in specific tissues and the identification of tissue-specific genes in an organism that is based on the cross-linking of a specific → RNA-binding protein with → messenger RNA (mRNA) by → formaldehyde, and the co-immunoprecipitation of the protein-mRNA complex with → antibodies directed against a FLAG → epitope sequence engineered into the RNA-binding protein. In short, the target tissue is first fixed with formaldehyde to cross-link → poly(A)$^+$-mRNA with e.g. → poly(A)-binding protein (PABP). PABPs are involved in → translation initiation and mRNA stabilization/degradation. The gene encoding PABP for mRNA tagging is engineered to contain a → downstream → in-frame FLAG coding sequence (encoding the FLAG tag, general sequence: H$_2$N-DYKDDDDK-COOH). This gene construct is transcribed from a tissue-specific → promoter in the target tissue, and the expressed protein can bind to cell-specific mRNAs and be co-immunoprecipitated with bound mRNA by an epitope-specific antibody. This procedure fractionates cell-specific mRNA from mRNAs of other (contaminating) tissues. After → immunoprecipitation, the mRNA-PABP complexes are dissociated by → sodium dodecylsulfate (SDS) at 65 °C for 30 minutes. Then the mRNA is isolated, converted to → cDNA, biotinylated → cRNA synthesized in vitro from this double-stranded cDNA (amplification step), and fragmented cRNA mixed with a → hybridization mixture and finally hybridized to an appropriate → microarray to identify the different mRNAs.

Messenger RNA translation state: The number of → messenger RNAs in a given cell at a given time that are actually translated into their cognate proteins. Since not all mRNA transcripts are also translated, and since proteins, not mRNAs determine the → phenotype of a cell, estimation of the messenger RNA translation state informs about the protein potential of the cell, and can be measured by → translation state array analysis.

MeST: See → methylated sequence tag.

Metabolic engineering: The use of → genetic engineering technology to transfer, stably integrate, and express foreign genes in a host organism to shift a metabolic pathway towards overproduction of its products, or to rechannel metabolites of a pathway into another one.

Metabolic fingerprint (biochemical fingerprint): The depiction of (preferably all) metabolites of a cell at a given time. Such fingerprints are established by extraction of the various metabolite classes (e.g. sugars, amino acids, dicarbonic acids, hydroxy acids, fatty acids, polyamines, to name few), their identification and quantification by gas or liquid chromatography coupled to mass spectrometry (GC-MS, LC-MS) or *nuclear magnetic resonance* (NMR) analyses. Highly correlated metabolites (e.g. amino acids or sugars) are then combined in socalled network clusters. Such metabolic network clusters are specific for an organism.

Metabolic fingerprinting: See → metabolic profiling.

Metabolic labeling: A technique for the identification and quantification of peptides and proteins from two (or more) complex samples A and B that is based on the *in vivo* incorporation of differentially labeled precursors (e.g. ^{14}N-labeled amino acids in sample A, ^{15}N-labeled amino acids in sample B, "^{15}N metabolic labeling") into the proteins and their discrimination by → mass spectrometry. Carbon isotopes can also be used for labeling ("^{13}C metabolic labeling"). The differentially labeled → proteomes can then be extracted from the cells and directly be combined in equal amounts (to warrant equal representation) for separation in the same → two-dimensional polyacrylamide gel. After staining of the separated proteins, the interesting spots can be excised, in-gel digested with e.g. trypsin, and the tryptic fragments analyzed by → *matrix-supported laser desorption/ionization* (MALDI) mass spectrometry and the resulting *peptide mass fingerprints* (PMFs) treated with appropriate software (as e.g. ProFound, Mascot, or MS-Fit). Since the isotope-labeled peptides are visible in the mass spectra as ion doublets, the relative protein masses in the original samples can directly be determined from the signal intensities.

Metabolic map: The graphical depiction of (preferably) all metabolites of a cell, showing their quantitative relationships among each other at a specific point of time.

Metabolic phenomics: Another vague term of the → omics era, describing the analysis, interpretation and prediction of genotype-phenotype relationships from genomic data. See → phenome, → phenomic fingerprint.

Metabolic profiling (chemical profiling, metabolic fingerprinting, metabolite profiling):

a) The isolation of (preferably) all metabolites of a cell, their separation (by e.g. liquid chromatography, capillary electrophoresis, gas chromatography, or →

matrix-assisted laser desorption/ionization) and identification (by e.g. matching of the mass of each compound to reference masses or using internal standards) to establish a metabolic map (an inventory of all metabolites of a cell at a given time), or the cataloguing of up- or down-regulated compounds as a result of intrinsic or environmental stimuli. Metabolic profiling allows to monitor entire pathways simultaneously.

b) In a more specific sense, metabolic fingerprinting encircles the identification of a sample on the basis of the profile (i.e. the pattern and concentration) of a selected series of metabolites that are indicative for specific metabolic pathways.

Metabolic selection: The influence of an animal's metabolic activity on its genome size. For example, metabolic selection for a more compact genome seems to occur in birds. Stronger fliers possess smaller genomes than weak fliers. Since flight demands a high metabolic rate, and a high metabolic rate in turn restricts cell size, genome size has to fit in a small cell. A similar correlation holds for bats.

Metabolome: The complete set of low molecular weight compounds (metabolites) in a given cell and its organelles at a given time. The thousands of metabolites (*E. coli*: about 1,200) are extracted, separated by e.g. two-dimensional thin-layer chromatography, and identified by various detection techniques. Conveniently the target cells are fed with ^{14}C-labeled precursors (e.g. ^{14}C-glucose), and the newly synthesized compounds extracted, separated, and detected by → phosporimaging. The result is termed a "metabolite profile". Compare → genome, → proteome, → transcriptome.

Metabolomics: The whole repertoire of techniques to study the → metabolome. It involves the identification, quantitation and interpretation of the complete set of metabolites of a cell at a given time ("metabolic fingerprinting"). The competing term "metabonomics" is virtually identical to metabolomics.

Metabolon: A series of tightly connected protein complexes (many of the proteins being enzymes) that catalyse the highly coordinated and cooperative processing of a substrate to a product (in some cases, an endproduct).

Metabonomics: The technologies to monitor changes of the → metabolome in response to stress.

Metagenome: The entirety of the nucleic acid material in a soil, deep sea, salt and sweet water, ruminant stomach, human digestive tract, including mouth, deep sea whale fall, acid mine drainage or rock sample, resembling the genomes of an extremely complex mixture of a natural, mostly bacterial community. See → environmental genomics, → metagenomics, → trash sequencing.

Metagenomic DNA: The total DNA isolated from a → metagenome.

Metagenomic library: A library consisting of → genomic DNAs from multiple organisms of a complex metagenomic sample (see → metagenome, → metagenomic DNA). Such libraries ideally comprise the complete genomic information of a microbial community with all its genetic diversity, and can be used for the isolation of novel genes or → gene clusters encoding pharmaceutically or industrially important proteins (e.g. cellulases, xylosidases, amylases, lipases/esterases, proteases, dehydratases, oxidoreductases and Na$^+$/H$^+$-antiporters). See → meta-proteome.

Metagenomics (environmental genomics): The analysis of the genomes of whole living communities in deep sea, salt and sweet water, soil or rocks, or also in the human digestive tract, including mouth. A synonym for → environmental genomics. See → environmental genetics, → metagenome.

Metal-chelate affinity chromatography: See → immobilized metal affinity chromatography.

Metal DNA: See → M-DNA.

Metallothionein: Any one of a series of highly conserved, low molecular weight, cysteine-rich proteins that bind heavy metals such as cadmium, zinc, copper, mercury, and others. See → metallothionein gene.

Metallothionein gene (MT gene): Any member of a small gene family that codes for the synthesis of → metallothioneins, cysteine-rich proteins with the potential to bind heavy metals (e.g. zinc). The promoter regions of these genes contain a highly conserved → consensus sequence of 15 bp (→ metal regulatory element), which causes activation of the adjacent genes in the presence of heavy metals. In the mouse, MT genes are selectively amplified in the presence of heavy metals. See also → heavy metal resistance.

Metal regulatory element (metal responsive element, MRE): A short (15 bp) sequence element in the → promoter region of → metallothionein genes that specifies → heavy metal resistance in animal and human cells. It is highly conserved (consensus sequence 5'-CTNTGCPuCPyCG GCCC-3') and occurs in multiple copies in a metallothionein gene promoter. The insertion of synthetic MREs into heterologous promoters (e.g. the HSV thymidylate kinase promoter) renders the adjacent gene inducible by heavy metals. See also → heavy metal resistance gene promoter.

Metal responsive element: See → metal regulatory element.

Meta-proteome: The entirety of peptides and proteins in a complex metagenomic sample. See → metagenome, → metagenomic DNA, → metagenomic library.

Metaproteomics: The whole repertoire of techniques for the isolation, detection and characterization of the complete → proteome from complex metagenomic samples. See → meta-proteome.

Meth-DOP-PCR: See → methylation degenerate oligonucleotide-primed polymerase chain reaction.

Methidium: An intercalating dye (see → intercalating agent), used for → DNA capture procedures.

Methotrexate (Mtx, amethopterin, 4-amino-10-methylfolic acid): An analogue of dihydrofolate that inhibits → dihydrofolate reductase and consequently purine synthesis. See also → methotrexate resistance.

Methotrexate

Methotrexate^r: See → methotrexate resistance.

Methotrexate resistance (methotrexate^r, Mtx^r): The ability of an organism to grow in the presence of the dihydrofolate analogue → methotrexate. The drug inhibits → dihydrofolate reductase (DHFR) and consequently purine biosynthesis. Resistance against methotrexate is usually based on the → amplification of the DHFR gene (→ gene dosage effect) but may also be a consequence of DHFR gene mutation. Methotrexate resistance is used as → selectable marker in cloning experiments with animal cells, but has also been used in plant cells.

Methylase: See → methyltransferase.

Methylase-limited partial digestion: The incomplete restriction of a DNA sequence by a particular → restriction endonuclease, caused by a simultaneously acting DNA modification methyltransferase that methylates cytosine residues within the → recognition site of the endonuclease. This technique is used to partially digest DNA in → agarose plugs for → pulsed field gel electrophoresis.

Methylated adenine recognition and restriction (Mrr) system (modified adenine recognition and restriction system): A series of → restriction endonucleases of *E. coli* that recognize DNA sequences containing methylated adenine residues (such as G^{N6m}AC and C^{N6m}AG). Compare → methylated cytosine recognition and restriction system.

Methylated cap: See → cap.

Methylated CpG island amplification (MCA): A technique for the preferential amplification of methylated CpG-rich sequences that is based on the → restriction of → genomic DNA with → restriction endonucleases with differential sensitivity to 5-methyl-cytosine, followed by → adaptor ligation to the resulting fragments, and subsequent amplification in a conventional → polymerase chain reaction (PCR) with adaptor-specific → primers.

Methylated CpG island amplification and microarray (MCAM): A variant of the conventional → methylated CpG island amplification (MCA) for the high-throughput genome-wide analysis of DNA methylation in combination with a → microarray platform. In short, → genomic DNA is first isolated from two (or more) contrasting samples (e.g. normal and tumor cells), methylated DNA is enriched and genome → complexity reduced by serial → restrictions of control (e.g. normal cells) and experimental DNA samples (e.g. from tumor tissues) with the → restriction endonucleases *Sma*I (eliminates unmethylated 5′-CCCGGG-3′ sites) and *Xma*I (leaves 5′-CCGG-3′ → overhangs in methylated sites), followed by → ligation of → oligonucleotide → adaptors to CCGG overhangs, catalyzed by → T4 DNA ligase, and amplification in a conventional → polymerase chain reaction (PCR) using specific MCAM → primers and → *Taq* DNA polymerase. The resulting amplicons, checked for their quality by electrophoresis in 1.5% agarose gels and representative of the methylated fraction of control and cancer cells, are labeled separately with the → fluorochromes → cyanin 3 (Cy3, as dCTP) and → cyanin 5 (Cy5, as dCTP), respectively, and the → Klenow fragment of DNA polymerase. The labeled fragments from control and tumor tissues are then combined in equimolar ratios, and co-hybridized onto a microarray slide. Image acquisition and data analysis finally identify methylated and non-methylated sequences (e.g. genes) by comparing fluorescence intensity values of Cy5 and Cy3 dyes for each pair of control and cancer samples. If control samples are labeled with Cy3, and tumor samples with Cy5, then laser scanning and false color imaging will depict red spots on the microarray as hypermetylated, green spots as hypomethylated, and yellow spots as metylated to the same extent in both samples.

Methylated-CpG island recovery assay (MIRA): A technique for the detection of methylated → CpG islands in normal and abnormal states of a cell that does not depend on sodium bisulfite conversion of → genomic DNA, followed by → PCR amplification of the target region (see e.g. → methylation-sensitive single nucleotide primer extension), but employs methyl-CpG-binding domain proteins (see → methyl-CpG-binding protein), such as methyl-CpG-binding domain protein-2 (MBD2), binding specifically to methylated DNA. In short, sonicated genomic DNA isolated from cells or tissues is incubated with a matrix containing glutathione-S-transferase-MBD2b conjugate in the presence of methyl-CpG-binding domain protein 3-like-1, a binding partner of MBD2 that increases the affinity of MBD2 for methylated DNA. After washing to remove unbound DNA fragments, specifically bound DNA is then eluted from the matrix, and gene-specific PCR reactions are performed to detect CpG island methylation.

Methylated cytosine recognition and restriction (Mcr) system (modified cytosine restriction system): A series of restriction endonucleases of *E. coli* that recognize DNA sequences containing methylated cytosine residues, and cleave them. Among these systems, Mcr A restricts the sequence $C^{5m}CGG$, Mcr B the sequence $Pu^{me}C$ (where three different cytosine modifications are recognized: 5-methylcytosine, N-4-methylcytosine, and 5-hydroxymethylcytosine).

Methylated DNA immunoprecipitation (MeDIP, methylcytosine immunoprecipitation, mCIP, Methyl-DNA immunoprecipitation): A technique for the identification of methylated CpG-rich sequences in a genome or in a specific DNA region by enrichment of the methylated fraction of a genome with a → monoclonal anti-methylcytosine → antibody raised against 5-methylcytidine (5 mC). In short, → genomic DNA is first isolated, purified, and fragmented by → sonication such that fragments of 300 to 600 bp are generated. This size range warrants an efficient immunoprecipitation. The resulting fragments containing methylated cytosines are denatured (e.g. by heat), and then precipitated with an antibody raised against → 5-methylcytosine. Denaturation is more efficient, because the antibody has a higher affinity for 5 mC in single-stranded DNA. The positive fragments are subsequently labeled with e.g. → cyanin 3 and co-hybridized with → cyanin 5-labeled input DNA to a → microarray, onto which 25-mer → oligonucleotides representing a → genome or a genomic region at 35 bp intervals (or even more dense) are spotted. Hybridization events localize regions with methylated cytosines on a genomic (or subgenomic) scale and establish high-resolution maps of the → methylome.

Methylated sequence tag (mST, MeST): A short DNA sequence isolated from a distinct → CpG island of a → promoter of a protein-encoding eukaryotic → gene, in which distinct cytosines are methylated at their C5. The methylation pattern of such a methylated sequence tag is an indicator for the reduced → expression of the adjacent gene and diagnostic for the state of at least some malignant tumors. For example, specific MeSTs are expected to allow early detection of colon cancer.

Methylation: The transfer of a methyl group from a methyl donor (e.g. → S-adenosyl-L-methionine) to a methyl acceptor molecule (e.g. a protein, RNA or DNA) by a → methyl transferase. → DNA methylation is described in more detail, see there, and also → restriction-modification system. For an example of RNA methylation, see → methylated cap. RNA methylation is a → post-transcriptional modification.

Methylation assay (DNA methylation assay): A technique for the detection of methylated nucleotides within → recognition sequences of → restriction endonucleases in genomic DNA, using methyl-sensitive endonucleases, or pairs of endonucleases recognizing the same sequence but differing in methylation sensitivity (→ heteroprostomers). For example, the endonucleases *Mbo* I and *Hpa* II recognize and cut the same cleavage site (5'-CCGG-3'). *Msp* I also recognizes this site, if the internal cytosine is methylated (i.e. 5'-CC^{m}GG-3'), whereas *Hpa* II does not. By comparison of the cleavage pattern obtained from the same genomic DNA with either *Msp* I or *Hpa* II differences in methylation of CCGG-sequences can be detected. Since methylation of specific bases in → promoters may influence their activity (see → DNA methylation), such methylation assays allow to correlate promoter methylation with transcription of the adjacent gene. See also the table "Methyl-sensitivity of restriction endonucleases" of the Appendix.

Methylation degenerate oligonucleotide-primed polymerase chain reaction (Meth-DOP-PCR): A combination between the → degenerate oligonucleotide primed polymerase chain reaction (DOP-PCR) and the → methylation-specific polymerase chain reaction (MSP) for the high-throughput analysis of the methylation status of multiple genes ("methylation profiling") in trace amounts of DNA (e.g. → circulating DNA). In short, → genomic DNA is first extracted from samples (e.g. serum from patients), treated with sodium bisulfite that deaminates unmethylated cytosine to uracil, while methylated cytosine is resistant to this treatment. The modified DNA is then amplified in two steps and at different → annealing temperatures. A first step employs two → degenerate oligonucleotide → primers at → low stringency ("low-stringency amplification", annealing temperature = 25 °C), whereas the second PCR step ("stringent amplification") occurs at → high stringency, again using the DOP primers of the first step. The DOP-PCR products are then subjected to MSP with two primers, one targeting unmethylated ("unmethylated Meth-DOP-PCR primer"; all G nucleotides are substituted with As), the other one methylated DNA ("methylated DOP-PCR primer"; only the Gs following Cs are substituted with A). The MSP-PCR products are finally resolved in 3% → agarose gels.

Methylation drift: Any change in the global (genome-wide) → cytosine methylation pattern during a physiological process. For example, DNA methylation generally decreases during aging in many tissue types, except immortal cells, and mammalian fibroblasts cultured to senescence increasingly lose DNA methylation. This drift is most probably caused by a progressive loss of → DNA methyltransferase (DNMT) efficiency or erroneous targeting of the transferase by co-factors (or both).

Methylation-free island: See → CpG-rich island.

Methylation frequency (MF): The frequency with which cytosyl residues in a sequence (mostly in → genes or → promoters, especially → CpG-rich islands) are methylated at their carbon atom 5. MF is defined as the number of individuals of a sample with methylated target sequences divided by the total number of individuals of the sample multiplied by 100. See → methylation index.

Methylation-independent polymerase chain reaction (MIP, methylation-independent PCR, M-PCR): A technique for the amplification of bisulfite-modified sequences regardless of their methylation status. In short, genomic DNA is treated with sodium bisulfite, then sheared, and the resulting fragments amplified in a conventional → polymerase chain reaction using methylation independent primers (MIP-primers). Such MIP primers are designed to avoid CpG dinucleotides or to replace Cs within CpGs by a mismatched base. MIP does not require the use of restriction endonucleases.

Methylation index (MI): A measure for the overall methylation rate of cytosyl residues in a sequence (mostly in → genes or → promoters, especially → CpG-rich islands) in individual samples or specimens, defined as the fraction of methylated sequences divided by the number of tested sequences. Compare → methylation frequency.

Methylation induced premeiotically (MIP): The extensive methylation of cytosyl residues in naturally or artificially duplicated DNA segments (regardless of their endogenous or exogenous origin) in the filamentous fungus *Ascobolus immersus*. MIP inactivates the genes located on the duplicated segments. Once established, both the C-methylation and gene silencing are stably maintained through vegetative and sexual reproduction, even after segregation of the duplicated segments. The minimum duplicate size for MIP-induced methylation is about 300–400 bp (i.e. smaller duplications are not methylated). MIP represents a special type of epimutation, and probably functions to shelter the genome against invasion by mobile elements and recombination of ectopic repeats that could be lethal.

Methylation interference (methylation interference assay, methylation interference analysis, methylation interference footprinting): A method to test the specificity of binding interaction(s) between a specific DNA sequence and a sequence-specific binding protein. In short, the DNA sequence is partially methylated *in vitro* at purine residues by dimethyl sulfate, mixed with a nuclear extract or a purified nuclear binding protein, and tested for its binding properties in a → mobility-shift DNA binding assay. Methylation of purine residues within the target DNA interferes with the binding of the specific protein that readily binds to the identical non-methylated sequence.

Methylation map: Any physical map, which depicts the location of methylated cytosines in a genome. Usually, → restriction landmark genomic scanning (RLGS) or bisulfite sequencing are used for the construction of a methylation map. Such maps show, for example that methylation of e.g. *NotI* restriction endonuclease recognition sites in the *Arabidopsis thaliana* genome are not evenly distributed along the chromosomes, but cluster near the centromeres. Only 35% of all *NotI* sites in the *A. thaliana* genome are methylated, and about half of the methylated *NotI* sites are within or close to repetitive DNA sequences.

Methylation-mediated gene silencing: The down-regulation of the → transcription of genes, whose → promoters carry cytosines with 5-methyl groups at strategic positions. These methylated cytosines may either sterically prevent the binding of → transcription factors to their respective recognition sites, or bind → methyl-CpG-binding proteins recruiting → histone deacetylases. In both cases, the adjacent gene is silenced.

Methylation pattern (histone methylation pattern): The specific distribution of methylated side chain residues in → histones within the → chromatin of eukaryotic cells that is continuously changing during the life cycle of a eukaryote. Methylation predominantly occurs on lysine and arginine residues in histone H3 and H4. For example, lysine residue 4 and 9 (K9) in histone H3 and lysine 20 in histone H4 are methylated by histone methyltransferase SU(VAR)$_{3-9}$ (in mammals) or Clr4 (in yeast). This methylated lysine is the only binding site for → heterochromatin protein HP$_1$ that is associated with silent heterochromatic regions of a genome. See → histone code.

Methylation protection: The masking of specific → restriction endonuclease → recognition sites (e.g. *Eco* RI sites) within a clonable → genomic DNA fragment or → cDNA by specific methylation of C or A residues using site-specific → methyltransferases (e.g. → *Eco* RI methylase). The protected DNA can then be modified e.g. by → linker tailing and restricted without internal → cuts. Compare → restriction-modification system.

Methylation-sensitive amplification polymorphism (MSAP): A variant of the conventional → amplified fragment length polymorphism (AFLP) technique that uses the → isoschizomers *Hpa* II and *Msp* I instead of *Mse* I as the frequently cutting → restriction endonuclease (the rare cutter *Eco* RI remains unchanged) in order to detect the extent and pattern of cytosine methylation in a target genome. In short, genomic DNA is isolated and divided into two reactions, one of which is digested with *Eco* RI and *Hpa* II (recognizes and cleaves the sequence 5'-CCGG-3', is inactive if one or both cytosines are fully methylated [both strands methylated] and cleaves the hemimethylated sequence [only one strand methylated]). The other part of DNA is restricted with *Msp* I (cleaves 5'-C^{5m}CGG-3, but not 5'-^{5m}CCGG-3') instead of *Hpa* II. Then → adaptors (*Eco* RI- and *Hpa* II- *Msp* I adaptor) are ligated to the restriction fragments using → T4 DNA ligase, and the products pre-amplified with adaptor-specific primers (no selective bases). The pre-amplified products are diluted and amplified with the *Eco* RI and *Hpa* II-*Msp* I primers in a → touchdown → polymerase chain reaction as described for → AFLP. The *Hpa* II-*Msp* I primer was end-labeled with [g-^{32}P]-ATP. The denatured fragments are separated on a denaturing → polyacrylamide gel, and the labeled fragments visualized by → autoradiography. See → methylation-specific polymerase chain reaction.

Methylation-sensitive high resolution melting (MS-HRM): A technique for the sensitive and high-throughput detection of → cytosine methylation in a target DNA. In short, methylated and unmethylated DNAs, respectively, acquire different sequences after bisulphite treatment. If these different sequences are amplified in a conventional → polymerase chain reaction (PCR), the amplification products differ markedly in their melting profiles. Therefore, MS-HRM allows to estimate the methylation level of unknown PCR products by comparing their melting profiles to those of PCR products with a known unmethylated to methylated template ratio.

Methylation-sensitive single nucleotide primer extension (Ms-SNuPE): A technique for the detection and quantitation of cytosine methylation at specific CpG sites in → genomic DNA. In short, genomic DNA is first treated with sodium bisulfite, which converts unmethylated cytosine to uracil, but leaves methylcytosine, since it is resistant to deamination. In the subsequent → polymerase chain reaction amplification of the target sequence, using → primers specific for bisulfite-converted template DNA, the uracil is relicated as thymine and the methylcytosine as cytosine. After → agarose gel electrophoresis of the amplification product and its isolation, it is used as template in a primer extension reaction using appropriate internal Ms-SNuPE primers (that terminate immediately 5′ of the single nucleotide under investigation), and ^{32}P-labeled dNTPs (either ^{32}PdCTP or ^{32}PdTTP). The radiolabeled products are than separated by denaturing 15% → polyacrylamide gel electrophoresis and visualized by → autoradiography or → phosphorimaging. The ratio of methylated versus unmethylated cytosine (C versus T) at the genomic CpG sites can then be determined. Ms-SNuPE avoids → restriction endonucleases and can be multiplexed.

Methylation single nucleotide polymorphism technique (MSNP technique): A technique for the high-throughput genetic and epigenetic profiling of mutations in target tissues or organs, revealing different DNA methylation patterns.

Methylation-specific digital karyotyping (MSDK): A technique for the quantitative genome-wide identification of methylated CpG sites (see → epigenome) that is based on the generation of short (21 bp) sequence → tags derived from specific genomic locations by a methyl-sensitive → rare-cutting → restriction endonuclease. In short, → genomic DNA is first isolated from a control and an experimental cell culture or tissue, digested with the rare-cutting methylation-sensitive restriction endonuclease *Asc*I (mapping enzyme), and → biotinylated linkers ligated to the resulting fragments. The linkered fragments are then restricted with *Nla*III ("fragmenting enzyme"), and the restriction fragments captured on → streptavidine-coated → magnetic beads. Onto these fragments, → linkers containing *Mme*I recognition sites are ligated, and *Mme*I used to release tags that are then dimerized ("ditags"), concatemerized, cloned into an appropriate → vector, and sequenced. Following *Mme*I digestion, the single tags ("monotags") can also directly be sequenced by any high-throughput → next generation sequencing technique (e.g. → sequencing by oligonucleotide ligation and detection [SOLiD], avoiding the tedious → concatemerization and → cloning steps. The sequenced tags can then be mapped to a genomic sequence to determine the location of *Asc*I sites and the nearest genes, and to compare the genome-wide CpG methylation pattern (i.e. the differentially methylated *Asc*I sites) of control and experimental tissue, respectively. The number of tags in an MSDK library reflects the methylation status of the mapping enzyme sites. See → digital karyotyping.

Methylation-specific oligonucleotide array: Any → microarray that allows the simultaneous analysis of → DNA methylation in high numbers of → candidate genes. In short, → genomic DNA is first treated with bisulfite that deaminates unmethylated cytosine to uracil, while methylated cytosine is resistant to this treatment. Then a specific → CpG island of interest is amplified by conventional → polymerase chain reaction techniques, converting the modified UG to TG and conserving the originally methylated dinucleotide as CG.

The amplified product is then labeled with → cyanin3 or → cyanin5 and hybridised to a microarray on a glass slide, containing → oligonucleotide probes that discriminate between the originally methylated and unmethylated CG dinucleotides, respectively.

Figure see page 927

Methylation-specific polymerase chain reaction (MSP): A technique for the detection of the extent and pattern of cytosine methylation in a target genome, more specifically the determination of the methylation status of any group of CpG sites within a → CpG-rich island without the use of methylation-sensitive → restriction endonucleases. In short, genomic DNA (usually employed in the µg or even in the ng range) is first denatured with 0.2 M NaOH, and unmethylated cytosines subsequently chemically converted to uracil, using the → bisulfite technique, which leaves the methylated cytosines unchanged. Then two types of → primers are employed

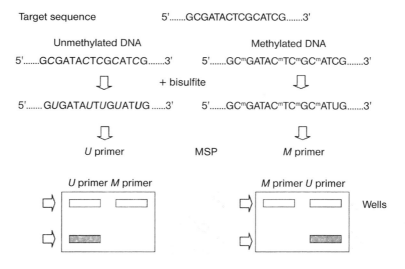

MSP

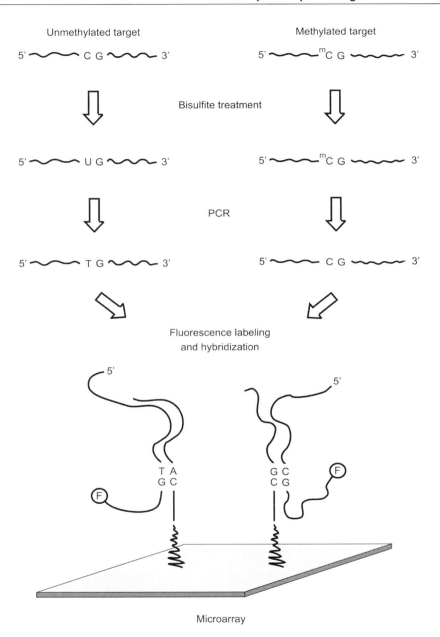

Methylation-specific oligonucleotide array

for amplification of unmethylated versus methylated DNA, respectively, using conventional → polymerase chain reaction (PCR) techniques. These primers are designed such that → mismatches are created that prevent → mispriming between primer and undesirable target sequences. The unmethylated sequence can only be amplified with the *U* primer set (*u*nmethylated sequence detection), whereas the methylated sequence can be amplified only with the *M* primer set (*m*ethylated sequence detection). After amplification, each PCR reaction is loaded onto a non-denaturing 6–8% → polyacrylamide gel, electrophoresed, stained with → ethidium bromide, and visualized under UV light. MSP allows the precise mapping of cytosine methylation in all CpG residues of a target DNA (e.g. CpG-rich islands), and avoids the frequent false positives that result from only partial digestion of target DNA with methylation-sensitive restriction enzymes. See → methylation-sensitive amplification polymorphism, → methylation-sensitive single nucleotide primer extension, → methylation-specific single base extension. Compare → methylation-sensitive amplification polymorphisms.

Methylation-specific single base extension (MSBE) assay: A technique for the simultaneous detection of DNA methylation at multiple CpG motifs in a genome. In short, → genomic DNA is first treated with sodium bisulfite and the modified DNA amplified in a conventional → polymerase chain reaction (PCR) using non-methylation sensitive → primers. The resulting PCR product is purified and a single base extended at the predetermined 5′-CpG-3′ site with primers complementary to either the methylated or unmethylated sequence. The extended products are then treated with shrimp alkaline phosphatase to remove unincorporated ddNTPs, and the dephosphorylated samples mixed with formamide. The → single nucleotide polymorphism is then analyzed by capillary electrophoresis on a DNA sequencer, and the DNA methylation of the target site inferred from the relative peak heights of the extended products generated from the methylated and unmethylated templates, respectively. See → methylation-sensitive amplification polymorphism, → methylation-sensitive single nucleotide primer extension, → methylation-specific polymerase chain reaction. Compare → methylation-sensitive amplification polymorphisms.

Methylation variable position: See → epigenetic marker.

Methyl-*b*inding *p*olymerase *c*hain *r*eaction (MB-PCR): A technique for the detection of the methylation status and level (high, intermediate, low) of a particular genomic region that uses a recombinant protein with high affinity for methylated CpG sequences to bind and retain relevant sequences, which then can be detected by gene-specific PCR in the same tube.

Methyl-CpG-*b*inding *d*omain column (MBD column): An affinity matrix with an attached DNA-binding domain of the → methyl-CpG-binding protein that is used for the isolation of DNA regions with methylated cytosines. DNA fragments with many mCpG sites own a higher affinity to the column than fragments containing less CpG sites. Therefore, the former are eluted at high salt (sodium chloride) concentrations, the latter at lower salt concentrations. The relative salt concentration reflects the overall methylation status of different DNA fragments with otherwise identical nucleotide sequence. MBD column chromatography suffers from insensitivity to small variations in methylation, and the fact that the methylation status of DNA fragments with identical or similar elution profiles can be heterogeneous.

Methyl-CpG-*binding* protein (MBD, CpG-binding protein, MeCP): Any one of a family of nuclear proteins (MBD1-4, Kaiso family, ZBTB4) that bind to methyl-CpG rich regions of a genome (see → CpG-rich islands) and mediate transcriptional silencing associated with → DNA methylation (see → methylation-mediated gene silencing). Knockdown of e.g. MBD1 therefore activates the target genes. For example, rat nuclear protein MeCP2 binds symmetrically to methylated CpG (^mCpG) sites in the genome, has a DNA-binding domain of 85 amino acids, and requires an A/T run adjacent to a methylCpG. MeCPs probably prevent the demethylation of methylated CpG residues and may thus be involved in long-term silencing of transcription of whole → chromatin domains, especially if located in → promoters. See → methyl-CpG-binding domain column.

Methylcytosine immuno-precipitation: See → methylated DNA immuno-precipitation.

Methyl-DNA immunoprecipitation: See → methylated DNA immunoprecipitation.

Methyl filtration (methyl filtration genome sequencing, methyl filtering): A somewhat misleading term for a technique to separate methylated from less or non-methylated regions of a genome. Most of the → repetitive genomic DNA (i.e. → retrotransposons, → satellite DNAs, → transposons) is heavily methylated, whereas → genic DNA is under- or unmethylated. For the separation of both classes of DNA, shotgun libraries of the target genome are established in genetically engineered bacterial strains that restrict methylated DNA. Thereby the library is enriched for non-methylated (genic) DNA, which can directly be sequenced to discover the genes. In short, genomic DNA is isolated and size-selected fragments generated by either → restriction or mechanical shearing. The relatively small fragments are then cloned in the methylation-restrictive *E. coli* host strains JM 101, JM 107, or JM 109. The enzyme encoded by gene *McrBc* of these strains restricts methylated DNA, requiring two 5'-Pu-^{m}C-3' dinucleotides separated by 40–80 bp for restriction. Therefore, methylated DNA is under-represented ("filtered") from these libraries, and as a consequence, → genic DNA (which is less or not at all methylated) is enriched severalfold.

Methyl filtration genome sequencing: See → methyl filtration.

Methyl *filtration* read (MFR): The sequence of an → insert of a clone from a → methyl filtration small-insert → genomic library in an *E.coli* host with a 5 mC restriction system (McrBC). This system prevents the propagation of clones carrying methylated inserts and achieves a 5–7-fold enrichment of genic sequences as compared to a control library in a normal *E. coli* host. Therefore, any methyl filtration read contains mostly sequences of the → genic space of a genome.

Methylglyoxal: A highly reactive α-oxoaldehyde, formed *in vivo* primarily from the glycolytic intermediates dihydroxyacetone phosphate and glyceraldehyde 3-phosphate that leads to glycation of amino acids in proteins and nucleotides in DNA. For example, in diabetes, hyperglycemia enhances the production of methylglyoxal that rapidly modifies proteins ar arginine residues to generate socalled *advanced glycation end* products (AGEs), as e.g. hydroimidazolone, and additionally glycate the corepressor protein mSin3A. Glycation diminishes the binding of mSin3A to a glucose-responsive GC box in the angiopoietin-2 → promoter, thereby increasing the expression of the angiopoietin-2 gene. Methylglyoxal is decomposed by glyoxalase I.

Methyl guanosine: 1-methyl guanosine, and N^2-dimethyl guanosine, both → rare bases.

Methyl inosine: 1-methyl inosine, a → rare base.

Methylmercuric hydroxide: The toxic chemical CH_3HgOH that reacts with imino bonds of uridine and guanosine in RNA and thus prevents the formation of secondary structures. The chemical is used to denature RNA completely before its electrophoresis in → agarose gels.

Methylome: The pattern of methylation of cytosyl (also adenyl) residues of a genome or parts of it (e.g. → promoters) at a given time. See → methylomics.

Methylomics: The whole repertoire of techniques for the quatitative determination of the methylation pattern of cytosyl or adenyl residues in target DNA. See → methylome.

Methylosome: A 20S protein complex consisting of methyltransferases and associated proteins that symmetrically dimethylate arginine residues in arginine- and glycine-rich domains of socalled Sm proteins. For example, the methyltransferase JBP1 produces methylated SmD1 and SmD3, which drastically increase their affinity for the *s*urvival *m*otor *n*euron (SMN) complex that in turn plays a decisive role in the assembly of → small *n*uclear (sn) RNA-protein core particles. See → pre-spliceosome, → spliceosome.

Methylphosphonate: A hydrophobic non-ionic nucleic acid analogue that contains nuclease-resistant methylphosphonate linkages instead of the naturally occurring negatively charged → phosphodiester bonds. Methylphosphonates form duplex hybrids with complementary DNAs by standard → Watson-Crick base pairing and are effective → antisense molecules that prevent virus and cellular → messenger RNA translation (antisense inhibition

Methylphosphonate

of herpes simplex virus replication, triplex-directed inhibition of chloramphenicol acetyltransferase mRNA expression in cell cultures, and inhibition of human collagenase IV mRNA expression). Methylphosphonate-RNA duplexes are not attacked by → RNase H.

Methyl-phospho switch: A special feature of the socalled → histone code that is based on the enzymatic phosphorylation of a serine and/or threonine residue (see → amino acid) next to a methylated lysine residue in → histones. The resulting methyl-phospho module prevents the interactions of methyl-binding proteins with their methylated histone target by repulsing the proteins. Phosphorylation can readily be reversed by a phosphatase, such that the original pattern of methylation of the histones persists. For example, heterochromatin protein 1 (HP1) that promotes → heterochromatin formation and consequently → gene silencing, recognizes and binds to methylated lysine at position 9 of histone H3 (H3K9) via a → chromodomain. This protein is associated with heterochomatin during most of the cell cycle, yet leaves the chromosomes with the onset of chromosome condensation. Dissociation is brought about by the phosphorylation of serine 10 (H3S10) by Aurora B kinase. Once the cell leaves the S phase of the cell cycle and prepares for division, all histones H3 are phosphorylated at S10.

MethylScreen: A technique for the sensitive detection and quantification of → cytosine methylation at specific genomic → loci that combines the action of methylation-sensitive (MSRE) and methylation-dependent → restriction endonucleases (MDRE) in single and double → digests. In short, → genomic DNA is first isolated, and separated into four equal aliquots. The first aliquot is incubated with an MSRE, as e.g. *Hha*I, *Aci*I or *Hpy*CH4IV that cuts within the region of interest (e.g. a → gene or → promoter), and the second aliquot is treated with an MDRE (e.g. McrBC). McrBC recognizes a pair of methylated cytosine residues in the context 5'-Pu^{m}C(N$_{40-2000}$)Pu^{m}C-3' (R^{m}CG), and cleaves within ~30 bp from one of the methylated residues. The MDRE quantifies the unmethylated molecular fraction. The third aliquot is mock-incubated (no enzymes present) and determines the total DNA amount in the reaction. The fourth aliquot is restricted with both the MSRE and the MDRE (double digest). Then the DNA from all four aliquots serves as → template in → fluorescence-based → quantitative PCR reactions with primers flanking the region of interest (using either → TaqMan assays, or → molecular beacons, or SYBR green). The DNA remaining after each treatment is thereby quantified and compared to the amount of DNA before the treatment. The inclusion of an MDRE eliminates false-positive reporting. Moreover, MethylScreen circumvents bisulfite conversion of DNA, and therefore can be started with nanogram quantities of template. See → methylation degenerate oligonucleotide-primed polymerase chain reaction, → methylation-sensitive high resolution melting, → methylation-sensitive single nucleotide primer extension, → methylation-specific digital karyotyping, → methylation-specific oligonucleotide array, → methylation-specific polymerase chain reaction, → methylation-specific single base extension.

Methyl single nucleotide polymorphism (methylSNP): Any methylation-dependent DNA sequence variation between two (or more) indidual → genomes, in which a specific → cytosine methylation status superimposes a → single nucleotide polymorphism. MethylSNPs can be converted into common SNPs of the C/T type by sodium bisulfite treatment of the DNA, which then can be subjected to conventional SNP typing.

Methyl tag: Any methyl (–CH$_3$)–group that is covalently attached to C5 of a cytosine base in the configuration 5′-CpG-3′ in DNA.

Methyltransferase (MTase, methylase, modification methylase): An enzyme that catalyzes the transfer of a methyl group from → S-adenosyl-L-methionine onto a substrate (e.g. a protein or nucleic acid). Methylation of DNA is described in more detail, see → DNA methylase, → DNA methylation, → DNA methyltransferase and → restriction-modification system, also → Dam methylase, → Dcm methylase, → *Eco* RI methylase, → heteroprostomer, → isoprostomer, → modification methylase. For an example of RNA methylation, see → cap.

MF: See → methylation frequency.

MFI: See → CpG-rich island.

M-FISH: See → *m*ultiplex *f*luorescent *in situ h*ybridization.

MFLP: See → microsatellite-anchored fragment length polymorphism.

MGB: See → *m*inor *g*roove *b*inding probe.

MGC: See → Mammalian Gene Collection.

M gene: An operational term for any DNA sequence that is **m**oderately homologous to → genic sequences deposited in the data banks, and therefore less likely to resemble a gene. M genes own confidence scores of <60%. See → H gene, → L gene.

mhpDAF: See → *m*ini-*h*airpin *p*rimed *D*NA *a*mplification *f*ingerprinting.

MI: See → methylation index.

MIAME: See → minimal information about a microarray experiment.

Micellar electrokinetic capillary chromatography (MECC): A variant of the → capillary electrophoresis technique that allows to separate uncharged molecules. In short, surface-active substances are first added to the separation buffer in concentrations above the *c*ritical *m*icelle *c*oncentration (CMC), leading to the formation of socalled micelles. In an electro-osmotic flow, negatively charged micelles migrate to the cathode, but slower than the running buffer. Molecules are distributed within and out of the micelles according to their lipophilic character, and different molecules are separated from each other by their different affinity towards the micelles.

MicroRNA chip (miChip, microRNA microarray): A variant of the conventional → LNA microarray (LNA array), onto which >1200 T$_m$-normalized → *l*ocked *n*ucleic *a*cid (LNA) capture oligonucleotides (usually 12–50 nucleotides long) complementary to microRNAs are immobilized (e.g. by photo-coupling procedures) with a spot-to-spot distance of 100–200 nm, covering all human, mouse and rat microRNA sequences annotated in miRBase 10.0. Such miChips discriminate between single nucleotide differences and therefore between closely related miRNA family members, and are used to profile the expression of mature miRNAs during e.g. developmental processes in a target organism.

micRNA: See → countertranscript.

Microalgal expression system: Any *Chlamydomonas reinhardtii* (Chlorophyceae) cell, into whose → genome one (or more) foreign gene(s) are stably inserted that can either be induced to express the encoded protein or synthesizes this protein constitutively, depending on the type of driving promoter (→ inducible versus → constitutive promoter). Since *C. reinhardtii* is a non-pathogenic and non-toxic organism for humans, the removal of endotoxins characteristic for e.g. *E. coli* vectors is unnecessary for clinical applications. Moreover, no risk of virus or → prion contamination exists. The favorable generation time of the alga (only 8 hours), the relatively short time span from → transformation to the production of the protein (some weeks), the well-established transformation protocols, the comparatively inexpensive cultivation, the expression of the foreign protein on chloroplast as well as cytoplasmic → ribosomes, the yield of the foreign protein of up to 1% of the algal dry matter, the potential secretion of the protein into the surrounding medium (allowing its simple isolation), the similarity of glycosylation patterns of *Chlamydomonas* proteins to mammalian proteins, and the excellent biological safety standards (algae are cultivated in self-contained systems, no release of transgenic algae into the environment, no viability of accidentially escaped strains outside the containment due to appropriate → mutations), no → antibiotic marker genes necessary for selection of transformants) recommend microalgal expression systems for the production of any mammalian protein.

Microamplification: The amplification of DNA from a single band of a polytenic chromosome. The target band is first microdissected from a polytenic chromosome preparation, the underlying DNA digested with the restriction endonuclease *Sau*3A, then oligonucleotide → adaptors ligated to the resulting fragments, and → primers complementary to these adaptors used to amplify the fragments in a conventional → polymerase chain reaction. See → microdeletion, → microcloning.

Microarray: Any microscale solid support (e.g. nylon membrane, nitrocellulose, glass, quartz, silicon, or other synthetic material), onto which either DNA fragments, → cDNAs, → oligonucleotides, → genes, → open reading frames, peptides or proteins (e.g. antibodies) are spotted in an ordered pattern ("array") at extremely high density. Such microarrays (laboratory jargon: "chips") are increasingly used for high-throughput → expression profiling. See → adenoviral siRNA kinome chip, → ADME array, → all-exon array, → antibody chip, → antibody array, → antigen array, → antigen microarray, → antisense genome array, → aptamer chip, → aptazyme array, → ASK chip, → autoantibody array, → autoantigen microarray, → BAC DNA microarray, → BAC microarray, → bead array, → bead-based array, → bioarray, → bioelectronic array, → biological array, → Brownian ratchet, → cancer cell profiling array, → cantilever array, → capillary chip, → capture array, → capture stretch microarray, → cDNA array, → cDNA expression microarray, → cDNA microarray, → cell microarray, → cell-based microarray, → cell biochip, → cell chip, → cell lysate array, → cell microarray, → cellular biochip, → cellular chip, → cellular microarray, → CGH chip, → chemical microarray, → chemiluminescent protein array, → chip, → chromosomal region expression array, → combinatorial protein array, → compartmented chip, → constitutional chip, → cryoarray, → cylindrical microarray, → dendrimer-based microarray, → designer microarray, → diagnostic microarray, → diffusion sorting array, → diversity array technology, → DNA array, → DNA chip, → DNA colony array, → DNA microarray, → double-stranded DNA microarray, → drug metabolism genotyping assay, → dynamic array, → electrochemical microarray, → electronic biochip, → electronic microarray, → electrophoresis chip, → enamel chip, → entropic trap array, → EST array, → evanescent resonator chip, → exon array, → exon junction microarray, → exon tiling array, → expression array, → fiber bead array, → fiber-optic DNA array, → filter array, → flow-through biochip, → flow-through chip, → fluorescent bead array, → force-based chip, → force-based protein biochip, → format I microarray, → format II microarray, → forward array, → forward-phase array, → 4D array, → 4D chip, → functional protein array, → gel pad array, → gene array, → gene chip, → gene expression microarray, → gene interaction array, → genome array, → genome chip, → genomic array, → genomic microarray, → genomic tiling array, → genomic tiling path microarray, → glycochip, → gold microarray, → haplotype chip, → high-density chip, → high density colony array, → high-density oligonucleotide array, → high-density protein array, → high-resolution microarray, → histological chip, → HLA chip, → human endogenous retrovirus chip, → human leucocyte antigen chip, → human single nucleotide polymorphism probe array, → hybridization array, → hydrogel-based microarray, → immobilized microarray of gel elements, → *in situ* array, → intein-mediated peptide array, → interaction chip, → intergenic array, → ion channel array, → isotope array, → lab-on-a-chip, → lentivirus-infected cell microarray, → liquichip, → live cell microarray, → living chip, → living microarray, → LNA microarray, → low density array, → low density chip, → lymphochip, → macroarray, → medium density chip, → membrane microarray, → microarray Western, → microcantilever array, → microchip, → microelectrode array, → microelectronic array, → microelectrophoresis chip, → microfluidics chip, → microRNA array, → microRNA microarray, → microsphere array, → microtube microarray, → miniarray, → miRNA array, → modular array, → modular microarray, → *Mu* array, → multiallergen chip, → multi-functional biochip, → multiplex hybridisation array, → multiplex microarray, → nanoarray, → nanoparticle chip, → nanowire sensor array, → non-living array, → nucleic acid microarray, → nucleic acid-programmable protein array, → nucleosomal array, → nylon macroarray, → oligonucleotide array, → oligonucleotide chip, → oligonucleotide microarray, → one-chip-for-all, → ordered array, → pathochip, → pathogen chip, → pathogen detection array, → pathway slide, → PCR array, → peptide array, → peptide chip, → peptide microarray, → phenotype array, → photoaptamer array, → phylogenetic array, → planar array, → planar waveguide chip, → PNA array, → polydimensional single nucleotide polymorphism microarray, → polydimensional SNP microarray, → population-specific array, → printed microarray, → programmable chip, → protein biochip, → protein chip, → protein domain array, → protein expression array, → protein function array, → protein *in situ* array, → protein interaction array, → protein microarray, → protein-protein interaction chip, → proteome array, → proteome chip, → proteome microarray, → proximal promoter array, → recombinant protein array, → retroarray, → retrochip, → retrovirus chip, → reverse capture microarray, → reverse format array, → RNA biochip, → RNA chip, → RNA expression microarray, → RNAi cell chip, → RNA microchip, → SELDI chip, → separation chip, → sequencing array, → single base extension tag array on glass slides, → single molecule array, → single nucleotide polymorphism chip, → sipper chip, → small molecule microarray, → splice array, → splice microarray, → splice oligonucleotide array, → splice variant monitoring array, → spotted array, → spotted microarray → subarray, → substrate chip, → suspension array, → *Taq*Man arrayÔ, → tandem array, → theme array, → tiling array, → tiling microarray, → tiling path DNA microarray, → tiling path microarray, → tiling resolution DNA microarray, → tiling resolution microarray, → tissue array, → tissue microarray, → tissue-specific microdissection coupled with protein chip array technology, → transcript array, → transfection micoarray, → transgene chip, → 2D/3D biochip, → ultramicroarray, → ultra-high density microarray, → universal array, → universal microarray, → universal protein array, → whole genome microarray, → whole genome oligonucleotide array, → whole genome tiling array, → whole proteome microarray. Compare → microarray architecture, → microarray noise. See color plate 4:

See color figure page 936

Microarray architecture: A laboratory slang term for the overall layout of all components of a microarray system, such as the design of the microarray itself, the hybidization chamber, the detector with all the

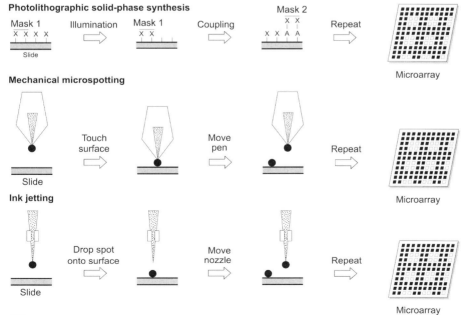

Microarray

Microarray

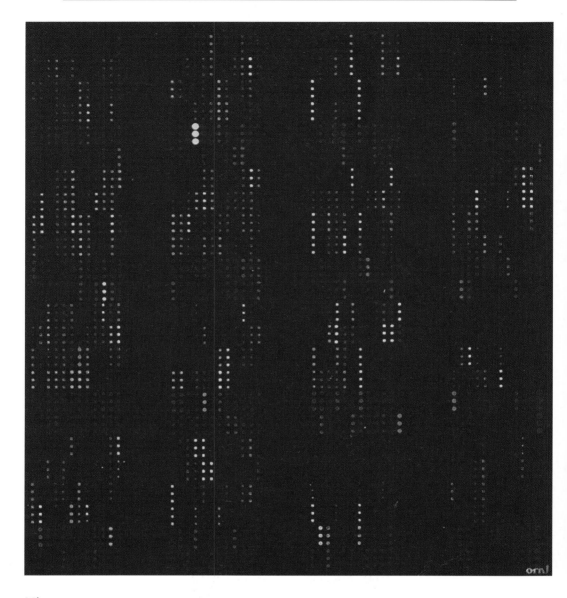

Microarray

A → medium-density → cDNA expression microarray image after → hybridization with two differently labeled cDNA preparations from intact and wounded leaves, respectively, of rice (*Oryza sativa*). One cDNA population was labeled with → cyanin 3, the other one with → cyanin 5. The → two color overlay shows → messenger RNAs of higher (green) or lower → abundance (red spots). Yellow spots resemble messages, that are not influenced by wounding.

See → expression array, → high-density chip, → low density chip, → medium density chip, → transcript array.

filters, light sources (e.g.lasers), optics and other hardwares.

Microarray-based cloning: See → transcript-based cloning.

Microarray density: See → density.

Microarray gene expression database group (MGED): A group of institutions that aims at defining specific guidelines for the submission of → microarray data. See → minimal information about a microarray experiment.

Microarray immunoassay: A variant of the conventional → immunoassay for the detection of → antibody-antigen interactions, in which the → antigens are spotted on a solid surface at low or medium density and reacted with antibodies from an individual, and the interactions detected by → enzyme-linked immunosorbent assay (ELISA). For example, in a specific case, different autoantigens, commonly used as diagnostic markers for autoimmune diseases such as systemic rheumatic disorders are immobilized on a → microarray and employed for the parallel detection of different types of autoantibodies. Less than one µl of patient serum autoantibody titers can be determined with high accuracy.

Microarray noise: An undesirable contribution of → microarray parameters such as background or substrate fluorescence or cross-reactivity of the probe to the readings of the fluorescence detection instrument. See → background subtraction, → dark current, → electronic noise, → optical noise, → sample noise, → substrate noise.

Microarray Western: Any → microarray, onto which target proteins (for example, in the form of cellular extracts) are immobilized that can be used to screen with → antobodies raised against specific proteins. If such a specific antibody recognizes and binds its cognate protein on the chip, the complex can be detected by a secondary antibody labeled with e.g. a → fluorochrome and active against the first antibody. The chip can then be scanned by a laser. Compare → Western blot, → Western blotting.

Microautoradiography: A variant of → autoradiography, which uses a liquid photoemulsion into which a sample (e.g. a tissue section whose RNA was labeled with ^{3}H-uridine) is embedded. The generated silver grains can be visualized on a sensitive film. See → macroautoradiography.

Microbial cell-surface display (cell-surface display): A technique for the display of peptides or proteins on the surface of bacterial or lower eukaryote cells that is based on the expression of a fused gene encoding N- or C-terminal sequences of so called carrier proteins (usually cell surface proteins or their fragments) and sequences encoding the target protein. In short, the sequence encoding the target peptide or protein (→ "passenger protein") are first fused to either the N- or C-terminus of the carrier, or inserted into the center of the carrier (→ sandwich fusion), cloned into an → expression vector and expressed in an appropriate bacterial host cell, which should be compatible with the displayed protein and deficient from cell wall-associated or extracellular proteases (as e.g. certain *E.coli, Bacillus* and *Staphylococcus* strains). Distinct strains of *Saccharomyces cerevisiae* are also used, because they are considered as safe (e.g. for food or pharmaceutical applications), and possess protein folding and secretory systems similar to other eukaryotes (e.g. mammals). The expressed → fusion protein is then transported to the membrane (preferentially to the outer membrane) and exposed on the surface of the cell. The carrier protein should possess an efficient → signal peptide (or transport signal) and a strong anchoring motif to prevent detachment of the fusion protein from the cell surface, should remain stable after fusion and be resistant towards proteases of the periplasmic space. For example, bacterial fimbriae, S-layer proteins, ice nucleation proteins and some *E. coli* outer membrane proteins (e.g. TraT) are such efficient carrier proteins (especially for immunostimulation and the development of recombinant vaccines). The passenger protein sequence also influences the efficiency of display and can even prevent it. For example, a passenger containing four phenylalanine residues is only inefficiently displayed on *Staphylococcus xylosus* cells, but their replacement by serine residues allows efficient display. Substantial improvement of the display system can be introduced by spacers of appropriate (experimentally proven) lengths that permit correct folding of both carrier and passenger proteins, prevent functional interference between both, or between passenger and cell surface. Microbial cell-surface display is widely applied for e.g. vaccine development (by exposure of heterologous epitopes on human commensal or attenuated pathogenic bacterial cells to elicit antigen-specific antibody responses), bioremediation (e.g. the development of efficient bioadsorbents for the removal of toxic chemicals or heavy metals from the environment), whole-cell biocatalysis (by e.g. immobilization of enzymes on the surface), biosensor design (by e.g. anchoring enzymes, receptors, or other signal-sensitive compounds for diagnostic or environmental purposes) and mutation screening (e.g the detection of single amino acid changes in target peptides after → random mutagensis). See → peptide display, → phage display, → ribosome display. Compare → differential display.

Microbial diagnostic microarray: See → diagnostic microarray.

Microbiome: The entirety of all microorganisms in a certain environment. For example, all the symbiotic and commensal, but also parasitic and temporal microorganisms that populate a human body (including skin and hair), are considered a microbiome. On the extreme, the mitochondria (in animals and plants) and the plastids (in plants) are also part of the microbiome. See → aggregate genome.

Microblot: A miniaturized → Western blot for the simultaneous detection of up to 12 → antibodies directed against 12 selected → antigens. In short, antigens (e.g. proteins from a patient) are first electrophoretically separated on → nitrocellulose membranes, the proteins stained with an all-protein stain, the resulting bands excised and stacked, the stack turned by 90°, paraffin-embedded and cut into 10 µm thin sections with a microtome. The slices are then mounted onto a second, solvent-resistant membrane, and this membrane deposited into wells of a microtiter plate and exposed to probe samples (e.g. human serum). The antibodies bound to the immobilized proteins are finally detected by an anti-human IgG peroxidase conjugate followed by precipitation of the peroxidase substrate.

Microcantilever: Any microfabricated silicon support, onto which a gold monolayer is deposited, which in turn serves as docking substrate for the covalent immobilization of synthetic 5'thio-modified oligonucleotides. Any hybridisation of an unlabeled → probe to the immobilized targets leads to a difference in surface tension between the functionalized gold and the non-functionalized silicium surface, which bends the microcantilever. The bending force can be transduced into measurable electric signals. See → cantilever array, → nanomechanical transduction.

Microcantilever array: Any → microarray that contains hundreds or thousands of → microcantilevers. Microcantilever arrays are used for the label-free detection of DNA-DNA-, DNA-RNA-, DNA-protein-, RNA-protein-, protein-protein-, and peptide protein interactions. See → nanomechanical transduction.

Microcell-mediated gene transfer **(MMGT):** A method for the transfer of single chromosomes from one mammalian somatic cell to another, using so-called microcells. In brief, donor cells are treated with colcemid to block mitoses. This leads to a reorganization of the nuclear membrane which engulfs single or small groups of chromosomes (micronuclei). Addition of cytochalasin B and centrifugation of these multinucleated cells produce microcells (micronuclei surrounded by plasma membrane) which can be fused to recipient normal-sized cells with the aid of → polyethylene glycol.

Microchip ("chip"):

a) A packaged computer circuitry ("integrated circuit") of minute dimensions that is manufactured from silicon and produced for program logic ("microprocessor chip") or for computer memory (memory or RAM chip).

b) Any miniaturized solid support (e.g. of nylon, nitrocellulose, glass, quartz, silicon or other synthetic material), onto which socalled target molecules are spotted at a low, medium or high density (see → microarray), or into which nanochannels are microfabricated (see → microfluidic chip). Such microchips are used for → microarrays.

MicroChIP: A variant of the conventional → *ch*romatin *i*mmuno*p*recipitation (ChIP) technique that is optimized to small samples (e.g. 10^4 to 10^5 cells). Since the amount of → chromatin DNA from such samples (e.g. biopsies, archival tissues) is very low, it has to be amplified by → *w*hole *g*enome *a*mplification (WGA) procedures. Do not confuse with → microchip.

Microchromosome: Any one of several → chromosomes of most (if not all) avian orders and some primitive vertebrates that has an average size of 12 Mb (smallest size: 1 Mb; largest size: 20.6 Mb) and a → gene content twice that of the → macrochromosomes. For example, the domestic chicken (*Gallus gallus*) genome consists of three chromosome size classes: five macrochromosomes (GGA 1–5), measuring from 50–200 Mb in size, five intermediate chromosomes (GGA 6–10) with sizes ranging from 20–40 Mb, and 28 microchromosomes (GGA 11–38) spanning from 3–12 Mb. Chicken microchromosomes account for only 18% of the total female genome, but harbour ~31% of all chicken genes. Moreover, the GC content, the number of → CpG islands, CpG sites and hypermutable CpG dinucleotides, the level of → cytosine methylation, the number of → stable gene deserts as well as the → recombination rate are all higher on microchromosomes (e.g. average recombination rate on microchromosomes: 6.4 cM/Mb;

intermediate chromosomes: 3.9 cM/Mb, and macrochromosomes: 2.8 cM/Mb. See also → human engineered chromosome.

Microcin: Any low molecular weight → colicin.

Microcloning: The → cloning of specific subchromosomal regions produced by microdissection (i.e. the removal of parts of a metaphase chromosome by physical means). Microdissection and microcloning procedures are used to generate → markers from specific chromosome regions that can serve as starting points to clone more extended regions of the chromosome. See → chromosome hopping, → chromosome walking.

Micrococcal nuclease (staphylococcal nuclease, micrococcus nuclease, nuclease S7, EC 3.1.31.1): An endonucleolytic enzyme from *Staphylococcus aureus* catalyzing the Ca^{2+}-dependent nucleolytic cleavage of → linker DNA between adjacent → nucleosomes in → chromatin. The enzyme is preferentially used to isolate nucleosome monomers (monosomes) and linker-free particles (→ core particles). The complete digestion of DNA by this nuclease leads to 3′ mononucleotides.

Micrococcus nuclease: See → micrococcal nuclease.

Microdeletion: Any → deletion in a → genome that resulted from the removal of only a few bases up to several kilobases of DNA, as opposed to the large deletions of several to many megabases or even whole chromosome arms as detected by e.g. cytogenetic techniques. Compare → microamplification. See → microinsertion.

Microdissection: A technique to fragment a chromosome by physical microsurgery (e.g. by a laser beam). Subchromosomal fragments can then be used to establish → subgenomic gene libraries.

Microdissection PCR: See → microdissection *polymerase chain reaction*.

Microdissection *polymerase chain reaction* (microdissection PCR): A method to amplify DNA fragments obtained by the dissection of specific regions of chromosomes. In brief, squashed chromosomes are microdissected, the fragment is extracted and the DNA of the fragment digested to completion (e.g. with Mbo I). Then → adaptor DNA sequences are ligated to the termini of the Mbo I-fragments. These adaptors serve as → primers for the → polymerase chain reaction that allows the amplification of the microdissected DNA in some 35–40 cycles to amounts of 100 ng or more.

Microdomain: A specific compartment of plasma membranes that harbors functionally related protein complexes serving as a platform for interaction(s) between the membrane and the cytoskeleton, especially protein trafficking. Microdomains are isolated by enriching plasma membrane fractions, detergent treatment, and floating detergent-resistant fractions in sucrose density gradients.

Microdrop *in situ* hybridization (MISH): A technique for the detection of mutations in DNA of isolated nuclei, chromosomes, or RNA in single cells, which uses microencapsulation of the test material in gel droplets, where the → fluorescence *in situ* hybridization takes place. Subsequently → flow cytometry allows to reveal gross aberrations in chromosomes.

Microelectrode array: See → microelectronic array.

Micro-electromechanical system sequencing machine (MEMS): An instrument for the → sequencing of DNA, consisting of a silicon chip that contains engraved in its surface all components necessary for a sequencing reaction (e.g. a thermal cycler for → cycle sequencing, purification of the product, electrophoresis and a detector).

Microelectronic array (MEA, microelectronic chip, microelectrode array, bioelectronic array): Any → hybridization array, onto which the target sequences (e.g. nucleic acids or proteins) are directed to specific locations by programmable sets of microelectrodes. Each microelectrode (diameter: 100 μm) is covered by a thin layer of → agarose and generates a controllable electric current that forces the target sequences to specific pre-programmed location on the chip surface. A variant of the conventional MEA works with electrogenic tissue (e.g. central or peripheral neurons, heart cells, muscle cells) that is grown on the array (and therefore also on the electrodes). The array is incubated with a test substance (e.g. a toxin, metabolite, or drug) and the reaction of the electrogenic tissue directly measured as changes in action potentials. For example, a myocyte monolayer is grown on an array and exposed to e.g. quinidine (causing arythmia *in vitro*). The extracellularly measured *f*ield *a*ction *p*otentials (fAPs) comprise all components of a classical cardiac action potential *in vivo*. The fAP duration, i.e. the interval between the maximum depolarization and maximum repolarization reflects the situation *in vivo*, and can therefore be used to directly measure drug effects onto the electrogenic tissue on the array. For other MEA designs consult www.multichannelsytems.com

Microelectrophoresis chip: A variant of the → microfluidics Chip that contains multiple channels for the sequencing of nucleic acids and the separation of amplified DNA fragments in short periods of time.

Microexon: Any extremely short → exon that encodes only one or up to a dozen amino acids.

Microfabricated fluorescence-activated cell sorter (μFACS): A disposable microfabricated device for the separation of fluorescently labeled cells that contains channels with diameters in the μm range, thereby reducing both volume of reagents and samples, and increasing speed of e.g. cell separation.

μFACS: See → microfabricated fluorescence-activated cell sorter.

Microfluidics: A series of technologies for the management and control of flow of minute volumes of liquids (and gasses) in miniaturized systems (e.g. a → microchip). These technologies encircle the production of glass or silicon chips ("microprocessors") with microchannel systems, appropriate valves, and devices to apply and move nanoliters of various samples and reagents in parallel, and the detection of interactions between target and probe molecules or the generation of products. See → microfluidic chip, → microfluidics technology.

Microfluidics chip (microfluidics-based chip): A microfabricated silicon chip, whose interior contains a network of micrometer channels, in which fluids can be moved by controlled pressure-driven flow. Microfluidic chips can be used to monitor cellular parameters (e.g. cell density, cell shape, apoptotic cells, transformed cells), allow highly parallel and high-throughput (HTP) analyses, and help to reduce the costs for chemicals and test materials. See for example → continuous-flow polymerase chain reaction, → flow-through biochip.

Microfluidics technology: The whole repertoire of techniques combining liquid chromatography and microfabricated chip application. Such glass, quartz, silicon, or plastic chips contain interconnected gel-filled microchannels for the molecular sieving of nucleic acid molecules. In short, each sample is transported from its well to a separation capillary system, into which it is injected. Separation of the nucleic acid molecules in the capillaries occurs according to their size. The differently sized nucleic acids are in-capillary loaded with an intercalating FLuorochrome (e.g. → ethidium bromide) and the complexes detected by fluorescence. Appropriate software plots fluorescence intensity versus time and displays the data as an electropherogram. See → microfluidics.

Microgel electrophoresis: See → single cell electrophoresis.

Microgene: Any gene that encodes a → microRNA.

Microgenomics: The whole repertoire of techniques to study the → genome, → transcriptome and → proteome of single cells on a microscale. Single cells are first isolated (by e.g. → laser microdissection) and the DNA, RNA or total protein extracted with techniques specifically adapted to extremely small quantities. For example, → messenger RNA is isolated from such cells, reverse transcribed into → cDNA, and the cDNA amplified by → *in vitro* transcription. The amplified cDNA is then labeled and used in → microarray experiments to detect cell-specific transcription or transcriptional responses of single cells to external stimuli (e.g. drugs). The term microgenomics is also employed for expression profiling of tissue sections.

Microhaplotype: Any → haplotype that comprises only two → single nucleotide polymorphisms (SNPs) within 10–20 base pairs of a particular → allele in → genomic DNA. Other alleles or alleles from another individual may possess a different microhaplotype. Therefore, microhaplotyping is one approach to → genotype various organisms (e.g. patients suffering from the same disease).

Microhomology: Any → homology between two nucleic acid strands (e.g. DNA-DNA, DNA-RNA, RNA-RNA) that is based on only one or few (1–4) complementary nucleotides.

Micro-indel: A more general term for any → insertion or → deletion (→ indel) that comprises only 20 bp or less. See → microdeletion, → micro-insertion.

Microinjection: A technique for the mechanical transfer of genes into living cells that uses a → micromanipulator consisting of a glass capillary with a blunt end ("holding capillary") to position the recipient cell and to hold it under slight suction, and a finely drawn glass capillary into which the injection fluid (e.g. a DNA solution) is sucked under slight vacuum ("injection capillary"). The injection capillary is then moved with mechanical or hydraulic devices ("chopsticks") towards the fixed cell, until it penetrates the cell membrane. It can also be directed to enter the nucleus (A). The injection capillary is now emptied and releases the solution (frequently mixed with an indicator dye such as red oil or the fluorescent Lucifer-Yellow). The recipient cell may also be glued to a glass surface with poly-L-lysine (B) or partly embedded in a thin layer of agarose (C). This technique of → direct gene transfer is comparably accurate and highly efficient, but needs experienced experimentors.

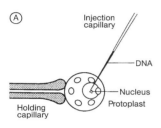

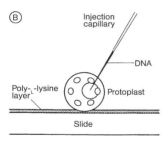

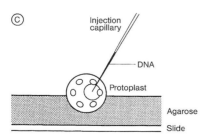

Microinjection

Micro-insertion: Any → insertion that comprises 20 bp or less. See → microdeletion, → micro-indel.

Microlesion: See → point mutation.

Micromachining: The patterning of glass or silicon surfaces such that three-dimensional microstructures are generated, as for example in → DNA microchip production.

Micromanipulator: An instrument for the injection of subcellular particles or molecules (e.g. DNA) into cells, see → microinjection. It also allows the isolation of single cells or → protoplasts.

Micromutation: See → point mutation.

Micron:

a) Micron (µ, µm): A unit of length, equivalent to 10^{-6} meter. Frequently used in contour-length determination of DNA (or RNA) molecules (1 m = 3 kb).

b) A family of short (393 bp) conserved → mobile elements with extensive sequence identity (90%) that are variants of the socalled → *m*iniature *i*nverted-repeat *t*ransposable *e*lements (MITEs), and almost exclusively occur within → microsatellites dispersed throughout the rice (*Oryza sativa*) genome. One of these elements, Micron 001 resides in a $(TA)_n$ microsatellite upstream of the rice phytochrome A (*phyA*) gene, and probably integrated into this microsatellite locus prior to the divergence of the two wild species *O. rufipogon* and *O. barthii* from a common ancestor. Microns possess 20 nucleotides long *s*ubterminal *i*nverted *r*epeats (SIRs), and the single strands own potential to form stable secondary structures via several internal repeats. About 100–200 copies of Micron-related sequences are present in the rice → nuclear, but not → chloroplast and → mitochondrial genomes. All Micron elements are flanked on both sides by microsatellite sequences consisting mainly of $(TA)_n$.

Micronucleus:

a) The smaller, generative nucleus of certain protozoan species (ciliatae) that contains the complete genome in the form of typical eukaryotic chromosomes with associated → histones, divides by mitosis, and is transcriptionally silent during asexual growth of the ciliate. However, it becomes active during sexual reproduction and is responsible for the genetic continuity of the protozoon ("germ-line nucleus"). Compare → gene-sized DNA, → macronucleus, → nuclear dimorphism.

b) Any one of several structures formed from acentric chromosome fragments and the nuclear membrane in cells treated with colcemid. See → microcell-mediated gene transfer.

Microplate array: Any → microarray that is printed in the base of a well in a → microtiter plate. A single well accomodates 100–300 spots, a whole 96-well plate then contains from 9,600 to 28,800 spots for targeted → expression or → mutation analysis (e.g. → single nucleotide polymorphism screening).

Microplate array diagonal gel electrophoresis (MADGE): A high-throughput microformat technique to separate minute amounts of proteins and nucleic acids. In short, a microplate format slot former with 9 mm pitch between wells is placed in an appropriate tray with the teeth upward. Then a → poly*a*crylamide (PA) gel is poured into the tray, and a sticky silane-coated glass plate placed onto the arrangement. After about 5 minutes, the glass plate is lifted with the open-faced microplate-compatible, 2 mm thick, 96-well PA gel attached to it. The array of wells is slightly turned on a diagonal angle (18.4° to the axis of the rows) such that the track lengths for electrophoresis are extended (i.e. the samples of a particular well are electrophoresed in between two wells of the next array beneath). The gel can be loaded with multi-channel pipettes, and running times be reduced to less than an hour.

Microplate-based PCR: See → microplate-based *p*olymerase *c*hain *r*eaction.

Microplate-based polymerase chain reaction (microplate-based PCR): A variant of the → polymerase chain reaction technique that allows the amplification of DNA sequences directly in lysed bacterial colonies or phage plaques without laborious DNA preparation. In short, bacterial colonies or phage plaques are transferred to the wells of a microplate using a toothpick. A microplate consists of a thin flexible polycarbonate mold that provides good thermal transfer properties (i.e. does not warp at high temperatures nor leach any potentially inhibitory organic compounds). Then a PCR reaction mixture containing deoxynucleotides, primers (→ amplimers) and → *Thermus aquaticus* DNA polymerase is added, and overlayered with light mineral oil. The bacterial cells or the phages are lysed and their DNA is denatured by heating. Then amplification cycles are started. This technique facilitates the rapid and simultaneous characterization of large numbers of clones and the production of single-stranded DNA templates for → Sanger sequencing, using appropriate primers (e.g. M13-based amplimers). If biotinylated nucleotides (see → biotinylated dATP, → biotinylated dUTP) are used for PCR amplification of the target DNA, it is possible to recover the amplified products easily. Following PCR, → streptavidin-coated magnetic beads are added to the microplate wells, again covered with mineral oil. The biotinylated strand is bound to the magnetic beads, the non-biotinylated strand eliminated (e.g. by alkali treatment) and the remaining single strand can be prepared for sequencing with e.g. fluorescently labeled → dideoxynucleotides directly in the microplate well.

Microprocessor: A nuclear multiprotein complex that processes long → *pri*mary *micro*RNAs (pri-miRNAs) into ~70 nucleotide miRNA precursors (→ pre-miRNAs) with fold-back (→ stem-loop) structures. Main components of the microprocessor are → Drosha, an RNaseIII-like enzyme, and → Pasha, a double-stranded RNA-binding protein. The pro-miRNA processing starts with the binding of DGCR8 to the junction between the rigid double-stranded stem and the 5′ and 3′ flexible single-stranded segments of the pri-miRNA. The correct positioning of DGCR attracts the processing center of Drosha to ~11 bp up the stem, where it introduces a staggered pair of → cuts into the RNA to produce the ~65 nucleotides long pre-miRNA. If DGCR8 binds at the loop end of the stem, Drosha is positioned such that it leads to unproductive cleavage and abortive "pre-miRNA". The correctly processed pre-miRNAs are then exported to the cytoplasm and subsequently cleaved by another RNaseIII-like enzyme called → Dicer to generate mature → miRNAs.

Microprotein: Any one of a series of naturally occurring peptides of a few dozen amino acids in length that possess extraordinary stability, a distinct and highly ordered tertiary structure and a good affinity for target proteins. For example, the socalled cystine knot microproteins are highly selective inhibitors of targets, e.g the microprotein EE-TI-II from *Ecballium elaterium* (a Mediterranean plant of the Cucurbitaceae) inhibits trysin efficiently, and the neuroactive conotoxins from deep sea snails of the genus *Conus* block ion channels in the membranes of neurons. Other microproteins exhibit hemolytic, antiviral, antimicrobial or uterotonic effects. Artificial microproteins can be derived from a native precursor (e.g. a cystine knot microprotein) by permutation. Libraries of permutated variants can then be screened for interaction partners by expressing them in *E. coli*, presenting them on the surface of the host cells, where interaction with binding ligands can be monitored by e.g.cytometric methods. The interacting complex can then be identified. Such selected microproteins are lead structures for the development of novel pharmaca.

Microproteomics: The whole repertoire of techniques to isolate, separate, identify and Characterize PRoteins and proteomes on a micro-or nano-scale.

Micro-*r*epresentational *d*ifference *a*nalysis (micro-RDA): A variant of the conventional → representational difference analysis that eliminates the high proportion of → ribosomal RNA by employing the → *p*henol *e*mulsion *r*eassociation *k*inetics *t*echnique (PERT) during the subtractive hybridisations.

Micro-ribonucleoprotein (miRNP, "RISC-like complex"): A 15S → ribonucleoprotein particle that consists of several proteins (e.g. the survival of motor neurons [SMN] proteins Gemin 3 [a DEAD-box RNA helicase], 4, 5 and 6, and the Argonaute protein and eukaryotic translation initiation factor eIF2C2 as major constituents) in a complex with at least 40 → microRNAs, ranging in size between 16 and 24 nucleotides. The complexity of microRNAs reflects the ability to recognize a wide range of diverse target RNAs for degradation via the → RNA interference pathway. MiRNPs are probably involved in the maturation and activity of microRNAs and → small temporal RNAs.

MicroRNA (miRNA, also tiny RNA): Any one of a class of hundreds (vertebrates: more than thousand) of ubiquitous, usually single-stranded, evolutionary conserved, 16–24 nucleotides long non-coding, regulatory, eukaryotic → RNAs that are processed in nucleo by the double-strand RNA-specific ribonuclease III Drosha from longer and normally polyadenylated and 5′-capped transcripts (pri-miRNAs, usually 70–171 nucleotides, in extreme cases up to 1 kb long) carrying a stem-loop structure (see → primary microRNA). Drosha, in concert with its cofactor DGCR8 that binds the junction between the double-stranded stem and the flanking single-stranded regions of the pri-miRNA, cuts the stem-loop at an 11 bp distance from the junction. The resulting hairpin RNAs (precursor miRNAs, "pre-miRNAs") are then transported to the cytoplasm by a transportin-5- ("exportin") dependent mechanism, where they are again trimmed by a second, double-strand RNA-specific ribonuclease called → Dicer. One of the two strands (→ active strand) of the resulting 19–23 nt long RNA is bound by a complex similar or identical to the → RNA-induced silencing complex (RISC) involved in → RNA interference (RNAi). The complex-bound single-stranded miRNA is targeted to and binds specific → messenger RNAs (mRNA) with complete or only partial sequence → complementarity (socalled "seeds"). The bound mRNA remains untranslated, resulting in reduced expression of the corresponding gene without degradation of the mRNA. MicroRNAs associate with proteins to form socalled → micro-ribonucleoprotein (microRNP) complexes. One of the proteins of this RNA-protein complex is the eukaryotic translation initiation factor eIF2C2, others are Argonaut, Gemin3 and 4 (components of the survival of motor neurons [SMN] complex). Some of the miRNAs (e.g. Lin-4 and Let-7) are also called → small temporal RNAs, because their mutational inactivation affects developmental timing in Caenorhabditis elegans. MicroRNAs inhibit the translation of target mRNAs containing 3′-untranslated region (3′-UTR) sequences with partial complementarity, and are probably involved in the development of spinal muscular atrophy, a hereditary neurodegenerative disease of (predominantly) children. The SMN complex is involved in the assembly and restructuring of diverse → ribonucleoprotein machines, as e.g. the → spliceosomal small nuclear RNPs (snRNPs), the → small nucleolar RNPs (snoRNPs), the → heterogenous nuclear RNPs (snRNPs), and the → transcriptosomes. MicroRNAs should not be confused with → short interfering RNAs, though the two RNA species are both generated by Dicer from longer precursors. However, siRNAs are not encoded by discrete genes, microRNAs are. Numerous miRNAs are encoded by → introns, and these miRNAs are different from the intergenic miRNAs, because they are transcribed by → RNA polymerase II and use specific spliceosomal components for their processing. In vertebrates, about 60% of the miRNA genes are expressed independently, 15% are expressed in clusters, and 25% are located in introns. Many of the more than 250 miRNA genes reside at chromosomal fragile sites associated with cancer in humans. A single miRNA possibly regulates multiple genes in processes like early development of *Caenorhabditis elegans*, apoptosis, fat metabolism, cell proliferation, cell differentiation, and cell death in *Drosophila melanogaster*, cell differentiation in *Arabidopsis thaliana*, brain development, chronic lymphatic leukemia, colon adenocarcinoma, Burkitt's lymphoma, and regulation of viral infection in humans. See → cell cycle RNA, → microRNome, → miRNarray, → microRNA*, → non-coding RNA, → short hairpin RNA, → short interfering RNA, → small RNA, → small endogenous RNA, → small non-messenger RNA, → small regulatory RNA, → small temporal RNA, → spatial development RNA, → stress response RNA, → tiny RNA.

MicroRNA array (miRNA array, microRNA microarray): Any solid support (e.g. a nylon membrane, a polymer chip, a N-hydroxysuccinamide glass slide), onto which 54–72 nucleotides long DNA → oligonucleotides → sense or → anti-sense to → microRNAs, or → *p*eptide *n*ucleic *a*cids (PNAs) and → *l*ocked *n*ucleic *a*cids (LNAs), containing e.g. a 5′-terminal C6-amino modified linker, or mixed LNA/DNAs as capture probes are spotted that can be hybridized to → total RNA or microRNA-enriched fractions (rich in RNAs below 60 nucleotides) of a cell, a tissue, an organ, or an organism, and used for the simultaneous analysis of miRNA → expression profiles. 2′-*O*-(2-meth*oxy*ethyl)-(MOE)- modified oligoribonucleotides are also employed for microRNA expression profiling, since they bind with high affinity and specificity to natural RNA. In short, total RNA including all small RNAs is first isolated, fractionated into different size classes by a special electrophoresis run, and the microRNA recovered. The miRNAs are then (A)-tailed by → poly(A)polymerase that incorporates *a*mino*a*llyl (aa)-UTP onto the 3′ end of the RNAs, resulting in microRNAs with a 3′-terminal U-tail. The amine-modified nucleotide reacts with activated → cyanine3 or → cyanine5 to fluorescently label the miRNAs. The fluorescent miRNAs are hybridized to the microRNA array that is finally processed using standard array scanners. The specificity of → hybridization is warranted by a series of oligonucleotides with three mismatches (G → C or C → A) on the array, which produce a significantly lower fluorescence signal as compared to their cognates. Additionally, synthetic 21-nt RNAs with sequences not corresponding to any miRNA, are spotted onto the array as a reference for normalization. MicroRNA arrays on membranes can also be hybridized to target RNAs that are γ^{33}P-dATP end-labeled by → T4 polynucleotide kinase. In this case, the membranes are shortly prehybridized, and overnight hybridized in the same solution containing the RNA probe. After hybridization, membranes are washed (e.g. with 2 × SSC/0.5% SDS at 37 °C), exposed to a phosphor storage screen, scanned by a Phosphor Imager, and hybridization signals quantified. In general, → hairpin-containing probes increase the specificity toward the target miRNA, and the stability of probe-target interactions. Among the many variations of microRNA arrays, a specific array provides both sequence and size discrimination, resulting in highly specific detection

MicroRNA cluster

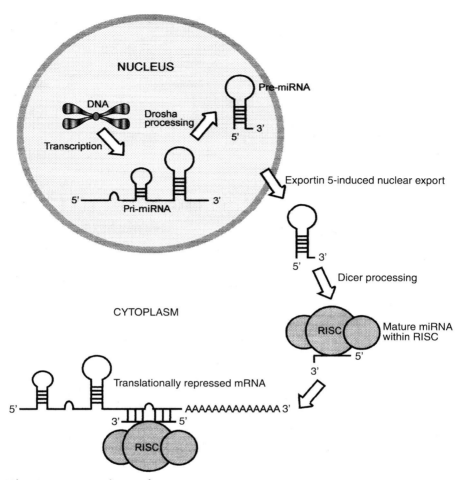

MicroRNA processing pathway (1) miRNAs are expressed in the nucleus as parts of long primary miRNA transcripts (Pri-miRNA) that have 5' caps and 3' poly(A) tails. (2) The hairpin structure that likely forms around the miRNA sequence of the Pri-miRNA acts as a signal for digestion by a double-stranded (ds) ribonuclease (Drosha) to produce the precursor miRNA (Pre-miRNA). (3) Exportin-5 mediates nuclear export of the pre-miRNAs. (4) A cytoplasmic dsRNA nuclease (Dicer) cleaves the pre-miRNA leaving 1–4 nt 3' overhangs. The single-stranded mature miRNA associates with a comlex that is similar, if not identical, to the RNA induced Silencing Complex (RISC). (5) The miRNA/RISC complex represses protein translation by binding to sequences in the 3' untranslated region of specific mRNAs. The exact mechanism of translation repression is still undefined.

of closely related mature miRNAs that may differ by only a single → nucleotide.

Figure see page 949

MicroRNA cluster: A laboratory slang term for any region of a → genome, in which genes encoding → microRNAs (miRNAs) are clustered. For example, in the human genome, such clusters comprise 35–40 genes that are co-transcribed by → DNA-dependent RNA polymerase II and span about 50 kb of genomic sequence.

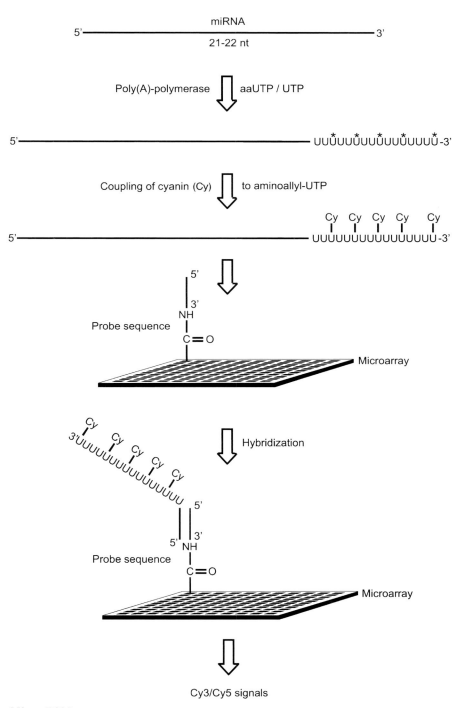

MicroRNA array

MicroRNA disease (miRNA disease): Any disease caused by the irregular → expression or mutation of → microRNA genes. For example, chronic lymphocytic leukemia (CLL) patients commonly exhibit a chromosomal abnormality at 13q14: the genes for miR-15 and miR-16 RNAs are deleted. Or, normal human colorectal mucosa expresses at least 28 different → microRNAs, of which two are down-regulated in adenocarcinoma and precancerous adenomatous polyps. Also, a chromosome rearrangement in the region containing the miR-155 gene seems associated with Burkitt lymphoma. Moreover, about 50% of the miRNA genes are located in common break-point regions, fragile sites, minimal regions of → loss of heterozygosity and minimal regions of amplification.

MicroRNA expression reporter vector ("miRNA vector"): Any → plasmid-based → cloning and → expression vector that allows the quantitative expression of → microRNA (miRNA)-encoding inserts in target cells. Such vectors each contain a multiple → cloning site for the insertion of miRNA-encoding sequences, a reporter gene (e.g. → luciferase gene) under the control of a strong → promoter (in mamalian systems: a CMV promoter) and a termination site, a ColE1 → origin for bacterial replication, and a pro- and eukaryotic → selectable marker gene (e.g. → ampicillin and → puromycin resistance genes, respectively), under the control of appropriate promoters. miRNA vectors can be used to identify miRNAs that bind to a target sequence of interest.

MicroRNA (*mir*) gene: Any one of a family of evolutionary conserved eukaryotic genes that encode → microRNAs, are partly arranged in tandem gene clusters, and are e.g. coexpressed in the germ line and early embryo of *Caenorhabditis elegans* and *Drosophila melanogaster*.

MicroRNA inhibitor: Any synthetic, chemically modified, single-stranded nucleic acid that specifically binds to, and inhibits endogenous → microRNA(s). Such inhibitors can be introduced into target cells via → electroporation of → transfection.

MicroRNA* (microRNA star, miRNA*): A less abundant small RNA, originally detected in *Caenorhabditis elegans* that derives from the arm of the hairpin of the → primary microRNA (pri-miRNA), the microRNA (miRNA) precursor, opposing the miRNA, and therefore pairs with the corresponding miRNA, leaving ~2 nucleotides 3'-overhangs. In animals, this miRNA:miRNA* duplex is generated by the sequential action of → Drosha and → Dicer RNaseIII endonucleases. Drosha cleaves near the base of the → hairpin stem in the pri-miRNA, liberating a 60–70 nucleotides fragment comprising the major part of the hairpin that is then cleaved near the loop by Dicer. The miRNA strand of the resulting miRNA:miRNA* duplex is then loaded onto the → silencing complex. Once within the silencing complex, the miRNA guides the posttranscriptional repression of the cognate → messenger RNA.

MicroRNome: The whole complement of genes in a → genome encoding → microRNAs.

MicroSAGE: A variant of the original → serial analysis of gene expression (SAGE) technique for the global analysis of gene expression patterns that requires only minute quantities of starting material (e.g. bioptic material or microdissections). MicroSAGE is run in a single →

streptavidin-coated PCR tube (to which the RNA or cDNA remains immobilized) from RNA isolation to the release of tags, thus avoiding step-by-step losses. Also, re-amplification of excised → ditags is reduced to only 8–15 cycles. In between different steps, enzymes from the previous reactions are removed by heat inactivation and disposal, so that after washing the reaction buffer and all ingredients for the next step can easily be added. MicroSAGE also uses total RNA rather than → polyadenylated RNA, because the poly(A)⁺-fraction is directly bound to the strepavidin-coated wall of the tube via a biotinylated oligo(dT)primer that also serves as primer in subsequent cDNA synthesis. See → SAGE-Lite, → SAR-SAGE.

Microsatellite (short tandem repeat, STR; repetitive simple sequence, RSS; simple repetitive sequence, SRS; simple sequence repeat, SSR; outdated: "CA-repeat"): Any one of a series of very short (2–10 bp), → middle repetitive, tandemly arranged, highly variable (hypervariable) DNA sequences dispersed throughout fungal, plant, animal and human genomes. For example, the microsatellite sequence $(TG)_n$ is present in $5-10 \times 10^4$ copies per human genome, spaced at intervals of 50–100 kb. Such microsatellites arise by → slipped-strand mispairing in combination with point mutations and → unequal crossing over of sister chromatids or homologous chromosomes during meiosis. See also → hypervariable region, → simple repetitive sequence, compare → minisatellite and → variable number of tandem repeats.

Microsatellite-anchored fragment length polymorphism (MFLP): A technique for the fingerprinting of genomes that is based on a combination of the → amplified fragment length polymorphism and → microsatellite-anchored primer technique. In short, → genomic DNA is digested with a → restriction endonuclease (e.g. *Mse* I) and an *Mse* I-adaptor ligated onto the restriction fragments. Then an *Mse* I-adaptor → primer and a microsatellite-anchor primer are used to amplify the intervening sequences. Usually over 100 fragments are amplified with MFLP, using conventional → polymerase chain reaction techniques, many of which are polymorphic between individuals. The sequence polymorphisms are partly caused by mutations in the *Mse* I site (→ restriction fragment length polymorphisms, RFLPs), in the microsatellite itself (→ variable number of tandem repeats, VNTRs), or in the internal sequence.

Microsatellite cluster: Any microsatellite-rich region of a → genome that contains at least 4 subregions with a high density of various → microsatellite repeat motifs. For example, in the *Arabidopsis thaliana* genome ~3500 such clusters exist, and ~30% of all microsatellites (preferentially of low GC content) are organized in such clusters.

Microsatellite expansion (triplet repeat expansion): The increase in numbers of a specific → microsatellite at a particular genomic locus. Such expansions probably occur at various locations in genomes, but in most cases the resulting → mutations (here: → insertion mutations) remain neutral, i.e. without phenotypic effect. However, in a series of human disorders such microsatellite expansion causes the onset of a disease. Some of the more important triplet expansion diseases are detailed:

Huntington's disease (HD) is an autosomal, dominantly inherited neurodegenerative disorder with uncontrolled movements (chorea), general motoric impairment, psychiatric abnormalities (personality changes) and dementia, which usually starts in the third or fourth decade of life and affects one in 10,000 individuals of European origin. The symptoms progressively worsen over the next 15–20 years and lead to death, associated by neuronal death and astrogliosis (especially in the caudate and putamen, but later on throughout the cerebral cortex). The underlying human gene IT 15 is 170 kb in length, consisting of 67 exons, located on chromosome 4 p 16.3, and mutated in HD chromosomes by a CAG microsatellite expansion at its 5′-end. The normal range of CAG repeats is from 6–34 triplets, in HD patients from 37 to over 100. An inverse correlation exists between age of disease onset and repeat length. IT 15 encodes a 348 kDa protein (huntingtin), which in HD suffered a polyglutamine expansion. The gene is widely expressed in human tissues, with highest expression levels in the brain. The huntingtin protein is localized in the cytoplasm.

The HD gene is highly conserved throughout vertebrates (murine-human sequence identity on the peptide level: 91%). The generally smaller (23 kb) homologous gene from the pufferfish *Fugu rubripes* contains all 67 exons, is highly conserved, and serves as model to decipher the disease mechanism(s). It is most probable that the disease is caused by a gain-of-function (e.g. the stimulation or inhibition of some unrelated target gene by the mutated protein, most likely a → transcription factor).

Moderate expansions of glutamine-encoding CAG repeats are also underlying other neurological disorders. The so-called **dentarubral-pallidoluysian atrophy** (DRPLA), a rare autosomal dominant disease with progressive dementia, epilepsy, gait disturbance and involuntary movements (chorea and myoclonus) is linked to gene CTG-B37 on the short arm of chromosome 12 that contains CAG repeats whose number in normal individuals ranges from 7–23, but expands to 49–75 repeats in DRPLA patients. Again, the number of CAG repeats is inversely correlated with the age of disease onset and is clearly associated with the severity of clinical symptoms.

Another disease, the so-called **spinocerebellar ataxia type 1** (SCA 1), an autosomal dominant disorder with ataxia, progressive motor deterioration and severe loss of cerebellar Purkinje neurons, is likewise caused by an expansion of CAG repeats within a gene. Also here, the age of disease onset and severity is highly correlated with the size of the CAG repeat island. Both the normal and expanded alleles are transcribed in lymphoblasts of SCA 1 patients. The SCA 1 gene product, ataxin-1, is localized in nuclei of neurons from various cortical regions, caudate, putamen, globus pallidus, pons, and dentate nucleus of the cerebellum, but predominantly in the cytoplasm of Purkinje cells. The expanded glutamine stretch of the mutant protein probably leads to a gain-of-function.

Other CAG microsatellite expansion diseases are spinobulbar muscular atrophy (SBMA), Kennedy's disease; affected is a gene encoding the androgen receptor, AR) and Machado-Joseph disease (MJD). However, other microsatellite motifs may also expand and lead to disorders, e.g. fragile X syndrome (FRAXA; caused by an expanding CGG repeat in a large open reading frame, which turns off the transcription of the adjacent gene), fragile XE mental retardation (FRAXE), and myotonic distrophy (MD, also dystrophia myotonica, DM, caused by an expanding CTG repeat in the 3′-untranslated region of an mRNA encoding a protein kinase). See → dynamic mutation, → microsatellite instability. Compare → loss of heterozygosity.

Microsatellite-initiating mobile element: See → mini-me element.

Microsatellite instability (MIN): The expansion or contraction of the number of → microsatellite repeats at a given locus of a → genome. For example, repeated CAG codons within an → open reading frame of the *H*untington's *d*isease gene (HD gene) on human chromosome 4p16.3 are stable, if the repeat number stays below a threshold of about 40 triplets. Above this threshold, the repeat number becomes instable. As a consequence of this microsatellite instability there is a significant probability that the length of the CAG island will increase when transmitted from one generation to the next. In such cases, the carriers with such a → dynamic mutation will develop a serious neurodegenerative disorder, Huntington's chorea. See → microsatellite expansion.

Microsatellite map: A → genetic map that is solely based on single-locus, codominant → microsatellite markers.

Microsatellite obtained from BAC (MOB): Any → microsatellite sequence that has been cloned into a → bacterial artificial chromosome (BAC) and recovered by either hybridization with microsatellite-complementary → probes, or amplification in a → polymerase chain reaction using → primers complementary to BAC sequences or sequences flanking the microsatellite. Do not confuse with → *mob*.

Microsatellite obtained using strand extension (MOUSE): A fast and effective method to enrich → genomic libraries for → clones containing → microsatellite sequences. In short, blunt-ended genomic → restriction fragments are first size selected (350–550 bp) on agarose gels, ligated to double-stranded, linearized → M13 vector DNA, and then transformed into electro-competent bacterial host cells. The bacteria are plated, → plaques develop, and M13 phage particles are eluted from the plate with → LB medium. Single-stranded M13 DNA is then isolated and used for the enrichment procedure, which starts by adding single-stranded → biotinylated microsatellites (e.g. $[CA]_{20}$). These anneal to the complementary microsatellites of the M13 clones and are extended by the → Klenow fragment, which displaces the newly formed strand from the template strand (which is removed by capture to → streptavidin-coated → magnetic beads). The new strand is eluted at 85 °C, made double-stranded by → primer annealing and extension with → *Taq* and → *Pwo* DNA polymerases (no strand displacement), and the resulting microsatellite-containing M13 DNA again transformed into bacteria. After this second round of amplification, the single-stranded DNA from the plaques can be isolated and sequenced. Then primers complementary to the microsatellite-flanking sequences are designed and used to amplify locus-specific → sequence-tagged *m*icrosatellite sites (STMS) from genomic DNA(s) in a conventional → polymerase chain reaction.

Microsatellite polymorphism (short tandem repeat polymorphism, STRP): Any difference in the number of → microsatellite repeat units at corresponding genomic loci in two (or more) different → genomes that can be detected by → sequence-tagged microsatellite polymorphism marker technology.

Microsatellite-primed polymerase chain reaction (MP-PCR; *inter-simple sequence repeat amplification*, ISSR; ISSR *amplification*; *inter-SSR amplification*, [1]SA; *single primer amplification reaction*, SPAR): A variant of the conventional → polymerase chain reaction that uses → microsatellite sequences as → primers to amplify regions of a genome located between two microsatellites on opposing DNA → strands. MP-PCR detects → polymorphisms in genomic DNA of different organisms of a population. See → anchored microsatellite-primed polymerase chain reaction, → inter-simple sequence repeat amplification, → minisatellite-primed amplification of polymorphic sequences, → simple repetitive DNA.

Microsequencing (protein microsequencing): A technique to increase the sensitivity of conventional → protein sequencing by two to three orders of magnitude into the picomole range. For example, in gas phase microsequencing the reagents of the → Edman degradation are carried by a stream of argon to the protein that is bound to a polybren film. This reduces the amount of solvents, reagents, and byproducts, which leads to increased sensitivity of the technique.

Microsphere array: See → suspension array.

Microspot: A technical term for any tiny area on a microarray that is less than 500 µm in diameter and contains a particular class of molecules (e.g. proteins, RNAs, oligonucleotides, DNAs) bound to the solid phase of the underlying chip material.

Microsynteny: The conserved order, sequence, and orientation of genes, conserved gene repertoire and conserved gene spacing (similar length of intergenic regions) in the range of about 100 kb in the genomes of closely related species. See → macrosynteny, → synteny.

microTAS: See → lab-on-a-chip.

Microtiter plate (MTP): A plastic plate with regularly arranged wells. The number of such wells ranges from 96–384 (and more), allowing to run reactions simultaneously, but physically separated and in minute volumes. This miniaturization reduces the costs for the incubation medium (e.g. PCR reaction mixture, culture medium).

Micro total analysis system: See → lab-on-a-chip.

Microtransponder: A light-powered silicon-based miniature radio-frequency (RF) transmitter of minute dimensions (250 × 25 × 100 µm) representing an integrated circuit composed of photocells, memory and an antenna, onto which 20–25 nucleotide long target sequences are covalently attached. During attachment, each transponder stores an identification number (ID number) for the attached sequence. Microtransponders allow to detect and identify large numbers of unique DNA sequences in one single assay. The microtransponder is first hybridised to fluorochrome-labeled probes, then the unbound (non-complementary) sequences are removed by washing, and the fluorophors attached to the hybridised probes excited by laser light. The microtransponders are then pumped through the flow chamber of a high-throughput scanner, where the fluorescence signal is detected, transformed to an RF signal, and assigned to a specific oligonucleotide on the transponders surface. The sequence information is stored in the electronic memory of the transponder. The microtransponder technology is used for the determination of DNA or RNA sequences.

Microtube microarray ("array tube"): Any → microarray that is embedded in the bottom of a 1.5 ml microcentrifuge tube. This configuration allows to perform the hybridisation, washing and blocking steps in one and the same tube, avoiding evaporation and contamination. The microarray can be made of either DNAs, oligonucleotides, or proteins that are either spotted or synthesized *in situ*. Interactions between the spotted molecules and probes can be visualized by either fluorochrome labeling of the probes and laser excitation, or non-fluorescently with gold-induced silver precipitation, in which a → biotinylated target is stained with a gold-strepavidin conjugate. The gold particle catalyses precipitation of silver particles that in turn can be detected by transmission imaging. Compare → DNA dip stick.

Microwell polymerase chain reaction (microwell-PCR, microPCR): A variant of the conventional → polymerase chain reaction that uses 96 or 384 2–3 µl microchambers in a supporting glass slide chip that are loaded by nanodispensors and sealed. Then a thermoelectric heating-cooling device is used for cycling in a flatbed block of a thermocycler. Usually, microwell-PCR is coupled to a gel chip electrophoresis, and the samples separated in ultra-thin gel layers and a combination between slab and capillary gel electrophoresis Moving lasers allow detection of bound fluorochromes during the run ("real time"). Micro-PCR reduces the amounts of chemicals (total reaction volume: 0.5–1.0 µl), and the amplification and electrophoresis times (15–20 minutes and 1.5 to 5 minutes, respectively).

Middle repetitive DNA (moderately repetitive DNA): A fraction of genomic DNA that – after denaturation – forms duplexes fairly late in a → C_0t analysis (i.e. reassociates at medium → C_0t values). It is composed of diverse sequences 100–500 bp in length which each are repeated from 100 to 10000 times (see e.g. → rDNA, → transfer RNA genes and → histone genes). Middle repetitive DNA also encompasses → microsatellite sequences.

MIDGE vector: See → *m*inimalistic *i*mmunogenically *d*efined *g*ene *e*xpression vector.

Midpoint *d*issociation *t*emperature (Td): A parameter for the characterization of the dissociation dynamics of oligonucleotides and their homologous target sequences. Td is defined as the temperature at which 50% of the originally bound, short (probe molecules dissociate from the membrane-bound target DNA within a specific time period and under specific conditions (e.g. special buffer composition).

MIL: See → minimum inactivation length.

Miller spreading: See → Miller spreads.

Miller spreads (Miller spreading, Miller technique): A method to prepare → chromosomes for electron microscopy, starting with the centrifugation of chromosomes isolated from lysed nuclei through 10% formalin in 0.1 M sucrose onto membrane-coated grids. These are then treated with a chemical to reduce the surface tension before being dried. Finally the specimens are stained with phosphotungstic acid and examined with the electron microscope.

Miller technique: See → Miller spreads.

Millicurie (mCi): The amount of a radioactive nuclide in which 3.7×10^7 disintegrations per second (dps) occur.

Millipore filter: The trade name of a series of filters with defined pore sizes ranging from 0.001 to 10 µm. Used to sterilize non-autoclavable solutions, or to trap nucleic acid precipitates.

Mimivirus: A double-stranded DNA virus (with 400 nm icosahedral capsid size the largest known virus) with a 1,181 Mb genome that belongs to the *Mimiviridae* family of *n*ucleo*c*ytoplasmic *l*arge *D*NA *v*iruses (NCLDVs) and resides in amoeba (e.g. *Acanthamoeba polyphaga*). Its genome contains 1262 putative → *o*pen *r*eading *f*rames (ORFs) representing numerous genes encoding RNAs and proteins for e.g. protein → translation (e.g. six → transfer RNAs, four → aminoacyl transfer RNA synthetases, peptide chain release factor eRF1, translation elongation factor eF-TU, and translation initiation factors 4E, SuI1 and IF-4A, a → helicase), → DNA repair (e.g a formamidopyrimidine-DNA glycosylase, a UV damage endonuclease, UvdE, a 6-*O*-methylguanine-DNA methyltransferase, MutS proteins, rad2 and rad50 homologs), both type I and II → DNA topoisomerases for DNA maintenance, the → chaperon Hsp70 (DnaK), the associated DnaJ protein, and many polysaccharide synthesis enzymes. Mimivirus also harbors four self-excising → introns in its → RNA polymerase encoding genes, but lacks any → ribosome component.

MIN: See → *m*icrosatellite *in*stability.

Min A min B mutant: An → *E. coli* → double mutant that divides into two cells of different size, a normal wild-type cell and a smaller → mini-cell.

Miniarray: Any solid support (e.g. glass slide), onto which a limited number (i.e. 500–800) of 50–70-mer → oligonucleotides, → PCR fragments, → cDNAs or → tags (see e.g. SuperSAGE) are spotted in triplicates together with appropriate control spots that is used for small scale → expression profiling. See → macroarray, → microarray.

Miniature genome: The → genome of various eukaryotes that is relatively small in comparison to other eukaryotes of comparable complexity, but nevertheless encodes all the functions vital for its carrier. For example, the marine chordate *Oikopleura dioica*, belonging to the class of larvaceans, owns a genome of only 65 Mb, with an average → gene density of 1/5 kb and a → gene number of 15,000. The miniature genome facilitates → replication and allows to cycle at extreme speed (2–4 dys at room temperature).

Miniature *i*nverted repeat *t*ransposable *e*lement (MITE): Any non-autonomous transposable element of about 0.3 kb flanked by 14 bp → terminal inverted repeats, and preferably occurring in → 3′ untranslated regions of genes from worms, insects, mammals, and plants. MITEs attain very high copy numbers in most genomes (from 1000 to over 15,000), possess no coding capacity, insert preferentially into non-coding regions of single or low copy sequences and exceptionally function as part of → transcription initiation or → polyadenylation sites. MITEs do not recognize any specific target site, but rather distinct secondary structures of the target DNA. See → heartbraker, → Ping, → Pong.

Miniature inverted repeat transposable element amplified fragment length polymorphism (MITE-AFLP): A variant of the conventional → amplified fragment length polymorphism technique (AFLP) that combines the power of AFLP with the abundance and relative diversity of → miniature inverted repeat transposable elements (MITEs) to generate → molecular markers for the estimation of genetic diversity and phylogenetic relationships within or among species. In short, → genomic DNA is first digested with *Mse*I, and → adaptors (sequence:5'-GACGATGAGTCCTGAG-3' and 5'-TACTCAGGACTCAT-3') ligated to the ends of the resulting fragments. Then a pre-amplification step follows with the → primer 5'-5'-GACGATGAGTCCTGAGTAA-3' and a primer complementary to a consensus domain of MITEs (e.g. *Hbr*: 5'-GATTCTCCCCACAGCCAGATTC-3'; *Pangrangja*: 5'-AARCAGTTTGACTTTG ATC-3', where R = A,G). For selective amplification, an *Mse*I selective primer in combination with a MITE primer is used. The resulting fragments are then processed according to the standard AFLP protocol.

Miniaturization: The reduction in the size of scientific instruments, and with it, the reduction in reagent volumes, reagent masses, frequently a reduction in time and cost of experimentation.

Miniaturized protein: A synthetic peptide-based model of a naturally occurring protein, which contains a minimum set of constituents necessary for an accurate reconstruction of a defined three-dimensional structure and a reproduction of a defined function. Such miniaturized proteins are model systems for structure-function relationship studies.

Mini-cells: Spherically shaped small cells which are continuously produced during the growth of specific mutant strains of bacteria (e.g. *E. coli*, see → min A min B mutant, or *B. subtilis*) and can be separated easily from normal-sized cells by → density gradient centrifugation. These mini cells do contain plasmids but not chromosomal DNA, are capable of RNA and protein synthesis and therefore serve to detect the expression of plasmid-borne genes and to characterize the proteins encoded by these genes without chromosomal background (*in vivo* transcription-translation system).

Minichromosome:

a) The circular 5.2 kb duplex DNA genome of → Simian virus 40 after its transfer into the host cell nucleus, where it becomes complexed with host cell → histones H2A, H2B, H3 and H4, and resembles a small chromosome.

b) A synonym for → artificial chromosome (see also → human engineered chromosome, → yeast artificial chromosome).

Minichromosome maintenance (Mcm): A comprehensive term for a series of nuclear proteins that probably function as replication licensing factors.

Minichromosome maintenance protein (MCM protein): Any one of a family of ATP-binding proteins regulating DNA replication such that it occurs only once per cell cycle. Expression of MCM proteins increases during cell growth and reaches a maximum in the transition phase from G1 to S.

Mini-exon: A synthetic → exon flanked by consensus 3' and 5' → splice sites that contains → open reading frames encoding short (e.g. 40–50 amino acids) peptides. In neither of the three open reading frames any → stop codon exists, and each reading frame encodes a peptide recognized by the same → monoclonal antibody. The 3' splice site includes a consensus branch point sequence, a polypyrimidine tract and the mandatory AG dinucleotide. The 5' splice site in turn carries the mandatory GT dinucleotide. Since exon-intron boundaries tend to map to the surface of the final protein product, the mini-exon peptide will be displayed on the surface of the protein, and be accessible for the antibody. The mini-exon can be inserted into the → introns of genes, so that the small encoded peptide appears in the protein encoded by the host gene. This protein can then be recognized by the antibody. See → mini-exon epitope tagging.

Mini-exon-derived RNA (medRNA): An untranslated sequence at the 5'-end of the mature → messenger RNA of trypanosomes, transcribed from a tandemly repeated mini-exon (see → exon). This sequence is ligated to the messenger RNA after the displacement of a mini-intron sequence ("minRNA"; see → intron) at the 5'-end by a process resembling → cis-splicing.

Mini-exon epitope tagging (MEET): A technique for the discovery of genes and their analysis, which is based on the insertion of a synthetic → mini-exon into → introns of a target gene, and permits detection of encoded proteins with the same → monoclonal antibody regardless of the intron class. The resulting protein is altered minimally by the mini-exon and can be analysed by functional assays such as → immunofluorescence or isolated using affinity-purification of the tagged protein.

Mini-gel: See → baby gel.

Minigene: A hypothetical precursor of a present-day → gene that was formed prebiotically as a small part of a nucleotide sequence with no biological information content. Minigenes were assembled during evolution and became present-day → exons, whereas the intervening "senseless" sequences still exist as → introns.

Mini hairpin: A small → hairpin-like DNA structure with a stem of only two, and a loop of only 3–4 nucleotides (d[GCGA AAGC]). Mini hairpins are compact structures with a high melting temperature, and are used for → mini-hairpin primed DNA amplification fingerprinting.

Mini-hairpin primed DNA amplification fingerprinting (mhpDAF): A variant of the → DNA amplification fingerprinting technique that employs → mini-hairpin primers to amplify genomic sequences in a conventional → polymerase chain reaction. The primers harboring a mini-hairpin at their 5'-termini and an arbitrary core of only 3 nucleotides at the 3'-termini allow to amplify multiple loci in DNAs from → plasmids, PCR-amplified fragments, → bacterial artificial chromosome and → yeast artificial chromosome clones, and small and big → genomes. In contrast to conventional DNA amplification fingerprinting with → arbitrary primers, mhpDAF enhances the detection of polymorphisms in target DNAs.

Mini-hairpin primer: An → oligonucleotide → primer that contains highly stable → hairpin-like structures with a short stem and a 3 nucleotide looped domain at its 5' terminus, and 8 nucleotides long 3' terminal stretches of arbitrary sequences. Such

Linear primer:

5´-G T A A C G C C-3´

Mini-hairpin primer:

```
    G
   / \
  A   C — G — 5´
  |   ‖
  A   G — C — G — C — C — 3´
   \ /
    A
```

Linear and mini-hairpin DAF-primers

primers select annealing sites during the primer-template screening phase of the amplification reaction much better than normal arbitrary primers, and are effectively anchored at their target sites (e.g. in genomic DNA). Mini-hairpin primers are used in → mini-hairpin primed DNA amplification fingerprinting.

Minilibrary (partial gene bank): A laboratory term for a → gene library that contains preselected and enriched → genomic DNA (genomic minilibrary) or → cDNA sequences (cDNA minilibrary). Such minilibraries contain only part of complex genomes or mRNA populations and are therefore easier to screen for target sequences than complete → genomic or → cDNA libraries. Their establishment, however, requires sequence information(s) and separation and enrichment procedures.

Minimal common region (MCR): Any genomic sequence that is lost or gained in → copy number variations from a variety of similar or (diagnostically) identical tumors (e.g. colon tumors) from different human individuals.

Minimal domain vector: Any → plasmid expression vector that contains only a truncated version of a cloned transcription → activation domain. For example, the VP16 activation domain from herpes simplex virus as part of the → tet-on/tet-off gene expression system plasmid contains repeats of a 13 amino acids tract that represents the functional core of the domain. Now, the → overexpression of VP16 can be deleterious, because it interacts with specific components of the transcription machinery. Therefore a truncated version of VP16 is less toxic than the full-length activation domain.

Minimal gene set: The minimal number of cellular genes that allows life, estimated for the genome of *Mycoplasma genitalium*. This organism seems to contain a → minimal genome made up of only 517 genes, about 265–350 of which are indispensable for life. *Bacillus subtilis* harbors a total of 4071 genes, of which 271 genes are essential for survival (under optimal conditions in the laboratory).

Minimal genome: The smallest set of genes that allows the replication of an organism in a particular environment (i.e. encode proteins for the catalysis of basic metabolic and reproductive functions). See → minimal gene set.

Minimal information about a microarray experiment (MIAME): A guideline for the publication of → microarray data that sets standards for good performance and reliable evaluation of microarray experiments, and their easy interpretation and independent verification (see the home page of the

Microarray Gene Expression Data Society: www.mged.org).

Minimalistic immunogenically defined gene expression vector (MIDGE vector): A linear double-stranded DNA vector for the transfer and expression of a gene in target cells (or, more precisely, nuclei) that contains only one single gene with its → promoter and → terminator sequences necessary for its → expression, and single-stranded → loops on either end for its protection from intra- and extracellular → nucleases (especially → exonucleases) and for the covalent introduction of peptide-, glycopeptide- or carbohydrate → ligands (for e.g. binding to cell surface receptor molecules, or directed transport to the nucleus or other cell organelles). MIDGE vectors circumvent the need for → selectable marker genes (e.g. → kanamycin/neomycin resistance genes) whose products may cause undesirable immune reactions in the target organism, and – in case of bacteria-derived sequences – may represent highly immuno-modulatory agents. More over, MIDGE vectors lack → origins of replication, since they are not propagated in bacterial hosts. MIDGEs can be targeted to specific cells by attaching specific peptide sequences to their end, and are vectors of choice for → genetic vaccination and somatic → gene therapy.

Minimal promoter: Any → promoter that consists only of the essential sequences for correct initiation of transcription of the adjacent gene (e.g. the → TATA-box and → cap site in TATA-box containing promoters, and 5′-PyPyAN[TA]PyPy-3′ in TATA-less promoters).

Minimal protein identifier (MPI): A collection of data that unequivocally identifies a specific protein from thousands of other proteins. Identification is based on a specific peptide map, generated by mass spectrometry, fragment ion spectrum (actually mass fingerprints of proteolytically [mostly tryptically] generated peptide fragments), and peptide fragment sequences. Nuclear magnetic resonance and X-ray crystallographic data can be used as additional identifiers.

Minimal RNA-induced silencing complex (minimal RISC): A variant of the → RNA-induced silencing complex (RISC) that is engineered to contain only a minimal number of proteins and therefore is only 160 kDa in size (native RISC: 500 kDa). This minimal RISC owns similar kinetic properties as the activated RISC in cell extracts, and serves to analyze e.g. the influence of RNA secondary structures on the recognition of RNA targets and target degradation.

Minimal tiling path (minimum tiling path): Any map or table showing the placement and order of a set of minimally overlapping clones (e.g. → bacterial artificial chromosome clones) that completely and contiguously cover a specific segment of DNA, a chromosomal region of interest, or a complete chromosome. Such minimal tiling paths are constructed from a genomic library by creating a physical map of the segment, region, or chromosome, and aligning selected clones using known markers. Sequences from these clones are then arranged into an unambiguous order based on their overlapping regions.

Minimal vector: Any DNA-based vector that only contains the sequences required for its maintenance.

Mini-me element (*m*icrosatellite-*i*nitiating *m*obile *e*lement): Any one of a group of highly abundant Dipteran → retroposons that contain two internal → proto-microsatellite regions (see also → microsatellite) with the potential to expand. Mini-me elements are flanked by 10–20 bp long, → inverted repeats (IRs) with the 3' repeat located sub-terminal (22–45 bp from the actual 3' end of the element). A partial duplication of the 5'-IR allows the formation of a → hairpin loop. A highly conserved 33 bp core region is flanked by both proto-microsatellites, the 3' proto-microsatellite consisting of $(TA)_n$ repeats, the 5' one of (GTCY), where Y is either C or T. The elements in different dipteran genomes vary in size from 500 to 1,200 bases, caused by → insertions or → deletions in a socalled variable region 3' of the 3' proto-microsatellite. These elements comprise about 1.2% of the *Drosophila melanogaster* genome and represent sources for new microsatellites. Basically two mechanisms generate these new repeats: (1) preexisting tandem repeats expand by an as yet unknown process, and sequences with high → cryptic simplicity are converted to tandemly repetitive DNA, and (2) the elements move to new genomic loci, where the new environment relaxes constraints on proto-microsatellites such that they expand more rapidly.

Minimum efficient processing segment (MEPS): The minimum length of a DNA sequence that is required for efficient homologous → recombination. MEPSs are 25–30 bp in *E. coli*, 50 bp in bacteriophage T4, about 250 bp in yeast (33 bp for → meganuclease-induced repeat recombination), and 250–400 bp in plants and cultured mammalian cells. Efficient intrachromosomal recombination in mouse L cells requires 134–232 bp of uninterrupted homology, and effective meiotic homologous recombination in humans is only possible with MEPS of 337–456 bp in length. The introduction of only one single basepair mismatch in a MEPS sequence substantially reduces recombination efficiency in prokaryotes, as do two single-nucleotide mismatches in 232 bp MEPS of human DNA. If the size of a previously functional MEPS is reduced below a certain threshold, recombination becomes very inefficient, or ceases.

Minimum inactivation length (MIL): The shortest length of an inhibitory → oligonucleotide (e.g. an → antisense oligonucleotide, → deoxyribozyme, → DNAzyme, → locked nucleic acid, → morpholino, → ribozyme) that achieves substantial target RNA inactivation (preferably at target concentrations prevailing within cells).

Minimum tiling set: The smallest number of clones that span the entire length of the DNA molecule from which they were derived for cloning. See → tiling path.

MiniPing: See → Ping.

Mini-prep (mini preparation): A small-scale method to extract and purify DNA and RNA from any source (e.g. phage, bacteria, plant, animal). Mini-preps are specifically adapted to small amounts of material (cells, tissues) and small volumes, and are therefore used to analyze → insert DNA in large numbers of transformants or cloning vectors.

Minipreparation: See → miniprep.

MiniSAGE: A variant of the conventional → serial analysis of gene expression (SAGE) technique for the analysis of global gene expression that capitalizes on the use of a single tube to perform → messenger RNA (mRNA) isolation, → reverse transcription of mRNA into → cDNA with a → biotin-labeled oligo(dT) → primer, enzymatic digestion of cDNA, binding of digested biotin-labeled 3'-terminal cDNA fragments to → streptavidin-coupled → magnetic beads, → ligation of → linker oligonucleotides containing recognition sits for a tagging enzyme to the bound cDNA fragments, and release of cDNA tags, with only one microgram of starting total RNA. MiniSAGE also reduces the amount of linker oligonucleotides in the ligation reaction, which minimizes their interference with SAGE → ditag amplification and increases the yield of SAGE ditags, and uses a → phase lock gel for the extraction of RNA from the original sample. See → microSAGE, → SAGE-Lite.

Minisatellite: Any one of a series of short (9–64 bp), usually GC-rich, → middle repetitive, tandemly arranged, highly variable (hypervariable) DNA sequences which are dispersed throughout the human genome (but also occur in animal and plant genomes) and which share a common 10–35 bp consensus or core sequence (core repeat unit, tandem repeat unit). The minisatellites show substantial length polymorphism arising from → unequal crossing-over that alters the number of short tandem repeats in a minisatellite, so that arrays about 0.1–20 kb in length are formed. Unequal exchanges may be favored by a recombination signal within the core sequence, especially since this core is similar to the *E. coli* recombination signal (→ chi sequence). A hybridization probe consisting of the core, repeated in tandem, can detect many highly polymorphic minisatellites simultaneously within a genomic digest and may therefore provide genetic markers for → linkage analyses (used in individual-specific → DNA "fingerprinting"). The theoretical probability that the same set of DNA fragments (the fingerprint) is identical in two human beings is so small that every human individual (except identical twins) is expected to have a unique pattern of bands detected with a minisatellite on autoradiograms. The mutation process acts preferentially at the 3' end of a minisatellite, so

Jeffrey's minisatellite core sequences (myoglobin gene)

Consensus	Origin
GGAGGTGGGCAGGAAG	myoglobin
aagGGTGGGCAGGAAG	clone 33.1
GGAGGTGGGCAGGAAX	clone 33.3
tGgGGaGGGCAGaAAG	clone 33.4
GGAGGYGGGCAGGAGG	clone 33.5
GGAGGaGGGCtGGAGG	clone 33.6
GGA-GTGGGCAGGcAG	clone 33.10
GGtGGTGGGCAGGAAG	clone 33.11
aGAGGTGGGCAGGtGG	clone 33.15
GGAGGTGGGCAGGAXG	core
GCTGGTGGGCTGGTGG	chi dimer

X = A or G   Y = C or T   - = deleted

that most of the sequence variability originates from here. In contrast, the 5′ end belongs to a low mutable region, and therefore stabilizes the repeat. The causes for this polarity are unknown.

An extraordinarily high minisatellite variation occurs in African populations (many groups with independent characteristic minisatellite patterns). The term minisatellite overlaps with → hypervariable region and → variable number of tandem repeats. See also → minisatellite-primed amplification of polymorphic sequences and → minisatellite variant repeat. Compare → microsatellite.

Minisatellite-primed amplification of polymorphic sequences (MAPS): A technique for the detection of sequence → polymorphisms in → genomic DNA of different organisms, in which single synthetic → minisatellite sequences are used as → primers in a conventional → polymerase chain reaction to amplify regions flanked by them. After amplification, the polymorphic bands can already be detected on → agarose gels with → ethidium bromide. Compare → DNA fingerprinting, → interspersed repetitive sequence polymerase chain reaction, → oligonucleotide fingerprinting.

Minisatellite variant repeat (MVR): A → minisatellite repeat sequence that differs from its neighbouring repeat(s) by only one or few → restriction endonuclease sites. This "interrepeat unit sequence variability" arises by mutations within certain repeats of a minisatellite. See also → variant repeat unit.

Minisequencing: See → pyrosequencing.

Mini-short tandem repeat (miniSTR): Any → microsatellite-containing repeat amplified with two flanking → primers positioned in the immediate neighbourhood of the microsatellite. Amplification in a conventioinal → polymerase chain reaction with these primers results in short amplification products, an advantage over longer products, if degraded samples (e.g. blood, tissue, semen) are only available for analysis.

miniSTR: See → mini-short tandem repeat.

Mini-Ti (mini-Ti-plasmid): A small derivative of the → Ti-plasmid of → *Agrobacterium tumefaciens* from which most of the → T-region has been deleted, except the → opine synthase gene and its promoter, a cloning site into which foreign DNA can be inserted, and the left and right → T-DNA borders. This plasmid replicates in *E. coli*, and may be conjugatively transferred into *A. tumefaciens*. The recipient can then transfer the modified T-region into wounded plant cells provided the vir functions are supplied in trans (e.g. by a Ti-plasmid carrying the → *vir* region).

Mini-Ti-plasmid: See → mini-Ti.

Minizyme: A synthetic oligoribonucleotide with hammerhead structure and → ribozyme activity.

Minocycline: The semisynthetic tetracycline derivative 2-(amino-hydroxy-methylidene)-4,7-bis (dimethylamino)-10,11,12a-trihydroxy-4a,5,5a,6-tetrahydro-4H-tetracene-1,3,12-trione that inhibits protein synthesis in Gram-positive and Gram-negative bacteria, thereby arresting their growth. This broad spectrum tetracycline antibiotic is active against many tetracycline-resistant strains of e.g. *Staphylococcus, Streptococcus* and *E. coli*.

Minor allele frequency (MAF): The frequency with which a minor → allele occurs in a given population, expressed as percentage of all alleles.

Minor base: See → rare base.

Minor gene: Any → gene, whose contribution to the expression of a particular polygenic trait is inferior to the contribution of another → major gene.

Minor groove: The indentation on the surface of a DNA → double helix molecule, formed by the sugar phosphate backbones and the edges of the base pairs (linked by → Watson-Crick base pairing forces) that contain the N3, (in → purines) or O2 atoms (in → pyrimidines). See → double helix, → major groove.

Minor groove binding probe (MGB probe, MGB ligand, minor groove binder): Any → oligonucleotide → probe that preferentially hybridizes to target sequences in the → minor groove of the DNA double helix. Such MGB probes (as e.g. dihydrocyclopyrroloindole tripeptide, DPI$_3$) can be used for e.g. the detection of → single nucleotide polymorphisms. The term is also used for → fluorochromes that bind to the minor groove (e.g. → HOECHST 33258 and variants, also → DAPI).

Minor spliceosome: A less abundant variant of the ubiquitous → spliceosome ("major spliceosome") that assembles on → splice junctions with a consensus sequence deviating from the canonical → GT-AG rule, and occurs in plants, metazoans, and humans, but not in yeast or *Drosophila*. The assembly of the minor spliceosome starts with the recognition of the aberrant splice sequence 5'-exon-AUAUCCUUU-3' of the → pre-messenger RNA by the di-snRNA U11/U12 (as opposed to the conventional U1/U2 binding to normal splice junctions). Therefore this special assembly pathway is called the U12-dependent pathway. The subsequent steps of the spliceosome formation are identical to the conventional assembly (i.e. recruitment of U4/U6 and U5). Minor spliceosomes harbor a set of specific proteins, but also share common proteins with the major spliceosomes (e.g. SF3b, G, F, E, D1, D2, D3, and B) and assemble on pre-mRNAs of e.g. ion channel protein-encoding genes.

Minus: Located → upstream of the → cap site. See → plus.

Minus strand (minus viral strand, –strand):

a) In a single-stranded DNA virus the strand complementary to the → plus strand, which can be transcribed into mRNA.
b) In a single-stranded RNA virus the non-coding strand which is copied by RNA-dependent RNA polymerase into translatable mRNA. Compare → plus strand, definition b.

Minus strand cDNA: See → antisense cDNA.

-10 sequence: See → Pribnow box.

-3/-1 rule: See → von Heijne rule.

Minus viral strand: See → minus strand.

MIP:

a) See → methylation-independent polymerase chain reaction.

b) See → *methylation induced premeiotically*.
c) See → *molecularly imprinted polymer*.

MIRA: See → *methylated-CpG island recovery assay*.

MIR fingerprinting: The establishment of genomic fingerprints using → primers derived from the 70 bp consensus sequence of a → *mammalian-wide interspersed repeat* (MIR) to amplify regions within the MIR element with conventional → polymerase chain reaction techniques. For example, the → forward primer 5′-ACCTTGAGCAAGTCACT-3′ and the → reverse primer 5′-GCCTCAGTTTCCTCAT-3′ are used to amplify MIR regions, the resulting amplification products separated by → polyacrylamide gel electrophoresis and visualized by either → autoradiography (if one of the primers was labeled with ^{32}P) or → silver staining. The MIR fingerprint patterns are species-specific and can be used for phylogenetic analyses.

miRISC: Any → RISC that contains → microRNA bound to its constituent Ago2 protein. See → siRISC.

miRNA: See → *microRNA*.

miRNarray: See → *microRNA array*.

miRNP: See → *micro-ribonucleoprotein*.

Mirtron: (pre-*miRNA*/*intron*): Any pre-microRNA (pre-miRNA) originating from a debranched intron that mimics the structural features of a pre-miRNA (e.g. contains short hairpins) and therefore enter the miRNA-processing pathway without Drosha-mediated cleavage. The mirtron pathway uses the action of the splicing machinery and lariat-debranching enzyme and merges with the canonical miRNA pathway during hairpin export by exportin-5, and both types of hairpins are subsequently processed by Dicer-1. This generates small RNAs that can repress perfectly matched and seed-matched target messenger RNAs. At least 14 such mirtrons are identified in *Drosophila melanogaster* and another four in *Caenorhabditis elegans*, some of which are selectively maintained during evolution. Mirtrons are an alternate source of miRNA-type regulatory RNAs.

Misactivated amino acid: Any activated amino acid (i.e. an amino acid condensed with ATP to yield an aminoacyl adenylate) that is erroneously transferred to the 3′end of a → *transfer RNA* (tRNA), although it is not the cognate amino acid of this tRNA. The individual aminoacyl-tRNA synthetases possess an editing mechanism to detect misactivated noncognate amino acids, which are then hydrolyzed before they can be incorporated into a polypeptide chain during ribosomal protein synthesis.

Mis-cleavage (Mc): A somewhat misleading term for the trypsin-catalyzed cleavage of some, but not all recognition sites within a target peptide or protein.

Miselongation: See → *misextension*.

Misextension (miselongation): The addition of bases onto the 3′ end of a → primer oligonucleotide that have no complementary counterparts in the → template strand. Such errors occur at a rate of 1 per 10,000–30,000 bases in a → DNA-dependent DNA polymerase-catalyzed reaction, but are corrected, if the enzyme possesses a 3′ →

5′ proofreading activity. Compare → misinsertion.

Misfolded protein: Any protein that adopts an abnormal, usually non-functional or even toxic three-dimensional configuration after its synthesis on the → ribosome. Misfolded proteins comprise a large fraction of the total proteins of a cell, and are primarily degraded by the → *u*biquitin-*p*roteasome *s*ystem (UPS). See → toxic protein.

MISH: See → *m*icrodrop *i*n *s*itu *h*ybridization.

Misinsertion: The incorporation of bases into a growing → polynucleotide chain (→ DNA or → RNA) that have no complementary counterparts in the → template strand. Such mismatched bases are normally excised by → mismatch repair systems and replaced by the matching bases. Compare → misextension.

Mismatch: See → base mismatch.

Mismatched primer: Any oligonucleotide → primer used in the → polymerase chain reaction that is not perfectly homologous to its template DNA. Despite such a mismatch, the primer can be used for amplification if its 3′ end is well matched to the template.

Mismatch extension: The ability of → DNA polymerases to use even a mismatched base at the 3′ end of a → primer to synthesize a new strand complementary to the → template strand. The DNA polymerases extend a → mismatch more slowly than a matched 3′ terminus, however, and different mismatches extend at different rates. For example, a G/T mismatch extends readily, whereas C/C, A/G, G/A, A/A and G/G mismatches extend less efficiently.

Mismatch gene synthesis: The *in vitro* synthesis of two single-stranded complementary oligodeoxynucleotides that differ in sequence by only one or a few bases. Upon reannealing, these mismatched bases

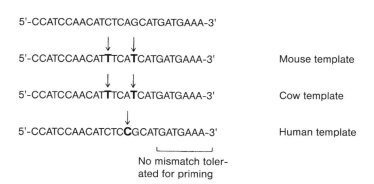

Cytochrome b primer from fish

5′-CCATCCAACATCTCAGCATGATGAAA-3′

5′-CCATCCAACAT**T**TCA**T**CATGATGAAA-3′   Mouse template

5′-CCATCCAACAT**T**TCA**T**CATGATGAAA-3′   Cow template

5′-CCATCCAACATCTC**C**GCATGATGAAA-3′   Human template

No mismatch tolerated for priming

↓ : Arrows denote mismatches in the primer core that does not interfere with amplification

Mismatched primer

cannot base-pair with a complementary partner. If such mismatched genes are cloned into a plasmid and transformed into a host cell, DNA repair processes will eliminate the mismatch, using both strands as templates. This leads to a population of two distinct genes, differing by only one or a few bases at a specific site.

Mismatch oxidation: A technique for the visual detection of → base mismatches in target DNA that relies on the attack of mismatches by $KMnO_4$ in the order $T>C\gg G>A$. The chemical oxidizes the double bonds of the bases in mismatches more efficiently than the matched bases, which are oxidized over time to generate a comparatively low background signal. Low oxidation (or none) leads to the development of a pink colour, high levels of oxidation produce a yellow colour. In short, the target fragment is first amplified by conventional → polymerase chain reaction (PCR), the resulting products heated and then cooled to form → homo- and → heteroduplexes, then $KMnO_4$ added, upon which the corresponding colour develops.

Mismatch repair (MMR; postreplication repair): The detection and replacement of incorrectly paired (mismatched) bases or small insertion/deletion (ID) mispairs in newly synthesized DNA. For example, in *E. coli* a → mismatch repair system consisting of 11 proteins, encoded by the genes *mutH, mutL, mutS, uvrD* and *uvrE*, screens the newly synthesized DNA strand for mismatched bases. Proteins MutS, MutL, and MutH recognize mismatches and incise the newly synthesized

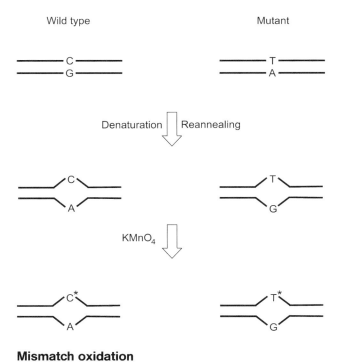

Mismatch oxidation

unmethylated DNA strand ("initiation"). The mispaired bases and a short region surrounding them are excised by one of four → exonucleases (Exo1, Exo VII, ExoX, or RecJ) that catalyze 5′ or 3′ excision from the DNA strand → break in concert with UvrD → helicase ("excision"). Finally, the → DNA polymerase III holoenzyme catalyzes the repair process, which is completed by → DNA ligase. This repair mechanism acts before the newly replicated DNA is methylated. Only after its completion the *de novo* synthesized strand is modified e.g. by → dam methylase according to the methylation pattern of the → complementary strand (→ maintenance methylation). In eukaryotes, the initial recognition of mismatches is accomplished by a complex of the two proteins MSH2 and MSH6 (MutSα), or to a limited extent by MSH2-MSH3 (MutSß), which binds to mismatched bases. A second complex, consisting of proteins MLH1, PMS2, PCNA (proliferating cell nuclear antigen), RPA (replication protein A), EXO1, HMGB1, RFC (replication factor C), and → DNA polymerase δ then joins the mismatch-MSH2/6 complex, and catalyzes excision and repair of the mismatch. The MutLα complex promotes termination of the EXO1-catalyzed excision upon mismatch removal by dissociating EXO1 from the DNA. DNA ligase I finally catalyzes the → ligation step. Hereditary deficiencies in the MMR system result in gene mutations and subsequent susceptibility to specific types of cancers, including hereditary colorectal cancer. Loss of the MMR leads to the socalled mutator phenotype that exhibits increased mutation rates.

Mispairing: See → base mismatch.

Mispriming: An undesirable artifact generated by the → annealing of → amplimers to non-target sequences and the extension of these amplimers by → *Thermus aquaticus* DNA polymerase in the → polymerase chain reaction. The generation of such artifactual products can be circumvented by the → hot start technique.

Misreading: See → mistranslation.

Missense mutant: A → mutant carrying one or more → missense mutations.

Missense mutation: Any gene mutation in which one or more → codon triplets are changed so that they direct the incorporation of amino acids into the encoded protein, which differ from the wild type (e.g. UUU, encoding phenylalanine, mutates to UGU, encoding cysteine). The replacement of a wild type amino acid by a missense amino acid in the mutant potentially produces an unstable or inactive protein. Compare → mistranslation, where a "wrong" amino acid is incorporated despite of a correct mRNA.

Missense single nucleotide polymorphism (missense SNP): Any → single nucleotide polymorphism that occurs in the coding region of a gene, and changes the amino acid sequence of the encoded protein. Such missense SNPs, if responsible for a functional change of e.g. a protein → domain, may cause diseases. See → silent SNP.

Missense SNP: See → missense single nucleotide polymorphism.

Missing contact analysis: See → DNA-protein interference assay.

MIST: See → multiple spotting technique.

Mistranslation (misreading): The incorporation of an incorrect amino acid into a nascent polypeptide inspite of the presence of an mRNA with the correct sequence. Mistranslation may becaused by the improper function(s) of the → tRNA, the → aminoacyl tRNA synthetases, or the → ribosome. Compare → missense mutation, where a mutated gene causes the transcription of an incorrect mRNA, which consequently directs incorporation of a "wrong" amino acid.

MITE: See → miniature inverted repeat transposable element.

MITE-AFLP: See → miniature inverted repeat transposable element amplified fragment length polymorphism.

***Mit*hramycin (MIT; also plicamycin, aureolic acid):** One of a series of acid oligosaccharide antibiotics produced by different strains of *Streptomyces*. For example, mithramycins A, B and C are synthesized by *S. argillaceus* and *S. plicatus*, mithramycin A being the dominant compound. Mithramycins bind to GC-rich stretches in the minor groove of double-stranded DNA, which can be quantified by the yellow fluorescence of the DNA-mithramycin complex. At the same time, mithramycins prevent RNA synthesis from the complexed DNA *in vitro* and *in vivo*.

Mitochip: See → mitochondrial genome chip.

Mitochondrial diseases: A variety of human diseases caused by a series of → mutations in → mitochondrial DNA (mtDNA). Since the proportion of mitochondria with mutations at specific sites in mtDNA varies after repeated cell divisions, mitochondrial diseases are highly variable.

For example, *L*eber *h*ereditary *o*ptic *n*europathy (LHON), characterized by optic atrophy, abnormal heart beat and neurological abnormalities, is a consequence of several mutations in five NADH dehydrogenase genes at different sites in the mtDNA. *M*yoclonic *e*pilepsy and *r*agged *r*ed *f*ibers (MERFF) disease, a variant of myoclonic epilepsy with shock-like convulsions, is caused by a → point mutation in the tRNA^{Lys} gene, whereas MELAS (*m*itochondrial *m*yopathy, *e*ncephalopathy, *l*actic *a*cidosis, and *s*troke-like episodes) is dependent on an A → G substitution in tRNA^{Leu} at nucleotide 3,243. The same mutation occurs in about 20% of patients with the recessive autosomal *p*rogressive *e*xternal *o*pthalmoplegia (PEO). The *K*earns-*S*ayre *s*yndrome (KSS) with progressive opthalmoplegia, pigmentary degeneration of the retina, and cardiomyopathy is associated with either a → deletion or a → base substitution in nucleotide 8,993 of the mtDNA, and the Pearson marrow-pancreas syndrome is caused by deletions in mtDNA affecting subunit 4 of NADH dehydrogenase, subunit 1 of cytochrome oxidase, and subunit 1 of ATPase. These are few examples of mitochondrial diseases that affect mostly male patients in humans. For example, more than 85% of all LHON cases affect human males.

Mitochondrial DNA (mtDNA): The circular → duplex DNA of → mitochondria, which is found in about 5–15 copies per organelle (exception: the 40 kb *Paramecium* mtDNA is linear). The mtDNA of most organisms is by far smaller than the → nuclear DNA (e.g. 17–101 kb in yeast, 15–20 kb in animals; from 200–2500 kb in flowering plants) and consists of two strands. The heavy (H) strand encodes two → ribosomal RNAs, 14 → transfer RNAs (tRNAs), and about a dozen proteins (e.g.

cytochrome b and cytochrome oxydase, seven subunits of NADH dehydrogenase, mitochondrial → ribosomal proteins, → elongation and termination factors, and ATP synthase), and is transcribed into a single → polycistronic RNA, which is subsequently processed into smaller units. The light (L) strand codes for eight tRNAs and one subunit of NADH dehydrogenase. Some genes overlap, and the → promoters are usually short. The → transcripts lack a typical → cap structure at the 5′-end, their → leader sequence is minimal, and in addition to the AUG → initiator codon, AUA, AUU and AUC serve as methionine start codons. The socalled → universal code is exceptionally varied in mammalian, but not plant mitochondrial DNA. For example, UGA (stop codon) translates into tryptophan, AUA (codon for isoleucine) into methionine, CUA (leucine) into threonine, and AGA and AGG (arginine) for stop. Some mitochondrial genes do not end in → stop codons, but instead their transcripts terminate in an A or UA that is extended by adenylation into a UAA stop signal. Most (and especially strategic) mitochondrial proteins are encoded by

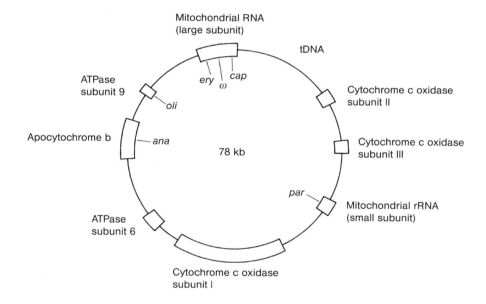

Simplified map of yeast mitochondrial DNA

Ana : Antimycin A resistance gene
Cap : Chloramphenicol resistance gene
Ery : Erythromycin resistance gene
Oli : Oligomycin resistance gene
Par : Paromomycin resistance gene
ω : Locus interfering with recombination of mitochondrial genes

nuclear genes, and a coordinated interaction between the mtDNA and the nuclear DNA is required for maintenance and division of mitochondria. For example, in mammals all mitochondrial ribosomal proteins are encoded by nuclear genes and, after their synthesis on cytoplasmic → ribosomes, are imported into mitochondria. The massive → horizontal gene transfer of mitochondrial genes into the nucleus over evolutionary times led to a reduction in mitochondrial autonomy ("semi-autonomy of mitochondria"). Some of the mitochondrial DNA sequences are promiscuous and move into other organelles (e.g. the chloroplast in plants, and the nucleus in all organisms) in surprisingly high frequencies (1×10^{-5} per cell generation).

Mitochondrial DNA depletion (mtDNA depletion): The complete removal of → mitochondrial DNA from a → mitochondrium, caused by environmental toxins interfering with mtDNA → replication, as e.g. *azido*thymidine (AZT) that inhibits DNA replication catalysed by the mitochondrial → DNA polymerase γ- mtDNA depletion can be a cause for recessive Mendelian human disorders. Do not confuse with → DNA deletion.

Mitochondrial DNA lineage (mtDNA lineage): A group of closely related and therefore mostly homologous mitochondrial DNA molecules.

Mitochondrial Eve (African Eve, "phylogenetic Eve"): A human female, who lived approximately 150,000–200,000 years ago in Africa and from whom everyone on this planet descended through the maternal line. In essence, all the maternal lines are connected to African Eve by their mitochondrial DNA (which is almost exclusively inherited maternally). Sequence comparisons of this mtDNA from many humans discovered that people tend to cluster into a small number of groups, each one defined by the precise sequence of their mtDNA. For example, in native Europeans seven such groups, four among Native Americans, nine among Japanese, and so on, exist. Each of these groups then traced back to just one woman, the common maternal ancestor of everyone in her group, or clan.

In present-day Europe, seven clans prevail ("seven daughters of Eve"):

Clan Ursula: The oldest of the seven native European clans, founded around 45,000 years ago by the first modern *Homo sapiens* established in Europe. About 11% of modern Europeans are direct maternal descendants of Ursula. The clan is particularly well represented in western Britain and Scandinavia. It is molecularly characterized by a C-T → transition at position 16,270.

Clan Xenia: second oldest of the seven native European clans. It was founded 25,000 years ago by the second wave of modern humans, *Homo sapiens*, who established themselves in Europe, just prior to the coldest part of the last Ice Age. Today around 7% of native Europeans are in the clan of Xenia. Within the clan, three distinct branches fan out over Europe. One is still largely confined to Eastern Europe, while the other two have spread further to the West into central Europe and as far as France and Britain. About 1% of Native Americans are also in the clan of Xenia. The mtDNA of this clan is characterized by a transition mutation at position 16,223.

Clan Helena: is by far the largest and most successful of the seven native clans with 41% of Europeans belonging to one of its many branches. It began 20,000 years ago with the birth of Helena somewhere in the valleys of the Dordogne and the Vezère, in south-central France. The clan is widespread throughout all parts of Europe, but reaches its highest frequency among the Basque people of northern Spain and southern France.

Clan Velda: is the smallest of the seven clans containing only about 4% of native Europeans. Velda lived 17,000 years ago in the limestone hills of Cantabria in northwest Spain. Her descendants are found nowadays mainly in western and northern Europe and are surprisingly frequent among the Saami people of Finland and Northern Norway. The mtDNA of this clan is characterized by a transition mutation T-C at position 16,298.

Clan Tara: includes slightly fewer than 10% of modern Europeans. Its many branches are widely distributed throughout southern and western Europe with particularly high concentrations in Ireland and the west of Britain. Tara herself lived 17,000 years ago in the northwest of Italy among the hills of Tuscany and along the estuary of the river Arno. The mtDNA of this clan is characterized by transition mutations T-C at position 16,126 and C-T at 16,294, respectively.

Clan Katrine: is a medium-sized clan comprising+ 10% of Europeans. Katrine herself lived 15,000 years ago in the wooded plains of northeast Italy, now flooded by the Adriatic, and among the southern foothills of the Alps. Her descendants are still there in numbers, but have also spread throughout central and northern Europe. The mtDNA of this clan is characterized by transition mutations of the T-C type at position 16,298 and 16,311, respectively.

Clan Jasmine: is the second largest of the seven European clans after Helena and is the only one originating outside Europe. Jasmine and her descendants, who now make up 12% of Europeans, were among the first farmers and introduced the agricultural revolution into Europe from the Middle East about 8,500 years ago. The mtDNA of this clan is characterized by transition mutations of the C-T type at positions 16069 and T-C type at position 16,126, respectively.

Clan Ulrike: is not among the original "Seven Daughters of Eve" clans, but with about 2% of Europeans is included among the numerically important clans. Ulrike lived about 18,000 years ago in the cold refuges of the Ukraine at the northern limits of human habitation. Though Ulrike's descendants are common today, the clan prevails mainly in eastern and northern Europe with high prevalence in Scandinavia and the Baltic states.

For an identification of the corresponding clan a 400 bp sequence between positions 16001 and 16400 of the → *Cambridge Reference Sequence (CRS)* is screened:

5'-ATTCTAATTT AAACTATTCT CTGTTCTTTC ATGGGGAAGC AGATTTGGGT
ACCACCCAAG TATTGACTCA CCCATCAACA ACCGCTATGT ATTTCGTACA
TTACTGCCAG CCACCATGAA TATTGTACGG TACCATAAAT ACTTGACCAC
CTGTAGTACA TAAAAACCCA ATCCACATCA AAACCCCCTC CCCATGCTTA
CAAGCAAGTA CAGCAATCAA CCCTCAACTA TCACACATCA ACTGCAACTC
CAAAGCCACC CCTCACCCAC TAGGATACCA ACAAACCTAC CCACCCTTAA
CAGTACATAG TACATAAAGC CATTTACCGT ACATAGCACA TTACAGTCAA
ATCCCTTCTC GTCCCCATGG ATGACCCCCC TCAGATAGGG GTCCCTTGAC-3'

Any mutation in this sequence (mostly transitions) allows to categorize any carrier into a clan. If the tester sequence is identical to the CRS sequence, then it belongs to the clan of Helena. Everyone in the same clan is therefore a direct maternal descendant of one of these clan mothers and carries her mtDNA within every cell.

See → Y-chromosome Adam.

Mitochondrial genetics: A branch of → genetics that focusses on the inheritance of mitochondrial traits (especially human disorders caused by mutations in → mitochondrial DNA), the isolation and characterization of the underlying genes, mutations in these genes, the establishment of → genetic and → physical maps of the mitochondrial DNA, mostly depending on whole genome sequencing, the interference of nuclear genes with mitochondrial functions, the characterization of mitochondrial mutants and the estimation of mutation rates. Mitochondrial genetics is complicated by the presence of multiple copies of mtDNA molecules in one and the same organelle that may differ in base composition.

Mitochondrial genome chip (mitochip): Any → microarray that contains the complete sequence of → mitochondrial DNA (e.g. from humans). Mitochips are used for the detection of somatic mitochondrial → mutations, which frequently are associated with different cancer types and serve as potential markers for an early diagnosis.

Mitochondrial import: The transport of peptides and proteins synthesized on cytoplasmic ribosomes across the two cooperating membranes of the mitochondrion into the interior of the organelle.

Mitochondrial proteome: The complete set of proteins in a mitochondrion, probably comprising more than 2,000 individual proteins, of which only about 30% are characterized. Mitochondrial proteins are potential drug targets, since a number of major diseases involve abnormal mitochondrial functions and therefore disfunctional mitochondrial proteins.

Mitochondrial transcription *ter*mination *f*actor (mTERF): A → leucine zipper DNA-binding protein that regulates the transcription termination of mitochondrial genes. For example, a specific mTERF binds to a 28 bp region at the 16S rRNA-leucyl tRNA genes boundary, promotes termination of 16S rRNA gene transcription, and thereby regulates the ratio of rRNA/mRNA in mitochondria.

Mitochondriomics: Another term of the → omics era that describes the whole set of genetic, genomic, transcriptomic, proteomic, and bioinformatic technologies to characterize the → mitochondrial genome, its genes, their expression patterns in various experimental situations, the encoded proteins, their functions, and their localization within the mitochondrium.

Mitochondrium (mitochondrion; Greek: mitos-thread, chondrion-grain): Any one of hundreds or thousands of cytoplasmic semiautonomous organelles of eukaryotic cells that is surrounded by a double membrane and carries a series of → mitochondrial DNA molecules encoding relatively few proteins needed for mitochondrial functions. Most of the mitochondrial proteins are encoded by nuclear genes, whose → messenger RNAs are translated on cytoplasmic → ribosomes. The translated proteins contain socalled signal peptides for their import into the organelle. Among the nucleus-encoded proteins are strategic proteins as e.g. a series of ribosomal proteins, porins, DNA and RNA polymerases, enzymes of the citrate cycle, subunits of the ATPase, cytochrome c oxidase and cytochrome bc_2 complex, to name only few. Main functions of the mitochondria are the electron transport chain with the coupled oxidative phosphorylation (generation of ATP), the citric acid cycle, and the oxidative degradation of fatty acids.

Mitofusin (Mfn): Any one of a family of large mitochondrial GTPase proteins that are integral components of the outer mitochondrial membrane. A large N-terminal region contains a GTPase domain, whereas the C-terminus projects into the cytoplasm. At least two homologs, termed Mfn1 and Mfn2, are present in mammals that together with Opa1 mediate mitochondrial fusions. Mfn complexes form in trans (i.e. between adjacent mitochondria). Multiple missense mutations in the *Mfn2* gene leads to the peripheral Charcot-Marie-Tooth type 2A syndrome, an inherited progressive neuropathy with muscular atrophy of hands and feet as symptoms of the affected individuals. In fibroblasts, the missense mutations cause severe aggregation of mitochondria in the cell. Likewise, mutations in the *Opa1* gene cause dominant optic atrophy (an inherited optic neuropathy). Experimental mice lacking either of the *Mfn* genes die as embryos.

Mitogen-activated protein kinase (MAPK): Any one of a series of highly conserved protein kinases with different protein substrate specificity that altogether function in the transmission of signals from the intra- and extra-cellular environment to the nucleus, where specific genes are turned on in response to the signal. Such socalled signaling cascades include MAPK or extra-cellular signal-regulated kinase (ERK), → MAPK kinase (MKK or MEK), and MAPK kinase kinase (MAPKKK or MEKK). MAPKK kinase/MEKK phosphorylates and activates its downstream protein kinase MAPK kinase/MEK, which in turn activates MAPK. For example, MAPKKK5, an 11 kinase subdomain protein of 1,374 amino acids is abundantly expressed in human heart and pancreas. The MAPKKK5 protein phosphorylates and activates MKK4 (synonyms: SERK1, or MAPKK4) *in vitro*, and activates c-Jun N-terminal kinase (JNK)/stress-activated protein kinase (SAPK) during transient expression in COS cells.

Mitogen-activated protein kinase kinase (MAP kinase kinase, MEK kinase): Any one of a family of protein kinases (e.g. MEK-1 to MEK-7) that activate MAP kinases by phosphorylation. For example, the prototype MEK kinase, MEK-1, specifically phosphorylates strategic threonine and tyrosine residues of the sequence H_2N-thr-glu-tyr-COOH in the MAP kinase protein.

Mitogenomics: The whole repertoire of techniques to identify specific characteristics of the → mitochondrial genome, as e.g. the full → nucleotide sequence, the → gene number, → gene order, → gene size, number of → non-coding regions (NCRs), compositional features and divergence of protein-coding genes, size and position of → intergenic spacers, GC content, and others.

Mitomycin C: An aziridine → antibiotic produced by *Streptomyces caespitosus* that cross-links complementary strands of a DNA duplex molecule, and thereby prevents DNA → replication and → transcription.

Mitomycin C

Mitosome: An organelle of anaerobic parasitic amoeba (e.g. *Giardia intestinalis* and *Trachipleistophora hominis*) that is surrounded by a double membrane and involved in the synthesis of iron-sulfur clusters. These clusters are assembled into cellular proteins. Mitosomes are considered to be reduced mitochondrial descendants that lost their genomes (either by → deletion or transfer to the nuclear genome of the host cell) and with them all functions characteristic for mitochondria (as e.g. respiratory chain). See → horizontal gene transfer, → hydrogenosome, → mitochondrium.

Mixed codon family: Any group of four → codons that share the first two bases, but code for more than one amino acid. For example, AAU, AAC, AAA, and AAG share the first two bases, but encode asparagin (AAU, AAC) and lysine (AAA, AAG) and therefore represent such a mixed codon family. See → unmixed codon family.

Mixed infection: See → marker rescue.

Mixed lineage leukemia gene (MLL gene): A gene on human chromosome 11 encoding a protein that recruites other proteins in the nucleus to form a huge protein machine responsible for → chromatin structure. Within the MLL gene a breakpoint region is located, at which DNA double-strand → breaks occur. During the subsequent repair processes erroneously chromosome → translocations occur. All known reciprocal translocations of the MLL gene (e.g. onto all chromosomes, but with high frequency onto chromosome 17) inevitably lead to *a*cute *l*ymphatic (ALL) or *a*cute *m*yelotic leukemia (AML).

Mixed oligonucleotide-primed amplification of cDNA (MOPAC): A technique for the isolation of genes of far-reaching homology (e.g. genes of a → gene family) by deducing → primers for → polymerase chain reaction from peptide sequences. Due to the → degenerate genetic code usually a series of primers differing in nucleotide sequence are generated from one peptide sequence and used to amplify the corresponding sequence out of a → cDNA library. Usually two degenerate pools of → oligonucleotides (all sequence variants which may encode the same set of amino acids according to the → degenerate code) to prime → first strand cDNA amplification. In short, mRNA is first reverse-transcribed into first strand cDNA, using → reverse transcriptase. Then two pools of oligonucleotides (pool 1: oligos complementary to all possible sequences encoding a particular tract of amino acids in the target protein; pool 2: oligos complementary to all possible sequences encoding another tract of amino acids in the same protein) are annealed as → amplimers to the first-strand cDNA. The amplified product can then be cloned into appropriate vectors and used as probe to screen → genomic or → cDNA libraries.

Mixed oligonucleotide probe (mixed oligo probe): A mixture of synthetic single-stranded oligo deoxynucleotides about 12–25 bases in length that differ from each other in one single base only. Mixed oligo probes are used to screen → genomic or

→ cDNA libraries for a gene whose protein product is known and whose sequence has been inferred from the corresponding amino acid sequence. However, the → codon bias does not allow to deduce the correct nucleotide sequence of the corresponding gene indubitably. The use of mixed oligos increases the probability that at least one perfectly matched oligonucleotide will detect the desired gene. Frequently, inosine is incorporated as the → wobble base, since it base-pairs with most of the other bases.

Mixed primer labeling: See → random priming.

Mixed target polymerase chain reaction (mixed target PCR): A conventional → polymerase chain reaction, in which two (or more) target → template DNAs are present and simultaneously amplified.

MLL: See → mixed lineage leukemia gene.

MLP: See → multilocus probe.

MLPA: See → multiplex ligation-dependent probe amplification.

MLST: See → multilocus sequence typing.

MM: See → mismatch, or → base mismatch.

MMA: See → multiplex messenger assay.

MMGT: See → microcell-mediated gene transfer.

M-MLVRT: See → moloney murine leukemia virus reverse transcriptase.

MMR: See → mismatch repair.

MMS: See → marker-mediated selection.

MN assay: See → *in vitro* micronuclei assay.

MNAzyme: See → multi-component nucleic acid enzyme.

MNP: See → multiple nucleotide polymorphism.

MOB: See → microsatellite obtained from BAC.

Mob (mob, mob **functions,** *mobilizing* **functions):** Two defined regions of a conjugative → plasmid, of which one encodes a mobilizing protein that specifically binds to the *mob* region, and induces a → nick in its so-called → *nic/bom* site (*nic* for *nick*; *bom* for *basis of mobility*). One of the *mob* functions (synthesis of a mob protein) can also be supplied in trans. If for example, one plasmid has lost this property (*mob*⁻), a second plasmid coding for a functional mob protein (*mob*⁺) may complement the *mob* functions so that the deficient plasmid can be transferred from one cell to another. In addition to the *mob* regions functional → *tra* genes are also necessary for plasmid transfer. See also → mobilization.

Mobile domain: Any typically compact, cystein-rich → domain of 30–130 amino acids in a → mosaic protein that is able to fold independently of other domains and is evolutionarily mobile, i.e. has spread during evolution and now occurs in many functionally unrelated proteins.

Mobile gene: Any gene that changes its location within a → genome, or moves from one genome to another genome within the same cell (e.g. from → mitochondria to the → nucleus, or vice versa, or from → chloroplasts to the mitochondria or the nucleus, in green plant cells), or from the genome of one organism to the genome of another organism (see → horizontal gene transfer).

Mobile genetic element: Synonym for → transposon.

Mobile group II intron: Any → intron that functions not only as → ribozyme, but also as → transposable element. Intron transposition is mediated by the intron-encoded DNA endonuclease, which is a → ribonucleoprotein particle containing both the intron RNA and the intron-encoded protein. The DNA target for intron transposition is about 31 nucleotides long, therefore highly specific, and primarily recognized by the intron RNA through → base-pairing.

Mobility-shift DNA-binding assay (band shift, band shift assay, DNA-binding assay, gel electrophoresis DNA-binding assay, gel mobility shift assay, gel retardation assay): A method to detect specific DNA-protein interactions that is based on an altered mobility of protein-DNA complexes during non-denaturing gel electrophoresis, as compared to free DNA. In short, the target DNA fragment is labeled (e.g. end-labeled with $\gamma\text{-}^{32}\text{P-dATP}$ using deoxynucleotidyl transferase) and incubated with a nuclear extract. Specific DNA-protein complexes are detected by low ionic strength → polyacrylamide gel electrophoresis and → autoradiography. The free fragment moves faster than the protein-DNA complex (which is retarded). Usually an excess of heterologous competitor DNA is added to saturate the more abundant, non-specific DNA-binding proteins. See also → electrophoretic mobility shift assay.

Mobilization:

a) The directed movement of a non-conjugative plasmid from one bacterium (donor) to another bacterium (acceptor) with the aid of a → conjugative plasmid of the donor.
b) The directed movement of chromosomal genes of one bacterium (donor) to another bacterium (acceptor) with the aid of a → conjugative plasmid of the donor.
c) The movement of → transposons (→ transposition) or → retrosequences.

Mock infection: A laboratory term for a fictive infection of cells with a → bacteriophage or → virus, In which the cells were either not exposed to the infectious agent or treated WIth killed agents, but otherwise processed as the truly infected cells.

Moderately affected Alzheimer disease DNA (MAD-DNA): The DNA from cells of the hippocampal region of patients with weak to moderate Alzheimer's disease symptoms. MAD-DNA shows a modified → B-DNA conformation with a small shoulder peak at 290 nm of its spectrum, binds more → ethidium bromide than → severely affected Alzheimer disease DNA, and exhibits an unusual biphasic melting profile with two T_M values of 54 °C and 84 °C.

Moderately repetitive DNA: See → middle repetitive DNA.

Modification:

a) Any change in a protein or nucleic acid molecule after its synthesis.
b) See → DNA methylation, also → restriction-modification system.

Modification enzyme: Any enzyme that modifies → DNA or → RNA. Typcial examples for modification enzymes are → bacterial alkaline phosphatase, → calf intestinal alkaline phosphatase, → DNA polymerase I, → DNase I, → exonucleases, → Klenow fragment, → MOdification methylase, → mung bean nuclease, → nuclease P1, → nuclease S1, → reverse transcriptase, → RNA polymerase, → RNase, → T4 DNA ligase, → T4 DNA polymerase, → T4 RNA ligase, → T7 DNA polymerase, → T7 RNA polymerase, → T3 RNA polymerase, → terminal transferase.

Modification gene: See → modifier gene (b).

Modification methylase (DNA modification methyltransferase, modification enzyme, EC 2.1.1.72): A bacterial enzyme that catalyzes the transfer of methyl groups from → S-adenosyl-L-methionine to specific positions of specific bases in DNA. Since the methylation of such bases within the recognition sequence of a → restriction endonuclease prevents the recognition process and thus the cleavage at this sequence, methylation protects bacterial DNA against own and foreign restriction enzymes (→ restriction-modification system). Some modification methylases with high specificity for distinct recognition sequences of particular endonucleases are used to protect internal recognition sequences of a DNA fragment which is to be cloned (e.g. → Eco RI methylase specifically methylates bases in the recognition sequence of Eco RI endonuclease). If for example, such a fragment contains Eco RI sites, any cloning into an Eco RI site of a

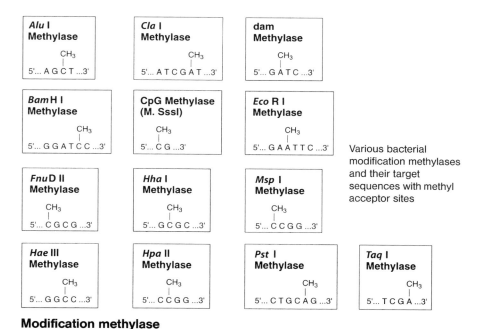

Various bacterial modification methylases and their target sequences with methyl acceptor sites

Modification methylase

→ cloning vector would be obsolete, since excision of the insert by *Eco* RI would inevitably destroy it. If, however, the *Eco* RI sites of the fragment are methyl-protected, then the use of *Eco* RI linkers and cloning into *Eco* RI sites of vectors becomes feasible. See also → heteroprostomers and → isoprostomers, → Dam methylase, → Dcm methylase, and compare → methyltransferase.

Modified adenine recognition and restriction system: See → methylated adenine recognition and restriction (Mrr) system.

Modified base:

a) Any nucleic acid base that is altered postsynthetically, e.g. by methylation.

See → DNA methylation, → restriction-modification system.
b) See → rare base.

Modified *c*ytosine *r*estriction (Mcr) system: See → methylated cytosine recognition and restriction system.

Modified deoxynucleoside-5'-triphosphate: Any one of a series of synthetic → deoxynucleoside triphosphates, into which an additional group is incorporated, as e.g. an amino, azido, or methyl group (see formulas). Such analogues are used for DNA-protein interactions and inhibition of DNA polymerases. See → modified ribonucleoside-5'-triphosphate.

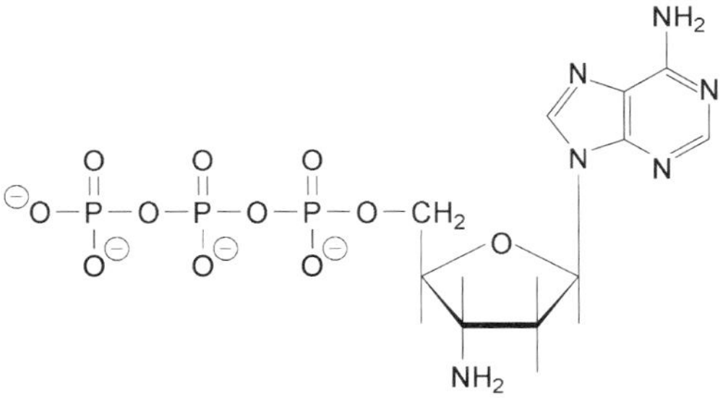

3'-Amino-2', 3'-ddATP

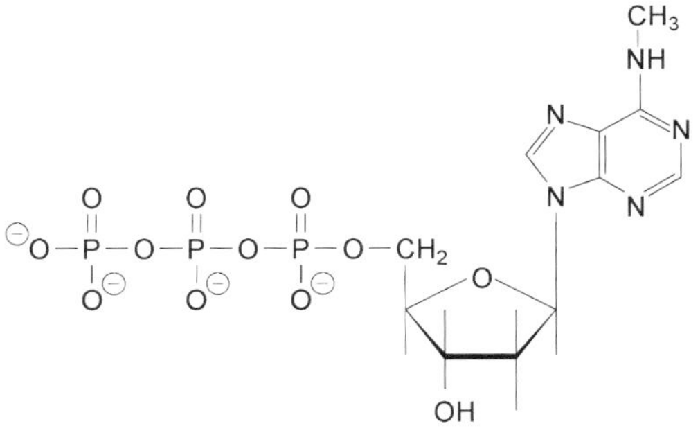

N^6-Methyl-2'-dATP

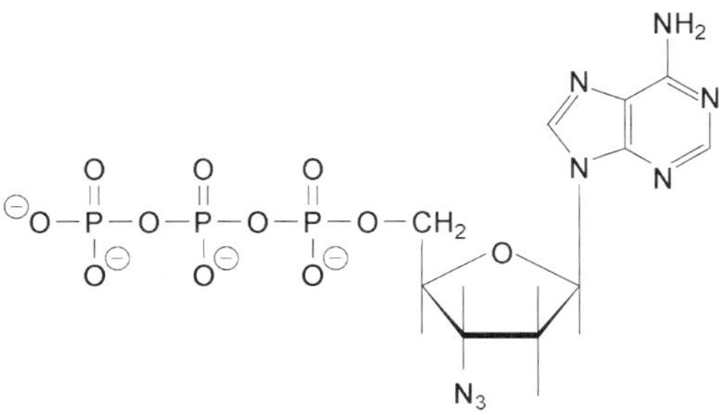

3'-Azido-2', 3'-ddATP

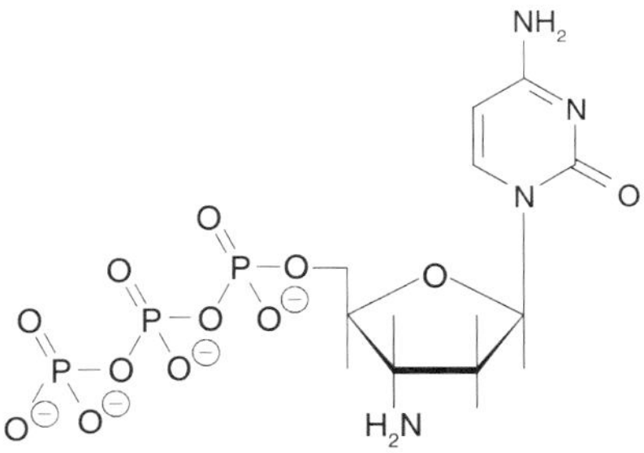

3'-Amino-2',3'-dideoxycytidine-5'-triphosphate

Modified deoxynucleoside-5'-triphosphate

Modified deoxynucleoside-5'-triphosphate

3'-Amino-2',3'-dideoxyguanosine-5'-triphosphate

3'-Azido-3'-deoxythymidine-5'-triphosphate

3'-Amino-3'-deoxythymidine-5'-triphosphate

3'-Azido-2',3'-dideoxycytidine-5'-triphosphate

3'-Deoxythymidine-5'-triphosphate

3'-Azido-2',3'-dideoxyuridine-5'-triphosphate

3'-Azido-2',3'-dideoxyguanosine-5'triphosphate

5-Bromo-2',3'-dideoxyuridine-5'-triphosphate

Modified deoxynucleoside-5'-triphosphate

Modified deoxynucleoside-5'-triphosphate

2'3'-Dideoxyadenosine-5'-O-(1-thiotriphosphate)

4-Thiothymidine-5'-triphosphate

2'3'-Dideoxycytidine-5'-O(1-thiotriphosphate)

2'-Deoxyadenosine-5'-O-(1-thiotriphosphate)

2-Thio-2'-deoxycytidine-5'-triphosphate

2'-Deoxycytidine-5'-O-(1-thiotriphosphate)

3'-Deoxy-5-methyluridine-5'-triphosphate

2'-Deoxyguanosine-5'-O(1-thiotriphosphate)

Modified deoxynucleoside-5'-triphosphate

Modified deoxynucleoside-5'-triphosphate

Thymidine-5'-O-(1-thiotriphosphate)

3'-Deoxy-5-methyluridine-5'-triphosphate

3'-Deoxyadenosine-5'-triphosphate (Cordycepin)

3'-Deoxyuridine-5'-triphosphate

3'-Deoxycytidine-5'-triphosphate

2',3'-Dideoxyadenosine-5'-triphosphate

3'-Deoxyguanosine-5'-triphosphate

2',3'-Dideoxycytidine-5'-triphosphate

Modified deoxynucleoside-5'-triphosphate

Modified deoxynucleoside-5'-triphosphate

2',3'-Dideoxyguanosine-5'-triphosphate

3'-Azido-3'-deoxythymidine-5'-O-(1-thiotriphosphate)

2',3'-Dideoxyguanosine-5'-O-(1-thiotriphosphate)

Thymidine-5'-O-(1-methyltriphosphate)

3'-Deoxythymidine-5'-O-(1-thiotriphosphate)

Guanosine-5'-O-(1-thiotriphosphate)

2',3'-Dideoxyuridine-5'-O-(1-thiotriphosphate)

Uridine-5'-O-(1-thiotriphosphate)

Modified deoxynucleoside-5'-triphosphate

Modified deoxynucleoside-5'-triphosphate

Adenosine-5'-O-(1-thiotriphosphate)

5-Prophynyl-2'-deoxycytidine-5'-triphosphate

Cytidine-5'-O-(1-thiotriphosphate)

5-Propynyl-2'-deoxyuridine-5'-triphosphate

8-Oxo-2'-deoxyadenosine-5'-triphosphate

5-Methyl-2'-deoxycytidine-5'-triphosphate

8-Oxo-2'-deoxyguanosine-5'-triphosphate

O^6-Methyl-2'-deoxyguanosine-5'-triphosphate

Modified deoxynucleoside-5'-triphosphate

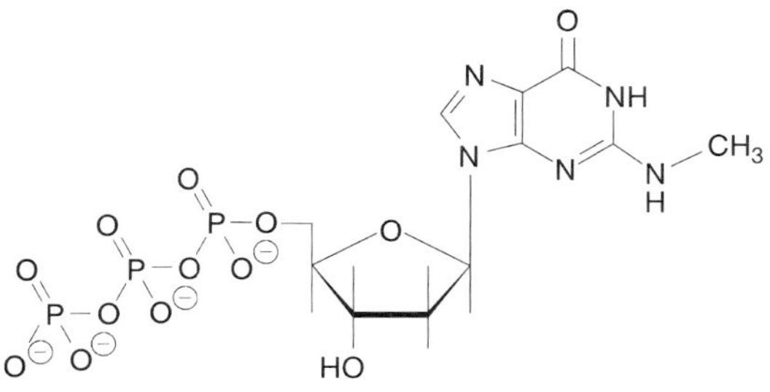

N²-Mehtyl-2'-deoxyguanosine-5'-triphosphate

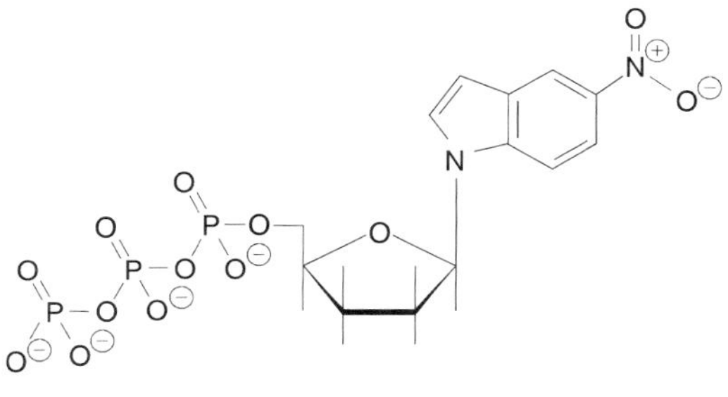

3'-O-Methyl-ATP

5-Nitro-1-indolyl-2'-deoxyribose-5'-triphosphote

3'-Amino-3'-dATP

2'-Azido-2'-dATP

Modified deoxynucleoside-5'-triphosphate

Modified ribonucleoside-5'-triphosphate: Any one of a series of synthetic → ribonucleoside triphosphates, into which an additional group is incorporated, as e.g. an amino, azido, or methyl group (see formulas). Such 2'-modified analogues prevent nuclease degradation of RNAs, and 3'-modified nucleotides inhibit RNA polymerases. See → modified deoxynucleoside-5'-triphosphate.

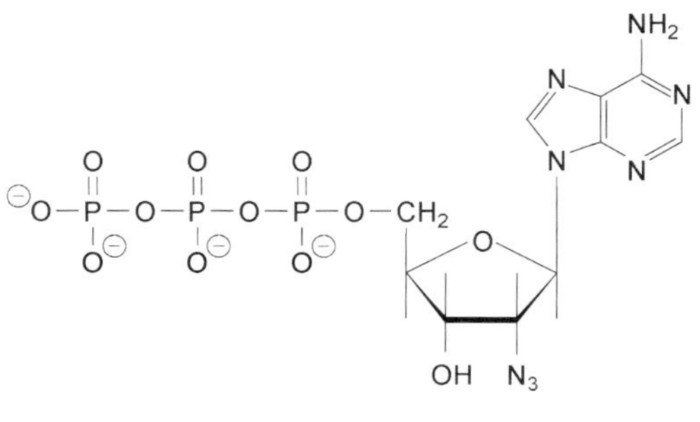

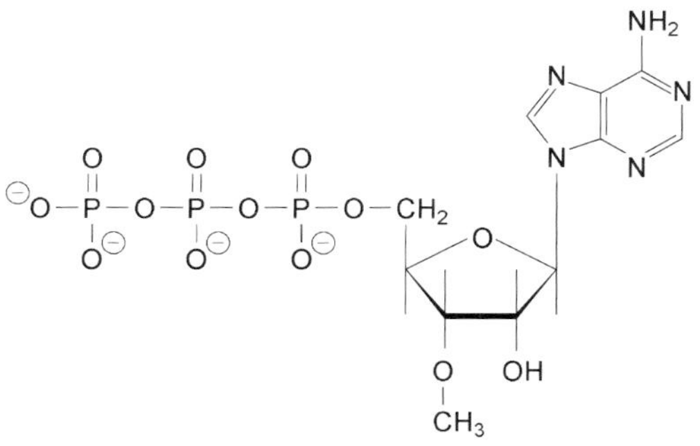

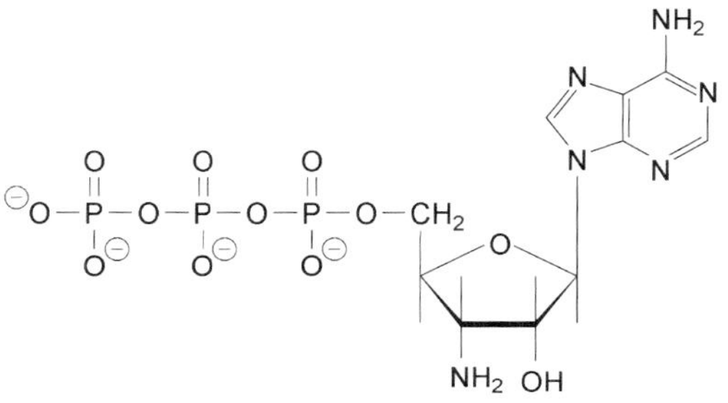

Modifier:

a) Any gene that modulates the phenotypic expression of one or more other genes.
b) Any DNA sequence motif that is located 5' upstream of a → promoter and either enhances (→ enhancer) or reduces (→ silencer) the rate of expression of a gene located downstream.

Modifier gene:

a) Any gene that either controls or affects → DNA methylation or → genomic imprinting.
b) Any mammalian gene (or → gene family) that modifies a trait encoded by another gene (or genes), and affects penetrance (i.e. the frequency with

which affected individuals occur among carriers of a particular genotype), → dominance, expressivity (i.e. the extent to which specific processes are influenced by a particular → genotype), and → pleiotropy. Modifier genes may be the cause for extreme phenotypes ("enhanced phenotypes"), less extreme ("reduced"), novel ("synthetic") or also wild-type ("normal") phenotypes. For example, a dominant modifier gene on human chromosome 7 reduces penetrance of a non-syndromic deafness gene (linked to *DFNB26*) on chromosome 4q31 that leads to hearing loss in the homozygotic state. However, several individuals in carrier families that are homozygous for *DFNB26* nevertheless hear normally: an effect of the modifier gene. Or, a modifier gene on chromosome 13q affects genes responsible for familial hypercholesterolaemia, an autosomal dominant trait (affecting one person in 500 and causing elevated cholesterol levels). Familial hypercholesterolaemia homozygotes often die of cardiovascular diseases. However, some individuals in afflicted families have low density lipoprotein (LDL) levels 25% lower than expected: again the effect of the modifier gene.

Modifier genes influence various phenotypes, preferentially in mice, rats and humans (e.g. gene *brachyura* [T] modifies tail length. [mouse], *Pax 3*^{Sp} suppresses spina bifida [mouse], and *Cfm* 1 a meconium ileus [mouse]).

c) Modifier of *mdg4*: A specific *Drosophila melanogaster* gene, of which both DNA strands are transcribed into different pre-messenger RNAs. These in turn are then ligated into one single mRNA that is translated into a protein. The modifier of *mdg4* corrupts the dogma that only one strand (see → antisense strand, → coding strand) encodes the mRNA of a gene.

d) Any gene that on its own has little (if any) detectable phenotypic effect, but can cause subtle or profound changes in the expression of the phenotype by mutation at another gene locus.

Modifier protein: Any peptide or protein that can be conjugated to a target protein and influences the activity and/or life-time of the target. For example, → sumoylation or → ubiquitinylation of substrate proteins are carried out by such modifier proteins.

Modular array (modular microarray, modular chip): Any → microarray that consists of several arrays separated by e.g. microfluidic hybridisation chambers. Each module can be separately used for specific experiments.

Modular microarray: See → modular array.

Modular vector: Any → cloning vector that is composed of a series of easily exchangeable → modules. For example, a modular plant transformation vector could contain the right and left → T-DNA border sequences, unique → restriction recognition sites for the insertion of → foreign DNA, a strong constitutive → promoter (e.g. the CaMV 35S promoter), a → Kozak consensus sequence, a → flag sequence, a → histidine tag, and a → 3' untranslated region as useful modules.

Module: Any DNA sequence that contains one or more conserved sequence → motifs and encodes a specific function (as e.g. the → TATA box as a module of → promoters of regulatable genes) or a specific domain of an RNA or a protein (as e.g. → exons as modules of genes).

```
A G C A C C C GGT  ACACTGTGTC  CT C C C G C T G C  A C C C A G C C C C  T T C A G C G C G A
                    GRE
G G C G T C C C G   A G G C G C A A G T  G G G C G G C C T T  C A G G G A A C T G  A C C G C C C G C G
                                                                                          AP-2
G C C C G T G T G C  A G A G C C G G G T  GCGCCCGGCC  C A G TGCGCGC  GGCCG G G T G T
                                              MRE 4              MRE 3
T T C G C C T G G A  G C C G C A A G TG  ACTCA G C G C G  G G G C G T G T G C  A G G C AGCGCC
                                    AP-1                                              MRE 2
CGGCCGGGGC  G G G C T T T T G C  A C TCGTCCCG  GCTC T T T C T A  G C T TATAA A C A
      Sp-1                              MRE 1
C T G C T T G C C G  C G C T G C A C T C  C A C C A C G C C T  C C T C C A A G T C  C C A G C G A A C C
                                              +1 ──────► RNA
C G C G T G C A A C  C T G T C C G A C  T C T A G C C G C C  T C T T C A G C T C  G C C ATGGATC
```

Modular architecture of the metallothionein gene promoter

Module-shuffled primer: Any → primer composed of six modules, each consisting of three or four nucleotides. All module-shuffled primers contain the same modules, but in different arrangements. Modules with three nucleotides carry a C at both their 5'- and 3'-termini, modules with four nucleotides a T at both ends. Therefore, the modules of module-shuffled primers are always connected by either C/T or T/C:

5'-CCC-TTCT-CAC-TGTT-CTC-TCAT-3'

or 5'-CAC-TCAT-CTC-TTCT-CCC-TGTT-3'

or 5'-CTC-TGTT-CCC-TCAT-CAC-TTCT-3'

Since the different module-shuffled primers differ only by the order, in which the otherwise identical modules are arranged, their sequences are unique, but their → melting temperatures are identical. Module-shuffled primers hybridise only with → complementary strands and have identical PCR amplification efficiency in conventional → polymerase chain reaction. A mixture of module-shuffled primers labeled with different → fluorochromes (e.g. → FAM, → HEX, → cyanin 5) is used for → module-shuffling primer PCR.

Module-shuffled primer polymerase chain reaction (MSP-PCR, module-shuffled primer PCR, multiplex PCR with colour-tagged module-shuffling primer): A variant of the conventional → polymerase chain reaction (PCR) for the comparative → gene expression profiling in different cells, tissues, or organs that employs specifically designed so called → module-shuffled primers to drive the analysis of several to multiple genes in one single reaction tube. In short, total RNA is first isolated, → poly(A)$^+$→ messenger RNA (mRNA) extracted and converted to double-strandede → cDNA by any conventional method (e.g. → RNA priming). The resulting cDNAs are then digested with a → four-base cutter → restriction endonuclease, leading to fragments averaging 256 bp. Three different → oligonucleotide → adapters corresponding to three different module-shuffled primers are then prepared and separately ligated to the restriction fragments from three different original cDNA preparations (i.e. from three different sources). The double-stranded adapters are designed such that after → ligation to the cDNAs, they form a Y-shaped end with one recessed strand, which avoids → priming by two module-shuffled primers. The adapter-ligated fragment populations are then mixed in equal quantities and serve as PCR → template. Each target cDNA fragment in the mixture is amplified with a primer pair consisting of one member of the module-shuffled primer mixture and a → gene-specific primer, where the module-shuffled primer discriminates between the sources of each amplified gene. PCR products are then analysed in a → DNA sequencer, and each fragment is identified by its specific electrophoretic mobility and the specific emission light wave length of its → fluorochrome. MSP-PCR circumvents the need for internal standards and a calibration curve, because the same genes from different sources are simultaneously amplified and directly measured and compared in one run. See → adapter-tagged competitive PCR, → enzymatic degrading subtraction, → gene expression fingerprinting, → gene expression screen, → linker capture subtraction, → preferential amplification of coding sequences, → quantitative PCR, → targeted display, → two-dimensional gene expression fingerprinting. Compare → cDNA expression microarray, → massively parallel signature sequencing, → microarray, → serial analysis of gene expression.

Molecular agriculture: See → molecular farming.

Molecular backcrossing: The shuffling (see → DNA shuffling) of a mutated DNA sequence with a specific function with a large molar excess of DNA fragments of highly related parental or wild-type sequence that is desirable as background, followed by selection of the mutant → phenotype with mostly parental DNA sequence. Only mutations necessary for the function are transferred into this preferred background. See → backcrossing.

Molecular beacon: A single-stranded → oligonucleotide that contains a → fluorochrome (e.g. → fluorescein, → TAMRA, → Cy3, → Cy5, → Texas red) at its 5'-terminus and a non-fluorescent quencher dye (e.g. [4(4-(dimethylamino)phenyl)azo]benzoic acid; DABCYL) at its 3'-terminus.

The sequence of such a molecular beacon is designed such that it forms a → hairpin structure intramolecularly, with a 15–30 bp probe region (complementary to the target DNA), and 5–7 bp long stem region (self-complementary). In this folded state the fluorochrome is quenched (i.e. any photon emitted by the fluorophore through exciting light is absorbed by the quencher [e.g. TAMRA], and emitted in the non-visible spectrum). After binding to a homologous target sequence, the beacon undergoes a conformational change forcing the stem of the hairpin apart, displacing the fluorochrome from the quencher, and abolishing the quenching (i.e. fluorescence occurs). Such molecular beacons are used for quantitation of the number of → amplicons synthesized during conventional → polymerase chain reaction, for the discrimination of → homozygotes from → heterozygotes, the detection of → single nucleotide polymorphisms, in situ visualization of → messenger RNA within living cells, and the simultaneous detection of different target sequences in one sample, if different fluorochromes with differing emission spectra are used. See → aptamer-beacon, → gene pin.

Figure see page 989

Molecular beacon aptamer (MBA): The combination of an → oligonucleotide → aptamer and a → molecular beacon that is used to (1) detect and (2) report target DNA, RNA or protein inside living cells. For example, the aptamer part of an MBA binds to a complementary nucleotide sequence or a site in a cognate protein. Binding induces a conformational change in the molecular beacon part, leading to a decrease in quenching and an increase in fluorescence light emission (if → fluorochromes are the reporter molecules) that can be detceted by a laser scanner and

Molecular beacon

allows to localize the DNA, RNA or protein target.

Figure see page 989

Molecular biology: A comprehensive term for a modern branch of biology, historically developing from physiological chemistry (in medical sciences) and biochemistry that tries to explain biological phenomena and processes on an atomic or molecular basis. Molecular biology engages physical, chemical, physico-chemical and conventional biological methods as well, and recently focussed on the structure and

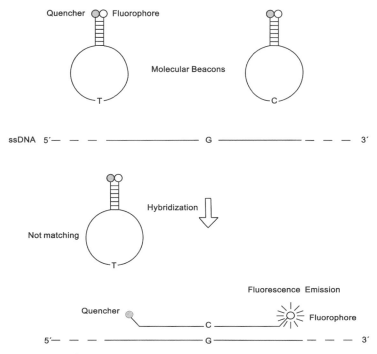

Molecular beacon

function of → chromosomes, the sequencing of whole → genomes, the DNA and its → replication and dynamic changes, → gene expression (with → microarrays), the → splicing processes and transport of RNAs into the cytoplasm, → RNA interference, → epigenetic phenomena as e.g. methylation and demethylation of cytosyl bases, the → histone code and the → ribosomal protein machine and → protein synthesis. See → molecular genetics, a specific field of molecular biology.

Compare → molecular genetics, a specific field of molecular biology.

Molecular breeding: The application of the whole repertoire of → genetic engineering and → molecular marker technologies for the improvement of fungi, plants, and animals.

Molecular brightness (q): The number of photons emitted by a → fluorochrome at a given excitation intensity. A change in molecular brightness as e.g. induced by a specific binding process (of a protein with a second protein, or a protein with its recognition sequence on a DNA) results in a change in total fluorescence intensity.

Molecular cap: Any non-nucleosidic ligands that can bind to correctly matched terminal base pairs of an → oligo- or → polynucleotide (e.g. DNA), but do not bind to → mismatches. Such caps are designed for the → 3'- and the → 5'-ends of a target DNA. For example, an anthraquinone residue can be covalently linked to the 2' position of the 3'-terminal nucleotide ("3'-cap"), and a trimethoxystilbene substituent functions as 5'-cap. Molecular caps increase

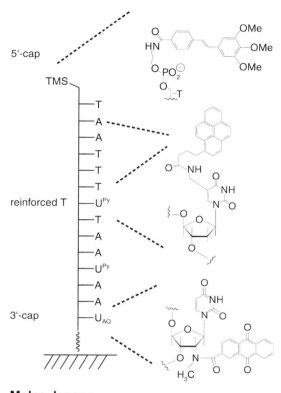

Molecular cap

the duplex stability of probes on DNA → microarrays and increase the sensitivity of the hybridization reaction (i.e. improve the detection of hybrids).

Molecular chaperone: See → chaperone.

Molecular clock: The hypothetical regular rate of nucleotide → substitutions or amino acid replacements over time. Considerations of a molecular clock are part of the study of rates of molecular evolution.

Molecular cloning: See → cloning.

Molecular combing: The stretching of cloned or native DNA molecules on a microscope slide. In short, the termini of solubilized DNA are bound to the silanated surface of a glass slide, the DNA solution covered with an untreated coverslip such that the drop spreads uniformly, and the solution evaporated. This procedure leads to the straightening and stretching of the bound DNA molecules, so that the labeled → probes can more easily be hybridized as compared to condensed chromosomes. See → chromosome stretching, → dynamic molecular combing.

Molecular docking: The prediction of the molecular geometry and binding affinity between two (or more) different molecules (e.g. a protein and a low molecular weight → ligand). It includes the calculation of the relative orientation and the conformational space of both the ligand and the protein, the geometric characterization of all potential molecular interaction(s), energy functions and possible torsion angles, and the major energy contributions (e.g. hydrogen and salt bridges, hydrophobic contacts). The process of molecular docking ("modelling") may start with e.g. the ligand, which is first dissected into smaller fragments. Each of these fragments is then placed into the active center of the protein (in case of enzymes) or a specific domain with a particular function. After finding the best fit, other fragments of the ligand are then successively incorporated in an energetically favourable way, unless the whole molecule is assembled and accommodated in the protein structure.

Molecular epidemiology: An ill-defined multidisciplinary approach to detect causative relationships between the genetic background (genes, mutated genes) and diseases, and its modification by racial and environmental parameters. Molecular epidemiology applies a series of different technologies spanning from the design of family studies for linkage analyses, the definition of factors influencing → linkage, localization of the relevant chromosomal regions or underlying gene(s) to the sequence variants of these genes, their frequencies in different populations, and their diagnosis. For example, the genetic causes of atherosclerosis, asthma, diabetes, schizophrenia or cancerogenesis and association with genetic markers (e.g. a mutation in the chemokine receptor gene as indicator for asthma bronchiale) are research areas of molecular epidemiology. Another research focusses on the distribution of mutant alleles in various populations. For example, the variants IVS10, I65T, E280K, and P281L are frequent in Spanish phenylketonuria patients, whereas R 408W and IVS12 prevail in Scandinavian patients.

Molecular farming (biofarming, molecular agriculture): The exploitation of plants for the production of peptides and proteins of biomedical, therapeutical and/or technical applications (e.g. → antibodies, industrial enzymes). For example, the first commercially available product of molecular forming was a recombinant avidin from chicken egg, produced by transgenic maize (*Zea mays* L.). Other products encircle a secretory IgA antibody against the causal agent of caries, *Streptococcus mutans* (tobacco), an antibody effective against herpes simplex virus (soybean), and a tumor-specific antibody (maize). Compare → gene farming.

Molecular forceps: See → DNA forceps.

Molecular fossil: Any DNA sequence that is phylogenetically old, but still a component of present-day genomes. For example, inactivated remnants of → retrotransposons with a low copy number in various eukaryotic genomes are such molecular fossils.

Molecular genetics: A branch of genetics that studies the structure and function of genes and genomes at the molecular level, using physical, physico-chemical and chemical techniques. Compare → molecular biology.

Molecular hybridization: See → hybridization.

Molecular imprinting: The generation of a polymeric mold around target molecules. Usually the polymer (e.g. → polyacrylamide) is synthesized from functional monomers that assemble around the target template. If the target is then removed, it leaves an impression on the polymer's surface, which can e.g. be used to selectively capture molecules that interact with the target molecule. For example, protein capture arrays are based on molecular imprinting. In this case, a glass or quartz slide, onto which target peptides are spotted, is coated with a mixture of monomers and cross-linkers that polymerises to a hydrogel layer over the peptides. When the film is removed, it carries the imprint of the immobilized peptides. See → molecularly imprinted electrosynthesized polymer, → molecularly imprinted polymer.

Molecularly imprinted electrosynthesized polymer (MIEP): A variant of a → molecularly imprinted polymer that is produced by growing the polymer around a template molecule on an electrode of a quartz crystal. The circulated charge controls the thickness of the polymer. For example, poly(o-phenylenediamine, PPD) in the presence of a low molecular weight template molecule (e.g. glucose) can electrochemically be polymerised into thin films (progressively covering the electrode), which can be used as a biomimetic sensor for glucose.

Molecularly imprinted polymer (MIP): Any synthetic support that mimics the three-dimensional shape of a target molecule and serves as an affinity capture mold largely resistant to biological degradation, acidity, extreme pH values, high temperature and other experimental conditions. Molecular imprinting starts with the polymerisation of functional monomers (e.g. methacrylic acid, MAA, and 2-(trifluoromethyl)acrylic acid, TFMAA) and crosslinkers (e.g. ethylene glycol dimethacrylate) in the presence of a polymerisation catalyser (e.g. 2,2'-azobisisobutyronitrile), a pore former (e.g. chloroform) and an imprinting molecule of choice (e.g. a low molecular weight compound such as a sugar or acid, or a peptide or protein as a template molecule) at 5°C under UV. Then the template molecule is removed from the resultant polymer network to leave a template-fitted cavity with template-selective binding capacity. MIPs are exploited e.g. for ligand-binding experiments, where they mimic the role of → antibodies in immunoassays. See → molecularly imprinted electrosynthesized polymer.

Molecular inversion probe assay: A technique for the high-throughput detection of → single nucleotide polymorphisms (SNPs) in → genomic DNA. In short, → padlock probes (containing two parts complementary to adjacent regions in the target DNA sequence, binding sites for PCR → primers, and socalled tag sequences for their capture on a solid support as e.g. a → microarray) are allowed to anneal directly to the genomic DNA. Hybridization of the two ends of the probes leaves a single-base → gap between the probe ends. The gap is filled by a single-base → primer extension reaction that can distinguish between the two SNP alleles, followed by circularization of the probe by joining its ends through → ligation. Excess → linear DNA is the removed from the reaction mixture by an → exonuclease. The circular probe containing primer-binding sites serves a → template for → inverse polymerase chain reaction (inverse PCR). The amplified inverted probes are subsequently cleaved, fluorescently labeled by either two or four → fluorochromes, and captured on a glass chip microarray carrying complementary tag sequences. Fluorescence is then excited by a laser and the fluorescence signals captured by a scanner.

Molecular marker (DNA marker): Any specific DNA segment whose base sequence is different (polymorphic) in different organisms and is therefore diagnostic for each of them. Such markers can be visualized by → hybridization-based techniques (e.g. → DNA fingerprinting, → restriction fragment length polymorphism) or → polymerase chain reaction-based methods (e.g. → DNA amplification fingerprinting, → random amplified polymorphic DNA, → sequence-tagged microsatellite sites). Ideal molecular markers are highly polymorphic between two organisms, inherited codominantly (i.e. allow to discriminate homo- and heterozygotic states in diploid organisms), distributed evenly throughout the genome and easily to be visualized. Moreover, molecular markers should occur frequently in the genomes, should easily be visualized, and be stable over generations. No single marker system fulfills all these criteria. Molecular markers (to which also protein markers, e.g. → isoenzymes belong) are used for the genotyping of single organisms, the detection of genetic variation(s) between organisms, the identification of hybrids, paternity testing, generally genetic diagnostics, and → genetic mapping. See → morphological marker.

Molecular medicine: A branch of medical sciences that employs techniques of → molecular genetics and → gene technology to unravel disease processes at the molecular level. Molecular medicine aims at diagnosing, preventing, treating and curing various human and animal diseases and also to develop animal models for human disorders. See → genetic medicine.

Molecular mimicry: See → molecular piracy.

Molecular motor (motor protein): Any one of a series of cellular proteins that decompose ATP to generate chemical energy for the production of physical force. Proteins transporting molecules or vesicles along the cytoskeleton, enzymes catalyzing DNA strand separation and DNA replication such as → helicases, → gyrases and → topoisomerases, and ATPases chanelling ions through membranes are such motor proteins. Intracellular molecular motors fall into three superfamilies with dozens of individual proteins, comprising myosins (moving along actin filaments), kinesins and dyneins (moving along microtubules). All motor proteins share a globular motor → domain, which undergoes a conformational change after ATP hydrolysis, thereby initiating motion along the associated filaments. For example, in muscle cells thick filaments of myosin pull on thin actin filaments, causing contractions. In neurons, kinesins use microtubules to transport vesicles with neurotransmitters through the cell to the axons. In the mucociliary elevator and sperm cells, dyneins connect microtubules in the cilia and flagella, respectively, thereby phasing their concerted bend, and kinesins and dyneins assist in mitotic spindle assembly, chromosome alignment and cytokinesis. Mutations in the genes encoding motor proteins may be lethal, but most of them lead to nonlethal defects. For example, the absence of myosins in cardiac muscles cause adult cardiac diseases (e.g. hypertrophic cardiac myopathy), deafness in mammals is incited by mutations in the gene coding for myosin VI that controls stereocilia movement in ear hair cells, and differences in coat color occur, when myosin V is mutated that transports melanocyte vesicles.

Molecular pathology: A scientific discipline that focuses on the interaction(s) between a pathogen and its host(s) on the molecular level.

Molecular phenotype (phenomic fingerprint): A vague term for the specific → transcript or protein expression pattern of a cell, or an organism (e.g. a bacterial cell) at a given time. Such fingerprints can be established by e.g. → expression microarrays or → protein chips.

Molecular piracy (molecular mimicry): The capture of host genes by pathogens (e.g. viruses or remainders of viruses as e.g. of herpes or pox viruses), or the acquisition of genes from cellular organelles and their incorporation into host genomes.

Molecular plasticity: The difference(s) in the spectrum of responses of different organisms to changes in the environment, usually detected by comparing the different → transcriptomes or → proteomes.

Molecular recognition force microscopy (MRFM): A variant of the → *atomic force microscopy* (AFM) that measures the strength of interaction(s) of two molecules (e.g. a receptor protein and its → ligand). In short, a silicium or silicium nitrite AFM tip is first functionalized with e.g. ethanolamine hydrochloride (in solution) or *aminopropyl- tri*ethoxysilane (APTES, in a gas phase). Functionalization creates a tip surface packed with primary amine groups, onto which long and elastic polyethylene glycol (PEG) chains are bound that possess one reactive end for amino groups, and therefore bind with this end to the tip. The other end carries another reactive group that binds proteins or other ligands. The ligand itself is attached to a support. Controlled movement then brings protein and ligand into close proximity, so that they react with each other. Subsequently the tip is slowly removed such that an increasing tractive force is imposed onto the protein-ligand complex that bends the AFM tip downwards. The increasing tractive force finally leads to the dissociation of the complex, recognized as spontaneous return of the tip into the neutral position. Such measurements allow to determine binding rates and affinities between two partners, and to collect structural information about the binding epitopes.

Molecular sensor: See → sensor probe.

Molecular sieving: See → gel filtration.

Molecular switch: Any supramolecular assembly that can switch from one to another conformation and back, thereby generating nanomechanical power. For example, a → DNA forceps represents such a molecular switch.

Molecular syringe: A tube-like structure formed by some Gram-negative bacteria pathogenic to plants and animals upon contact with the host cell that allows to secrete virulence factors directly into the cell. These factors are translocated from the bacterial cytoplasm to the host membrane or even into the host cytoplasm, passing three membranes consecutively. In some cases, the socalled secretion system III is assembled upon host cell contact, and a syringe-like structure formed that spans both inner and outer bacterial membranes. The system contains a translocator capable of channeling virulence proteins into the host cell. As an example, protein EspD of *enteropathogenic E. coli* (EPEC), as part of a pore, inserts into the host cell membrane (here: small intestinal mucosa cells), and is thought to catalyze such translocation processes.

Molecular tweezers: See → DNA forceps.

Molecular weight: The sum of the weights of all atoms of a molecule.

Molecular weight marker: See → molecular weight standard.

Molecular weight standard (molecular weight marker): A mixture of different peptides, proteins or nucleic acid fragments with known molecular mass that are used for the calibration of the molecular weight of proteins or nucleic acid molecules, after their separation by → gel electrophoresis. Compare → binning marker, see → marker.

Molecular writing: The graphical description of the surface of a material (e.g. a protein, DNA) by → atomic force microscopy.

Molecule: A complex of two or more identical or non-identical atoms that has a specific chemical property or properties different from those of the constituent atoms.

Moloney murine leukemia virus **(M-MLV)** *reverse transcriptase* **(RTase), RNase H minus (EC 2.7.7.49):**

a) A genetically modified → M-MLV reverse transcriptase, from which the → RNase activity has been removed. It is used to synthesize → cDNA, in → RT-PCR, RNA sequencing and → filling in 5' overhangs and → primer extension.

b) A single polypeptide enzyme from Moloney murine leukemia virus (M-MLV) that catalyzes the synthesis of a DNA strand from single-stranded RNA or DNA as a template requiring a → primer. The enzyme lacks → endonuclease activity, but has low → RNase H activity. It is used for full-length cDNA synthesis from large mRNAs (up to 10 kb) using oligo(dT) primers annealing to the poly(A) tail of the mRNA, and for → filling-in 5'overhangs. The enzyme is also available as recombinant DNA product completely devoid of RNase H activity, so that it no longer attacks the primer-poly(A) hybrid or the RNA-DNA hybrid arising from the reverse transcription. See → M-MLV RT, H minus.

Monoallelic expression: The predominant or exclusive → transcription of only one of two → alleles of a gene in diploid organisms. Usually → mutant alleles are differentially methylated at cytosine residues, buried in → chromatin with posttranslational → histone modifications, and are not (or aberrantly) expressed. Monoallelic expression, for example, is characteristic for some genes in less improved old hybrids of corn (*Zea mays* L.), and can be detected by e.g. → allele-specific amplification or → allele-specific polymerase chain reaction. See → allelic expression, → biallelic expression.

Monobromobimane: A → fluorochrome for the labeling of proteins that reacts with cystein residues. The compound is directly added to protein-containing gels, and the derivatized proteins can be visualized as fluorescent turquoise bands under UV light. Monobromobimane does not allow to label a series of cystein-free proteins (as e.g. myoglobin, concanavalin A, or cytochrome b5).

Monocistronic mRNA (monogenic mRNA): Any messenger RNA that codes for only one single polypeptide chain (in contrast to a polycistronic mRNA that codes for more than one protein).

Monoclonal: See → monoclonal antibody.

Monoclonal antibody (mAb, "monoclonal"): Any one of a population of immunoglobulins originating from one single clone of plasma cells, and therefore consisting of structurally and functionally identical antibodies. In the organism, monoclonals are produced by tumorous cells of the immune system (myelomas). However, it is experimentally possible to fuse such myeloma cells with activated antibody-producing plasma cells (B lymphocytes) to socalled hybrid (→ hybridoma) cells that are immortal, grow permanently *in vitro*, and produce and secrete practically unlimited amounts of identical mAbs. In short, animals (e.g. mice, rats, sheep) are first immunized with the specific → antigen, then the required cells (about 10^8 lymphocytes) are isolated from the spleen, and fused with myeloma cells that carry a defective gene encoding an enzyme of the nucleotide metabolism such that they can selectively be removed after cell fusion. This fusion can be performed with either p*o*ly*e*thyleneglycol (PEG), viruses, electrofusion or also laser fusion. A fusion is a relatively rare event (10^{-4}), and the fused cells have to be selected by HAT (H: hypoxanthine, A: aminopterine, T: thymidine). The fusion products usually are distributed into microtiter plates with 96 wells such that about 5×10^4 cells are contained in one well (half a spleen original material). One week later the supernatants of the fused cells are tested for the secreted mAb by e.g. → enzymelinked immunosorbent assay. Since several hybridoma clones with different specificities may reside in one well, the cells have to be cloned either by limited dilution or → flow cytometry. For mass production, the mAb-encoding cells are either injected into and grown in a mouse peritoneum ("Ascites technique"), synthesized by → transgenic plants (e.g. tobacco) or → transgenic animals (sheep, cow), or grown in various types of bioreactors (e.g. spinner flasks, stirred tank fermenter, airlift fermenter, hollow fiber reactor). Monoclonal antibodies can also tailored by → genetic engineering. For example, murine antibodies can be "humanized" by coupling the variable region of the murine antibody to the Fc region of a human antibody, creating a chimeric antibody. After integration of the hypervariable regions (*c*omplementarity *d*etermining *r*egions, CDRs) of the murine antibody into a human antibody a fully humanized antibody is regenerated (CDR-grafted or reshaped monoclonal antibody) that still retains its binding specificity. Antibody-encoding DNA can also be tailored by → site-directed mutagenesis. For example, mutations in the hypervariable regions or the Fc region can improve important properties of the resulting antibody (e.g. binding affinity, target specificity, biological half-life time). Monoclonal antibodies can also be expressed in *E. coli* ("coliclonal antibodies"), but cannot be glycosylated by the bacterium. Therefore these antibodies may be immunogenic. Generally, monoclonal antibodies are important tools for the identification and characterization of peptides and proteins, and are also components of → antibody arrays. See → catalytic antibody.

Monoclonal aptamer: Any synthetic → aptamer generated by e.g. → *s*ystematic *e*volution of *l*igands by *ex*ponential enrichment (SELEX), whose sequence is fully known, and that specifically reacts with a particular target molecule. These high-affinity binding aptamers compete with antibodies for *in vitro* diagnostics, biosensors, affinity resins and pharmaceutical applications. See → polyclonal aptamer. Compare → monoclonal antibody, → polyclonal antibody.

Monocuts: Laboratory slang for two (or even more) fragments of defined size arising through cleavage of → lambda DNA with a → restriction endonuclease that cuts only once.

Monogenic mRNA: See → monocistronic mRNA.

Monogenic trait (Mendelian trait): A (usually phenotypic) feature of an organism that is controlled by only one single gene. See → polygene, → polygenic trait. Compare → multifactorial trait.

Monomer:

a) Any basic unit from which polymers are made (e.g. amino acid monomers are polymerized by peptide bond formations into the polymer protein).
b) A subunit of a supramolecular, multimeric complex (e.g. a protein consisting of different or identical polypeptide chains).

Monomeric red fluorescent protein (mRFP): A variant of the → red fluorescent protein from the coral *Discosoma* spec. that is generated *in vitro* by → directed molecular evolution. In short, the dimerization of the → green fluorescent proteins (GFPs) of *Aequorea victoria* and *Renilla*, and the obligate tetramerization of the red fluorescent protein from *Discosoma striata* (DsRFP) largely prevent their use as genetically encoded → fusion tags. These inadequacies can be circumvented by a monomeric fluorescent protein. One approach towards a monomeric protein starts with the disruption of dimeric subunit interfaces (i.e. first AB, then AC) in the DsRFP by insertion of arginines, which cripples the protein. However, red fluorescence can be rescued in some subunits by random and directed mutagenesis (see → site-specific mutagenesis). The resulting monomeric protein owns a lower extinction coefficient, reduced quantum yield, and weaker photostability as compared to DsRFP, yet shows similar fluorescence brightness in living cells. Since the excitation and emission peaks of mRFP are shifted by 25 nm (584 and 607 nm, respectively), it can be combined with green fluorescent protein for the construction of fusion proteins and multicolor labeling. A series of mRFP variants with improved properties (e.g. new colors, increased tolerance of N- and C-terminal fusions, higher extinction coefficients, quantum yields and photostability) are available that were engineered by *in vitro* directed evolution and known as e.g. mBanana (Excitation maximum: 540 nm; emission maximum: 553 nm), mCherry (Excitation maximum: 587 nm; emission maximum: 610 nm), mHoneydew (Excitation maximum: 487/504 nm; emission maximum: 537/562 nm), mOrange (Excitation maximum: 548 nm; emission maximum: 562 nm), mStrawberry (Excitation maximum: 574 nm; emission maximum: 596 nm), and mTangerine (Excitation maximum: 568 nm; emission maximum: 585 nm).

Mononucleotide editing: A variant of → RNA editing in the mitochondria of the slime mould *Physarum polycephalum* and several other members of the phylum Myxomycota (e.g. *Stemonitis* and *Didymium*), which is characterized by the insertion of mononucleotides in RNAs relative to their mtDNA template. The most commonly inserted mononucleotide is cytidine, although a number of uridine mononucleotides are inserted at specific sites, whereas adenosine and guanosine are not at all inserted. See → dinucleotide editing, → transfer RNA editing.

Monosome:

a) One mRNA-ribosome complex.
b) Any → chromosome that has no homologous counterpart.
c) A single → nucleosome.

MOPAC: See → *m*ixed *o*ligonucleotide-*p*rimed *a*mplification of *c*DNA.

MOPS buffer (*m*orpholino-*p*ropane *s*ulfonic acid buffer): A synthetic zwitterionic buffer with a pK_a of 7.2 widely used in biochemical experiments.

Morbid map (*morbid*ity map): A laboratory slang term for a diagram showing the chromosomal location of genes associated with a particular disease.

mORF: See → *m*erged *o*pen *r*eading *f*rame.

Morgan unit (M): A measure for the relative distance between two genes on a chromosome, or, concomitantly, for the frequency of → recombination between two genetic markers. One Morgan corresponds to the length of a chromosome in which, on average, one recombination event (a → cross-over or a → chiasma) occurs each time a gamete is formed. One Morgan is equivalent to a crossover value of 100%, a *centi*Morgan (cM) corresponds to 1% crossover value and to 0.01 Morgan.

Morphodoma technique ("morphogenics"): A technology for the *in vivo* generation of → mutations and new → mutant organisms. The → replication of DNA prior to mitosis is accompanied by the introduction of wrong bases (see → polymerase infidelity). Now a series of postreplicative repair processes screen for replicative errors and eliminate them (e.g. → *mism*atch *r*epair systems, MMR). The Morphodoma technique is based on the reversible inhibition of MMR that is mediated by the activity of an → allelic variant of the human *PMS2* gene ("morphogene", identified in patients suffering from hereditary nonpolyposis colorectal cancer). The encoded protein antagonizes MMR and thereby permits naturally occuring mutations to be transmitted to offspring (at frequencies 1000 times greater than normal). This progeny can be screened for desirable mutations (i.e. gene mutations with new functions). For example, an → expression vector carrying the morphogene is transfected into → antibody-producing cells. The morphogene is expressed, and stably expressing cell lines ("Morphodoma cells") grown for a series of generations (up to 30). During this passage genome-wide mutations accumulate. The genetically diverse pool of cells is then single-celled by limited dilution, and the single cell clones propagated for two weeks. Then subclones are identified that express antibodies with new properties (e.g. higher affinity), or produce higher antobody titers. These elite cell clones are "cured" (i.e. they lose the expression vector by intragenic homologous recombination; the cells are "morphogene null"). This is achieved by adding a prodrug that is enzymatically converted to a cytotoxic molecule in the presence of a morphogene-linked negative selection marker gene. All cells expressing the morphogene therefore will die, whereas those cells will survive that lost the marker gene naturally.

Morpholino antisense oligonucleotide (MASO): Any → antisense oligonucleotide containing morpholinos; that is used for e.g. → gene silencing. To that end, MASOs

Morpholino oligonucleotide ("morpholino"): Any non-ionic → oligonucleotide with a backbone different from the → phosphodiester backbone of DNA or RNA. The individual elements of a morpholino oligonucleotide are derived from a → ribonucleoside, whose ribose ring is opened with $NaIO_4$. Subsequently the ring is closed and a nitrogen atom introduced by treatment with NH_3, and the two hydroxy groups removed by $NaCNBH_3$. The bases are protected during this proce-

are transferred into target cells via → electroporation or → microinjection, bind to complementary target sequences by → Watson-Crick base-pairing and silence the expression of genes. Or, alternatively, they block the → spliceosome by hybridizing to → splice junctions in → pre-messenger RNA, and additionally inhibit the → translation initiation complex by hybridizing to the first 25 bases of coding sequence in the → 5'-untranslated region. See → morpholino oligonucleotide.

Phosphodiester bond   **Morpholino bond**

Ribonucleoside to morpholino transformation

B=Thymine and base-protected adenine, cytosine and guanine

Protection and activation of morpholino subunit

Preparation of morpholino oligonucleotides

dure. Morpholinos are resistant to exo- and endonucleases, do not interact with e.g. proteins non-specifically, because they are not charged, and are therefore extremely stable in biological systems. Additionally, morpholinos possess excellent aqueous solubility and are not toxic. They require about 14 to 16 bases minimum target sequences.

Morphological marker: Any easily identifiable trait (e.g. eye or flower color) that is characteristic for an individual. Morphological markers (slang: "morphos") can be placed on socalled → genetic maps, where they identify → linkage to other markers or traits and help to tag the underlying gene(s). See → molecular marker.

Morphome: The description of all anatomical and histological structures, the organ and body architecture, and their structural relationship in an intact organism.

Mosaic gene: See → split gene.

Mosaic protein: Any protein that is composed of a series of discrete → domains, where each domain (or a set of different domains) has a specific function for the overall activity of the protein. For example, hemostatic proteases carry large extensions N-terminal to their serine protease domains. These extensions consist of a number of discrete domains with defined functions as e.g. substrate recognition, binding to phospholipid membranes or interaction(s) with other proteins. The majority of metazoan mosaic proteins are extracellular or membrane-bound, and are considered as indicators for the evolution of multicellularity. Compare → mosaic gene.

Motif-primed PCR: A variant of the conventional → polymerase chain reaction that uses → primers complementary to conserved DNA sequence motifs important for gene function and regulation (e.g. parts of → promoters, → consensus sequences for → DNA-binding proteins, or regulatory domains of gene families).

MOUSE: See → microsatellite obtained using strand extension.

Movement protein: A protein encoded by certain plant viruses that is necessary for the spread of the virus throughout the infected plant. The gene encoding a movement protein can be engineered such that it codes for only part of a movement protein. The engineered gene can then be inserted into the target plant genome, where it is expressed. The truncated defective protein then competes with the native movement protein, leading to an inhibition of the movement of the virus. Genes encoding movement proteins can be isolated from a variety of viruses, e.g. luteovirus (pr17), tobamovirus (p30), potexvirus (TGB), or gemini virus (BC1).

MP: See → transport protein.

M-PCR: See → methylation-independent polymerase chain reaction.

MP-PCR: See → microsatellite-primed polymerase chain reaction.

M-PVA: See → magnetic polyvinyl alcohol microparticle.

M_r: Abbreviation for relative → molecular weight.

MRE: See → metal regulatory element.

MRE11-Rad50-Nbs1 complex (MRN): A multiprotein complex of eukaryotic cells that recognizes ("senses") → *double-strand breaks* (DSBs) in DNA directly, binds to them, unwinds the adjacent → double-stranded DNA, using the energy of ATP, recruits → *ataxia-telangiectasia mutated* (ATM), dissociates the ATM dimer, activates it and thereby starts the repair of the break. Cells of patients suffering from the Nijmegen breakage syndrome (NBS) or *ataxia telangiectasia-like disorder* (ATLD) express mutated Nbs1 (nibrin) and/or Mre11 (DNA-binding, 3′,5′-exonuclease) proteins, respectively, and, as a consequence, show decreased levels of ATM substrate phosphorylation. MRN is therefore essential for the maintenance of genomic stability. See → ATM- and Rad3-related (ATR) checkpoint pathway.

MRFM: See → molecular recognition force microscopy.

mRFP: See → monomeric red fluorescent protein.

MRI: See → magnetic resonance imaging.

M RNA (medium RNA): One of the three linear single-stranded RNAs of the tripartite genome of Tospoviruses (family: Bunyaviridae) that is about 5 kb in length, is associated with the nucleocapsid proteins, and encodes a socalled nonstructural protein and the two viral membrane glycoproteins G1 and G2. The terminal sequences of M RNA carry complementary repeats 65–70 nucleotides long, which allow to form a quasi-circularized (pseudo-circular) molecule. See → L RNA, → S RNA.

mRNA: See → messenger RNA.

mRNA display: See → messenger RNA display.

mRNA initiation site: See → cap site.

mRNA profiling: See → messenger RNA profiling.

mRNA-protein fusion: See → messenger RNA display.

mRNP: See → messenger ribonucleoprotein.

MRP: See → mapped restriction site polymorphism.

Mrr system: See → methylated adenine recognition and restriction system.

MS: See → microsatellites.

MSA:

a) See → multiple sequence alignment.
b) See → multiple substrate array.

MSAP: See → methylation-sensitive amplification polymorphism.

MsDNA: See → multicopy single-stranded DNA.

MSNT (1-mesithylene-2-sulfonyl-3-nitro-1,2,4-triazole): A → coupling reagent used in → chemical DNA synthesis.

MSP: See → methylation-specific polymerase chain reaction.

MS-PCR: See → mutagenically separated polymerase chain reaction.

MSSCP: See → multiplex single-strand conformation polymorphism.

Ms-SNuPE: See → methylation-sensitive single nucleotide primer extension.

mST: See → methylated sequence tag.

MTase: See → methyltransferase.

mtDNA: See → *mi*tochondrial DNA.

mTERF: See → *m*itochondrial *t*ranscription *termination factor*.

MTF: See → *m*embrane-associated *t*ranscription *f*actor.

MT gene: See → *m*etallothionein gene.

M13: A → filamentous phage of *E. coli* (→ coliphage) containing a circular, single-stranded DNA molecule of 6.407 kb ("plus-strand"). Filamentous phages infect only *E. coli* strains with F pili (containing → F factors), where they adsorb and invade the host cell. The latter is not lysed but grows at a slower rate. Infected cells may thus be recognized as → plaques. As soon as the ssDNA of the phage enters the cell, it becomes converted into a double-stranded → replicative form which multiplies rapidly until an accumulating phage-encoded single-strand-specific DNA

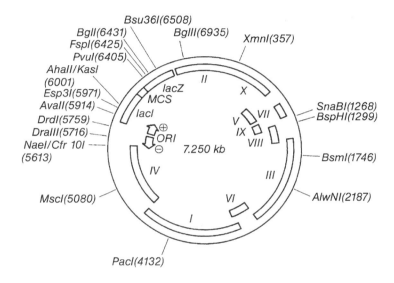

**Simplified map of M13mp18
(with unique restriction sites)**

I-X : Viral genes (transcribed clockwise)
ORI : Origin of DNA replication
⊕⊖ : Plus and minus strand replication
MCS : Multiple cloning site

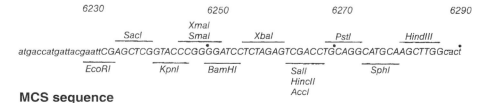

MCS sequence

binding protein prevents the synthesis of the complementary strand. From then on, only single stranded phage DNA is produced, which is packaged at the host's cell membrane into → capsid proteins which replace the ssDNA-binding proteins. Finally the complete phages are released from the host.

A multitude of M13 derivatives have been developed as → cloning vectors (M13mp series) which contain the E. coli *lac* regulatory region (see → *lac* operon) and the coding sequence for the α-peptide of → β-galactosidase together with either single cloning sites (in M13mpl, for Ava II, Bgl I, Pvu I), symmetrical → polylinkers (e.g. M13 mp7), or asymmetrical polylinker regions (e.g. M13 mp8, mp10, mp11, mp18, mp19). Insertion of foreign DNA into these cloning sites will interrupt the sequence coding for the α-peptide of the β-galactosidase gene. The defective gene does not allow the conversion of the indicator dye → X-gal into its blue derivative upon induction of the *lac* operon by → IPTG so that the phages containing an insert can be easily selected as white plaques. The single-stranded phage DNA can be obtained in bulk quantities, and used directly for large-scale → sequencing of DNA.

M13mp cloning vector: See → M13.

MTMR: See → multi-target multi-ribozyme.

MTN blot: See → multiple tissue Northern blot.

MTP: See → microtiter plate.

MTS: See → membrane-translocating sequence.

Mtx: See → methotrexate.

Mtx^r: See → methotrexate resistance.

μ:

a) μm, mu: See → micron.
b) See → Mu phage.

Mu: See → *Mu* phage.

***Mu* (mutator):** Any one of a class of transposable elements in the maize (*Zea mays* L.) genome that increases the frequency of mutation of various loci by more than an order of magnitude. *Mu* elements, present in the genome in 10–100 copies, comprise maximally 2 kb and are flanked by 200 bp → inverted repeats with adjacent 9 bp → direct repeats. Basically two size classes prevail, of which the shorter ones are derived from longer ones by internal → deletions. *Mu* elements transpose by a replicative mechanism, and can also occur in circular extrachromosomal state (e.g. *Mu*1 [1.4 kb] and *Mu*1.7 [1.7 kb]). Methylation of inserted *Mu* sequences prevents their transposition and stabilizes the mutation, whereas less than complete methylation leads to transpositional activity. See → *Mu* phage, → mutator gene.

***Mu*-AFLP:** See → mutator amplified fragment length polymorphism.

***Mu* array:** A glass chip or a nylon membrane, onto which thousands of → mutator transposon flanking regions are spotted at high density, and which serves to identify specific genes with mutator insertions. The various mutator flanking regions are isolated from individual *Mu* active plants with the → mutator amplified fragment length polymorphism technique, so that each spot on the array represents the *Mu* flanks of an individual plant. Hybridization of these arrays with e.g. → cyanin-labeled or radiolabeled gene probes (e.g. → cDNAs) identifies plants with *Mu* insertions in specific genes.

MudPIT technology: See → multidimensional protein identification technology.

MUG:

a) 4-*methyl*-*umbelliferyl*-β-D-galactopyranoside; MUGal: A colorless chromogenic substrate for β-galactosidase which is converted into the strongly fluorescent 4-methyl umbelliferone (MU) after cleavage, and used in → enzyme-linked fluorescent assays.

b) 4-*methyl*-*umbelliferyl*-glucuronide: A fluorogenic, synthetic substrate for → β-glucuronidase.

MUGal: See → MUG, entry a.

Muller's ratchet: The more rapid accumulation of mutations in asexually propagated genomes as compared to sexually propagated genomes.

Multiallergen chip: Any solid support (e.g. glass, quartz, silicon), onto which synthetic peptides or proteins are spotted that represent the spectrum of → allergens causing allergic reactions in sensitive individuals. Though about 20,000 sources for allergens exist, only about 50 allergenic molecules lead to allergenic responses (e.g. the Betv1, the main allergic protein of the birch tree, *Betula* spp.). These 50 allergens can be detected and monitored via the multiallergen chip through their interaction (binding) with chip-bound molecules.

Multibranch loop (bifurcation loop, junction loop): Any region, in which three (or more) helices form a closed loop. See → bulge loop, → hairpin loop, → internal loop, → tetraloop.

Multicolor fluorescence *in situ* hybridization (multicolor FISH): A technique to identify several specific sequences of intact chromosomes simultaneously by → hybridization with different nucleic acid → probes, each of which is labeled with a specific → fluorochrome (e.g. probe A with → fluoresceine isothiocyanate, probe B with → rhodamine B isothyocyanate, probe C with a coumarin derivative), and each of which detects a specific chromosomal site. The simultaneous use of several differently labeled probes in one single → *in situ* hybridization experiment generates multicolor chromosome pictures.

Figure see page 1006

Multicolor fluorescence microscopy: The microscopic visualization of different (up to eight) proteins in a single living cell, their subcellular localization, co-localization, movement and interaction(s) by the differential labeling of the proteins with fluorochromes of non-overlapping excitation and emission spectra.

Multicolor *polymerase chain reaction* (multicolor PCR): A variant of the conventional → polymerase chain reaction that allows to simultaneously amplify several different → template DNAs (or several different regions of one template) in a single reaction tube by using e.g. complementary → primers labeled with different compatible fluorochromes (e.g. → FAM and → HEX, or → TET, or → Cy3, or → TAMRA, or → Texas Red, or → Cy5).

Multicolor spectral karyotyping: A variant of the → spectral karyotyping, which com-

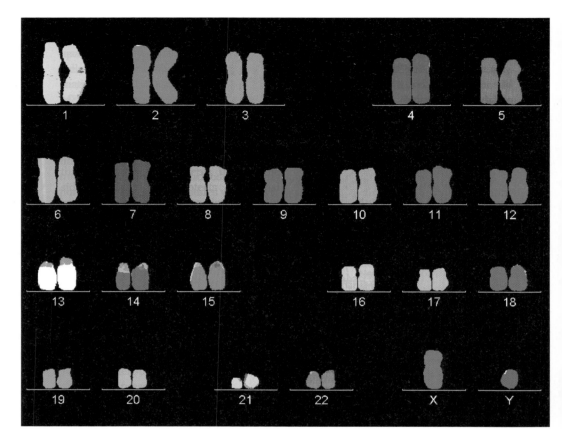

Multicolor fluorescence *in situ* hybridization (multicolor FISH)

Multicolor karyotype of a male human, visualized by → *in situ* hybridisation of a mixture of 24 whole → chromosome libraries labeled with combinations of five → fluorochromes to metaphase spreads. All 24 chromosomes are simultaneously detected by a single hybridisation. Note the → translocation onto chromosomes 13 and 14.
(Kind permission of Dr. Ilse Chudoba, MetaSystems, Altlußheim, Germany)

bines visualization of emitted fluorescence light from chromosome-specific fluorescent-labeled probes (hybridized to a metaphase spread) through a triple band pass filter, sent through an interferometer and imaged with a charge-coupled device (CCD) camera. The interferogram generated for each pixel is analyzed by Fourier transformation, and the measured spectra converted to display or classification colors. This technique allows to identify each chromosome in a metaphase spread after → fluorescence *in situ* hybridization with chromosome-specific probes labeled with different → fluorochromes in a single experiment.

Multicomponent nucleic acid enzyme (MNAzyme): Any non-protein enzyme composed of at least two (or more) short → oligonucleotide components ("partzymes"), which combine in a self-assembly process in the presence of one or more socalled MNAzyme assembly facilitator(s) to form a catalytically active multi-component nucleic acid (DNA) enzyme. Such MNAzymes are used for the detection, identification and/or quantification of one or more target nucleic acids in a → quantitatice RT-PCR assay, since MNAzyme complexes cleave generic oligonucleotide probes between → fluorophore and → quencher dyes generating a fluorescent signal. Since the MNAzyme approach requires target-specific binding of two partzymes and two PCR → primers, it has four levels of specificity, much greater than that of → TaqMan and → molecular beacon methods, which have only three levels of specificity, conferred by two primers and one target-specific probe. See → DNAzymes, → ribozymes.

Multicopy inhibition: The reduction in → transposition frequency of a single-copy → transposon 10, resident in the host chromosome, by the interference of a → multicopy plasmid carrying an IS 10-R sequence (IS 10-R is the right IS element flanking Tn 10, and encodes a → transposase that mediates transposition). The mechanism of interference involves pairing of the start region of the transposase mRNA and a short complementary RNA (→ antisense RNA) transcribed from the opposite IS 10 strand. As a consequence, translation of the transposase message is impaired.

Multicopy plasmid (high copy number plasmid): A → plasmid that is present in bacterial cells in copy numbers greater than one per chromosome, because it is under → relaxed control. Its copy number can therefore spontaneously or artificially be increased (spontaneously to 10–100, artificially to 20–40000 copies; compare →

amplification). See also → multicopy inhibition, compare → low copy number plasmid.

Multicopy single-stranded DNA (ms DNA): Any one of a family of small (from 48 to 163 nucleotides) single-stranded DNAs of Gram-negative bacteria that are present in hundreds of copies per cell ("multicopy"). ms DNAs are apparently always – associated with small (i.e. from 49 to 119 bases), single-stranded RNA molecules that are joined to the 5'-end of the DNAs by 2', 5' phosphodiester linkages at a specific internal guanosine base in the RNA (sequence context 5'- AGC-3'), and protein(s) to form an extrachromosomal nucleoprotein complex. The ms DNAs of different bacteria vary considerably in both their DNA and RNA strands, which nevertheless share more or less conserved secondary structures: they fold into stable → stem-loop structures, all RNAs contain the G residue (forming the branched, bond with the DNA), and a small RNA-DNA hybrid region forms between the 3' ends of both strands. ms DNA is encided by a → retron in the bacterial chromosome, which contains *msd* (the gene for ms DNA), *msr* and *ret* (the gene for reverse transcriptase) driven by an upstream → promoter. This retron is transcribed into a long → messenger RNA that is translated to produce → reverse transcriptase, which then uses the upstream *msr-msd* region of the mRNA as the → template as well as the → primer for ms DNA synthesis. An other locus, located → upstream of *msd* (designated *msr*) encodes the RNA strand of the ms DNA complex. Many stream of *msd* (designated *msr*) encodes the RNA strand of the ms DNA complex. Many retrons (in *E. coli*: Ec 48, Ec 67, Ec 73, Ec 83, Ec 86, and so on) are associated with → prophages of the P2 family. The function(s) of ms DNA is still inknown.

Multicopy tag sequence: Any → flag sequence that is reiterated up to 10 copies per tag, so that its detection in the corresponding protein is facilitated, since it produces stronger signals on e.g. → Western blots. The copy number of any multicopy tag has to be optimized, as it should not interfere with the function of the tagged protein. See → epitope tag, → epitope tagging.

Multidimensional protein identification technology (MudPIT): A combination of liquid chromatography of peptides derived from complex mixtures of proteins and the determination of their masses by → mass spectrometry. In short, the isolated protein mixture is first denatured and then digested with proteases as e.g. trypsin or Lys-C (or chemically). The resulting peptides are separated on a multidimensional fused silica capillary column (for example, a column packed with a reversed phase and subsequently with a strong cation exchanger). After their application onto the capillary, the peptides are first loaded onto the cation exchanger, and a small part washed into the reversed phase using a KCl salt step gradient. Then the peptides are eluted from the reversed phase with a reversed phase gradient directly into an → electrospray ionisation mass spectrometer. Up to 20 cycles of washes with increasing salt concentrations are needed for a complete elution of all peptides. In a high-throughput format of MudPIT, some 100,000 different mass spectra can be established in a mere 24 hours. These spectra are processed with data bank searches. For example, MudPIT allows the identification of 1.500 proteins derived from the → proteome of *Saccharomyces cerevisiae*, including low- and high-abundance proteins, proteins with extreme pI values and molecular weights, and membrane proteins. The technology therefore is a tool to produce a comprehensive view on the → proteome of an organism.

Multidomain protein: Any protein that is composed of a set of discrete, structurally and functionally independent modules cooperating to achieve the overall function of the protein. Multidomain proteins most probably evolved by the fusion of two (or more) ancestral single domain peptides or by domain duplication or domain swapping. See → supradomain protein.

Multidrug resistance (MDR): The indifference of bacteria against two (or more) different antibiotics. MDR is based on mutations in genes encoding e.g. ABC transporters, and presently represents a serious medical problem.

Multi-epitope imaging: A technique for the simultaneous detection and imaging of many different proteins in a tissue section that capitalizes on the sequential *in situ* interactions between fluorescently labeled specific → antibodies and their epitopes. Multi-epitope imaging of specific classes of proteins illustrate the localizations of these classes within a cell, and their changes in different phases of the life cycle or after environmental challenges.

Multi-epitope ligand cartography (*multi-epitope ligand kartographie*, MELK): A technique of → topological proteomics that allows to simultaneously determine both the cellular abundance of about 50 specific proteins and their cellular or subcellular localization in a single cell (→"whole cell protein fingerprinting"). MELK produces three-dimensional distribution patterns of proteins by first reacting a specific fluorescent → antibody with its cognate target protein *in situ*, recording the → fluorescence (and with it, the protein localization and quantity), then bleaching out the → fluorochrome, repeating the process with a second fluorescent antibody directed against a second protein, and so on. The signals can be visualized under a microscope, and are collectively compiled into a single panoramic view of the cell with a resolution of about 100 nm. Since only fixed cells can be used, real-time imaging of the target proteins in a living cell is not possible.

Multi-exon deletion: The → deletion of more than one (usually two or three) → exons from a → multi-exon gene. For example, exons 1–4 of the → breast cancer gene 2 (BRCA2) are frequently deleted in patients with breast cancer. See → single exon deletion.

Multi-exon gene (multi-exonic gene): Any → gene that contains more than one → exon. See → single-exon gene. Compare → multi-intronic gene.

Multifactorial analysis: The identification of the relative contributions of two (or more) genes and environmental factors to the expression of a distinct → phenotype.

Multifactorial trait: A (usually phenotypic) feature of an organism that is controlled by more than one, in extreme cases up to ten different genes and at least one, but normally many environmental factors. See → polygene, → polygenic trait (multigenic trait).

Multi-functional biochip **(MFB):** An integrated chip that allows the simultaneous detection of DNA-DNA, DNA-RNA, DNA-protein, and protein-protein interactions. The different "chips-on-a-chip" each contain an integrated circuit electro-optical system produced by the socalled *c*omplementary *m*etal *o*xide *s*ilicon (CMOS) technology. The MFB is supplied with photodiode sensor arrays, electronics, amplifiers and all necessary elements for analysis, so that nucleic acid hybridisations can be detected side by side with protein-antibody interactions.

Multi-functional phagemid: Any → phagemid that has been engineered to serve several functions at the same time (e.g. permits DNA cloning, double- or single-stranded sequencing of cloned inserts, *in vitro* mutagenesis, *in vivo* and *in vitro* transcription).

Multi-functional plasmid: Any → plasmid that has been engineered to serve several functions at the same time (e.g. permits → DNA cloning, → Sanger sequencing of cloned inserts, → *in vitro* mutagenesis, and → *in vitro* transcription).

Multifunctional protein ("moonlighting protein"): Any protein that catalyses several enzymatic and non-enzymatic reactions by using different → domains or binding sites, engaging in different multi-protein complexes, or being active in different cellular locations or different organs of an individual. For example, protein PMS2 is part of a proofreading system in nuclei of mammals that removes → mismatched nucleotides from newly synthesized DNA strands. The same protein is, however, increasing → mutation rates of immunoglobulin genes in B cells about 100,000-fold over normal cells. This → hypermutation allows the B cell carrier to produce an immense repertoire of → antibodies against a wide range of external → antigens. Here, PMS2 removes nucleotides from the parental rather the new DNA strand, upon which a nucleotide complementary to the mismatch is inserted. As a consequence, the mutation Is fixed. The protein distinguishes parental from newly synthesized strand by the presence of methyl groups on cytosyl residues of the parental DNA. Another example for a moonlighting protein is phosphoglucoisomerase (PGI), which catalyses the conversion of glucose-6-phosphate to fructose-6-phosphate in the glycolytic pathway, acts as a signalling molecule in B cell maturation and as nerve growth factor, is a stimulator of differentiation of myeloid leukemia cells and represents a migratory mediator for cancerous cells. This "catalytic promiscuity" is wide-spread in enzymes of intermediary metabolism.

Multigene analysis: The simultaneous determination of the → expression patterns of hundreds, thousands or even hundred thousands of genes in a particular cell, tissue, or organ at a given time, as opposed to the analysis of the expression of only a single gene or a few genes. Multigene analysis can be performed with → microarrays (see → cDNA array, → cDNA expression array, → expression microarray, → transcript array) and high-throughput profiling techniques such as → massively parallel signature sequencing or → serial analysis of gene expression.

Multigene family (gene family): A set of closely related genes originating from the same ancestral gene by duplication and mutation processes (see → gene amplification). They may either be clustered on the same chromosome (e.g. genes coding for ribosomal RNAs, see → rDNA) or be dispersed throughout the genome (e.g. → heat shock protein genes). Most of the members of such multigene families retain a far-reaching homology in the coding region, but are divergent in the → intron and → promoter regions. See for example → histone genes. Compare also → gene battery, definition b: the genes are related and contiguous in a specific chromosomal region; compare → supergene family: related genes with limited homologies.

Gene Family	Organism	Number of Genes	Clustered (C) Dispersed (D)
Actin	Yeast	1	–
	Slime mold	17	C & D
	Drosophila	6	D
	Chicken	8–10	D
	Human	20–30	D
Tubulin	Yeast	3	D
	Trypanosome	30	C
	Sea urchin	15	C & D
	Mammals	25	D
α-Amylase	Mouse	3	C
	Rat	9	D
	Barley	7	C
β-Globin	Human	6	C
	Lemur	4	C
	Mouse	7	C
	Chicken	4	C

Multigene family

Multi-gene shuffling (gene-family shuffling, family shuffling, multiple gene shuffling, multiple gene family shuffling): A variant of the → DNA shuffling technique, in which many homologous genes from related organisms are used for creating diversity. The resulting shuffle libraries contain novel chimeras that differ in many positions. For example, a single cycle of family shuffling of the four cephalosporinase genes from *Citrobacter freundii*, *Klebsiella pneumoniae*, *Enterobacter cloacae* and *Yersinia enterocolitica* resulted in a mutant enzyme that differs by 102 amino acids from the *Citrobacter*, by 142 amino acids from the *Enterobacter*, by 181 amino acids from the *Klebsiella*, and by 196 amino acids from the *Yersinia* enzyme. In contrast, three rounds of → single gene shuffling yielded only four amino-acid substitutions.

Multigene transformation: The simultaneous transfer of multiple genes into a target organism and their (preferably linked) integration into its genome. For example, up to 14 different genes have been transformed into the genome of rice plants by → biolistic gene transfer techniques, were mostly genetically linked, stable over several generations, and almost all transcribed. Multigene transformation allows to engineer durable traits into target organisms (e.g. resistance to a pathogen will be more effective and lasting, if several resistance genes are involved).

Multigenic trait: See → polygenic trait.

Multi-intronic gene: Any → gene that is composed of more than one → intron. See → multi-exonic gene.

Multi*locus* probe (MLP): Any → repetitive DNA sequence that allows to detect two or more, in extreme cases a multitude of loci in a genome, using → labeling and → hybridization techniques. Compare → single locus probe.

Multi*locus* sequence typing (MLST): A variant of the conventional multi*locus* enzyme electrophoresis (MLEE) that involves the sequencing of internal fragments of a series of → house-keeping genes (as e.g. *abcZ* transporter gene, glucose-6-phosphate dehydrogenase [*gdh*], phosphoglucomutase [*pgm*], polyphosphate kinase [*ppk*], 3-phosphoserine aminotransferase [*serC*], adenylate kinase [*adk*], shikimate dehydrogenase [*aroE*], pyruvate dehydrogenase subunit [*pdhC*]) to detect allelic variations between strains of bacterial pathogens. In short, chromosomal DNAs of the test strains are first isolated, and gene fragments of a size between 400 and 600 bp (for convenient sequencing) amplified by conventional → polymerase chain reaction using gene-specific → primers. Both strands of the resulting products are sequenced, the sequences compared, and used to establish socalled sequence *t*ypes (STs), i.e. the allelic combination at each locus. Related STs are grouped in socalled *c*lonal *c*omplexes (CCs). Special software packages allow a comparison of the resulting STs with STs from a world collection of bacterial isolates to precisely describe the target population structure. Since many loci are involved, the typing identifies multilocus sequence types.

Multimer: A supramolecular complex consisting of two or more identical or non-identical subunits (monomers). For example, a protein molecule, made up of two or more individual polypeptide chains.

Multi-microRNA hairpin vector: A → cloning and → expression vector (e.g. a → TOPO cloning or → lentivirus vector) that contains two (or more) synthetic genes encoding two (or more) different → microRNAs. The microRNA gene sequences are usually derived from naturally occurring miRNA genes, placed *in tandem* on the vector, each one separated from the next one by an artificial → linker. Their expression is controlled by a strong → promoter (e.g. the *cytomegalovirus* [CMV] promoter), leading to the synthesis of a → transcript that folds into two (or more) independent → hairpin structures. These hairpins can be cleaved into two (or more) independent miRNAs. This arrangement produces disproportionally large amounts of mature small RNAs and, as a consequence, leads to a more efficient → gene knockdown as compared with a single miRNA construct. The different miRNAs may target the same gene, or also different genes.

Multi-nucleotide polymorphism: Any polymorphism between two → genomes that is based on the exchanges of several adjacent nucleotides, as compared to the → single nucleotide polymorphism.

Multi-pass protein: Any protein that spans a plasmamembrane more than once, i.e. contains more than one → transmembrane helix. See → single-pass protein.

Multiple alignment:

a) The → alignment of two (or more) nucleic acid sequences, into which → gaps are introduced such that residues with common features (e.g. pyrimidines versus purines) and/or ancestral residues are ordered in the same vertical line. The most widely used program for multiple alignments is ClustalW.

b) The comparison of the amino acid sequences of many proteins of a protein family or the nucleotide sequences of many genes of a gene family to identify homologous regions.

Multiple allele: Any one of a series of alternative → alleles of a single gene.

Multiple allelism: The occurrence of more than two alleles of a genomic → locus in a population.

Multiple arbitrary amplicon profiling (MAAP): See → arbitrarily amplified DNA.

Multiple bidirectional transcription: A variant of the → bidirectional transcription process that produces one → sense transcript and two smaller → antisense transcripts. See → single bidirectional transcription.

Multiple cloning site (MCS): Synonym for → polylinker.

Multiple-copy single-strand DNA (msDNA): A → satellite DNA of myxobacteria and some natural isolates of *E.coli* that contains a single-stranded DNA branching out from an internal guanosyl residue of an RNA molecule by a unique 2', 5'- → phosphodiester linkage. Both DNA and RNA possess considerable internal base-pairing. The synthesis of msDNA requires → reverse transcriptase.

Multiple deletion strain: Any bacterial strain, whose → genome is artificially reduced in size by the → deletion of various non-vital sequences not necessary for the experimentor. For example, a total of 43 genomic regions are removed from *E. coli* strain MG1655, including → prophages, phage remnants, the → *lac* operon, recA,

→ restriction modification genes, large K-islands, flagellar and chemotaxis-related genes and → mobile elements together with → IS elements and → recombination *hot spots* (RHSs). These deletions reduced the genome size from the original 4,639,221 to 3,930,956 bp (reduction: 15.27%). The strain still grows like the parental wildtype strain, and also retains most other physiological characteristics, but the → mutation frequency caused by IS elements drops to zero. Multiple deletion strains are stable hosts for library construction of small and large → inserts, respectively, and for large-scale protein expression.

Multiple displacement amplification (MDA): A variant of the conventional → rolling circle amplification technique, which allows to amplify whole genomes directly from biological samples as e.g. blood or tissue cultured cells in a single tube. MDA is an isothermal amplification reaction, as such does not require any heating or cooling steps compared to e.g. the → polymerase chain reaction, and differs from rolling circle amplification in that linear → genomes can be amplified. In short, genomic DNA and random, exonuclease-resistant hexamer → primers at high concentrations (50 mM) are first mixed, and the template amplified with the highly processive, strand-displacing bacteriophage φ29 DNA polymerase at 30°C into amplicons more than 10, sometimes 100 kb in length. The polymerase is tightly bound to the template and therefore able to replicate through difficult primary and secondary structures of the DNA. After an initial priming step, branched amplification generates additional single-stranded templates that in turn serve for primer binding and extension: an exponential cascade of branched amplification ensues ("secondary priming"). Since the polymerase displaces downstream product strands, it creates the templates for multiple concurrent and overlapping rounds of replication. At the end of the reaction (usually after 18 hours), most of the amplified strands are converted to double-stranded DNA and the reaction stopped by heating to 65°C. MDA yields up to 20–30 µg product DNA, starting from only 1–10 copies of genomic DNA. The MDA process is highly reliable, since φ29 DNA polymerase has an error rate of only 1 in 10^6–10^7. MDA-amplified genomic DNA is used for a whole series of genomic techniques such as → chromosome painting, genotyping of → single nucleotide polymorphisms, → RFLP analysis, → cloning, → subcloning, and DNA sequencing.

Multiple exon skipping: The removal of multiple → exons from a → pre-messenger RNA, usually *in tandem*. For example, 11% of all → exon skipping events in humans involve the removal of several to many exons. See → single exon skipping.

Multiple *f*luorescence-based PCR-SSCP (MF-PCR-SSCP): A sensitive mutation detection technique that combines conventional → polymerase chain reaction with two (or more) primers labeled with different → fluorochromes and → single-strand conformation polymorphism analysis. In short, the target sequence is first amplified using forward and reverse primers labeled with two different dyes, respectively (e.g. FAM [blue] and JOE [green]) at their 5'-ends. Amplified products are then heat-denatured, mixed with an internal DNA size marker labeled with a third dye (e.g. ROX [red]), and run in temperature-controlled non-denaturing → polyacrylamide gels in an automated DNA sequencer. Mutations are detected as positional shifts of two-coloured peaks in the electropherogram. The technique allows to diagnose single base exchanges and → loss of heterozygosity, and works without radioactivity.

Multiple *f*luorescent *in situ* *h*ybridization (multiple FISH): A variant of the conventional → fluorescent *in situ* hybridization procedure, in which several different → probes are each labeled with a different → fluorochrome, synchronously hybridized to a target DNA (or RNA) *in situ* and simultaneously detected in the sample. See → multiplex labeling.

Multiple gene disruption: The → insertion of DNA sequences into two or more genes within the same → genome with the result of a → knock-down or complete → knock-out of all the genes. The function(s) of all the disrupted genes in concert is then deduced from a changed → phenotype. For example, the knock-out of only one strategic gene of the parasite *Plasmodium berghei* (e.g. gene *UIS3*, "*u*pregulated in *i*nfectious *s*porozoites gene *3*"), whose encoded protein is necessary for the establishment of the parasite in the human body, is not sufficient for a long-term and efficient protection against malaria. In fact, sporozoites lacking *UIS3* do not fully develop the liver cycle, but enduring resistance of a host is only expected from the disruption of more genes, so that the parasite cannot replace them all in a short time period.

Multiple gene DNA shuffling: See → DNA shuffling.

Multiple gene family shuffling: See → multi-gene shuffling.

Multiple genes: See → polygene.

Multiple gene shuffling: See → multi-gene shuffling.

Multiple hit: See → superimposed substitution.

Multiple *n*ucleotide *p*olymorphism (MMP): Any polymorphism between two (or more) → genomes that is based on more than one → single nucleotide polymorphism. For example, many human diseases probably are caused by single base exchanges at strategic sites of several genes (e.g. coding for functional → domains of different proteins) that are not present in the wild type genomes and act in concert to cause a disease. These altogether are multiple nucleotide polymorphisms.

Multiple overlapping primer PCR: A variant of the → PCR *in situ* hybridization, using → primers with overlapping sequences to generate large amplification products that do not diffuse away from their original location in cells or tissue specimens.

Figure see page 1015

Multiple promoter usage: The occupation of two (or more) → transcription factor IID-binding sites within one single → promoter by → TFIID. The corresponding → transcripts usually differ by their → 5'-UTRs or first → exons. For example, the promoter of the human *WEE1* gene contains two TFIID-binding sites, corresponding to the 5'-ends of two distinct → *m*essenger RNAs (mRNAs). Each of these mRNAs encodes a distinct protein. One codes for a full-length WEE1 protein, the shorter one only for its kinase domain. The shorter transcript is more abundant in the G_0 phase, the longer one is highly transcribed in both G_0 and S phases of the cell cycle, respectively.

Multiple recognition site (multiple recognition sequence): An infelicitous term for the occurrence of more than one → restriction site recognized by a specific →

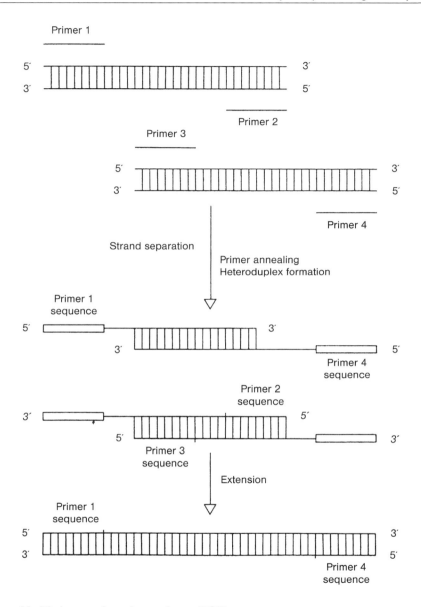

Multiple overlapping primer PCR

restriction endonuclease. For example, the endonuclease *Acc* I recognizes the four sequences 5'-GTAGAC-3', 5'-GTATAC-3', 5'-GTCGAC-3', and 5'-GTCTAC-3', and cleaves 3' of the T residue.

Multiple sequence alignment (MSA): The iterative search for → homologs of a protein of interest in the → proteomes of selected species with → algorithms such as PSI-BLAST, and their sequence

→ alignment using a multiple-sequence alignment tool as e.g. CLUSTALW, MUSCLE, MAFFT, T-Coffee, ProbCons and SATCHMO. The selection of an alignment method is dependent on available computer resources and the size and evolutionary divergence of the dataset, both of which affect alignment accuracy. For large or divergent datasets, MAFFT and MUSCLE are recommended, because they are computationally efficient.

Multiple spotting technique (MIST): A technique for the simultaneous screening of multiple → analytes that interact with peptides, → antibodies or other proteins immobilized on a → protein chip surface. In short, in a first step the target proteins in minute volumes (e.g. 0,2–0,6 nl) are serially spotted onto the same spot of a poly-L-lysine coated glass → microarray. In a second step, a reaction buffer containing glycerol or polyethylene glycol together with the analytes (e.g. antibodies) is again robotically spotted in extremely small volumes onto the immobilized proteins. After a distinct reaction time the slides are rinsed and then incubated with → cyanin5-labeled anti-antibodies, and an interaction between target protein and antibody detected by laser-induced → fluorescence. MIST allows to quantify analyte concentrations as low as 400 zeptomoles (equivalent to 240,000 antibody molecules).

Multiple substrate array (MSA): Any solid support (e.g. a glass slide), onto which a multitude of microspots, each containing a specific protein are printed. Such MSAs are incubated with cells (usually 5,000 – 10,000 cells), non-adherent cells are washed off, and the adherent cells detected by e.g. staining with appropriate dyes and scanning with e.g. a confocal laser. MSAs therefore allow to monitor cell populations for the adhesion potential of individual cells, and can simultaneously be tested for a multitude of substrates (proteins) and culture conditions.

Multiplet: Any single band on a → DNA fingerprint gel that contains two (or more) different DNA molecules. This undesirable comigration of different sequences can be resolved by isolation of the band from the gel, the cloning of the different DNAs, and their sequencing.

Multiple tissue Northern blot (MTN blot): A ready-to-hybridize → Northern blot that contains poly(A)$^+$-RNAs (→ polyadenylated RNA) from a series of tissues of an organism, separated by denaturing → agarose gel electrophoresis and blotted onto a nylon membrane. Equal amounts of RNA per lane allow to detect tissue-specific expression of genes.

Multiplex amplifiable probe hybridization (MAPH): A technique for the simultaneous detection of varying copy numbers at 40–100 → genomic loci, including complete → deletions at some loci. In short, → genomic target DNA is first isolated, then denatured, and irreversibly immobilized on a nylon membrane by UV irradiation. A mixture of amplifiable → probes in the size range of 140–600 bp, each flanked by identical → primer-binding sequences and recognizing a unique region in the genome (e.g. a → gene), is the hybridized to the filter-bound DNA. Subsequent stringent washing removes all non-specifically bound probes. The washed membranes are then transferred to an amplification mixture, and the probes amplified in a concentional → polymerase chain reaction (PCR), using the common ^{33}P 5'-endlabeled primer pairs. The amplification products are then separated by → denaturing polyacrylamide gel electrophoresis, and radioactivity detected by → autoradiography. Since the sizes of the various amplicons are different from each other, the various gene-specific fragments can easily be discriminated from each other. By varying the composition of the probe set, different genomic regions can be targeted (as e.g. → subtelomeric regions, specific chromosomes).

Multiplex DAF: See → multiplex DNA amplification fingerprint.

Multiplex DNA amplification fingerprint (multiplex DAF): A variant of the → DNA amplification fingerprinting technique that uses at least two or multiple primer oligodeoxynucleotides to generate → amplification fragment length polymorphisms.

Multiplex dosage pyrophosphorolysis-activated polymerization (MD-PAP): A variant of the → pyrophosphorolysis-activated polymerization (PAP) technique for the detection of large heterozygous chromosomal → deletions and gene → duplications that employs 30 nucleotides long → oligonucleotides blocked at their 3′-end by a → dideoxynucleotide (ddNTP) not extendable by → DNA polymerase. When such blocked oligonucleotides are specifically and completely anneal to a complementary target → template, pyrophosphorolysis removes the blocking ddNTP in the presence of pyrophosphate (PPi). After removal of the blocking ddNTP, DNA polymerase can now extend the activated oligonucleotide. Blocked oligonucleotides can be multiplexed in solution, because no primer-dimers can be formed and false priming is excluded. The amplification products are then electrophoresed through a → denaturing polyacrylamide gel.

Multiplexed array: See → multiplexed microarray.

Multiplexed microarray (multiplexed array, multiplex microarray): Any → microarray support (e.g. glass, quartz or polypropylene slide or microtiter plate) that either contains multiple microarrays in a planar arrangement (each physically separated from the others, as on a slide), or in the wells of the plate. Such a geometry allows parallel printing and hybridization as well as processing of multiple samples against hundreds or thousands of genes for → expression profiling under identical conditions. This geometry increases reproducibility of the hybridization results and reduces costs. See → compartmented microarray, → multiplex hybridization array.

Multiplex-endonuclease genotyping approach amplified fragment length polymorphism (MEGA-AFLP): A variant of the conventional → amplified fragment length polymorphism (AFLP) technique, which employs the AFLP protocol, but uses four (or more) → restriction endonucleases (e.g. *Eco*RI, *Bgl*α, *Bcl*I and *Mun*I) in two series of digestions, each in combination with one pair of → adapters/primers. This technique allows robust and stringent → polymerase chain reaction (PCR) conditions for subsequent amplification. See → three-endonuclease amplified fragment length polymorphism.

Multiplex fluorescent *in situ* hybridization (M-FISH): A technique for the simultaneous detection and discrimination of all different chromosomes in a metaphase spread by different colorization. In short, chromosome-specific DNA libraries are first labeled with distinct combinations of → fluorochromes. Then the different, specifically labeled chromosomal DNA libraries are hybridized onto spreads of metaphase chromosomes (or cell nuclei), and the individual fluorochromes detected by epifluorescence microscopy (using all filters to excite all the fluorochromes in the sample) coupled to a *c*ooled *c*harge-coupled *d*evice (CCD) camera. The different fluorescence signals allow to unequivocally assign a specific fluorogram (i.e. colour) to a specific chromosome.

Multiplex hybridisation array: A general term for any → microarray that contains hundreds or thousands of individual genes, cDNAs or oligonucleotides and is used to simultaneously detect DNA-DNA-, DNA-RNA- or DNA-oligonucleotide-interactions by hybridisation (i.e. by → Watson-Crick base pairing).

Multiplexing: A technical term from electronics that describes the mixture of many different signals at the start of an electronic circuit, which are separated from each other later on.

Multiplex labeling: A variant of the conventional → *fluorescent in situ hybridization* (FISH) and multiplex fluorescent in situ hybridization (M-FISH) technique that allows to simultaneously visualize several (up to 50) → messenger RNAs in a single cell. In short, aminoallyl-UTP is incorporated into all the various → probes. The aminoallyl linker of each probe serves as docking site for a specific → fluorochrome. Each probe therefore is labeled with a different fluorophore, where each fluorophore owns different properties (e.g. different excitation wave lengths, different emission spectra, different fluorescence intensity). After hybridization, the probes (and with them their targets) can be visualized side by side in the target cell.

Multiplex *ligation-dependent probe amplification* (MLPA): A technique for the detection of → mutations, more precisely → exon duplications and → deletions, deletions of whole genes, → single nucleotide polymorphisms, or chromosomal aberrations (in e.g. tumor cell lines or samples). In short, MLPA starts with the → hybridization of target-specific → probes to denatured and fragmented → genomic DNA (usually 20–100 ng). Each probe consists of two → oligonucleotides A and B that bind to adjacent nucleotides of the target sequence via their 50–70 nucleotides long DNA-*b*inding *s*equence (DBS) at the 3'-end. Oligonucleotide A additionally contains a flanking universal *p*rimer-*b*inding *s*equence (PBS), whereas in oligonucleotide B DBS and PBS are separated by a stuffer fragment of variable length (*variable fragment*, VF). If both oligonucleotides hybridize to the target DNA, they can be covalently ligated by a thermostable → DNA ligase (e.g. the mismatch-sensitive, NAD$^+$-requiring ligase-65). The resulting, usually 130–480 bp long strand can then be amplified by conventional → polymerase chain reaction (PCR), using one fluorescently labeled, and another non-labeled → primer directed to the PBSs. Since all ligated probes share identical 5'-end sequences, they can be amplified with only one single primer. The difference in length of the different probes allows their separation and quantification in high-resolution capillary gel electrophoresis (or also 6.5% → polyacrylamide gel electrophoresis). In case the target sequence is deleted, the ligation is prevented, and the fragment cannot be amplified by the universal primer. Should the target DNA be absent in both homologous chromosomes, the corresponding fragment cannot be detected. If the target sequence is deleted in only one of the → alleles, then the peak

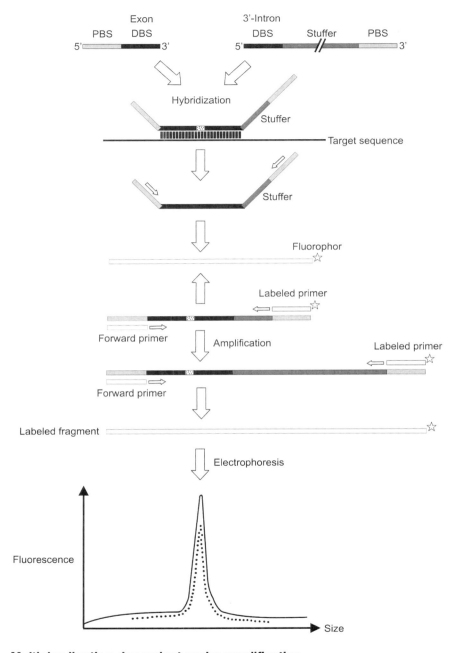

Multiplex ligation-dependent probe amplification

area of the eluting fragment is reduced to about 50% of the control. Up to 40 probes with different stuffer lengths (or sequences) and targeting at 40 different genes can simultaneously be run in a single reaction. See → multiplex amplifiable probe hybridization.

Multiplex messenger assay (MMA): A technique for the simultaneous analysis of the expression of many different genes in a cell, tissue, organ, or organism. In short, → cDNA clones corresponding to known genes are picked from → replica plates, spotted onto nylon-based filters in duplicate, and fixed onto the membrane, which can contain up to 50,000 such cDNAs ("high-density filter"). Then complex → probes are prepared separately from cell A and B, respectively, by → reverse transcription of their total RNAs, which are either labeled radioactively or fluorochromated (perferably with two different → fluorochromes with varying emission spectra, so that they can be detected simultaneously). Each of these probes is then hybridized to the high-density array filter, the hybridization signals detected by → autoradiography or → fluorography, quantified, and socalled hybridization signatures established. These signatures reflect the expression profiles of the corresponding genes in the two target tissues. See → cDNA expression array.

Multiplex PCR: See → multiplex polymerase chain reaction.

Multiplex polymerase chain reaction (multiplex PCR): A variant of the → polymerase chain reaction technique that uses at least two, frequently many → primer oligodeoxynucleotides to either amplify different stretches on a target DNA molecule or different genomic loci simultaneously. Multiplex PCR requires optimization of the → annealing temperature, the concentrations of *Taq* DNA polymerase, nucleotide triphosphates, $MgCl_2$ and the temperature profile of the PCR reaction.

Multiplex-QEXT: See → multiplex quencher extension.

Multiplex quencher extension (multiplex-QEXT): A single-step technique for the simultaneous real-time detection and quantification of several different → single nucleotide polymorphisms (SNPs) that is based on the direct measurement of fluorescence changes in a closed tube. In short, the target DNA (e.g. a gene) is first amplified by specific → primers in a conventional → polymerase chain reaction (PCR), the amplified fragment treated with shrimp → alkaline phosphatase and → exonuclease I to inactivate the nucleotides and to degrade residual PCR primers. Then different → probes detecting different SNPs in the amplified fragment are 5′-labeled with different reporter → fluorochromes (e.g. one is labeled with → 6-FAM, the second one with → TET, the third one with → HEX, the fourth one with → Texas Red, the fifth one with → cyanin 5, and so on. These probes are subsequently extended by a single → TAMRA-labeled dideoxy cytosine (TAMRA-ddCTP), if the respective SNP alleles are present. TAMRA may function as → fluorescence acceptor (quencher-based detection) or donor (→ fluorescence resonance energy transfer (FRET)-based detection), depending on the 5′-fluorescent reporter. The extension generates increased reporter fluorescence, a result of → fluorescence resonance energy transfer (FRET), if TAMRA serves as energy donor. If TAMRA functions as energy acceptor, then the reporter fluorescence is quenched. See → multilocus sequence typing.

Multiplex ratio: The number of genetic loci that can simultaneously be detected by a specific → molecular marker technique in a single experiment. For example, → amplified fragment length polymorphism techniques display many different loci (their number depending on the → genomic DNA and the amplification → primers used, among others), whereas → sequencetagged microsatellite site methods detect only one single locus.

Multiplex sequencing (Church sequencing): A → DNA sequencing method that allows the determination of base sequences in 10–50 different DNA fragments synchronously. In short, the various fragments are first cloned into → plasmid vectors that differ from each other by sequences flanking the cloning site (so that each cloned fragment has specific unique border sequences). All the different inserts are then excised from the plasmid vector, using a → restriction endonuclease that cuts outside all border sequences. The pooled inserts are then sequenced using the → Sanger sequencing procedure. After electrophoresis of the various resulting fragments, their transfer to and immobilization on membranes the sequences belonging to each of the original fragments are detected by sequential hybridization with synthetic → oligonucleotides that are complementary to the sequences flanking the inserts. After hybridization to one such oligonucleotide and autoradiography, the probe is stripped off and the membrane reprobed with another oligonucleotide to detect sequences from another insert. In this way, some 10–50 different DNA fragments can be sequenced synchronously.

Multiplex sequencing of paired-end ditags (MS-PET): A variant of the conventional → gene identification signature (GIS) technique for the ultra-high-throughput analysis of → transcriptomes and → genomes, in which the 5′- and 3′-signatures of each full-length transcript are simultaneously extracted, covalently linked into socalled → paired-end ditag (PET) → concatemers for high-throughput sequencing and accurate demarcation of → transcription unit boundaries in assembled genome sequences. In the MS-PET procedure, a modified paired-end ditagging (PET) method was combined with one of the high-throughput sequencing technologies (see → fiber-optic reactor sequencing, → four-five-four sequencing, → second generation sequencing) to simultaneously sequence up to 400,000 (or more) dimerized PET (diPET) templates, with an output of nearly half a million PET sequences in a single 4 h machine run. See → chromatin immunoprecipitation-PET (ChIP-PET), → polony multiple analysis of gene expression.

Multiplex single-strand conformation polymorphism (M-SSCP): A variant of the → single-strand conformation polymorphism detection technique that allows to discover multiple mutations within one gene in one single approach. The technique works with → 5′ endlabeled forward → primers and unlabeled reverse primers and → polymerase chain reaction amplification. The amplification products are electrophoresed in thin → polyacrylamide gels and detected by autoradiography. Small changes in base composition (e.g. → deletions, additions, or duplications) appear as a slightly different band position as compared to a reference band. With M-SSCP it is, for example, possible to simultaneously screen for deletions within → exons, and a variety of base substitutions within a multi-intronic gene (e.g. the human dystrophin gene with 79 exons).

Multiplex transcription factor assay: A technique for the simultaneous detection of the activation of many (up to 20) → transcription factors that combines → nuclease protection with a bead-based assay. In short, the transcription factors in a sample are first reacted with biotinylated target DNA probes (each containing a single-stranded socalled capture sequence and a double-stranded DNA motif recognized by a specific transcription factor), and then a → nuclease is added. If the transcription factors bind to their cognate sequence motifs, these are protected from digestion. Now, fluorescence-encoded beads covalently bound to single-stranded DNA are incubated with the mixture. The single-stranded DNA bound to the beads hybridizes with the single-stranded tails of the biotinylated DNA probes. Finally, the beads are reacted with a → strep*a*vidin-R-*p*hyco*e*rythrin (SAV-RPE) conjugate that binds to the → biotin associated with the beads. After washing to remove unbound label, the beads are analyzed with a → bead array. Multiplex transcription factor assays allow to simultaneously and routinely detect the activation of e.g. AP-2, CREB, EGR, HIF, NF-B, NF1, NFAT, PPAR, SRE, and YY1 in a single sample.

Multiplex walking (oligomer walking): A technique for the → sequencing of long DNA stretches. In short, the DNA is first restricted with different → restriction endonucleases, the various fragments are subjected to the reactions for → chemical sequencing and then processed as for → multiplex sequencing. After sequence determination of one fragment → oligonucleotides complementary to its 5'- or 3'- terminus can be synthesized and used as → probes to "walk" to the adjacent fragments. Compare → primer-directed sequencing.

Multiplex Western blotting: A variant of the conventional → Western blotting technique that allows the simultaneous detection of many proteins, separated by → polyacrylamide gel electrophoresis, blotted onto a suitable membrane, and reacted with specific (preferably monoclonal) → antibodies labeled with different → fluorochromes.

Multiprobe *R*Nase *p*rotection *a*ssay (multiprobe RPA): A variant of the → RNase protection assay, which allows to simultaneously detect and quantify many → messenger RNAs. In short, a multiprobe template set is first established, consisting of a series of defined → cDNA fragments, each cloned into a plasmid that encode specific → antisense RNAs. Such a set could e.g. contain cDNAs involved in cytokine expression during T-cell-mediated immune response (IL2, IL4, IL5, IL9, IL10, IL13, IL14, IL15, IFNg, together with the house-keeping genes L32 and GAPDH as controls). The corresponding antisense RNAs are synthesized by → T7 RNA polymerase, labeled with ^{32}P-UTP, and purified by phenol-chloroform. The antisense probes are then overnight hybridized to total RNA from the target cell under highly stringent conditions. Then → RNase A and → RNase T1 are added that digest single-stranded RNA. These enzymes are removed by proteinase K treatment, the double-stranded RNAs recovered by phenol-chloroform extraction, and the different RNA hybrids resolved by denaturing polyacrylamide gel electrophoresis. Subsequent → autoradiography or → phosphorimaging allows to analyze the transcripts both qualitatively and quantitatively (band intensity). Multiprobe RPA is about 50–100-fold more sensitive than → Northern blotting analysis.

Multiprotein complex (MPC): Any cellular complex consisting of several to many different proteins. For example, → ribosomes, → transcriptosomes, → spliceosomes and → nuclear pore complexes are such MPCs.

Multiregional evolution: A model for the evolution of modern humans. It claims that all human populations living today originate in their various continents ("multi-regions") with archaic human populations continuously linked by → gene flow. Genetic evidence, e.g. the fact that global genetic diversity is a subset of the diversity found in Africa, disfavors the multiregional evolution theory. Instead, the alternative "Out-of-Africa" model is presently widely accepted.

Multisite mutation: Any mutation that either involves alteration of two or more contiguous nucleotides, or occurs repeatedly at many loci in a given genome.

Multisite polyadenylation: A process whereby → messenger RNAs (mRNAs) encoded by a particular gene are cleaved and polyadenylated at various positions 3'-downstream of the canonical → poly(A) addition signal 5'-AATAAA-3'. Many such cleavage sites share the common nucleotide sequence 5'-PyPyA-3' (e.g TTA, or CCA). Multisite polyadenylation influences posttranscriptional regulation of e.g. the life-time of the mRNA.

Multispecies conserved sequence (MCS): Any DNA sequence that is conserved in multiple vertebrate species.

Multi-target multi-ribozyme (MTMR): A → construct, in which two (or more) different → ribozyme sequences with different substrate specificities are cloned into a → vector, separated by → linker sequences that contain the specific ribozyme cleaving sites. Once introduced into a target cell, the autocatalytic cleavage *in cis* releases the different ribozymes that in turn cleave their target → messenger RNAs *in trans*.

Mung bean nuclease (EC 3.1.30.1): A single-strand specific, Zn-containing → nuclease from mung bean (*Phaseolus aureus*) sprouts that catalyzes the degradation of single-stranded DNA or RNA molecules into deoxy- or ribonucleoside 5' monophosphates. The enzyme does not attack → double-stranded DNA, or DNA-RNA hybrids, unless very large amounts of enzyme are used. Mung bean nuclease can be used for the trimming of single-stranded overhangs produced by → restriction endonucleases, for the removal of single-stranded regions in DNA hybrids, for the cleavage of → fold-back DNA in → cDNA synthesis, and for transcript mapping.

MUP (4-methyl-umbelliferyl-phosphate): A fluorogenic substrate for → alkaline phosphatase used in → enzyme-linked fluorescent assays.

***Mu* phage (*Mu*, phage *Mu*, m, bacteriophage *Mu*, also → mutator phage):** A → temperate bacteriophage with transpositional properties that infects enterobacteria (e.g. *E coli, Salmonella typhimurium, Erwinia,* and *Citrobacterium freundii*) and consists of a 60 nm icosahedral head and a 100 nm tail containing base plates, spikes and fibers. Its 37 kb DNA consists of two double-stranded stretches of 33 kb (α) and 1.7 kb (β), respectively, separated by a 3 kb single-stranded G-loop (specifies host range) that is flanked by 5 bp → direct repeats and contains a transcriptional → enhancer ("internal activation sequence") of about 100 bp in length. It can exist in a linear and a circular form, and harbors coat protein genes, the *c* gene (encoding a → repressor preventing lysis), the genes *ner* (negative regulation of → transcription), *A* (→ transposase), *B* (→ replication), *cim* (controls immunity, i.e. superinfection), *kil* (kills host in the absence of replication), *gam* (encoded protein protects the phage DNA from → exonuclease V), *sot* (stimulates → transfection), *arm* (amplifies replication), *lig* (encodes → ligase), *C* (positive regulator of morphogenetic genes), and *lys* (necessary for → lysis, which liberates from 50 to 100 phage particles). The 75 kDa → transposase binds to the two termini and the enhancer, and cuts *Mu* at the 3' end. Transposition itself is mediated by a complex of nucleoproteins, the → transposome. Transesterification at the 3' OH integrates *Mu* into the host → genome, but the 5' ends of *Mu* are still attached to the old flanking DNA by a socalled strand transfer complex (STC). After nucleolytic cleavage of *Mu* from these old flanks, the gaps are repaired, and transposition is complete. Transposition may occur at about 60 different sites in the host chromosome, whereby inactivation of host genes (→ insertional mutagenesis) or chromosomal rearrangements are caused. Both → mutations manifest themselves as an altered → phenotype.

Muramidase: See → lysozyme.

Mutagen (mutagenic agent): Any physical or chemical agent that increases the frequency of → mutations above the spontaneous background level. Such mutagenic agents include ionizing irradiation, UV irradiation, alkylating compounds and → base analogues. See also → mutagenesis.

Mutagenesis: The induction of → mutations in DNA, either in the test tube (see → *in vitro* mutagenesis) or *in vivo*, e.g. by irradiation (irradiation mutagenesis), chemicals (→ chemical mutagenesis) or by the → deletion, → inversion or insertion of DNA sequences (→ insertion mutagenesis). See also → interposon mutagenesis, → transposon mutagenesis. Compare also → mutator gene, → site-specific mutagenesis.

Mutagenesis assistant program (MAP): A software program that allows to predict the amino acid exchanges of a given protein introduced by 19 different mutagenesis techniques and to develop strategies for → directed evolution.

Mutagenesis in aging colonies (MAC): The increased mutation rates induced by stress in aging colonies of bacteria (e.g. *E. coli*). MAC is characteristic for each bacterial strain, varies greatly, increases as a consequence of oxidative stress and carbon source starvation, is dependent on a downregulation of → mismatch repair systems, and requires the activity of the *cyaA* (encoding adenylate cyclase) and *crp* genes (encoding cAMP receptor protein). For an estimation of the extent of MAC, the frequency of mutations conferring resistance to an → antibiotic (as e.g. → rifampicin) is measured in 1day- versus 7day-colonies. Rifampicin resistance is conferred by mutations in a single gene, *rpoB*. MAC is considered as a gentic strategy for improving survival rate after/under stress.

Mutagenic: Capable of inducing → mutations.

Mutagenic agent: See → mutagen.

Mutagenically separated polymerase chain reaction (MS-PCR): A technique for the detection of → point mutations in a known DNA sequence, which relies on conventional → polymerase chain reaction. It allows to amplify normal and mutant → alleles of a gene simultaneously in the same reaction, using allele-specific → primers of different lengths. Additionally, the allele-specific primers differ from each other at several nucleotide positions and therefore introduce new and discriminating mutations into the allelic PCR products (thereby reducing cross-reactions between amplification products during the PCR process). Since both products possess different lengths, MS-PCR "separates" both amplified alleles that can then be identified by → agarose gel electrophoresis and → ethidium bromide staining.

Mutant: An organism harboring a mutant gene whose expression changes the phenotype of the organism. See → mutation.

Mutant allele-specific amplification (MASA): Any one of a series of → polymerase chain reaction-based techniques, allowing the specific amplification of an → allele that has undergone a → mutation (e.g. a → deletion, → insertion, → inversion, → transition, → transversion). MASA techniques are presently employed in clinical screening and diagnosis.

Mutant analysis by PCR and restriction enzyme cleavage (MAPREC): A technique for the detection of → point mutations in coding → genomic DNA. In short, RNAs of wild-type and mutant are first isolated, and → reverse transcribed into → cDNAs using a random hexanucleotide → primer. The cDNA serves as template for → asymmetric PCR with primers specific for the gene of interest, more precisely the gene region in which the mutation occurs. The primer is designed such that it creates a → recognition site for the → restriction endonuclease MboI (5'-GATC-3') in the wild-type target DNA. The mutant sequence will give rise to e.g. a *Hinf*I restriction site. An excess of sense polarity primer ensures that the product is predominantly single-stranded DNA. Now the second strand is synthesized, using a labeled antisense primer (→ biotin labeling), the double-stranded product digested with *Mbo*I, and the resulting restriction fragments separated by → polyacrylamide gel electrophoresis. After → Southern blotting or → vacuum blotting and fixation of the fragments onto the blotting membrane, the fragments are visualized with → streptavidin-conjugated alkaline phosphatase. The wild-type sequence will be fragmented, the mutant sequence will remain uncut.

Mutated promoter: Any → promoter sequence, into which a → mutation(s) is (are) introduced naturally or artificially. Such mutations may not at all affect the binding of → transcription factors to their cognate sequence motifs, but can also lead to either a stronger → affinity of the transcription factor to its binding sequence, or the partial or total loss of binding of the transcription factor. See → promoter-up mutation.

Mutation: Any structural or compositional change in the DNA of an organism that is not caused by normal segregation or genetic recombination processes. Such mutations may occur spontaneously, or may be induced by → mutagens such as ionizing radiation or alkylating chemicals. The change of a nucleotide base for example, may cause the conversion of one → codon into another one. It is silent, if the codon change does not cause any detectable phenotypic change (if e.g. both codons stand for the same amino acid, see → codon bias). See also → mutagenesis, → mutation breeding, → mutation rate.

Mutational cold spot: See → cold spot.

Mutational load: See → genetic load.

Mutation analysis: The detection and characterization of a → mutation in DNA, e.g. → deletion, → insertion, → inversion, → mismatch mutation, → point mutation, → translocation. Out of a multitude of techniques for mutation analysis, see → allele-specific hybridization, → amplified restriction fragment length polymorphism, → arbitrarily primed PCR, → arbitrary primer technology, → arbitrary signatures from amplification profiles, → base excision sequence scanning, → capillary electrophoresis hybridization, → chimeric oligonucleotide-directed gene targeting, → cleavase fragment length polymorphism, → cleaved amplified polymorphic sequence, → digested random amplified microsatellite polymorphism, → direct amplification of minisatellite DNA, → dynamic allele-specific hybridization, → forensically informative nucleotide sequencing, → heteroduplex analysis, → inter-retrotransposon amplified polymorphism, → methylation-sensitive amplification polymorphism, → methylation-specific PCR, → methyl filtration, → microsatellite-primed PCR, → minisatellite-primed amplification of polymorphic sequences, → multiple fluorescence-based PCR-SSCP, → mutagenically separated PCR, → mutant allele-specific amplification, → mutant analysis by PCR- and restriction enzyme cleavage, → mutator amplified fragment length polymorphism, → MutS mismatch detection, → PCR clamping, → PCR-ligation-PCR mutagenesis, → polymerase chain reaction, → restriction fragment length polymorphism, → primer-specific and mispair extension analysis, → random amplified microsatellite polymorphism, → random amplified polymorphic DNA, → retrotransposon-microsatellite amplified polymorphism, → reversed enzyme activity DNA interrogation test, → selective amplification of polymorphic loci, → semi-specific primer technology, → sequence-based amplified polymorphism, → sequence-specific amplification polymorphism, → single nucleotide polymorphism, → single-strand conformation analysis, → temperature modulated heteroduplex analysis.

Mutation breeding: The development of plants with improved characteristics (e.g. resistance against pathogens or environmental stress, increased agricultural productivity) through physically or chemically induced → mutations. Since such mutations are totally at random, no directed genetic change is possible. This method is still used but will be replaced by directed genetic engineering in future. But see → targeting induced local lesions in genomes.

Mutation cluster region (MCR): Any region of a → gene or → genome, where various types of mutations are present at a higher frequency than in the rest of the genome. MCRs represent extended → hot spots of mutations.

Mutation delay: The time lag between a → mutation event and its phenotypic expression. For example, recessive mutations may only be apparent, if they become homozygous.

Mutation detection electrophoresis (MDE) gel: A gel made of modified → polyacrylamide with slightly hydrophobic properties that selectively alters the electrophoretic mobility of → heteroduplexes such that even single mismatched bases in one kb of duplex DNA can be visualized by a mobility shift.

Mutation rate (μ): The number of → mutations occurring per unit DNA (e.g. → kb or a → gene) per unit time.

Mutator: See → mutator gene, → Mu.

Mutator amplified fragment length polymorphism (MuAFLP; amplification of insertion mutagenized sites, AIMS): A variant of the conventional → amplified fragment length polymorphism (AFLP) technique to screen, isolate and characterize → insertions of → Mu elements (or → T-DNA in plants) and their flanking sequences in target genomes, and to screen for sequence polymorphisms in the DNA flanking inserted mutator copies. In short, genomic DNA is restricted with Mlu I (that cuts within mutator → long terminal repeat sequences) and a four-base cutter (e.g. Mse I or Bfa I), then biotinylated Mlu I and four base cutter → adaptors are ligated to the corresponding ends of the restriction fragments, the biotinylated fragments captured on streptavidine beads and the fragments including insertion sequences amplified by linear → polymerase chain reaction of only 12–15 cycles (to minimize PCR artifacts) using → primers complementary to Mu sequences and the four-base cutter adaptor (the latter primer is labeled with ^{32}P). Then the amplified fragments are electrophoretically separated in → polyacrylamide gels, and Mu insertions detected by → autoradiography. If a primer labeled with a → fluorochrome is used for MuAFLP-PCR, the number and sizes of the amplified Mu flanks can be quantitatively determined by an automated fluorescence reader. Since Mu preferentially inserts into → genes, MuAFLP allows to detect insertions that lead to → gene knock-out mutants. See → Mu array.

Figure see page 1028

Mutator gene (mutator): Any gene (mut gene) that increases the rate of spontaneous → mutations of one or more other genes. Such mutators may themselves originate from normal genes by mutation. If for example, a gene is mutated whose product normally functions in DNA repair or replication, the mutant protein encoded by the mutated gene may introduce multiple errors (that is mutations) during these processes. High mutator gene activity probably increases evolutionary adaptation in bacteria.

Mutator phage: Any phage that is able to increase the rate of mutation in its host cell (e.g. the → Mu phage).

Mutator polymerase: A mutated, nucleus-encoded mitochondrial γ- DNA polymerase that gives rise to the accumulation of → frame-shifts, → point mutations, and → deletions in the → mitochondrial genome. One of the mutations converting the wild-type DNA polymerase to the mutator polymerase is a point mutation that changes a highly conserved tyrosine at position 955 (part of the binding pocket responsible for selection of deoxyribonucleotides against ribonucleotides) to cysteine ($Y_{955}C$). This simple base exchange does neither change the catalytic rate nor the intrinsic 3′ 5′- exonuclease proofreading activity, but decreases the fidelity of DNA replication, which in turn leads to the mtDNA mutations. These mutations cause a series of diseases. For example, the *progressive external ophthalmoplegia* (PEO) is the consequence of several kb long deletions primarily between short, direct repeats of 10–13 base pairs. These deletions are associated with point mutations caused by T.dTMP mispairing that occurs 100 times more frequent with mutator as compared to wild-type γ-poly-

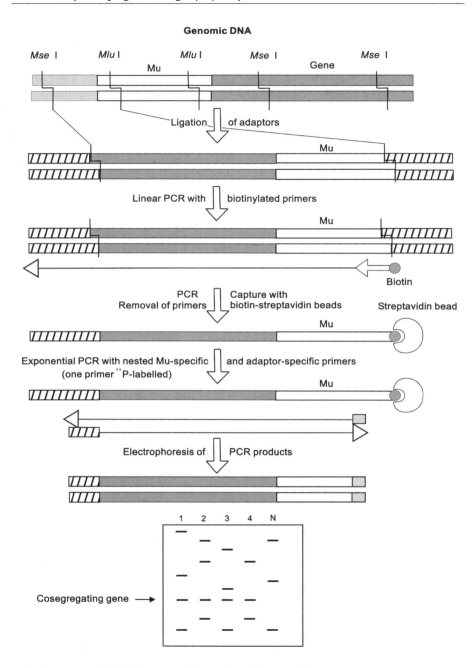

Mutator amplified fragment length polymorphism

merase. The disease appears in patients at the age of 30–40, and causes a weakness of muscles in general, the eye muscles in particular. As a consequence, the muscles moving the eye (especially the lateral rectus) deteriorate gradually, so that the patients can only follow a moving object by turning their heads. The muscle weakness is a result of impaired electron transport chain activity (depletion of ATP). Another cause of PEO is a mutant mitochondrial → helicase encoded by gene *twinkle*.

Mutator strain: Any *E.coli* strain that carries → mutations in one or several DNA repair pathways, and therefore has a higher random mutation rate as compared to the wild type. Typically, the *mut*D (deficient in 3′-5′-exonuclease activity of DNA polymerase III), *mut*T (unable to hydrolyze 8-oxodGTP), and *mut*S (error-prone mismatch repair) alleles are present in such mutator strains, and consequently the spontaneous mutation rate is increased from 50 to as much as 5000 times in triple mutants. If an → insert (e.g. a → gene) is maintained in a → plasmid of such a mutator strain, it also suffers random mutations at an increased rate (e.g. one base change per 2000 nucleotides per generation).

Mutein (*mut*ated pro*tein*): Any protein that is encoded by a mutated gene and therefore has an amino acid sequence different from the → wild-type protein encoded by a non-mutated gene. Muteins frequently are the result of → genetic engineering. For example, native insulin exists as a hexamer in solution that only slowly dissociates to release the physiologically active monomer. The targeted exchange of specific amino acids at the C-terminus of the insulin B-chain results in muteins ("insulin lispro", "insulin aspart"), whose hexamers dissociate more rapidly,

so that the insulin effect starts earlier after administration. Alternatively, another sequence change in the B-chain leaves the insulin ("insulin glargin") soluble during injection. However, at the injection site it precipitates as microcrystals, representing an insulin depot, from which the active monomer is released very slowly. Result: a relatively constant insulin level over a longer time period.

Muton: The smallest unit of a gene that may undergo → mutation (equivalent to one base pair of DNA).

Mut S: Any one of a family of *E. coli* methyl-directed → mismatch repair enzymes that recognize and bind to mismatched bases in target DNA. Mut S is part of a system for the correction of replication errors. See → Mut S mismatch detection.

Mut S mismatch detection: A technique for the detection of single base → mismatches in a target DNA that exploits the affinity of → Mut S to recognize and bind mismatched bases. In short, the target DNA is first amplified, using appropriate, radioactively endlabeled → primers and conventional → polymerase chain reaction techniques. The PCR products are then heat-denatured and re-annealed, which results in four different DNA duplexes (in case the target DNA is heterozygous at locus A [A/a]: two homoduplexes [AA and aa]), and two heteroduplexes [Aa and aA]). If the heteroduplexes contain e.g. a single base-pair mismatch, the added Mut S protein will bind to this mismatch, and the mutant allele can be detected by → mobility-shift DNA-binding assays in → polyacrylamide gels with subsequent → autoradiography.

Figure see page 1030

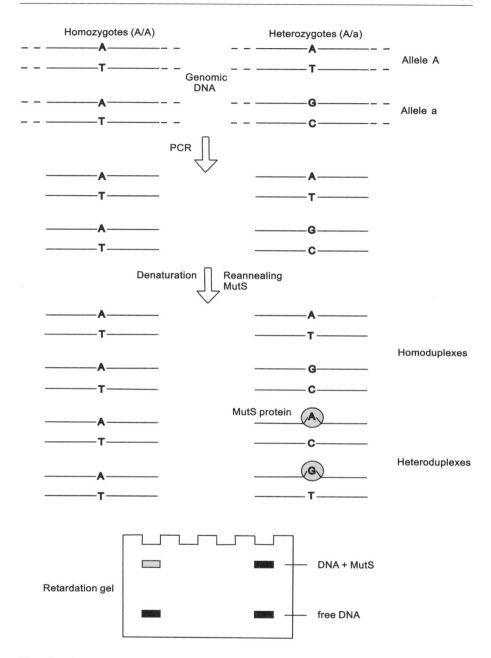

Mut S mismatch detection

Mutually exclusive splicing (ME splicing, mutually exclusive alternative splicing): A variant of the conventional → alternative splicing of → exons from a → pre-messenger RNA to produce the mature → messenger RNA that warrants the selection of only one single exon from an array of two (or more) tandemly arranged exons of the same type. ME splicing is specific for genes with → alternate exons such as e.g. the 61.2 kb *Dscam* gene of *Drosophila melanogaster*, harboring an array of potential alternative (mutually exclusive) exons: exon 4 has 12, exon 6 a total of 48, exon 9 altogether 33, and exon 17 two almost identical alternatives. This means that each mRNA contains one of twelve possible alternatives from exon 4, one of 48 alternatives from exon 6, one out of 33 from exon 9, and one of two from exon 17. ME splicing requires the interaction between a conserved 36 nucleotide long sequence element located ~250 nucleotides → upstream of a specific exon (the socalled docking sequence, "docker"), and an intronic socalled selector sequence ("selector"), preceding each exon variant. Every selector sequence can base-pair with the docker over a stretch of ~27 nucleotides, and this base-pairing interaction is specific for each exon such that only one single exon variant will be spliced. For example, the *Dscam* gene of *Drosophila melanogaster* harbors an exon 6 cluster of 48 alternative exons (denoted as 6.1, 6.2 ... 6.48). The docking site (consensus sequence: 5'-AAATTGAAAACTGCCTGAAT-GTTGGGATAGGGTACTC-3') is located in the → intron → downstream of → constitutive exon 5 and upstream of the first exon 6 variant (6.1), consists of 66 nucleotides, and is 90–100% identical in 10 other *Drosophila* species (the central 24 nucleotides are invariant in all species). The relatively conserved selector sequence (consensus sequence: 5'-TTTAACTT-TTACGGACTTACAACCCTA-3'), located in the intron upstream of 6.1 is partly complementary with exon 6.1 and will form a → duplex with it. Therefore only 6.1 (and no other of the 47 exon 6 variants) will be selected. This docking site:selector sequence interaction is thought to inactivate a splicing → repressor on the downstream exon, and consequently activates the splicing of the downstream exon 6 variant to exon 5. Subsequently, the exon joined to exon 5 can only be spliced to constitutive exon 7, because the remaining exon 6 variants are still repressed by the splicing repressor.

MVR: See → minisatellite *v*ariant *r*epeat.

Mx-rMx: A family of interacting → transposons of corn (*Zea mays* L.), belonging to the *hAT* superfamily (along with → *Ac/Ds* and *Bg/rBg*). The hAT superfamily is named after the autonomous elements *h*obo in *Drosophila melanogaster*, *Ac* in *Zea mays*, and *Tam3* in *Antirrhinum majus*. The 3.731 kb long autonomous *Mx* ("*m*obile element induced by *X*-rays") is flanked by 13 bp → *t*erminal *i*nverted *r*epeats (TIRs) and causes an 8 bp duplication of the target site (5'-CACTACAC-3'). *Mx* carries three → exons encoding a 674 amino acids protein homologous to the *Ac* → transposase that, however, cannot transactivate *Ds* excision. The transposase in turn contains three → domains (hAT1-3) that are highly conserved among the members of the *hAT* transposon superfamily. *rMX* ("*r*esponder to *Mx*") comprises 0.571 kb, contains 13 bp imperfect TIRs and 15 copies of the direct hexanucleotide repeat sequence 5'-CCCGAA-3' within the subterminal 170 bp at either end, upon transposition causes an 8 bp → *t*arget *s*ite *d*uplication (TSD), is non-autonomous (i.e. does not harbor a transposase gene), and together with *Mx* forms a classical family of interacting transposable elements (as e.g. the *Ac/Ds* element family of *Zea mays* L.). The *rMx* insertion site (5'-GTGGAGGA-3') is located in the second exon of the bz gene, close to its 3'-end, and the presence of *rMx* causes the somatic instability of *bz-x3m*, a mutation from an X-irradiated stock of maize, expressed as spotted kernel phenotype.

MYB domain: A region of about 52 amino acids in socalled → MYB proteins that occurs either single or as two or three repeats, respectively, and binds to specific address sequences in target DNA. Each repeat adopts a → helix-turn-helix conformation, which binds in the → major groove of the target. Single MYB domains possess longer C-terminal helices (as e.g. the human telomeric protein hTRF1) than the repeated MYB elements, but all contain regularly spaced tryptophan residues (three per repeat) that contribute to a hydrophobic cluster.

MYB protein: Any one of a large and diverse class of DNA-binding proteins that either function as transcriptional activators, possibly also as repressors, or as structural proteins (as e.g. telomeric G-rich sequence-binding MYB proteins containing a socalled telobox). Common and characteristic feature of all MYB proteins is the occurrence of either one (e.g. StMYB1), two (e.g. ZmMYBC1), or three structurally conserved → MYB domains (e.g. C-MYB) that mediate specific binding to address sequences in target DNA (consensus sequence: 5'-PyAAC (G/T)G-3'. For example, in plants MYB proteins regulate secondary metabolism (e.g. the maize genes ZmMYBC1, ZmMYBPL and ZmMYBP, the snapdragon genes Am-MYBROSEA and Am-MYBVENOSA, and the *Petunia* gene PhMYBAN2), cellular development (e.g. the *Arabidopsis thaliana* gene AtMYBGL1, involved in trichome differentiation), meristem formation (e.g. AtMYB13 and AtMYB103) and the cell cycle (e.g. AtCDC5). The DNA-binding domain(s) of MYB proteins is (are) located close to the amino terminus, and the transcriptional → activation domain lies C-terminal of this binding sequence.

Mycophenolic acid (mycophenolate): The antibiotic 6-(4-hydroxy-6-methoxy-7-methyl-3-oxo-5-phthalanyl)-4-methyl-4-hexenoic acid from *Penicillium stoloniferum* that is an immuno-suppressive agent preventing rejection in organ transplantation, because it inhibits inosine monophosphate dehydrogenase, the enzyme controlling the rate of synthesis of guanosine monophosphate necessary for purine synthesis and the proliferation of B and T lymphocytes (i.e. inhibits lymphocyte proliferation and antibody production).

***Mycoplasma laboratorium* ("Synthia"):** A synthetic → minimal genome ("chromosome") of 580,000 bp in size, developed from the → genome of the living *Mycoplasma genitalium*, transferred into a genome-free cell (most probably of *Mycoplasma capricolum*) and in fact being the first partly synthetic organism. The minimal genome contains 381 protein-coding and 43 RNA-encoding genes, about 101 genes less than the wild-type *M. genitalium* that altogether allow the organism to perform metabolism, growth and replication.

Mycostatin: See → nystatin.

Myeloma cell line: A tumor cell line that originates from a single lymphocyte and produces only one defined immunoglobulin.

N

N:

a) Abbreviation for a Ny base in DNA (e.g. → A, → T, → G, → C) or RNA (e.g. A, → U, G, C). Synonym for → X.
b) Number of chromosomes in a haploid set.

N-ᶠMet: See → N-formylmethionine.

NAB: See → nucleic acid biotool.

N-acetyl deblocking aminopeptidase (Ac-DAP): A thermostable, CoCl2-activated exo-aminopeptidase from *Pyrococcus furiosus* catalysing the removal of N-terminal acyl-type blocking groups in proteins and peptides. For example, the tocacco mosaic virus coat protein contains an acetylated N-terminal amino acid residue, which cannot be directly analysed on protein sequencers based on Edman degradation. Therefore, the N-acetylaminoacid has to be released, which is done by Ac-DAP. Since the N-terminus of a commercially available Ac-DAP is acetylated, its amino acid cannot be determined by the conventional Edman degradation. Therefore, the sequence of only the target peptides or proteins is analysed.

NACS™ (*nucleic acid chromatography system*): An ion-exchange resin that is used to separate DNA or RNA from contaminating low-molecular weight substances (e.g. nucleotides, or sulfonated and sulfated polysaccharides which are components of most → agaroses, and inhibit many enzymes used in recombinant DNA experiments, e.g. → restriction endonucleases). During chromatography, nucleic acids are bound to the resin in low-salt conditions. Extensive washings remove all contaminations, and the nucleic acids can then be eluted with a high-salt buffer.

NAHR: See → non-allelic homologous recombination.

Naked DNA: Any → DNA that is devoid of all proteins, with which it is normally associated in the nucleus. See → chromatin.

Naked eye polymorphism (NEP): Any difference between two closely related organisms that can be detected visually (i.e. by the naked eye). See → visual marker.

Nanoarray: Any solid support (e.g. a gold-coated glass chip), onto which dots of peptides, proteins, oligonucleotides or DNAs are spotted via e.g. → dip-pen nanolithography in arrays of 100 nm (or less) diameter and 100 nm (or less) distance between spots (see → ultra-high density microarray). Interactions between the probes and target molecules on a nanoarray are detected by atomic force microspcopy.

Nanobarcode particle: Any encodeable, machine-readable, sub-micron particle manufactured in a semi-automated process by electroplating inert metals such as gold, nickel, platinum, or silver into metallized templates. These templates define the particle diameter, and are dissolved, thereby releasing the resulting striped nano-rods.

Nanobody: See → heavy-chain antibody.

Nanocavity trap: A staphylococcal α-hemolysin transmembrane → β-barrel that is engineered to accommodate two different cyclodextrin (CD) adapters (e.g. βCD, and hepta-6-sulfato-βCD) within its lumen such that they serve as *cis* and *trans* gates of a nanocavity of several thousand of cubic Å volume. Organic molecules (e.g. → oligonucleotides) can be pulled into this cavity by an electric potential and kept there for hundreds of milliseconds. The trapped molecules shuttle back and forth between the adapters, before leaving the cavity. Nanocavity traps alter the magnitude and selectivity of ion flux (conductivity) in a transmembrane potential, and the build-in adapters bind guest molecules that block or reduce conduction by the pore.

Nanocircle: See → DNA nanocircle.

Nanocrystal antenna: Any metal (preferentially gold) covalently linked to a biomolecule (DNA, RNA, or protein) that can inductively be heated by an alternating magnetic field and allows to selectively and reversibly control the function of the biomolecule carrier (e.g. the activity of an enzyme).

Nanodroplet: A small volume (100–200 nl) droplet, in which cellular reactions can be simulated and analyzed without the problems in liquid cultures (e.g. caused by diffusion of all molecules). For example, the influence of small effectors on protein-protein interactions can be analyzed, if nanodroplets containing defined media, yeast cells and beads carrying photochemically releasable effectors (e.g. galactose) are employed. The yeast cells harbor → two-hybrid system vectors (one carrying a gene fused to the LexA DNA-binding domain, the other one to a transcriptional activation domain proximal to a promoter containing LexA binding sites upstream of a *URA3* reporter gene. Both constructs are repressed by glucose and activated by a shift to galactose in the medium. The *URA3* expressing cells are killed in the presence of 5-fluoroorotic acid (5-FOA). If, however, a small molecule disrupts the intracellular protein-protein interaction, proximity of the activation domain to the promoter is diminished, and *URA3* gene transcription reduced: cells can survive in the presence of 5-FOA.

Nano-electrospray ionization (nano-ESI) mass spectrometry: A technique for the determination of the masses of biological molecules in solution that is based on the conversion of the liquid analyte into a fine mist of droplets by an electric field imposed onto a gold-plated borosilicate needle (nanoflow capillary) containing the analyte. In short, the analyte solution in the tip of this nanoflow capillary (inner diameter: 1–10 μm) is first exposed to the field (1.5–2.0 kV), which causes an electric stress onto the liquid. As a consequence, the analyte liquid develops an electric double layer, leading to charge accumulation at the surface. This destabilizes the meniscus of the solution such that a jet of droplets with an excess of positive (or negative) ions is formed at the end of the capillary (the socalled "Taylor cone"). The solvent evaporates from these initial droplets in a desolvation gas (N2) atmosphere, progressively smaller droplets are generated, and finally a single, multiply charged molecule is left, which is analyzed in a → time-of-flight (TOF) mass analyzer. In the TOF-MS, the ions are accelerated through a fixed potential into a field-free drift region. Low mass ions achieve a higher velocity than high mass ions, so that the mass of an ion can be deduced from the amount of time needed to reach the detector. Finally, the mass-to-charge ratio (m/z) of the ion is recorded.

Nanogenome: Any genome of a living organism, whose size is among the smallest genome sizes known. For example, the genome of the 400 nm Archaean *Nanoarchaeum equitans* measures about 0.49 Mb (a tenth of the size of the *E. coli* genome, and smaller than the genome of the bacterium *Mycoplasma genitalium*).

Nanomechanical transduction: The transformation of forces, generated by DNA-DNA, DNA-RNA or DNA-oligonucleotide hybridizations or protein-protein and protein-ligand interactions into nanomechanical responses of a microfabricated silicon support, on which the interactive processes take place. In short, one side of silicon → cantilevers is first covered with a monolayer of gold. Then synthetic, 5′ thio-modified oligonucleotides are covalently immobilized on the gold surface in a monolayer. Cantilever I may carry e.g. 12-mers, cantilever II e.g. 16-mers, and so on. These cantilever arrays are then placed in a liquid cell and equilibrated in hybridization buffer, after which a complementary 12-mer is injected. Hybridization between the 12-mer and the matching oligonucleotides on the cantilever surface leads to a difference in surface stress between the functionalized gold and the non-functionalized silicium surface, which bends the cantilever. The degree of bending is recorded by an optical beam deflection technique. The transduction of Watson-Crick hybridization into surface stress is triggered by electrostatic, steric, and hydrophobic interactions (e.g. the charge density in the sugar-phosphate backbone of the oligonucleotides and their counter-ions during hybridization is increased, as is the packing of the oligonucleotides on the cantilever surface. These changes lead to repulsion and produce compressive surface stress. If cantilever II is now hybridized to a complementary 16-mer oligonucleotide, the same process is repeated and leads to a deflection of cantilever I. This technique allows to detect single base mismatches, which are translated into (minimal) nanomechanical responses. Since the hybridized oligonucleotides can be dissociated from their cantilever-bound counterparts (by e.g. 30% urea solution), the same array can be re-used repeatedly.

Protein-protein recognition can also be translated into nanomechanical forces. For example, one cantilever can be loaded with → protein A, the second one with bovine serum albumin as a reference. Then immunoglobulin G (IgG) is bound specifically – through its constant region – to protein A, leading to a deflection of the corresponding cantilever.

Nanomechanical transduction neither requires labeling of probe oligonucleotides or proteins nor optical excitation (by e.g. laser light) or external probes, and can be expanded to a high-throughput format (by e.g. parallel organization of 1000 or more cantilevers).

Nanoparticle-based bio-barcode assay: A technique for the indirect detection of protein analytes that uses a pair of paramagnetic beads with capture reagents (→ antibodies or → oligonucleotides, respectively) for the target protein. One reagent is a gold *nano*particle (NP) with a diameter from 13–30 nm and functionalized with hybridized oligonucleotides ("bio-bar codes") and polyclonal antibodies to recognize the target protein, the other one a 1 μm diameter polyamine *magnetic*

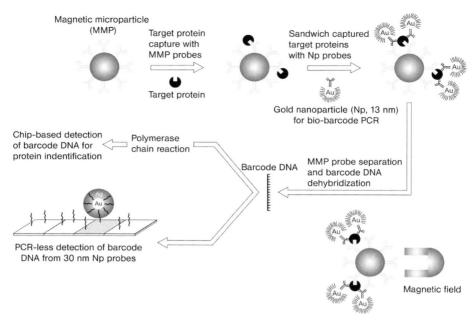

Nanoparticle-based bio-barcode assay

microparticle (MMP, containing an iron oxide core) with a bound target-specific monoclonal antibody, which both bind and sandwich the analyte. The MMPs with the bound analyte are then captured by a magnet, unreacted gold NPs washed off with ultrapure water, and the gold spheres with the bound analyte retained. Each gold particle carries unique identifier oligonucleotides that are released from the beads by heating to 55 °C, and ultimately detected as barcodes, using any DNA detection technique (e.g. gel electrophoresis, fluorophore labeling and detection, or scanometry). Therefore, not the analytes are detected themselves, but the bar code. See → nanoparticle-based DNA detection.

Nanoparticle-based DNA detection: A technique for the detection of specific DNA sequences and mutations in these sequences (e.g. → single nucleotide polymorphisms), in which → nanoparticle probes are hybridized to single-stranded DNA templates. Upon hybridization with complementary sequences, an extended polymeric network in two and three dimensions is formed. Whereas the non-hybridized particles (in case of gold, Au) have a red color, the hybridization-induced polymerization to aggregates changes the color to purple. This color change is brought about by the shortening of interparticle distances to less than an average gold particle diameter by the hybridization process. This shift is attributed to the → surface plasmon resonance of the Au. Therefore, the hybridization event can be monitored without radioactive label. Nanoparticle-based DNA detection can also be applied to DNA arrays on glass supports (→ DNA chips), where detection of a hybridization event is mediated by the reduction of silver ions to metallic silver through hydroquinone on the surfaces of the gold nanoparticle. Since this process can simply be scanned by a flatbed scanner, the procedure is also called scanometric DNA array detection.

Nanoparticle chip: A special type of a → microarray, in which the planar surface of conventional glass slides is replaced by silica nanoparticles, whose 4π-geometry allow a very high packaging with probes. At the same time, nanoparticle chips reduce the background signal intensity. Nanoparticle chips are used for nucleic acid hybridizations or the fabrication of protein arrays. See → nanoarray, → planar chip.

Nanoparticle probe: Any oligonucleotide → probe that is covalently bound onto the surface of gold, silver or platinum particles of a diameter of 5–10 nm. The noble metal particles contain N-propylmaleimide substituents that can be selectively coupled to sulfhydryl groups (thiol capping with S-trityl-6-mercaptohexylphosphoramidite) at the 3′ends of single-stranded DNAs. Nanoparticle probes are used for → nanoparticle-based DNA detection.

Nanoplex (*nanometer complex*): A complex of nanometer-sized particles (e.g. beads) and DNA (e.g. plasmid DNA) that can be used for → direct gene transfer by e.g. → particle gun techniques.

Nano-*polymerase chain reaction* (nano-PCR): A variant of the conventional → polymerase chain reaction that employs reaction volumes in the nanoliter range. Aside of saving reaction components, shorter diffusion routes and a highly efficient thermotransfer (governed by the big surface-to-volume ratio) make nano-PCR much faster than the conventional µl PCR. Therefore it is also called "fast PCR" or "rapid PCR".

Nanopore processing: A technique for the detection of single nucleic acid molecules, based on the blockage of ionic current in a single α-hemolysin pore channel caused by the traversing RNA or → ssDNA. Such pores of 1.0–1.2 nm internal diameter are formed by the self-assembly of the 293 amino acid staphylococcal α-hemolysin polypeptide into lipid bilayers. The duration and amplitude of the current blockage is related to both the concentration of the nucleic acid and its chain length. Also, → purine and → pyrimidine nucleotides produce distinct current blockades, and hybridization can be detected directly, since → dsDNA cannot penetrate the nanopore.

Nanopore sensing: The label-free electrical detection of specific DNA sequences by → hybridization of a target DNA (e.g. an → oligonucleotide) to a → probe DNA immobilized in the nanopore of a solid support (e.g. an alumina membrane). Such nanopores change their ionic conductance when they become partially blocked, as is the case when a DNA hybrid is formed within the pore. This change results in a spike in impedance, from which the concentration of the probe DNA (or any analyte interacting with the target DNA) can be measured. The pore diameter in e.g. alumina membranes is kept small by hydrothermal sealing (anodized aluminum is placed in boiling water, which leads to shrunken pores). Therefore anodized alumina membranes can be engineered to contain a multitude of parallel nanopore arrays for high-throughput sensing, in which each nanopore harbors a different single-stranded DNA probe. See → nanopore processing.

Nanopore sequencer: A high-throughput → DNA sequencing machine, based on the movement of the analyte DNA through a channel membrane protein (e.g. bacterial α-hemolysin) anchored in a lipid bilayer membrane (usually Teflon horizontal bilayers) that separates two solutions. When a voltage is applied across the membrane, charged DNA molecules migrate through the 1.5 nanometer pore ("nanopore") of the channel protein, and each base identifies itself with a characteristic electrical current, which in turn is measured by a detector. For example, one of the nanopore sequencers is based on α-hemolysin, a 33 kDa protein from *Staphylococcus aureus*, of which 7 subunits self-assemble in a lipid bilayer to form a nanopore, narrowing from 26 to about 15 nm. This pore remains open at neutral pH and high ionic strength, and the diameter at its narrowest point (about 1.5 nm) allows single-stranded DNA (but not double-stranded DNA) to pass through the pore. As the ssDNA enters the pore, it blocks the ionic current transiently and it does so specifically for each base. Therefore each base can be identified by a specific ion current. In the future, synthetic nanopores will certainly replace the α-hemolysin central pore for → nanopore sequencing. The socalled ion-beam sculpting technique allows to fabricate 5 nm pores in e.g. silicon nitride membranes, which shares many of the properties with α-hemolysin. Or, nanopores are made of carbon nanotubes that bind DNA and orient the bases such that they are optimally positioned for a transit. The nanopore sequencer concept is still in a developmental stage.

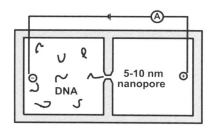

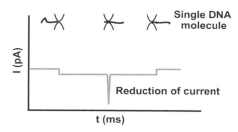

Nanopore sequencer

Nanopore sequencing: The determination of the sequence of bases of a single DNA molecule in a → nanopore sequencer.

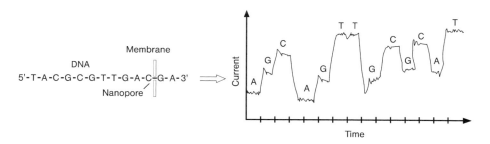

Nanopore sequencer

Nanos: Greek for dwarf, gnome.

Nanotag: Any supramolecular assembly of a rationally designed branched DNA with either covalently or also non-covalently bound intercalating, mostly fluorescent dyes. The loading of the fluorochromes follows the "nearest neighbour exclusion principle", i.e. one intercalating molecule locates between every other base pair. Such nanotags frequently exhibit very intense fluorescence.

Nanotechnology: A collective term for a series of techniques designed to study atoms and molecules at the nanometer level ("nano-scale") together with the theoretical background. For example, scanning tunneling microscopy, STM, scanning the tunneling current, and → atomic force microscopy, AFM, scanning the repulsive atomic forces between a sample and a probe, thereby producing high-resolution surface topographies of proteins and nucleic acids are nano-technologies.

Nanowire sensor array: A variant of an electronic biochip that detects interactions of biomolecules (especially protein-protein interactions) by electrical current. In short, silicon nanowires are deposited at specific locations on a specially designed chip surface, then the nanowires coated with the catcher molecule (e.g. a receptor protein or an → antibody) and the analyte solution added. If this solution contains a protein recognized by the receptor, or an → antigen specific for the antibody, this protein will bind, which leads to a change in electrical conductance of the receptor- or antibody-conjugated wire. The extent of the shift in conductance is proportional to the concentration of the bound ligand. Nanowire sensor arrays allow to detect → biomarkers in the subpicomolar range, even in very complex natural mixtures as e.g. sera. See → electronic microarray.

NAP: See → nucleoid-associated protein.

NAPPA: See → nucleic acid-programmable protein array.

Naptonuon (*non-apt*ative *nuon*): Any → potonuon that disintegrates over evolutionary times by random nucleotide changes ("nonaptation") and is lost.

Narrow range immobilized pH gradient strip: See → ultra-zoom gel.

NAS:

a) See → network-attached storage.
b) See → nonsense-associated alternative splicing.

NASBA: See → *n*ucleic *a*cid *s*equence-*b*ased *a*mplification.

NASBH: See → *n*ucleic *a*cid *s*canning *b*y *h*ybridization.

Nascent polypeptide: A chain of amino acids linked together via peptide bonds that is being formed (in statu nascendi) and still attached to the 50S (bacteria) or 60S (eukaryotes) ribosomal subunit through a tRNA molecule.

Nascent RNA, nascent DNA: A chain of nucleotides linked together via → phosphodiester bonds that is being formed (*in statu nascendi*).

NAT: See → *n*atural *a*ntisense *t*ranscript.

National Center for Biotechnology Information (NCBI): A unit of the National Library of Medicine (which in turn is part of the US National Institutes of Health, Bethesda, Md.) that is organized in (1) an Information Resources Branch (data acquisition, storage and distribution), an Information Engineering Branch (data control) and the Computational Biology Branch (data research). NCBI operation is based on more than 500 CPUs, serves millions of accesses day by day, and distributes a terabyte of data per day. NCBI databases are, for example, Gen Bank and Pub Med. Gen Bank alone contained about 33 billion bases and 25 million entries at the end of 2002.

URLs:

(1) NCBI Bookshelf
http://www.ncbi.nlm.nih.gov/entrez/query.fcgi?db=Books

(2) NCBI GeneRif
http://www.ncbi.nlm.nih.gov/LocusLink/GeneRIFhelp.html

(3) NCBI LocusLink
http://www.ncbi.nlm.nih.gov/LocusLink/index.html

(4) NCBI Reference Genomes
http://www.ncbi.nlm.nih.gov/entrez/query.fcgi?db=Genome

(5) NCBI RefSeq
http://www.ncbi.nlm.nih.gov/LocusLink/refseq.html

(6) US National Center for Biotechnology Information
http://www.ncbi.nlm.nih.gov/

National Center for Genome Resources (NCGR): An independent non-profit life science research institution located in Santa Fe (New Mexico, USA), in which the Genome Sequence Data Base (GSDB), a publicly available relational database of human genome sequences was developed. NCGR als designed bioinformatic tools as e.g. Gene-Xlite, ISYS and XGI. Web page: http://www.ncgr.org

National Human Genome Research Institute (NCGRI): A branch of the National Institute of Health (NIH) that led the Human Genome Project and develops novel technologies for DNA sequencing (e.g. the 1000 US$ genome). Web page: http://www.nhgri.nih.gov

National Institutes of Health Guidelines: See → NIH Guidelines.

Native chemical ligation: A technique for the synthesis of small proteins (up to 15 kDa) or protein domains, in which an N-terminal cysteine-containing peptide is chemically ligated to a second peptide possessing an α-thioester group with the resulting formation of a native peptide bond at the ligation junction. See → expressed protein ligation.

Native DNA: A double-stranded DNA molecule with intact hydrogen bonds between all its base pairs.

Native protein nanolithography (NPNL): A technique for the spotting of native (or nearly native) proteins onto a chip surface on a nanometer scale that exploits the cantilever tip of an → atomic force microscope analogous to the → dip-pen nanolithography (DPN). NPNL allows to detect protein-protein interactions between two proteins or within whole protein complexes under quasi-physiological conditions.

Native ultraviolet fluorescence detection (native UV detection): A technique for the detection of proteins without any labeling. Proteins contain amino acids with aromatic side chains (as e.g. tryptophan, tyrosine) that absorb ultraviolet light and emit fluorescence light. For an excitation of this native fluorescence two different UV laser sources can be used (e.g. a Ti:SA laser at 280 nm, and a Nd:YAG laser at 266 nm), whose light is focused through a spherical lens. The light then penetrates a quarz plate that covers a polyacrylamide gel, in which the target proteins were previously separated electrophoretically. A bandpass filter suppresses both the excitation and background light, so that only the emitted native fluorescence is monitored by sensitive CCD camera. Native UV detection is as sensitive as the conventional → silver staining of proteins.

NAT pair: Any two → messenger RNAs that form → sense-antisense complexes. See → natural antisense transcript.

Natural antisense miRNA (nat-miRNA): Anyone of a series of naturally occuring → antisense → microRNAs. The precursors of many of these nat-miRNAs carry large → introns that are critical for nat-miRNA evolution and the formation of functional miRNA → loci. See → natural antisense transcript.

Natural antisense transcript (NAT): Anyone of a series of naturally occuring → antisense → messenger RNAs in pro- and eukaryotic organisms. NATs are able to form double-stranded RNAs with → sense transcripts and therefore function in the regulation of → pre-mRNA splicing, → alternative splicing, control of → translation, the degradation of target RNA ("turnover"), RNA stability and trafficking (the transport of mRNA from the nucleus into the cytoplasm), → RNA interference, → genomic imprinting, → X chromosome inactivation, or → RNA editing. At least 2500 human genes are also transcribed into the corresponding antisense variants. Changes in antisense transcription have been implicated in pathogenesis, such as cancer or neurological diseases. In maize, as example for plants, more than 70% of all genes are transcribed in both sense and antisense transcripts that tend to be inversely expressed. Frequently, NATs anneal to → 3′-UTRs. See → cis-NAT, → NAT pair, → trans-NAT.

Natural competence: The ability of many bacteria to take up DNA from their environment, which is dependent on socalled competence → regulons (e.g. the Haemophilus influenzae regulon is composed of 25 genes in 13 → transcription units). Such regulons are conserved in many related pathogenic bacteria including Escherichia coli and Vibrio cholerae, a fact indicative for a widespread competence to ingest DNA.

Natural gene transfer: See → indirect gene transfer.

Natural plasmid: Any plasmid which has not been constructed in vitro for cloning purposes. Natural plasmids described in this book are for example the → colicin factor; → F factor, → Dictyostelium discoideum, → Dictyostelium giganteum, → Dictyostelium mucoroides and → Dictyostelium purpureum plasmid; → pSC 101, → pMB 9, → resistance factor, → RP1, → two micron circle.

Natural selection: An evolutionary process, during which individuals carrying a distinct gene possess a greater fitness than those without this gene.

Natural transformation: The modification of the genome of a cell by the active uptake of free DNA from the environment and its integration into the recipient's genome. In nature, DNA is liberated from pro- and eukaryotic cells via autolysis or excretion (in case of bacterial cells, also by bacteriophage-induced lysis), and may accumulate to relatively high concentration in the soil or water (marine ecosystems: 50 mg/l). If this DNA appears in the environment, potential acceptor bacteria acquire → competence, interact with the foreign DNA, take it up and integrate it, most frequently via → homologous recombination (divergence of 10–20% is not tolerated). See → transformation (b).

NBD: See → SA cluster.

NC: See → nitrocellulose.

NCBI: See → National Center for Biotechnology Information.

NCGR: See → National Center for Genome Resources.

NC-MiC: See → human engineered chromosome.

ncRNA: See → non-coding RNA.

NcSNP: See → non-coding single nucleotide polymorphism.

NE (negative element): See → silencer.

Nearest-neighbor frequency analysis: See → nearest-neighbor sequence analysis.

Nearest-neighbor sequence analysis (nearest-neighbor frequency analysis): A method for the characterization of DNA molecules that is based on the estimation of the relative frequencies with which pairs of each of the four bases lie next to one another. Any deoxyribonucleotide can be covalently bound to any one of the three others or to a nucleotide of the same type by its 3′ or 5′ hydroxyl group to form a dinucleotide molecule. Since there are 4 different deoxyribonucleotides (dATP, dCTP, dGTP, dTTP), the formation of 16 dinucleotides is possible. The frequency with which each of these combinations occurs is characteristic for a particular DNA. It can be determined by incubating a DNA template with *E. coli* → DNA polymerase and the four deoxyribonucleotides, one of which is labeled with ^{32}P at the α (innermost) phosphate position. The α-^{32}P then links the labeled nucleotide with its nearest-neighbor nucleotide. After synthesis the isolated DNA is digested with → micrococcal nuclease and spleen → phosphodiesterase to yield deoxyribonucleoside 3′ monophosphates. The ^{32}P is now attached to the 3′ carbon atom of the neighboring nucleoside (see scheme). The four deoxyribonucleoside 3′ monophosphates are separated by paper electrophoresis and their radioactivity is measured. This measure gives the frequency with which the originally labeled nucleotide has been bound to the other nucleotides. By using all four α-^{32}P-labeled deoxyribonucleotides in repeated nearest-neighbor analyses the frequency of all 16 dinucleotides can be determined.

***Near-isogenic line* (NIL):** A genotype, usually derived from repeated backcrossing, which differs from another genotype by only one or a few genes.

***Nearly identical paralogs* (NIPs):** Any pair of paralogous genes that share >98% sequence identity. Approximately 1% of all maize genes have a NIP, and both genes are expressed, frequently to different extents. NIPs arise by local duplication events, but also through transposon-mediated duplications.

***Near-upstream element* (NUE):** A DNA sequence that is located 4–40 nucleotides upstream the → poly(A) signal site in many plant and some plant virus genes and functions as part of a → termination signal complex in → transcription. NUEs of different genes contain a different core sequence (5'-AAUAAA-3' in the → cauliflower mosaic virus 35S gene complex; 5'-AAUGAA-3' in the zein gene of maize; 5'-AAUGGAAUGGA-3' in the ribulose bisphosphate carboxylase/oxygenase gene of pea). See → far-upstream element.

Nebulization: A simple method to fragment high molecular weight DNA by passing it through a small hole of a device used for inhalation directly onto a plastic hemisphere, where it is broken and dispersed in the surrounding plastic tube. The higher the applied pressure, the smaller the fragments.

Necroptosis: A non-apoptotic programmed cell death patrhway that leads to a necrosis-like cell decay. Necroptosis can be induced by tumor necrosis factor-α (a ligand for the Fas/TNFR family) and a pan-caspase inhibitor. Compare → apoptosis.

***Negative cofactor* (NC):** Any nuclear protein that disturbs the cooperation between the different proteins of the → transcription initiation machinery and thereby leads to an inhibition or complete cessation of transcription initiation. For example, NC2 of *Saccharomyces cerevisiae*, highly conserved from yeast to man, binds as dimer to the → DNA already loaded with → TATA-box-*b*inding *p*rotein (TBP), and prevents the recruitment of the general → transcription factors → TFIIA and → TFIIB. Result: no initiation of transcription of the adjacent gene. See → positive cofactor.

Negative control: Any experimental control element that provides little or no signal or result, irrespective of the results obtained from the actual experimental components. For example, a negative control on an → expression microarray consists of e.g. the → cDNA from a foreign gene that is therefore not active in the test organism (e.g. a human gene in a plant microarray experiment). If total and fluorescence-labeled cDNAs of a test organism are hybridised to the array with these negative control cDNAs, they will not produce a signal (e.g. a → fluorescence signal) notwithstanding the reaction of the other cDNAs spotted onto the array. Negative controls are necessary for a test of the function of the array. See → positive control.

Negative design: A variant of the → computational protein design that detects and eliminates undesired properties of the designed amino acid chain.

Negative dominance: The dominance of a mutated → allele over the wild-type allele by e.g. stronger → transcription of the former.

***Negative element* (NE):** See → silencer.

Negative gene control: The termination of gene expression by the binding of a specific → repressor protein to → operator sites upstream of the coding region of many genes which prevents the simultaneous binding of RNA polymerase. See → inducible gene, → inducible operon, for example → *lac* operon.

Negative regulator: A molecule that turns off → transcription or → translation. See → negative cofactor.

Negative selection: A procedure for the isolation of → transformants, in which detection is based on the loss of one or more specific functions. For example, an → insertion of a DNA fragment into the coding sequence of a → selectable marker gene of a vector inactivates this gene (→ insertional inactivation). Transformants can therefore be selected by the absence of marker gene function.

Negative supercoiling: The coiling of a covalently closed circular DNA (→ cccDNA) duplex molecule in a direction opposite to the turns of its double helix (i.e. in a left-handed direction). Compare → positive supercoiling, → supercoil.

N-end rule: The correlation between specific N-terminal amino acids of a protein and its *in vivo* half-life time. The N-end rule is based on a degradation signal, coined N-degron that consists of a destabilizing N-terminal amino acid residue and an internal lysine (or lysines), where the substrate-linked polyubiquitin chain is anchored. The ubiquitinylated protein substrate is then degraded by the 26S proteasome. Three types of destabilizing residues exist. The tertiary class of destabilizing residues consists of N-terminal asparagines and glutamines that are deamidated by N-terminal amidohydrolases to the secondary destabilizing residues aspartic acid and glutamic acid. These in turn are conjugated to arginine (one of the primary destabilizing amino acids) by *ATE-1*-encoded arginine-tRNA transferases (R-transferases). These in turn recognized by ubiquitin ligases (E3 enzymes) of the socalled N-end rule pathway. N-end rule pathways exist in both eu- and prokaryotes. In prokaryotes (as in eukaryotes) N-terminal phenylalanine, leucine, tryptophan, and tyrosine are primary destabilizing residues that are recognized directly by ClpAP, an ATP-dependent proteasome-like protease. Secondary destabilizing residues of prokaryotes are N-terminal arginines and lysines that are conjugated to either phenylalanine or leucine by *aat*-encoded leucine/phenylalanine-tRNA-protein transferase (L/F transferase).

Neocentromere: An ectopic → centromere that originates from non-centromeric regions of chromosomes. The human genome harbors at least 50 such neocentromeres that all lack typical centromeric sequences as e.g. satellite repeats, but still bind CENP-A, assemble a functional kinetochore, and exhibit mitotic stability. Plant neocentromeres (e.g. from maize or rye) appear during meiosis rather than mitosis, occur on chromosomes with normal centromeres, and form terminal → heterochromatin → domains (socalled knobs) composed of → tandem repeats differing from repeats of regular centromeres.

Neocentromere-based human minichromosome: See → human engineered chromosome.

Neofunctionalization: The acquisition of a novel, beneficial function by a duplicated gene (see → gene duplication) in evolutionary times, which is preserved by natural selection. The gene copy with the original function is retained. See → nonfunctionalization, → subfunctionalization.

Neoisoschizomer: Any → isoschizomer that cleaves at a position different from its prototype (i.e. the first restriction endonuclease sample of this type isolated). See → neoschizomer.

Neomycin (Nm): A broad-spectrum antibacterial aminoglycoside → antibiotic from *Streptomyces fradiae* that binds to the 30S subunit of bacterial ribosomes and causes severe miscoding, inhibits initiation factor-dependent binding of fMet-tRNA and transpeptidation in pro- and eukaryotes, and blocks translocation. It is effective against a wide range of Gram-negative (e.g. *E. coli*) and most Gram-positive bacteria. See → neomycin resistance, → neomycin resistance gene, → neomycin sensitivity.

Neomycin

Neomycin phosphotransferase: See → aminoglycoside phosphotransferase.

Neomycin resistance (Nm^r): The ability of an organism to grow in the presence of neomycin, an → aminoglycoside antibiotic from *Streptomyces fradiae*. See also → neomycin resistance gene, → neomycin sensitivity.

Neomycin resistance gene (Nm^r gene): A gene (*neo*) from → transposon 5, → transposon 601, and from *transposon 903* that encodes an → aminoglycoside phosphotransferase (APH I and II, respectively). These enzymes phosphorylate neomycin and related aminoglycoside compounds, and inactivate them. The neomycin resistance genes can be ligated to eukaryotic → promoters and transferred to eukaryotic cells, where their expression leads to neomycin resistance of the host. The neomycin resistance genes can be used as dominant → selectable markers in bacteria, fungi, animal and plant cells.

Neomycin sensitivity (Nm^s): The inability of an organism to grow in the presence of → neomycin, an aminoglycoside antibiotic from *Streptomyces fradiae*. Compare → neomycin resistance.

Neoschizomer: Any one of a subset of → isoschizomers that recognize the same DNA sequence motif and bind there, but cleave at different positions from the → prototype (i.e. the first discovered example of the corresponding isoschizomer). So, *Aat*II (recognition sequence 5'-GACGT↓C-3') and *Zra*I (recognition sequence 5'-GAC↓GTC-3') are neoschizomers of one another, whereas *Hpa*II (recognition sequence 5'-C↓CGG-3') and *Msp*I (recognition sequence 5'-C↓CGG-3') are isoschizomers, but not neoschizomers. See → neoisoschizomer.

NeoY: A chimeric chromosome of *Drosophila miranda*, generated by a → fusion of autosome 3, containing about 2800 genes, with the original Y chromosome some 2 million years ago. Whereas the original Y chromosome is completely heterochromatic, the fused autosome 3 is only partially heterochromatic, and therefore is continuously eroded. It suffered a series of → point mutations, massive accumulation of → insertions, → deletions, and a → duplication of several kb in length. Most of the insertions represent → retrotransposons (e.g. *TRIM*, the *YSY4* insertion, and *TRAM*, the *YSY5* insertion) that are trapped by the suppression of → recombination. The genes are at least partially duplicated (e.g. the larval cuticle protein-encoding gene *Lcp2*), but also inactivated by mutations. The neoY chromosome of *Drosophila miranda* therefore allows to study the molecular events of → genetic erosion.

NEP:

a) See → naked eye polymorphism.
b) See → non-exon probe.
c) See → nucleus-encoded polymerase.

NES: See → nuclear export signal.

NEST: See → nuclear expressed sequence tag analysis.

Nested oligo procedure: See → nested primer polymerase chain reaction.

Nested on chip (NOC) polymerase chain reaction (NOC-PCR): A variant of the conventional → polymerase chain reaction technique that combines the advantages of → DNA chip technology (as e.g. parallelism, speed and automation) with the specificity and simplicity of liquid phase PCR in a single carrier system. In short, socalled → nested primers (i.e. oligonucleotides complementary to sequences within an → amplicon) are immobilized on a solid support (glass or plastic chip). Each one of these primers (P3) contains a specific nucleotide sequence characteristic for e.g. a polymorphism within a gene. Now the target sequence (e.g. an → exon of a gene of interest) is amplified in the liquid phase around the chip, using a specific primer pair (P1, P2) targeting conserved regions. The resulting amplification products will bind to the chip-bound primer P3 only if the 3' terminal base of P3 is complementary to the corresponding base in the amplicon. In this case, an amplification of primer P3 on the chip takes place, the amplification products are covalently anchored on the chip, and the non-covalently bound molecules are washed away. If the PCR reaction runs with → biotin-labeled nucleotides, the amplified products on the chip can be detected by → streptavidin-cyanine 5 conjugates and appear as fluorescent spots. NOC-PCR allows e.g. the discrimination between different alleles of socalled *human leucocyte antigen* (HLA) genes on e.g. so called → HLA chips. See → nested primer polymerase chain reaction.

Nested PCR: See → nested primer polymerase chain reaction.

Nested primer: Any → primer whose sequence is complementary to an internal site of a DNA that has been amplified with other primers in a conventional → polymerase chain reaction (PCR). Such nested primers are used to re-amplify the target sequence at sites different from the original primer sites and thereby increase the specificity of the amplification reaction.

Nested primer polymerase chain reaction (nested PCR, nested oligo procedure): A modification of the → polymerase chain reaction which improves the yield of specific target sequences. During normal PCR, genomic DNA is denatured and annealed with an excess of two oligonucleotide → amplimers which bind to sequences just up- and downstream of the target DNA. These amplimers are then extended using thermostable → DNA polymerases. The DNA is again denatured, annealed to the same oligonucleotides and extended in a second cycle. This procedure is repeated some 20–30 times. Since the polymerase reads beyond the target DNA, a population of fragments arises, the lengths of which exceed that of the target DNA. In order to reduce the PCR to the target DNA, a second set of amplimers ("nested oligos") is annealed to sequences within the target DNA. After 20–30 cycles of PCR from these new primers, only amplified target DNA accumulates.

Nested RAP-PCR: See → nested RNA arbitrarily primed polymerase chain reaction.

Nested RNA arbitrarily primed polymerase chain reaction (nested RAP-PCR): A variant of the → RNA arbitrarily primed polymerase chain reaction (RAP-PCR) for the detection of differential gene expression and the partial → normalization of the RNA fingerprint to → messenger RNA abundance. In short, total RNA is first isolated from two contrasting samples (e.g. normal and tumor tissue), and a 10–18-mer → primer of arbitrary sequence used to prime the → first strand, and a → nested primer with one, two or three additional arbitrary chosen → nucleotides at the 3'-end of the first primer employed for → second strand cDNA synthesis. The primers are labelled with P^{32} or P^{33}. In a particular example, these primers have the following sequences:

5'-CCACACAGAAACCCACCA-3'
5'-CACACAGAAACCCACCA**G**-3'
5'-ACACAGAAACCCACCA**GA**-3'
5'-CACAGAAACCCACCA**GAG**-3'

The mixture of cDNAs is then amplified by conventional high-stringency → polymerase chain reaction, the amplification products electrophoresed in 4–6% → polyacrylamide gels containing 50% urea, and the gel wrapped in plastic and autoradiographed. If RNAs from two tissues are used for comparison, tissue-specific fingerprints are produced. Differences between such fingerprints arise from differently expressed genes. Fingerprint bands of interest can be isolated from the gel, cloned, and used as a kind of → expression-tagged sites. See → differential display reverse transcription polymerase chain reaction.

Nested transposon: Any → transposon that inserted into another transposon of the same (or also different) type. Such nested transposons frequently form whole sets of transposons, covering large areas of a genome. The → insertion sites vary from insertion event to insertion event, which can be exploited to generate → insertion-site-based polymorphic markers.

NET: See → nuclear envelope transmembrane protein.

Network-attached storage (NAS): A specialized server attached to a local area network and using a streamlined operating and file system that is employed to extract data from a database ("capture") and serve files to clients. For a better performance, the NAS system can be combined with the → SAN system ("NAS-SAN combo").

Network segregation: A somewhat misleading term for the preference of distinct proteins to interact with other distinct proteins such that groups of interacting proteins are formed within a cell.

Neurogenetics: A branch of → genetics that focusses on the relationship(s) between genes and neuronal function(s) and disfunction(s) on the molecular level. Major research areas of neurogenetics are the development of diagnostic and therapeutic tools for hereditary diseases that afflict the nerve system. For example, a mutation of the *L1* gene (one of a series of genes encoding diverse proteins as e.g. L1, CHL1 [*close homologue of L1*], NrCAM and neurofascin that represent the socalled L1 family) leads to the socalled CRASH syndrome (symptoms are hydrocephalus and mental retardation). The function of this gene can be deciphered with → knockout mouse mutants and their anatomical and molecular analysis.

Neurogenome: The total number of genes expressed in both the central and peripheral nervous system at a given time. See → neurogenetics, → neurogenomics.

Neurogenomics: The whole repertoire of techniques for the identification, isolation and characterization of preferably all genes involved in the various functions of the central and peripheral nervous systems (see → neurogenome) and their mutant forms, especially if they cause neuronal disorders or simply changes in behaviour. Neurogenomics still experiments with animal (frequently mouse) models. See → neurogenetics, → neuroproteome, → neuroproteomics.

Neuron-*re*strictive silencing factor (REST): A vertebrate → transcription factor binding to the highly conserved 21 bp so called repressor element 1 (RE1, also known as neuron-restrictive silencing element, NRSE), to which it recruits various → histone-modifying and → chromatin-remodelling complexes, and thereby represses transcription of RE1-containing target genes. These include many genes necessary for terminally differentiated neuronal function, such as synapse formation (*SYN1*), neurotransmitter secretion (*SNAP25*) and signalling (*CHRM4*). REST is also necessary for the regulation of the voltage-gated calcium channel subunit gene *CACNA1H*, whose encoded protein mediates transduction of electrical signals into cellular responses in calcium signalling of normal heart function in mouse. The *CACNA1A* gene encoding Cav2.1 is highly expressed in Purkinje cells of the cerebellum, and mutations in *CACNA1A* are responsible for a number of cerebellar disorders including migraine, epilepsy and ataxias. Duplication of functional RE1s, principally located within or beside transposable elements (TEs), is widespread. The vicinity of TEs suggests transposon-mediated duplication as a mechanism of evolutionary expansion in the REST regulon. The greatest number of RE1s are located in the introns of genes, and intronic RE1s are rather uniformly distributed within their target genes. However, chromosome regions exist, where the RE1 density is markedly lower or higher than the corresponding gene density. For example, on chromosome 1, a particular gene-rich region at the tip of the p-arm is highly enriched for RE1s, while another region at the centromeric end of the q-arm is markedly void of RE1s. The distribution of RE1s in the human genome is therefore non-random. Some REST target genes contain pairs of RE1s arranged *in tandem* that probably recruit REST at even low concentrations, or simultaneously recruit multiple REST complexes to a target gene. Mutations and insertion of RE1s played important roles in vertebrate brain evolution.

Neuropharmacogenomics: A branch of → pharmacogenomics that uses the whole repertoire of → genomics, → transcriptomics, and → proteomics technologies to identify genes and/or mutations in genes involved in neurological disorders and to design and develop new drugs to control such diseases. See → oncopharmacogenomics.

Neuroproteome: The complete set of peptides and proteins expressed in the central and peripheral nervous system at a given time. See → neurogenetics, → neurogenome, → neurogenomics, → neuroproteomics.

Neuroproteomics: The whole repertoire of techniques to characterize the → neuroproteome in molecular detail. See → neurogenetics, → neurogenome, → neurogenomics.

Neurospora crassa: A haploid saprophytic Ascomycete fungus that has a 40 Mb genome harbouring about 10,000 genes, grows as a mycelium, and exists in two mating types. Fusion of nuclei from two different mating types ("karyogamy") is followed by meiosis and post-meiotic mitosis with the production of eight ascospores that are arranged linearly in the ascus. This arrangement allows the identification of the various products of meiotic divisions and renders *Neurospora crassa* an ideal organism for genetic studies. Among others, such studies led to the formulation of the "one gene-one enzyme" concept. Transformation of *N. crassa* is possible, and → shuttle vectors have been constructed for the transfer of genes between e.g. *E. coli* and *N. crassa* that increase the → transformation frequency by a factor of 10. *N. crassa* represents the model organism for ascomycetes for e.g. studies on light perception, circadian rhythm, and analysis of differentiation processes.

Neutral allele: Any → allele of a gene, whose expression does not contribute to a → phenotype. Compare → contributing allele. See → protective allele.

Neutral DNA: An infelicitous laboratory slang term for any DNA that does not contain genes.

Neutral insertion: The → insertion of a → nucleotide or → oligonucleotide into a coding sequence of a → gene without changing the function of the encoded protein. See → insertion mutation, → neutral mutation.

Neutral mutation: Any → mutation that has no selective advantage or disadvantage for the organism in which it occurs, for example a mutation in a → cryptic gene or other → non-coding DNA.

Neutral substitution: An exchange of one (or more) amino acid(s) in a protein without any change of its function.

Next generation screening (NGS, NGS™): A variant of the classical screening for → single nucleotide polymorphisms (or other mutations) that uses high-density glass → microarrays, onto which thousands of → PCR-amplified single loci or gene fragments of patients are immobilized. Each spot on such microarrays corresponds to a single → locus of a particular patient and contains the specific → allele of this patient. These NGS microarrays are hybridized with synthetic, fluorescently labeled, allele-specific → oligonucleotides complementary to the disease alleles, and the → hybridization event detected by laser excitation of the corresponding → fluorochrome. Three signal intensities identify healthy (weak signals), carrier (intermediate signals) and disease (strong signals) genotypes. The use of multiple fluorochromes (e.g. cyanin 5 and cyanin 3) allows the screening of samples from up to 10,000 different patients on a single NGS array and the screening for 12–20 disease loci, whereas the classical microarray format only permits to screen one patient per chip. NGS therefore determines the → genotypes of multiple patients in a single test, and is used for blood typing, HLA analysis, forensic medicine, and research into hereditary hearing loss and infectious diseases.

Next generation sequencing (NGS; next generation sequencing technology): A generic term for novel DNA and RNA sequencing technologies with the potential to sequence a human genome for 100.000, or even only 1.000 US $ that are not based on the conventional → Sanger (→ dideoxy) sequencing procedure. Next generation sequencing relies on extremely high throughput procedures, mostly based on massively parallel reactions, as e.g. in → sequencing by oligonucleotide ligation and detection (SOLiD™), where each run produces at least 40 million reads, covering 1 billion bases. Next Generation Sequencing technologies fall into two broad categories: → clonal cluster sequencing, and → single molecule sequencing.

NF-I: See → CAAT-box transcription factor.

NF-κB (nuclear factor κB): Any one of a series of mammalian → transcription factors that regulate the activity of diverse genes such as genes involved in the immune response, the control of cell growth, cell differentiation, and → apoptosis. The five members of the NF-κB family, p65/RelA, c-Rel, Rel B, p50/p105 (NF-κB1) and p52/p100 (NF-κB2) share the highly conserved Rel homology domain (RHD, or Rel homology region, RHR) responsible for DNA-binding, dimerization and interaction with IκB, and form various homo- and heterodimeric complexes. For example, the NF-κB p50-p65 dimer is induced by cytokines, lipopolysaccharides (LPS) and T cell activation signals. The interaction of dimers containing the p65 or c-rel NF-κB subunits with e.g. IκBα or IκBβ, also IκBε, p105 and p100, masks the → nuclear localization signal of NF-κB, so that these dimers accumulate in the cytoplasm and can no longer bind their cognate DNA sequences (e.g. the 10 bp sequence 5'-GGGACTTTCC-3' in the immunoglobulin κ light chain enhancer of B lymphocytes). A strong → nuclear export signal in IκBα further augments this effect. Several agents such as TNFα, IL-1 or LPS stimulate the phosphorylation of N-terminal serines (e.g. serines 32 and 36 in IκBα) of the inhibitory IκB molecules, which leads to their → ubiquitinylation and degradation by the → proteasome. Thereby the unmasked NF-κB can be translocated into the → nucleus, where it binds to the → enhancer or → promoter regions of target genes and activates their → expression. The kinase catalyzing the phosphorylation of IκB (IκB kinase, IKK) is activated in response to TNF, IL-1 or other stimuli and also promotes the phosphorylation of the p105 and p100 precursors, leading to the production of p50 and p52, respectively. IKKα also controls → histone H3 phosphorylation of NF-κB-dependent promoters, thus positively contributing to gene expression. The Rel/ NF-κB signalling pathway is deregulated in a variety of different cancers, especially human lymphoid cancer. Cells of this cancer type own mutations or amplifications of genes encoding NF-κB transcription factors. In most cancer cells, NF-κB resides in the nucleus and is constitutively active, which protects the cells from apoptotic cell death.

NFR: See → nucleosome-free region.

NGS:

a) See → next generation screening.
b) See → next generation sequencing.

N-ᶠMet: See → N-formylmethiorine.

NF-1: See → nuclear factor 1.

NF1/CTF: A sequence-specific → DNA-binding protein that recognizes 5'-ATTTT GGCTTGAAGCCAATATG-3' and represents an initiation factor for → adenovirus DNA replication.

N-formylmethionine (N-ᶠMet): A derivative of the amino acid methionine that carries a formyl group at its terminal amino group and functions as starter amino acid in bacterial polypeptide synthesis. Since N-formylmethionine lacks a free amino group it is "blocked" that is, it can only form a peptide bond at its carboxy terminus. Thus it can only be the first amino acid of a polypeptide but cannot be incorporated into the growing chain.

Methionine (Met)   N-formylmethionine (fMet)

N⁴-hydroxycytidine (NHC): A modified ribonucleoside analog that is a weak alternative substrate for the → RNA-dependent RNA polymerase of viruses and therefore is employed as anti-pestivirus and anti-hepacivirus agent.

NG: See → nitrosoguanidine.

NHGRI: See → National Human Genome Research Institute.

NHP: See → non-histone protein.

nh-plot: A two-dimensional diagram depicting the variation of number of residues per helical turn (n) and the axial rise per residue (h) of a regular DNA double helix as a function of a pair of internal variations of the chain.

Nic/bom region: See → bom region.

Nick: A break in one 5'-3' phosphodiester bond in one of two strands of a DNA duplex molecule. Compare → cut, → break. See also → nick translation.

Nickase: A general term for an enzyme that introduces nicks (single-stranded breaks) in DNA duplex molecules. See for example → nick translation.

Nick-closing enzyme: A synonym for → DNA topoisomerase I.

Nicked circular DNA: See → open circle.

Nicking: The introduction of → nicks into one strand of a double-stranded DNA molecule.

Nicking-closing enzyme: See → DNA topoisomerase I.

Nicking enzyme: Any one of a series of enzymes that induce a → nick (a → single-strand break) into a DNA → double helix and relaxes e.g. → supercoiled covalently closed circular DNA (→ cccDNA). Nick enzymes are involved in e.g. → rolling circle amplification and → conjugative plasmid transfer ("relaxases").

Nick translation: The replacement of nucleotides in double-stranded DNA by radioactively labeled nucleotides using the nicking activity of → DNase I and the polymerizing activity of E. coli → DNA polymerase I. In short, → nicks are introduced into the unlabeled ("cold") DNA duplex molecule by a limited digestion with DNase I to generate 3'-OH termini. Then E. coli DNA polymerase I is added that starts a DNA replication reaction at the 3' hydroxy terminus of each nick and simultaneously removes nucleotides from the 5' side (5'-3' exonuclease activity of E. coli DNA polymerase I), thus extending the nick. Since at least one of the four nucleotides needed for the reaction is labeled (for example α-^{32}P deoxynucleotide triphosphates), the original nucleotides in the duplex DNA molecule are replaced by labeled nucleotides. In the case of radioactive labeling a probe with high specific

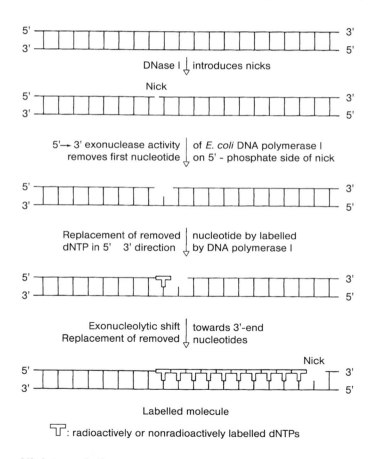

Nick translation

activity is generated that can be used in hybridization experiments. It is however also possible to use non-radioactively labeled nucleotides (see → non radioactive labeling). Compare → random priming.

***Nif* genes (nitrogen-fixation genes, nitrogen-fixing genes):** A set of about 20 genes organized in an → operon in the genome of certain soil bacteria, notably *Rhizobium* that encode subunits of nitrogenase and other enzymes catalyzing the fixation of atmospheric nitrogen (N_2) into ammonia NH_4^+ and nitrate (NO_3^-), a process called nitrogen fixation. For example, in *Klebsiella*, a cluster of 20 *nif* genes is organized in eight co-ordinately regulated operons. In *Bradyrhizobium japonicum* (a symbiont of soybean), the *nif* genes are arranged in → clusters. Cluster 1 contains genes *nif* D, *nif* K and *nif* H (encoding components I [molybdoferredoxin] and II [azoferredoxin] of nitrogenase), genes *nif* B, *nif* E, *nif* N (encoding proteins for the biosynthesis of the FeMo coenzyme), and gene *nif* S (the gene product stabilizes the structure of the nitrogenase complex). Cluster 2 harbors *nif* A (encodes an activator of the expression of the *nif* genes).

NIH Guidelines (National Institutes of Health Guidelines): A compilation of recommmended security measures (see for example → containment) in recombinant DNA experiments, obligatory for laboratories which are funded by NIH grants. These guidelines provided the basis for the establishment of similar guidelines in countries other than the US.

NIL: See → *near-isogenic line*.

9-β-D-ribofuranosyladenine: See → adenosine.

NIPs: See → nearly identical paralogs.

Nitrocellulose (NC; cellulose nitrate): A nitrated cellulose whose fibers can be used for the production of membrane filters, see → nitrocellulose filter.

Nitrocellulose filter (NC filter, cellulose nitrate filter): → Nitrocellulose fibers in the form of membranes with defined pore size (e.g. 0.45 µm). These filters selectively trap dsDNA or DNA RNA hybrids, but no single-stranded molecules. Single-stranded DNA or RNA may, however, be fixed to NC filters by → baking. Such blots can then be used in → Northern or → Southern blotting experiments.

Nitrogen-fixation genes: See → *nif* genes.

Nitrogen-fixing genes: See → *nif* genes.

Nitrosoguanidine (NG): An alkylating mutagenic chemical that adds methyl groups to many positions of all four bases in DNA, notably onto the oxygen at position 6 of guanine (leading to O^6-alkyl guanine). This causes mispairing with thymine and principally results in GC → AT transitions in subsequent rounds of replication. Example:

N-Methyl-N'-nitro-N-nitrosoguanidin

NLS: See → *nuclear localization signal*.

NLS receptor: See → *nuclear localization signal receptor*.

Nm: See → neomycin.

NMD: See → nonsense-mediated messenger RNA decay.

N-Methyl*ant*hraniloyl nucleotide (mant nucleotide): Any ribose-modified → nucleotide that contains the → fluorophore N-methylanthraniloyl linked to the 2' or 3' carbon of the sugar moiety. Mant nucleotides resemble natural nucleotides in their protein-binding properties and are therefore used for the sensitive detection of conformational changes in a nucleotide-binding protein, protein-protein- and protein-ligand-interactions.

2', 3'-O-(N-Methylanthraniloyl)-adenosine-5'-triphosphate

Nm^r gene: See → neomycin resistance gene.

Nm^s: See → neomycin sensitivity.

NOB: See → SA cluster.

NOC-PCR: See → nested on chip polymerase chain reaction.

Node: A specific point on a strand of supercoiled DNA, where it loops back and contacts itself.

NOGD: See → *n*on-*o*rthologous *g*ene *d*isplacement.

Noise: A laboratory slang term for → background.

Nomadic gene: See → jumping gene.

Non-aligned segments: Regions within various → homologous genes (or proteins) that cannot be aligned (i.e. do not share → homology).

Non-allelic homologous recombination (NAHR): Any → cross-over with subsequent genetic → recombination mediated by DNA base → mispairing, and resulting in → duplication and/or → deletion of DNA sequences.

Non-annotated expressed gene (NAE): Any one of a class of genes, for which either a tag (e.g. → SAGE or → SuperSAGE tag), an → *e*xpressed *s*equence *t*ag (EST) or a → cDNA is present in an organism's → transcriptome, but whose sequence has not been identified as coding in a sequenced genome of an organism. Most frequently NAEs reside in inter-genic regions. See → annotated expressed gene, → annotated non-expressed gene.

Nonautonomous controlling element: A defective → transposon that can transpose only with the aid of a second, autonomous element of the same type.

Non-autonomous sequence element: Any one of a series of genomic sequences that do not function or move by themselves (autonomously), but require the assistance of other sequences ("helper sequences") for function or movement.

Non-bi-directional cluster: See → SA cluster.

Non-canonical amino acid: Any one of a series of synthetic amino acid analogues that can be incorporated into proteins and change their physico-chemical properties. For example, selenomethionine (surrogate for methionine), olefinic and acetylenic methionine anlogues, p-chlorophenylalanine, bromo-, iodo-, azido-, cyano- and ethynyl substituents, p-fluoro-phenylalanine, p-aminophenylalanine (p-NH$_2$-Phe), p-acetylphenylalanine, O-methyl-L-tyrosine, aminobutyrate (surrogate for valine), trifluoroisoleucine and hexafluoroleucine, 4-, 5- and 6-aminotryptophan, β-selenol [3, 2-β]-pyrrolyl-alanine (surrogate for tryptophan) and azidohomoalanine are such non-natural amino acids. The incorporation of non-canonical amino acids is a means to selectively and globally label proteins for e.g. nuclear magnetic resonance (NMR) studies (as e.g. monitoring domain-domain interactions and conformational changes through the incorporation of ^{19}F-tryptophans). Moreover, the replacement of hydrophobic amino acids by their fluorinated counterparts can substantially stabilize hydrophobic folded regions. An increase in protein stability towards thermal and chemical → denaturation can be achieved by an increase in the extent of fluorination through incorporation of e.g. trifluoro- or hexafluoroleucine. Fluorinated amino acids also control protein-protein interactions. Introduction of 4-aminotryptophan into the two tryptophan positions of the → enhanced cyan fluorescent protein (ECFP) of *Aequorea victoria* shifts the emission maximum 69 nm beyond the most red-shifted mutant protein known. The engineered → autofluorescent protein is coined "gold fluorescent protein" (GFP), is more thermostable and less susceptible to aggregation, but suffers from quantum yield as compared to the wild-type protein. In principle, substitution of any one of the 20 natural amino

Non-canonical amino acids

acids in proteins by non-canonical amino acids expands the → genetic code. See → non-canonical amino acid.

Non-classic nuclear localization signal (ncNLS): A sequence of mainly positively charged amino acids in proteins necessary for their translocation from cytoplasm into the nucleus that binds directly to members of the importin β family nuclear receptors (and does not depend on an importin α-adaptor protein (as e.g. the → classic nuclear localization signal). The adapter-independent ncNLS-containing proteins encircle the → transcription factors CREB, Fos, and Jun, the retroviral proteins Rev and Tat of HIV-1, → ribosomal proteins L5 and L23a, the core → histones, and Smad-3.

Non-coding DNA: Any DNA that does neither encode a polypeptide nor an RNA. Non-coding DNA is a major constituent of most eukaryotic genomes, and includes → introns, → spacers, → pseudogenes, → centromeres, and most → repetitive DNA.

Non-coding exon (non-coding first exon): Any → exon that has no coding function (i.e. whose sequence does not contribute to the amino acid sequence of a protein). For example, 16 kb downstream of the → prion protein-encoding *Prnp* gene of mice a second *Prnp*-like gene, called *Prnd* and encoding the Dpl protein (German: doppel, for double) is located. These two genes are separated by an intergeneic space containing two intergene exons with no coding function(s). The term is also used for any exon that is present in the → pre-messenger RNA, but excised and skipped such that its information does not appear in any protein. See → coding exon.

Non-coding first exon: See → non-coding exon.

Non-coding RNA (ncRNA, non-protein-encoding RNA, non-protein-coding RNA, npcRNA): Any → ribonucleic acid that does not encode a protein and can therefore not be annotated by a search for → open reading frames. For example, → microRNAs, → ribosomal RNAs, → 7SL-RNAs, → small nuclear RNAs, → small nucleolar RNAs, → small interfering RNAs, → small temporal RNAs, → telomerase RNAs, → transfer RNAs, → Xist-RNAs are such ncRNAs.

Non-coding RNA gene: Any one of a series of genes encoding → non-coding RNA (ncRNA) that in turn does not encode a protein and can therefore not be annotated by a search for → open reading frames. In the human genome, such non-coding RNA genes encode at least 497 → transfer RNAs, 150–200 each of 18SrRNAs, 28SrRNAs, 5.8SrRNAs and 5SrRNAs (see → ribosomal RNA), 16 U1, 6 U2, 4 U4, 1 U5, 44 U6, 1 U7 and 1 U12 → small nuclear RNAs (snRNAs), 1 → telomerase RNA, 1 → Xist RNA, 69 C/D → small nucleolar RNAs (snoRNAs) and 15 H/ACA snoRNAs. See also → microRNAs, → 7SL-RNAs, → small interfering RNAs, → small temporal RNAs.

Non-coding sequence (NCDS): Any → DNA sequence that does not encode an → RNA or a → protein, as opposed to → coding sequences (e.g. a gene). Major NCDSs in eukaryotic genomes are → microsatellites, → minisatellites, → repetitive DNA, → retrotransposons, → satellites.

Non-coding single nucleotide polymorphism (ncSNP): A misleading term for any → single nucleotide polymorphism that occurs in a non-coding region of the genome (e.g. an → intron). NcSNPs are the most frequent types of SNPs in eukaryotic organisms. See → anonymous SNP, → candidate SNP, → coding SNP, → copy SNP, → exonic SNP, → human SNP, → intronic SNP, → non-synonymous SNP, → reference SNP, → regulatory SNP, → synonymous SNP.

Non-coding transcription: A laboratory slang term for the → transcription of genes into RNAs ("non-coding RNAs") that are not translated into proteins. Such → non-coding RNAs (ncRNAs), or non-protein-encoding RNAs, and non-protein-coding RNAs, npcRNAs are, for example, → microRNAs, → ribosomal RNAs, → 7SL-RNAs, → small nuclear RNAs, → small nucleolar RNAs, → small interfering RNAs, → small temporal RNAs, → telomerase RNAs, → transfer RNAs, and → Xist-RNAs, to name few.

Non-cohesive end: See → blunt end.

Non-conjugative plasmid (non-selftransmissible plasmid): Any → plasmid that does not contain all functions necessary for its own intercellular transmission by → conjugation (e.g. lacks the → tra genes).

Non-contact spotting (non-contact printing): The deposition of target oligonucleotides, → cDNAs, DNAs, peptides or proteins on solid supports ("chips") of glass, quartz, silicon or nitrocellulose by an electrically induced discharge of the solution from the pin onto the surface of the chip. The pin does not come into physical contact with the solid support. See → contact spotting.

Non-contiguous translation: The relatively rare → translation of a → messenger RNA, during which part of the message are skipped. For example, during translation of the message derived from bacteriophase T4 gene 60 about 50 nucleotides are skipped.

Non-conversion: A laboratory slang term for the inability of bisulfite to convert all methylated cytosines in a DNA to uracil. Non-conversions result from non-denaturation, are drawbacks in → bisulphite genomic sequencing and lead to reading artefacts.

Non-covalent protein delivery (peptide-mediated non-covalent protein delivery): A technique for the introduction of peptides or proteins into eukaryotic cells, using short synthetic peptides as transient carriers, which dissociate from the cargo protein after crossing the plasma membrane. For example, the 21 amino acid long peptide Pep-1 consists of a hydrophobic, tryptophan-rich motif (targeting the cell membrane and interacting with proteins hydrophobically), a hydrophilic lysine-rich domain derived from the → Simian virus 40 large → T antigen → nuclear localization sequence (that improves intracellular delivery of the peptide vector), and a spacer separating both. The peptides or proteins associate with Pep-1 through non-covalent hydrophobic interactions and form stable complexes, in which each protein is interacting with many Pep-1s (e.g. a 30 kDa → green fluorescent protein is complexed with 12-14 Pep-1 molecules). Once inside the cell, the Pep-1 and cargo rapidly dissociate ("decaging"), and the cargo can then translocate to its proper intracellular compartment. Proteins of up to 500 kDa and whole protein-DNA complexes can be rapidly delivered by this non-covalent protein delivery process.

Non-degenerate code: Any code in which the information is written in one specific sequence of symbols. In molecular biology, the genetic code is non-degenerate, if only one → codon specifies one amino acid.

Non-disjunction: The phenomenon that homologous chromosomes or sister chromatids do not separate at meiosis or mitosis, which leads to the formation of aneuploid cells.

Non-essential gene: Any gene that is not necessary for the survival, or, in another version for the fertility of an organism.

Non-exon-overlapping bi-directional cluster: See → SA cluster.

Non-exon probe (NEP): Any one of tens of thousands of 36 nucleotides long → oligonucleotide probes on a → microarray that is complementary to → intronic or → intergenic regions. Such oligonucleotides are synthesized on a glass substrate by e.g. → maskless array synthesis (MAS) and hybridized to → cDNA labeled with a → fluorochrome to determine the expression status of the underlying sequences. See → exon probe, → splice junction probe.

Nonfunctionalization: The prevention of an acquisition of a novel and beneficial function of a duplicated gene (see → gene duplication) by degenerative mutations. See → neofunctionalization, → subfunctionalization.

Non-functional polymorphism: Any sequence → polymorphism that has no consequences for the function of a protein and is therefore selectively neutral. Compare → functional polymorphism. See → intronic single nucleotide polymorphism, → non-coding single nucleotide polymorphism.

Nongenic DNA: The non-coding part of a → genome, mainly consisting of → microsatellites, → minisatellites, → retrotransposons, → satellite-DNA, → transposons, and in eukaryotes varying from about 3.0×10^6 to 1.0×10^{11} bp.

Non-histone protein (NHP): Any one of a large group of mostly acidic nuclear proteins of eukaryotes. These proteins serve enzymatic functions (e.g. → DNA and → RNA polymerases, → DNA methylases, RNA → processing enzymes), transport functions (e.g. RNA-binding proteins), regulatory functions (e.g. → transcription factors and → high mobility group proteins) and structural functions (e.g. → nuclear lamins).

Non-homologous end-joining (NHEJ): A mechanism ("pathway") for the repair of → double-strand breaks (DSBs) in DNA that rejoins the two ends of this break. NHEJ frequently leads to error-prone repair of DSBs, because the ends are only incompletely processed. Non-homologous end joining is catalysed by the concerted action of ligase IV, Xrcc4, Ku70 and 80, DNA-PKcs, Artemis and Nej1/Lif2 in rodent cells. See → homologous recombination, → single-strand annealing.

Nonhomologous random recombination (NRR): The random → recombination of DNA fragments in a length-controlled manner without the need for sequence → homology. For example, NRR is used to evolve DNA → aptamers that bind → streptavidin. Aptamer development starts with two parental sequences of modest affinity towards streptavidin, and repeated cycles of NRR evolve aptamers with 15- to 20-fold higher affinity. Therefore, NRR enhances the effectiveness of nucleic acid evolution. See also → error-prone PCR, → systematic evolution of ligands by exponential enrichment (SELEX).

Non-homologous recombination: See → illegitimate recombination.

Non-homologous synapsis: The indiscriminate association of non-homologous chromosomes during meiosis. Normally, only homologous chromosomes pair with each other, assisted by proteins of the socalled synaptonemal *c*omplex (SC). However, in certain mutants, the homolog pairing is not functioning. For example, in maize the *poor homologous synapsis* (*phs*) 1 gene encodes a protein that coordinates chromosome pairing, recombination and synapsis. A simple mutation in the gene results in the synthesis of a mutated PHS1 protein that fails to form chiasmata. Non-homologous synapsis leads to a random segregation of chromosomes.

Non-ionic detergent: A → detergent with an uncharged hydrophilic head-group that may be used to solubilize membrane proteins without their denaturation. Non-ionic detergents are for example the Tritons (see → Triton X-100), and octyl glucoside.

Non-linear splicing: See → rearrangements or repetition in exon order.

Nonliving array (chemical array): A polyethylene support or cellulose membrane, on which peptides or proteins are systematically arranged for high-throughput screening of oligonucleotide-protein, protein-protein, or protein-ligand interactions. The peptides can be synthesized on the polyethylene matrix by a *fl*uorenyl-*m*eth*o*xy*c*arbonyl (Fmoc) amino acid protection technique in a C- to N-terminus direction, the side chains and → α-amino groups being protected between consecutive cycles. Similarly, peptides can be synthesized on the cellulose membranes ("spot synthesis"), except that the hydroxyl groups of the cellulose can be derivatized by Fmoc-β-alanine groups, and the peptide arrays be synthesized via the cellulose-bound alanine (subsequent to its deprotection). The array size (= number of bound peptides per area unit) can be increased substantially by the combination of solid-phase synthesis with photolithographic techniques. For example, photolabile protective groups such as *n*itro*v*eratryl*o*xy*c*arbonyl, NVOC) on the growing peptide chain are selectively removed by light passing through a mask, similar to masks used for oligonucleotide synthesis (see → photolithography, → DNA chip). Nonliving arrays can be screened for e.g. chemical reactivity with low molecular weight ligands (e.g. pharmaceutically interesting compounds and their derivatives), or interactive peptides, proteins, RNAs, or oligonucleotides. See → living array.

Non-LTR retrotransposon: Any autonomous → retrotransposon that lacks → long terminal repeats. For example, the → LINE-1 or L1 retrotransposons are non-LTR retrotransposons. Typical full-length retrotransposons of this class are 4–6 kb in length and usually possess two → open reading frames, one encoding a nucleic acid-binding protein, the other one an → endonuclease and a → reverse transcriptase. L1s consist of a 5′-untranslated region (→ 5′-UTR), containing an internal →

promoter, the two → open reading frames (ORFs), a 3'-UTR, and a → poly(A) signal followed by a → poly(A) tail (A_n). L1s are usually flanked by 7–20 bp → target site duplications (TSDs). Some non-LTR retrotransposons integrate at specific sites in the host genome (e.g. R1 and R2 of *Drosophila melanogaster* insert at specific → ribosomal RNA gene locations), others insert at a very large number of genomic sites (e.g. mammalian elements, whose endonuclease prefers to cleave DNA at the short consensus sequence 5'-TTT/A-3'). Non-LTR retrotransposons are major drivers of host genome evolution.

Non-Mendelian inheritance: See → cytoplasmic inheritance.

Non-nuclear gene: Any gene that is localized outside of the nucleus in a eukaryotic cell. For example, chloroplast genes (in plants) and mitochondrial genes (in plants and animals) are such non-nuclear (organellar) genes. See → nuclear gene.

Non-nucleosomal histone: Any → histone protein that is not part of a → nucleosome, but exists free in the cytoplasm or nucleus in a so called histone pool.

Non-orthologous gene displacement (NOGD): The replacement of a gene encoding a protein with a particular function by a non-orthologous (unrelated, or distantly related), but functionally analogous gene during evolution.

Non-overlapping code: A → genetic code that specifies only as many amino acids as are triplets arranged in linear sequence. For example, the sequence UUUCCCUUU encodes only phenylalanine (UUU), proline (CCC) and phenylalanine (UUU). Compare → overlapping genes.

Non-overlapping FRET pair: Any pair of → fluorochromes, whose emission spectra do not overlap, but can nevertheless be used for → fluorescence resonance energy transfer (FRET) experiments. Normally, FRET between two fluorophores occurs only, if the emission spectrum of the socalled donor overlaps the excitation spectrum of the acceptor fluorophore. However, also non-overlapping FRET pairs can be employed for such experiments, except that both the donor and acceptor have to come into close vicinity to each other. The excited fluorophore then transfers the energy to the acceptor ("quencher"), and no photons are emitted.

Non-overlapping natural antisense transcript (NOT): Any → natural antisense transcript (NAT) with a sequence completely different from the corresponding → sense RNA transcribed from the same → locus.
    See overlapping *cis*-natural antisense transcripts (*cis*-NATs)

Non-palindromic cloning: The use of recombinant DNA techniques to propagate a DNA sequence inserted into non-complementary (non-palindromic) → cloning sites of a → cloning vector (→ non-palindromic vector). Non-palindromic sites on the vector can be generated by the ligation of non-palindromic → linkers to the termini of a linearized vector molecule. Non-palindromic cloning prevents the self-ligation of the vector molecules and the concatemerization of linkers. Thus dephosphorylation of the vector termini, and additionally any methylation or cutback steps, are superfluous. Compare → linker tailing.

Figure see page 1061

Non-palindromic vector: A → cloning vector that carries non-complementary

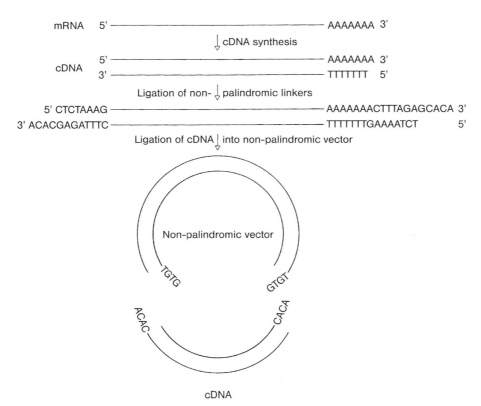

Non-palindromic cloning

(non-palindromic) termini at the cloning site. Such vectors allow → non-palindromic cloning.

Nonpenetrance: The absence of → expressivity of an → allele. See → complete penetrance, → incomplete penetrance.

Nonpermissive cell: A cell in which a particular virus cannot produce progeny viruses, because it is not capable to complete DNA replication (→ abortive infection).

Nonpermissive condition: A condition that does not allow the survival of a → conditional lethal mutant.

Nonprocessed pseudogene: Any one of two (or more) → pseudogenes originating from a common ancestor either by → gene duplication or, less frequently, by → unequal crossing over that may or may not be functional. Examples for the functionality of such pseudogenes are the genes encoding the hormone renin (*Ren1* and *Ren2*). Some mouse strains (as well as humans) own only one single copy of Ren1, expressed primarily in the kidney and, at a very low level, in the submaxillary glands, whereas other mouse strains contain the closely linked *Ren2* gene that encodes an isoform of rennin in submaxillary glands only. Both genes originate from an ancestral gene by → tandem duplication, but have different promoters.

Nonprocessive transcription: Any gene → transcription, whose → initiation occurs normally, but with inefficient → elongation. The transcription complex pauses and is rapidly released from the template, leading to an accumulation of short, non-polyadenylated RNAs, and only rarely full-length → messenger RNAs. See → processive transcription.

Nonproductive base-pairing: The imperfect pairing of bases in DNA that are not complementary to each other and therefore cannot form hydrogen bonds. For example, A-A, A-G, A-C, G-A, G-G, G-T, C-C, C-A, C-T, T-G, T-C, and T-T are such nonproductive base pairs, which e.g. destabilize hybrids and reduce the melting temperature of a hybrid through reducing the force of interaction between the two strands. See → productive base-pairing.

Non-productive infection: See → abortive infection.

Non-protein-coding RNA: See → non-coding RNA.

Non-protein-encoding RNA: See → non-coding RNA.

Non-radioactive labeling (chemical labeling): The introduction of nonradioactive groups into a DNA duplex molecule by → nick translation, → random priming, or → endlabeling. The introduced chemical compound (e.g. → biotin, → digoxigenin) can be detected by either colorigenic or luminogenic techniques (see → DNA detection system). See also → sulfonated DNA detection.

Nonribosomal peptide synthesis (NRPS): The synthesis of short peptides between two and 48 amino acid residues, most of them antibiotics (e.g. actinomycin D, bacitracin, bleomycin, cephalosporins, cyclosporins, erythromycin, penicillins or vancomycin) on non*r*ibosomal *p*eptide *s*ynthetases (NRPSs) of actinomycetes, bacilli, and filamentous fungi that consist of iterated modules. Each module activates a specific amino acid through a pair of closely coupled domains: a 550 amino acid *a*denylation domain (A domain) produces an *a*mino*a*cyl-O-*a*denosine *mo*nop*h*osphate (aa-O-AMP) that is then covalently tethered in a thioester linkage to the *p*hos*p*ho*pan*tetheinyl prosthetic group (ppan) of the neighboring 80 amino acid long thiolation domain (T; also peptidyl carrier protein, PCP). The peptidyl chain grows directionally in incremental steps of elongating acyl-S-enzyme intermediates. Peptide bond formation and chain translocation occur each time an upstream donor peptidyl-S-pantetheinyl is attacked by a downstream acceptor aminoacyl-S-pantetheinyl nucleophile. This process is catalysed by the 450 amino acid long socalled condensation domain (C), and terminated by a special termination domain (Te) that releases the peptide and sometimes induce it to form a loop (that never occur in proteins synthesized on cytoplasmic ribosomes). Some synthetases are multimeric complexes, others are single massive proteins. For example, the cyclosporin synthetase is composed of 11 modules comprising 15,281 amino acid residues (1,7 MDa). The resulting peptides can be linear, or adopt nonlinear (e.g. heterocyclic) and iterative configurations (e.g. branched peptide backbones), and include unusual amino acids (e.g. D-amino acids, methylated variants of the standard amino acids, nonproteinogenic, hydroxylated and glycosylated residues, of which more than 300 are known). The incorporation of D-amino acids requires the epimerisation domain (E domain, 450 amino acids) that catalyzes the racemisation of the C-terminal amino acid and transforms it from the L- into the D-enantiomer, until an equilibrium is reached. For an reactivation of the NRPS, the linear peptide has to be cleaved from the terminal ppan cofactor. Cleavage in most of the cases is catalyzed by C-terminal *thioesterase* (TE) domains, and follows two different pathways. One leads to a linear product via a hydrolytic cleavage (examples: bleomycin, vancomycin, and the penicillin precursor ACV), the other one to a cyclic molecule via intramolecular cyclization (examples: bacitracin and surfactin). Most of the modifications are performed during synthesis of the peptide, others are added postsynthetically. In each case, no genetic code and no ribosomes are involved. Probably, functional enzymes can also be synthesized by NRPS. For example, the 60 amino acids enzyme LPXTGase that cleaves the socalled LPXTG motif of many bacterial cell surface proteins (a prerequisite for an attachment of these proteins to the cell's surface), contains about 30% of unusual amino acids (not accepted by ribosomes). Therefore, this enzyme is most likely synthesized by NRPs. NRPs can also be rearranged by a cut-and-paste process to produce new combinations (i.e. new proteins). For example, the A subunit recognizing the amino acid leucine and the T subunit joining this leucine to the growing peptide chain can be replaced by A and T subunits specifically recognizing and joining other aminio acids. See → polyketide, → ribosome. The abbreviation NRPS stands also for nonribosomal peptide synthetase.

Non-selective polymerase chain reaction (NS-PCR): A variant of the conventional → polymerase chain reaction that allows to construct high quality → cDNA libraries employing sequence independent → primers. In short, → polyadenylated RNA is isolated, reverse-transcribed by → reverse transcriptase primed by oligo(dT), and the first strand cDNA is oligo(dC)-tailed at its 3′ terminus using → terminal transferase. Then the mRNA template is removed by → RNase H, and the resulting single-stranded cDNA amplified by → *Taq* DNA polymerase using an oligo(dT) primer complementary to the original poly(A) tail (e.g. 5′-GGGGCTCGAG [T_{16}]-3′), and an oligo(dC) primer complementary to the oligo(dC) tail (e.g. GGGGAATTC[G_{11}]-3′). Both primers contain an *Eco*RI → restriction site at their 5′-termini to facilitate subsequent cloning. The cDNA libraries obtained with NS-PCR are usually representative, i.e. contain each cDNA sequence at least once.

Non-selftransmissible plasmid: See → non-conjugative plasmid.

Nonsense-associated alternative splicing (NAS; nonsense-associated altered splicing): A relatively rare intranuclear → splicing process initiated by reading frame-sensitive recognition of → premature termination codons (PTCs) in certain → messenger RNAs (mRNAs) during translation. For example, the exon encoding the hypervariable VDJ region of human *T* cell receptor b (TCR-b) mRNA is a result of gene rearrangements, which probably generates PTCs. NAS leads to the accumulation of the corresponding pre-mRNA, the increased use of potentially alternative, but normally latent → splice sites, and reduced normal splicing of the PTC-containing mRNA. Alternatively, mutation of → exonic splicing enhancer (ESE) sequences also leads to the use of unusual splice sites by the spliceosome, because ESEs are targets for proteins defining pre-mRNA splice sites, and any mutation in such ESEs compromises splicing. See → nonsense-mediated messenger RNA decay, → nonstop messenger RNA decay.

Nonsense codon: Synonymous with → stop codon, see also → nonsense mutation.

Nonsense-mediated mRNA decay (Nonsense-mediated decay, NMD): The destruction of eukaryotic → messenger RNAs (mRNAs) containing → frameshift or → nonsense mutations that would otherwise lead to the synthesis of truncated and thus non-functional proteins. All mRNAs are first monitored for errors that would encode potentially deleterious proteins ("RNA surveillance"). During their exit from the nucleus to the cytoplasm, they are recruited for NMD by the shuttle protein Upf3p (in yeast), if they cannot be translated along their full length. In this case they will remain in a transition complex (i.e. associated with mRNP proteins and Upf3p), which triggers their decay. First, Upf3p forms a binary Upf3p-Upf2p complex ("recruitment complex"), then a transient bridge between recruitment and termination complexes (mediated by Upf1p in yeast). Finally, Upf1p-associated ATP-dependent 5′ → 3′ RNA/DNA helicase activity unwinds the faulty RNA in the 5′ → 3′ direction and induces a topology change that exposes the 5′ → cap, making it accessible to the decapping enzyme Dcp1p. Once decapped, the mRNA is fully degraded by Xrn1p from the 5′ end. NMD requires active → translation. Without NMD or similar processes (see → non-stop messenger RNA decay), the eukaryotic cell would produce truncated and most probably non-functional proteins. See → nonstop messenger RNA decay.

Nonsense mutation: Any mutation in a coding sequence that converts a sense codon into a nonsense codon (a → stop codon) or a stop codon into a sense codon. As a consequence, the encoded protein will either be truncated (premature termination) or too long which in turn hampers or abolishes protein function. See also → nonsense suppression, → amber mutation, → ochre mutation and → opal mutation.

Nonsense suppression: A secondary mutation occurring at a chromosomal site separate from the site of a nonsense mutation and correcting the phenotype associated with the latter. See for example → suppressor gene, also → suppressor mutation.

Nonsense suppressor: A → tRNA that is mutated in its → anticodon and recognizes a nonsense (→ stop) codon so that the synthesis of a specific polypeptide can be extended beyond the stop codon. As a consequence, the nonsense codon is ignored (suppressed).

Non-specific cross-hybridization: Any → hybridization occurring between two (or more) DNA sequences (e.g. on a → microarray) that do not share significant sequence similarity. Non-specific hybridization is a potential source for errors in → expression array experiments. See → specific cross-hybridization.

Non-specific transduction: See → transduction.

Nonstop decay: See → nonstop messenger RNA decay.

Nonstop messenger RNA decay (non-stop decay): A cellular process that eliminates eukaryotic messenger RNAs that do not possess → termination codons. Such mRNAs ("non-stop mRNAs") are degraded by the → exosome, a highly conserved complex of 3′ 5′-exonucleases. Compare → nonsense-mediated messenger RNA decay.

Nonstop transcript ("nonstop messenger RNA"): A laboratory slang term for a → messenger RNA that does not contain any → stop codon. Such transcripts are usually labile and removed by → nonstop messenger RNA decay.

Non-synonymous sequence change: Any alteration in the nucleotide sequence of a coding region that changes the amino acid sequence (and possibly the function) of the encoded protein. See → non-synonymous single nucleotide polymorphism, → synonymous sequence change.

***N*on-synonymous single *n*ucleotide polymorphism (non-synonymous SNP, nsSNP):** Any → single nucleotide polymorphism that occurs in a coding region of a eukaryotic gene and changes the encoded amino acid. NsSNPs may cause the synthesis of a non-functional protein, and therefore be involved in diseases. See → anonymous SNP, → candidate SNP, → coding SNP, copy SNP, → exonic SNP, → gene-based SNP, → human SNP, → intronic SNP, → non-coding SNP, → promoter SNP, → reference SNP, → regulatory SNP, → synonymous SNP.

Non-synonymous/synonymous mutation rate ratio: The ratio of → non-synonymous versus → synonymous mutations in a genome over evolutionary times, expressed as d_N/d_S (or ω). Negative selection is characterized by $d_N/d_S < 1$, no selection (H_0) by $d_N/d_S = 1$, and positive selection (H_A) by $d_N/d_S > 1$.

Non-templated nucleotide addition: The addition of 1–3 nucleotides (preferably dATP) onto the 3′-terminus of blunt-ended → duplex DNA substrates, catalyzed by a series of pro- and eukaryotic → DNA polymerases (e.g. → *Taq* DNA polymerase, yeast DNA polymerase I, avian retrovirus AMV → reverse transcriptase). Although occuring at a slow rate, the non-templated addition of dATP onto substrate DNA may result in +1 (or +2) → frameshift mutations *in vivo*. *In vitro* this addition is exploited by → TA cloning. See → terminal deoxynucleotidyl transferase.

Non-transcribed spacer: A DNA sequence that separates tandem copies of an expressed gene or an expressed transcription unit, but is not transcribed itself. See for example → rDNA.

Non-viral retroposon (non-viral retroelement, non-viral retrotransposable element): A → transposable element that transposes via an RNA intermediate, but does not contain → long terminal repeat sequences. Usually non-viral retroposons carry sequences with homology to → reverse transcriptase and → poly(A) tracts at their 3′ end, and are frequently truncated at their 5′-termini. Extensive modifications of some of these elements led to their inability to move. The elements probably originate from escaped → messenger RNAs. For example, *Ty* in *Saccharomyces cerevisiae*, *Cin* in *Zea mays*, *Ta* and *Tag* in *Arabidopsis thaliana*, *Tnt* in *Nicotiana tabacum*, *D, F, FB, Fw, G, "Doc", I, "Jockey"* and its incomplete variants called "*sancho I*", "*sancho II*" and "*wallaby*" are such non-viral retroposons, to name few.

Nopaline (N-α-[1,3-dicarboxylpropyl]-L-arginine): An amino acid derivative that is synthesized in plant cells transformed by the soil bacterium → *Agrobacterium tumefaciens* (e.g. strain C58). This bacterium, after contact with wound-exposed plant cell walls, transfers part of a large plasmid (→ Ti-plasmid) into the plant cell where it is integrated into the nuclear DNA. A gene (*nop* gene) close to the right border of the transforming DNA (→ T-DNA) encodes the enzyme → nopaline synthase that synthesizes nopaline from α-ketoglutarate and L-arginine. Nopaline is an → opine. It cannot be used by the host plant cell, but is secreted and serves as a carbon, nitrogen and energy source for agrobacteria possessing *noc* (*n*opaline *c*atabolism) genes on their Ti-plasmid (see → genetic colonization). See also → nopaline synthase gene.

$$HN$$
$$\diagdown$$
$$C - NH - (CH_2)_3 - CH - COOH$$
$$H_2N \diagup \qquad\qquad\qquad\quad |$$
$$\qquad\qquad\qquad\qquad\qquad NH$$
$$\qquad\qquad\qquad\qquad\qquad |$$
$$\qquad\qquad HOOC - (CH_2)_2 - CH - COOH$$

Nopaline synthase (Nos, nopaline synthetase): An enzyme present in → crown gall tumor cells and encoded by the *nop* gene of the → T-DNA originating from the → Ti-plasmid of → *Agrobacterium tumefaciens*. Nopaline synthase catalyzes the synthesis of the unusual amino acid → nopaline from L-arginine and α-ketoglutarate.

Nopaline synthase gene (NOP gene, *nop* gene): A gene encoded by the → T-region which is part of the → Ti-plasmid of → *Agrobacterium tumefaciens*. *nop* encodes the enzyme → nopaline synthase and is only expressed in transformed plant cells (see → crown gall). The *nop* gene is frequently used as a → reporter gene in plant transformation experiments, its → promoter (*Pnop*) and → termination sequences (3't nop) are incorporated in plant transformation vectors. See also → nopaline.

Nopaline synthetase: See → nopaline synthase.

Nopalinic acid (N²-[1,3-D-dicarboxypropyl]-L-ornithine; ornaline): An amino acid derivative that is synthesized in plant cells transformed by the soil bacterium *Agrobacterium tumefaciens*. Nopalinic acid belongs to the so-called → opines. See also → crown gall.

$$H_2N - (CH_2)_3 - CH - COOH$$
$$\qquad\qquad\qquad\quad |$$
$$\qquad\qquad\qquad NH$$
$$\qquad\qquad\qquad |$$
$$HOOC - (CH_2)_2 - CH - COOH$$

***nop* gene:** See → *nop*aline synthase gene.

NOR: See → *n*ucleolus *o*rganizer *r*egion.

NoRC: See → nucleolar remodeling complex.

norgDNA: See → nuclear insertion of organellar DNA.

Normalization:

a) The process of dotting → messenger RNAs from → housekeeping genes (e.g. → the ubiquitin gene sequence) onto → hybridization membranes such that a hybridization with labeled → cDNAs from different cells, tissues, or organs will produce consistent hybridization signals for all dots. The strength of these signals – as quantified by → autoradiography or → phosphorimaging – serves as internal standard for quantifying the relative → abundance of other transcripts in e.g. → Northern hybridization.

b) The equalization of the concentrations of various transcripts present in a cell at extremely different levels (e.g. single copy or "rare" or "least abundance" versus abundant or "highly abundant" or "most abundant" RNAs). Since the difference between single copy and highly abundant messages is more than 10^5 in most cells, any cloning of cDNAs will inevitably lead to an overrepresentation of clones from strongly expressed genes, whereas least abundant messages probably escape cloning. Normalization balances the otherwise unequal representation of the various messages in a cDNA library by reducing the proportion of highly expressed mRNAs with concomitant enrichment of rarely expressed messages. An efficient technique for normalization, → phenol emulsion reassociation technique, involves the amplification of cDNA, its precipitation with ethanol and resuspension in hybridization solution containing 8% phenol (reduces the aqueous phase and increases the rate of hybridization). Vigorous shaking leads to a mixing of the phases. The resulting emulsion then allows hybridization of abundant cDNAs. Subsequently chloroform-isoamyl alcohol extraction and desalting is performed, and the single-stranded cDNAs (representing single-copy mRNAs) enriched by → restriction of double-stranded CDNAs (representing abundant mRNAs). The efficiency of normalization can be monitored by the loss of distinct bands (overrepresented cDNAs) and an increase of the background smear in → ethidium bromide-stained agarose gels (normalization of previously underrepresented messages).

c) A procedure for the statistical elimination of differences in → microarray experiments that are caused by sampling (i.e. differences in RNA quantity and quality). Usually total RNA, a set of → house-keeping genes, or the means across all house-keeping genes are employed for normalization. For human microarray experiments, genes expressed at constant levels across samples and during developmental processes as well as experimental manipulations are selected, as e.g. genes encoding β-actin, glyceraldehyde-3-phosphate dehydrogenase, 18S and 28S ribosomal RNAs, transferrin receptor, β-glucuronidase, β-2-microglobulin, phosphoglycerate kinase and hypoxhantine phosphoribosyl transferase, to name some. However, hardly any house-keeping gene is in reality constantly expressed in any tissue, so that the optimal set of control genes has to be evaluated for each experimental condition and tissue separately.

Northern blot (RNA blot): A nitrocellulose or nylon membrane, onto which RNA molecules are transferred from a gel by e.g. capillary action and fixed by → baking or → cross-linking. Such blots can be hybridised to radioactively labeled → probes, and specific RNAs detected by → autoradiography. A Northern blot is the result of → Northern blotting. Compare → Southern blotting, → South-Western blotting, → Western blotting.

Northern blotting (Northern transfer, RNA blotting): A gel → blotting technique in which RNA molecules, separated according to size by → agarose or → polyacrylamide gel electrophoresis, are transferred directly to a → nitrocellulose filter or other matrices by electric or capillary forces (Northern transfer). Single-stranded nucleic acids may be fixed to the nitrocellulose filter by → baking and are thus immobilized. Hybridization of specific, radioactively or non-radioactively labeled, single-stranded probes to the immobilized RNA molecules (Northern hybridization) allows the detection of individual RNAs out of complex RNA populations. See also → multiple tissue Northern blot. Compare → Southern blotting, → South-Western blotting, → Western blotting. See → Northern blot.

Northern transfer: See → Northern blotting.

NOS:

a) See → nopalin synthase.
b) *Nitric acid oxide synthase* (Nos; L-arginine-NADPH:oxygen oxidoreductase, EC 1.14.23.39): A homodimeric hemeprotein that catalyzes the conversion of L-arginine into nitrogen monoxide (NO) and L-citrulline, consuming molecular oxygen and NADPH$_2$.

Notch protein: Any one of several conserved receptor proteins spanning the membrane of vertebrate cells such that a part of it extends inside and another part outside the membrane. Ligand proteins (Delta, Serrate, and Lag-2 or DSL families) binding to the extra-cellular domain induce proteolytic cleavage of the Notch protein directly outside the membrane (catalyzed by the ADAM-family metalloprotease TACE, for tumor necrosis factor alpha converting enzyme), and the activity of the socalled γ-secretase that cleaves the remaining part of the Notch protein just inside the inner part of the cell membrane of the Notch-expressing cell. This releases the intracellular domain of the Notch protein (ICN), which then moves to the nucleus, where it regulates gene expression by activating the transcription factor CSL (for CBF-1, suppressor of hairless, and Lag-1). CSL is a highly conserved protein comprised of N- and C-terminal Rel homology domains and a central trefoil domain binding specific sequences in DNA. The Notch signaling pathway is important for cell-cell communication during embryonic and adult life, as e.g. in neuronal function and development, stabilizing arterial endothelial fate and angiogenesis, regulating crucial cell communication events between endocardium and myocardium during both the formation of the valve primordial and ventricular development and differentiation, influencing binary fate decisions of cells that must choose between the secretory and absorptive lineages in the gut, to name few. Notch signaling is often repressed in cancer cells, and faulty Notch signaling is implicated in many diseases including T-cell acute lymphoblastic leukemia (T-ALL), multiple sclerosis (MS), cerebral autosomal dominant arteriopathy with sub-cortical infarcts and leuko-encephalopathy (CADASIL), among many others. Mutations in one (or more) of the four mammalian *Notch* genes (*Notch1-4*) cause various diseases. For example, germ-line → loss-of-function mutations in *Notch1* lead to congenital aortic valve disease, loss-of-function mutations in Notch-ligand JAGGED1 occur in Alagille syndrome (pleiotrophic developmental abnormalities), and somatic → gain-of-function mutations in *Notch1* are characacteristic for more than 50% of human *T* cell *a*cute *l*ymphoblastic *l*eukemias (T-ALL).

Notch signaling pathway: A highly conserved cell signaling system present in most multi-cellular organisms, whose crucial components are the Notch trans-membrane receptor proteins (Notch receptors, see → Notch proteins). Vertebrates possess four different Notch receptors, referred to as Notch1 to Notch4. A Notch receptor is a hetero-oligomer composed of a large extracellular domain associated with a short extracellular region in a calcium-dependent, non-covalent interaction, a single trans-membrane pass, and a small intracellular region. Signaling depends on the binding of a ligand to the Notch receptor protein, which promotes two proteolytic processing events and liberates the intracellular domain that enters the nucleus to form a complex with regulatory proteins (e.g. CSL, an acronym for CBF-1 in mammals, Suppressor of Hairless in *Drosophila*, and Lag-1 in *Caenorhabditis*). After complex formation, a third protein, Mastermind (MAM, MAML, lag-3) binds with high affinity, which in turn requires the ankyrin (ANK) domain of the Notch intracellular domain (NICD). This aggregate then regulates (activates) Notch-responsive genes. The Notch signaling pathway is involved in proliferation, stem cell niche maintenance, cell fate, differentiation, and cell death.

Not I library: See → chromosome linking clone library.

Novel gene: Any gene that has not been known before its detection by e.g. → genomic sequencing. The term is misleading, since a novel gene is not really novel (as e.g. a → synthetic gene might be), but normally a component of a genome for millions of years.

NPA: See → nuclease protection assay.

npcRNA: See → non-coding RNA.

NPNL: See → native protein nanolithography.

N-protein: A protein of the → lambda phage (and other → coliphages) that binds to specific sequences of the phage genome (*nut* sites, *N-u*tilization sites), prevents *rho*-dependent termination of leftward early transcription and induces the expression of adjacent genes. The gene for this anti-terminator protein (gene *N*) is transcribed during the early phase of infection (→ early gene).

NPT; *n*eomycin *p*hospho*t*ransferase: See → aminoglycoside phosphotransferase.

NR: See → *n*uclear *r*eceptor.

nRNA: See → *n*uclear RNA.

nrRNA: See → *n*uclear *r*egulatory RNA.

NS-PCR: See → *n*on-*s*elective *p*olymerase *c*hain *r*eaction.

nt: Abbreviation for *n*ucleo*t*ide(s).

NT: See → *n*uclear *t*ransplantation.

n*Taq*: Abbreviation for the *n*ative form of → *Thermus aquaticus* → DNA polymerase. Compare → r*Taq*.

N-terminal end (N-terminus; amino terminus, amino terminal end): The terminus of a protein where the amino (NH_2) group does not form part of a peptide bond. Polypeptide synthesis starts at this end. Compare → N-formylmethionine.

N-terminus: See → N-terminal end.

NTP: Abbreviation for any ribonucleoside-5′-triphosphate (e.g. ATP, CTP, GTP, TTP, or UTP).

Ntp: See → base pair.

NTT: See → nuclear transportation trap.

N²-di-methylguanosine: A → rare base.

Nu body: The equivalent of a → nucleosome in electron microscopic pictures of negatively stained Miller spreads.

Nuclear cage: See → nuclear lamina.

Nuclear chromosome scaffold: See → nuclear lamina.

Nuclear dimorphism: The presence of two differently sized nuclei in one and the same cell. For example, ciliates possess one or more socalled micronuclei and macronuclei. The smaller micronuclei harbor typical eukaryotic chromosomes with associated histones, divide by mitosis, and are transcriptionally silent during asexual growth of the ciliate. However, they are active during sexual reproduction and responsible for the genetic continuity of the protozoon ("germ-line nucleus"). The macronucleus in turn actively transcribes its genes during asexual growth, replicates during asexual reproduction, but is destroyed and re-formed during sexual reproduction. Therefore, macronuclei do not transmit genetic information to sexual offspring.

Nuclear DNA (nDNA): The DNA that is located within the nucleus of eukaryotic cells, in contrast to the DNA of mitochondria (mtDNA) or chloroplasts (cpDNA). See → mitochondrial DNA and → chloroplast DNA.

Nuclear envelope (NE, nuclear membrane): The double-membrane boundary of nuclei in eukaryotic cells. The outer lipid membrane (outer nuclear membrane, ONM) forms a continuum with some parts of the endoplasmic reticulum (ER), whereas the inner membrane (inner nuclear membrane, INM) functions in the organization of → chromatin (e.g. by anchoring → looped domains). Both membranes are perforated by complex pores (→ nuclear pore) that consist of a central channel and a peripheral layer of proteins, and mediate import and export processes.

Nuclear envelope transmembrane protein (NET): Any protein that is associated with the → nuclear envelope, and contains at least one, or more transmembrane domain(s).

Nuclear export sequence: See → nuclear export signal.

Nuclear export signal (NES; nuclear export sequence): A glycin- or leucine-isoleucine-rich domain in proteins that are synthesized in the → nucleus and exported into the cytoplasm of a cell. NESs are potential address sites where proteins (e.g. receptor proteins) bind and assist in the nucleo-cytoplasmic exportation process. In the Rev protein from the pathogenic human T-cell leukemia virus type 1 (HTLV-1), the NES consensus sequence is:

leu-X_{2-3}-phe/ile/leu/val/met-X_{2-3}-leu-X-ile/val.

Also, 5S rRNA is channeled into the cytoplasm after complexing with → transcription factor TF IIIa that contains an NES. See → nuclear localization signal.

Nuclear expressed sequence tag analysis (NEST): A technique for the identification of transcribed (active) genes in the nucleus of eukaryotic organisms. In short, nuclei are first labelled with → fluorochromes (e.g. via direct binding of the fluorophore to nuclear DNA, or indirectly with → autofluorescent proteins), isolated by → flow cytometry ("flow sorting"), lysed, and the released → poly(A)⁺-RNA captured on oligo(dT)-linked → magnetic beads. The captured RNA is then reverse transcribed into → cDNA, restricted with → four base cutters (restriction enzymes with a 4bp restriction recognition site), resulting in 3′-fragments bound to the beads. Then → linkers are ligated to the fragments, the fragments amplified via conventional → polymerase chain reaction techniques, using linker-complementary primers, and the resulting amplicons separated on → sequencing gels, which display characteristic expression profiles of the cells, tissues, organs or organisms of interest.

Nuclear factor: See → transacting factor.

Nuclear factor I: See → CAAT-box transcription factor.

Nuclear factor of activated T-cells **(NFATC):** Any cytosolic protein that is a component of the DNA-binding transcription complex of activated T cells. This complex consists of a pre-existing cytosolic component that translocates to the nucleus upon T cell receptor (TCR) stimulation, and an inducible nuclear component. Proteins belonging to this family of → transcription factors play a central role in inducible gene transcription during immune response. Some NFATCs are molecular targets for immunosuppressive drugs such as → cyclosporin A.

Nuclear factor 1 **(NF-1):** Any one of a large family of eukaryotic → transcription factors that recognize specific address sites and bind to DNA. The tremendous diversity within the NF-1 family is a consequence of the presence of multiple genes. The diversity of encoded proteins originate from → alternative splicing and heterodimerization.

Nuclear focus: Any one of some 500 to 10,000 compartments within the → nucleus of a eukaryotic cell that contains high concentrations of → DNA-dependent RNA polymerases and probably represent storage vesicles for these enzymes. See → processing bodies, → splicing speckles.

Nuclear gene: Any gene that is localized in the nuclear genome of a eukaryotic cell. See → non-nuclear gene.

Nuclear genome: The entire → genetic material of the → nucleus of eukaryotic cells. Synonym of → genomic DNA.

Nuclear halo: An artificial structure generated through the lysis of nuclei and the spread of the DNA as loops. These loops protrude from a central scaffold (→ nuclear lamina) that appears as a halo (gr.-lat.: zone of diffuse light around a light source).

Nuclear import: The process of transporting proteins from the cytoplasmic space into the nuclear space. Nuclear import of such proteins proceeds via several pathways. For example, proteins carrying the classical → nuclear localization signal are bound by an → importin (karyopherin) a/b1 heterodimer that docks at the → nuclear pore complex. The docked protein is then translocated into the nucleus in an energy-dependent step requiring a set of proteins, including nuclear transport factor 2 (NTF2), the GTPase Ran, and a nuclear pore protein designated nucleoporin p62. Certain RNA-binding proteins are imported by importin b2, some ribosomal proteins by importin b3. See → nuclear transport.

Nuclear insertion of organellar DNA (norgDNA): A comprehensive term for any DNA sequence in the → nuclear (genomic) DNA of a eukaryotic cell that is originating from an organelle (e.g. a mitochondrion, and in green plants additionally from a plastid, as e.g. a chloroplast) and has been transferred from the organelle by → lateral DNA transfer. See → nuclear mitochondrial DNA segment, → nuclear plastid DNA.

Nuclear lamin: A family of interrelated polypeptides that are the constituents of the → nuclear lamina network and fall into three major types: the neutral A- and C-lamins, and the acidic B-lamins (molecular weight range from 62–69 kDa). Less frequently occurring lamins belong to the D and E categories. The lamins are structurally related to the intermediary filaments, assemble to 10 nm filaments *in vivo*, and possess the typical → coiled coil-configuration of two intertwined α-helices. They consist of a short N-terminal domain ("head"), an α-helical rod domain and a long C-terminal domain ("tail"). During nuclear division the lamina disintegrates with concomitant strong phosphorylation of lamins. Specific mutations in nuclear lamina genes cause a variety of human hereditary diseases (→ laminopathies). For example, a single base exchange in the gene encoding lamin A leads to the use of a → cryptic splice site in the → pre-messenger RNA. Consequence: a shorter lamin A is synthesized that does not function correctly. The underlying mutation therefore is the cause for the Hutchinson-Gilford Progeria Syndrome (HGPS), an extremely rare disease leading to severe premature aging. Nuclear lamins and lamin-associated proteins are ubiquitous in metazoans, but absent in yeast and plants. Lamin-dependent complexes are formed by integral inner nuclear membrane (INM) proteins such as emerin and MAN-1. In cells lacking lamins, many of the proteins are not fixed to the → nuclear envelope (NE), but instead drift throughout the NE/ER network.

Nuclear lamina (fibrous lamina, karyoskeleton, nuclear cage, nuclear matrix, nuclear scaffold, nuclear chromosome scaffold): A filamentous meshwork located between the inner nuclear membrane (see → nuclear envelope) and → heterochromatin, which consists of lamins and lamin-associated proteins, and provides potential attachment sites for → chromatin and cytoplasmic intermediate filaments. It is involved in many nuclear activities, as e.g → DNA replication, → RNA transcription, nuclear and → chromatin organization, cell cycle regulation, cell development and differentiation, nuclear positioning and → apoptosis.

Nuclear localization sequence: See → nuclear localization signal.

Nuclear localization signal (NLS; nuclear localization sequence): A cluster of basic amino acids (usually containing a proline of glycine, for example the sequence proline-lysine-lysine-lysine-arginine-lysine-valine, PKKKRKV of the SV 40-like NLSs) in proteins larger than 40 kDa that directs their targeted import into the nucleus. Such NLSs have been identified in a series of yeast, *Drosophila*, amphibian, mammalian and plant proteins, and vary in amino acid sequence (smallest consensus sequence: $K_K^R X_R^K$). The NLS is not proteolytically removed after translocation of the linked protein, so that the protein retains the capacity to enter the nucleus repeatedly (e.g. after each cell division). Basically, two arrangements of NLS exist in import proteins. Single-cluster NLS consist of one single NLS sequence, bipartite NLS are composed of two interdependent domains with short intervening sequences that act synergistically, but can also independently, yet less effectively direct proteins into the nucleus. Three main NLS categories can be found:

1. Simian virus 40-like NLSs contain short tandem stretches of 6–8 basic amino acids with either a proline or glycine (PKKKRKV), and occur also in e.g. a transcription activator protein of maize (*Zea mays*).
2. Mating type a2-like NLSs consist of short hydrophobic regions that contain one or more basic amino acids (KIPIK or MNKIPIKDLLNPG).
3. Bipartite NLSs (nucleoplasmin-type NLS) are a combination of two regions of basic amino acids separated by a spacer of approximately ten amino acids, and are ubiquitous.

Proteins smaller than 40–60kD may also diffuse through nuclear pores. Larger proteins definitely require ATP and at least one NLS to traverse pores. Compare → nuclear export signal.

Nuclear localization signal receptor (NLS receptor; NLS-binding protein, NLS-BP): A protein that recognizes → nuclear localization signals, interacts with them, and directs the corresponding protein to nuclear pores.

Nuclear matrix: See → nuclear lamina.

Nuclear matrix protein enzyme-linked immunosorbent assay (NMP-ELISA): A technique for the *in vitro* detection and quantitation of specific nuclear matrix proteins from injured, dying, or dead cells. Specifically, NMP-Elisa detects the socalled *nu*clear *m*itotic *a*pparatus protein (NuMA) or its fragments that arise after an encounter with a toxic chemical or a pathogenic organism. This 240 kDa protein is restricted to the nucleus during interphase, but redistributed and concentrated at the spindle apparatus during mitosis. If cultured cells are injured or going to die, NuMA is released into the culture medium, where its concentration can be estimated by a detector → antibody. NuMA levels are positively correlated with *in vitro* cell death.

Nuclear membrane: See → nuclear envelope.

Nuclear mitochondrial DNA segment (Numt, pronounced "new mite"; nuclear pseudogene of mitochondrial origin): Any

Species	Protein	Motif
Xenopus laevis	Nucleoplasmin	KRXXXXXXXXXXKKKK
Homo sapiens	Glucocorticoid Receptor	RKXXXXXXXXXXRKXKK
Homo sapiens	Androgen Receptor	RKXXXXXXXXXXRKXKK
Homo sapiens	p53 Protein	KRXXXXXXXXXXKKK
Simian virus 40	SV40 T-Antigen	PKKKRKV

K = Lysine   P = Proline   R = Arginine   X = Any Amino Acid

NLS

→ pseudogene within the → nuclear genome that has high sequence similarity to mitochondrial sequences and therefore most probably originates from → mitochondrial DNA. For example, hundreds of Numts, representing mitochondrial genes (→ ribosomal RNA genes, → transfer RNA genes) or sequences from the control region (CR), are present on all chromosomes in the human genome. Some of the Numts encompass about 80% of the complete mitochondrial genome. Sequence similarity between human Numts and their mitochondrial counterparts comes close to 99%. The sizes of Numts range from 130 to 1,700 base pairs in chicken, where they altogether comprise only 0.0008% of the nuclear genome. The → horizontal gene (or DNA) transfer from mitochondria to the nucleus occurs via a DNA intermediate, and the transferred sequences are preferentially integrated into repeat-rich, but gene-poor regions. See → nuclear plastid DNA.

Nuclear mitotic apparatus (NuMA): A matrix protein that is concentrated in unfertilized meiotic and fertilized mitotic cells (centrosomal NuMA) and is involved in mitotic spindle pole assembly.

Nuclear periphery: A sub-nuclear → domain encompassing the inner → nuclear membrane, the → nuclear pore complex (NPC) and the peripheral part of the → nucleoplasm that contains silenced loci (e.g. the immunoglobulin genes in hematopoietic progenitor cells of B lymphocytes, regions of transcriptionally inactive → chromatin of → telomeres, and → constitutive heterochromatin) in some, but also transcriptionally active genes in other cell types. In yeast, the *INO1* (encoding an enzyme involved in phospholipid biosynthesis), *HSP104* (encoding a chaperone), *HXK1* (encoding a hexokinase), *SUC2* (encoding an invertase), *GAL1* (encoding a galactokinase), *GAL2* (encoding a hexose transporter), *GAL10* (coding for a glucose epimerase) and mating pheromone-induced genes, if highly expressed, locate to the nuclear periphery. Many other genes are recruited to the nuclear periphery upon activation, and physically interact with the NPC via the → nucleoporins (in yeast Nup2, Nup60, Nic96, Nup116, and the myosin-like proteins Mlp1 and Mlp2). Recruitment in some cases also requires the SAGA complex (Spt-Ada-Gcn5 acetyltransferase), a transcriptional co-activator altering gene expression by acetylating → histones in the → chromatin of → promoters of target genes that physically associate with the NPC. Localization of genes to the NPC is sufficient to activate → transcription in some genes, localization at the nuclear periphery in general promotes → transcription initiation and is a heritable → trait in some organisms, e.g. yeast (and therefore represents a novel epigenetic feature). A modified → histone, histone H2A.Z, plays a key role in maintaining the peripheral localization of active genes.

Nuclear plastid DNA (NUPT): Any DNA fragment that originates from the → genome of a plastid (e.g. a → chloroplast), has been transferred into the nucleus and integrated into → nuclear DNA (see → horizontal gene transfer). Large nuclear genomes contain more NUPTs than smaller genomes. NUPTs are frequently clustered and mixed with → nuclear mitochondrial DNA segments (Numts), possibly as a result of their → concatemerization before → integration. Original → insertions of NUPTs are large, but decay into smaller fragments with diverging sequence over evolutionary times.

Nuclear pore (nuclear pore complex, NPC, "porosome")

Nuclear pore (nuclear pore complex, NPC, "porosome"): A cylindrical channel through the → nuclear envelope that mediates cytoplasmic-nuclear and nuclear-cytoplasmic exchange of various molecules ("traffic"). A pore complex consists of a ring of eight globular subunits (annular granules) of 100–250 Å in size, arranged in a symmetrical, octagonal pattern at each side of the nuclear envelope. These rings border a circular hole of 900 Å in diameter and about 120 nm in length (*Dictyostelium discoideum*). From the ring at the cytoplasmic side a series of eight, irregularly formed filaments protrude into the cytoplasm. The nuclear ring consists of eight filaments that unite distally into a ring-like structure ("distal ring") such that a cage-like complex results ("nuclear basket"). A series of 800 to 1000 → nucleoporin proteins are more or less symmetrically distributed at both the cytoplasmic and nuclear sides. A central plug/transporter (CP/T) with a variable size, shape and position within the central pore represents cargo proteins in transit. About 100 to

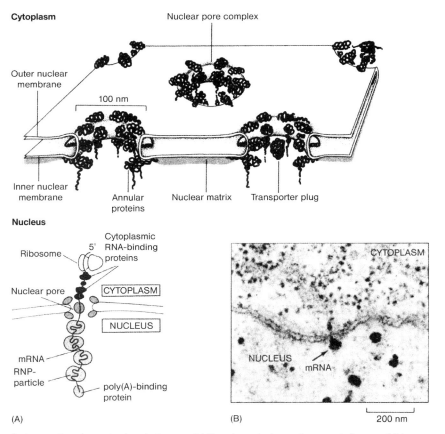

Scheme of nucleo-cytoplasmic transport (A) and an electron microscopic image of this process (B)

Nuclear pore

more than $5·10^7$ pores per nucleus may exist, their number varying with the metabolic state of the nucleus or the cell. Each pore complex catalyzes the transport of more than 1,000 molecules per second. Proteins over 40 kDa have to carry a → nuclear localization or → nuclear export signal to be transported through the pore complex, smaller molecules or ions diffuse "passively".

Nuclear pore complex protein (NUP): Any one of the 800 to 1000 proteins tightly bound to the → nuclear pores. Also called nucleoporin.

Nuclear positioning: The anchoring of a nucleus at a more or less fixed position within a eukaryotic cell. Nuclear positioning is dependent on specific proteins. For example, the cytoplasmic protein ANC-1, encoded by the *anc-1* (nuclear *anc*horage) gene in *Caenorhabditis elegans*, consisting of mostly coiled regions with a nuclear envelope localization domain (the KASH domain) and an actin-binding domain, simultaneously reacts with both another protein (UNC-84) at the nuclear envelope and with actin in the cytoplasm. Therefore it anchors nuclei by tethering the nuclear periphery to the actin cytoskeleton, creating a bridge across the nuclear envelope. Mutations in the *anc-1* gene leads to freely floating nuclei (normal cells: nuclei are located in specific regions).

Nuclear processing of RNA: See → post-transcriptional modification.

Nuclear proteome: See → nucleoproteome.

Nuclear pseudogene of mitochondrial origin: See → nuclear mitochondrial DNA segment.

Nuclear receptor (NR): Any one of a family of ligand-activated → transcription factors that bind to cognate *r*esponse *e*lements (REs) and induce the transcription of target genes. The ligand-dependent → transactivation by NRs is mediated by an *a*ctivation *f*unction motif (AF-2) which is present in the *l*igand-*b*inding *d*omain (LBD) of the receptor and functions via *t*ranscriptional *i*ntermediary *f*actors (TIFs). All nuclear receptors have at least four different domains that are differentially conserved between the subfamilies: the DNA-binding C-domain (a → helix-loop-helix or → zinc finger conformation), the ligand-binding/dimerization domain, the A/B transactivation domain, and the socalled hinge (D) domain. The various nuclear receptors bind different, mostly hydrophobic ligands such as dioxin, ecdysone, retinoic acid, steroids, thyroid hormones, and vitamin D, and form distinct complexes with → heat shock protein 90 that assists in domain-folding for ligand binding. The ligand-nuclear receptor complex directly acts upon the DNA, and therefore links extracellular signals to transcriptional response(s). Nuclear receptors (e.g. the estrogen receptor) regulate complex events in early embryogenic development, cell differentiation, and homeostasis.

Based on C-domain sequences and structural data, nuclear receptor genes fall into three subfamilies: subfamily I encodes *ear1* subgroup, retinoic acid and thyroid hormone receptors, subfamily II the orphan receptor genes (orphan: a nuclear receptor, for which no ligand has yet been identified), and subfamily III, harboring the steroid hormone receptor genes.

Nuclear regulatory RNA (nrRNA): Any one of a series of RNAs retained within the nucleus of a eukaryotic cell that act as → riboregulators or serve structural functions. For example, the X-chromosome-encoded → Xist RNA and its → anti-sense transcript → Tsix are such nrRNAs.

Nuclear reprogramming: The conversion of a differentiated adult cell to a mitotically active pluripotent cell through an erasure of epigenetic modifications (e.g. → histone modifications like methylations) and their re-setting by as yet unknown factors of an embryonic cell. For example, *human embryonic stem cells* (hESCs), if fused with somatic cells (e.g. fibroblasts) to generate heterokaryons, are able to re-program the fibroblast nuclei such that they become mitotically active. So, the original fibroblast nucleus expresses genes associated with pluripotency, while fibroblast-specific genes are repressed. Moreover, the ESC-fibroblast hybrid acquires properties of an hESC (e.g. the capacity of self-renewal over many passages, and the ability to differentiate into a variety of cell types). *OCT4*, a pluripotency-specific gene, repressed in mature fibroblasts by → promoter methylation, becomes unmethylated and active. hESCs most probably contain re-programming factors that catalyze the nuclear re-programming.

Nuclear retention: The blockage of export of newly synthesized RNA within the → nucleus of a eukaryotic cell. For example, the 8 kb → transcript of the *mouse cationic amino acid transporter 2* (mCAT2) gene, called CTN-RNA (for CAT2 *transcribed nuclear RNA*) harbors three → inverted repeats of → SINE origin within the → 3'-untranslated region (3'-UTR) that are each folded into a specific three-dimensional structure (an imperfect → stem-loop). This fold together with an ADAR (*adenosine deaminase acting on RNA*)-catalyzed A-to-I → editing in its 3'-UTR retains the transcript within the nucleus. Since multiple → inosine residues in an edited RNA interact with a protein complex comprised of PSF, p54^{nrb}, and matrin3, the CTN-RNA is fixed in so called → paraspeckles. Upon stress, however, CTN-RNA is cleaved at its 3'-UTR to produce the protein-coding mCAT2 → messenger RNA that is exported into the cytoplasm and translated into the mCAT2 protein, a cell-surface receptor for arginine, necessary for the synthesis of stress *n*itric *o*xide (NO).

Nuclear retention phenotype: Any mutant cell, whose nuclear DNA suffered one (or more) mutation(s) that lead to a defective → poly(A)$^+$-RNA transport out of the → nucleus. Such mutations frequently hit genes encoding → polyadenylation factors that are no longer functional. Consequence: the non-polyadenylated RNAs are retained within the nucleus, i.e. the cell is a mutant with a nuclear retention phenotype.

Nuclear RNA (nRNA): Any RNA that either remains within the nucleus after its synthesis, or is exported into the cytoplasm only after → processing. For example, heterogeneous nuclear RNA (hnRNA), including the primary transcripts of many genes (e.g. pre-mRNA, pre-tRNA, pre rRNA), occurs only in the nucleus. The processed transcripts (e.g. mRNA, tRNA, rRNA) are associated with specific proteins and transported into the cytoplasm.

Nuclear RNA *interference* (nuclear RNAi, RNA-mediated heterochromatin for-mation): A nuclear surveillance process that controls epigenetic gene regulation in eukaryotic organisms and the exclusion of foreign nucleic acids (e.g. → retrotransposons, → transposons). For example, the initial step in nuclear RNAi in *Schizosaccharomyces pombe* requires bidirectional →

transcription of a target → locus, or transcription of → inverted repeats (IRs), and the resulting formation of primary → double-stranded RNA (dsRNA). This dsRNA is then cleaved by the → RNaseIII-type endonuclease → Dicer to produce → small interfering RNAs (siRNAs) that are subsequently incorporated into the → RNA-induced initiator of transcriptional gene silencing (RITS) complex via → argonaute protein1 (Ago1), guiding the RITS to complementary sites of the genome. After RITS binding to these sites, the complex recruits chromatin-modifying proteins (e.g. → histone H3 lysine 9 [H3K9] methyltransferase [Clr4]). The ensuing methylation of H3K9 stabilizes binding of RITS to → chromatin, and RITS can now interact with → RNA-dependent RNA polymerase complex (RDRC, consisting of the RNA-dependent RNA polymerase itself, the putative → helicase Rdp1, and Cid12 associated with RNA → polyadenylation). This interaction leads to the production of secondary dsRNAs and amplifies the silencing signal. These processes trigger → heterochromatin assembly. It is yet unknown, what mechanism determines whether siRNAs initiate → RNA interference or nuclear RNA interference. The nuclear RNAi of plants and mammals differs from *S. pombe*. For example, more proteins with specific functions compose the nuclear RNAi pathway in *Arabidopsis thaliana*, i.e. it contains a fourth → DNA-dependent RNA polymerase (Pol IV, transcribing methylated DNA and being guided to [hemi]methylated DNA by 24 nucleotides long siRNAs) and the RNA-methylating enzyme HEN1 that stabilizes → microRNAs by methylating their 3′-overhangs. All eukaryotes engage small RNAs for the establishment of *de novo* DNA methylation pattern and/or the maintenance of epigenetic marks. Nuclear RNAi is additionally involved in the regulation of developmental genes, contributes to accurate chromosome segregation during cell division, and may engage specific processes such as DNA methylation and/or RNA amplification.

Nuclear run-off transcription assay: See → run-off transcription.

Nuclear scaffold: See → nuclear lamina.

Nuclear space: A synonym for the → nucleus of a eukaryotic cell.

Nuclear speckle (interchromatin granule cluster, IGC): A nuclear → domain that is involved in the assembly and modification and/or storage of the → pre-messenger RNA-processing machinery. The → poly(A)$^+$-RNAs in the IGCs are not transported into the cytoplasm. See → paraspeckle.

Nuclear transfer: See → nuclear transplantation.

Nuclear translation: The synthesis of proteins from → messenger RNAs (mRNAs) within the nucleus (*in nucleo*) of a eukaryotic cell. Actually → transfer RNAs, certain translation factors and → ribosomes (most probably not functioning prior to their export into the cytoplasm) are present in the nucleus, nuclear translation sites overlap with RNA polymerase II transcription sites (i.e. transcription and translation of the resulting messenger RNAs are probably coupled), and mRNA translation in the nucleus reportedly reaches 10–15% of the total cellular protein synthesis, nuclear translation is still not unequivocally proven and therefore controversial.

Nuclear transplantation (NT, nuclear transfer): The → microinjection of nuclei (or pronuclei) from one embryo into a second embryo, or the transfer of an isolated nucleus from one cell into the enucleated cytoplasm of another cell. For example, nuclei from frog blastomeres can be transferred into enucleated oocytes, where the nuclei are reprogrammed to a zygotic state. See → alternative nuclear transfer.

Nuclear transport: The import and export of molecules across the → nuclear membrane. The passage may be facilitated by specific proteins, may depend on a guide sequence (e.g. a → nuclear localization sequence), and may preferentially use → nuclear pores. See → nuclear import.

Nuclear transportation trap **(NTT):** A technique for the identification of → cDNAs (or genes) encoding nuclear transport signals (e.g. → *nuclear localization signals*, NLSs) that is based on a yeast selection system. The NTT consists of two components: (1) A yeast expression plasmid for → nuclear export signal (NES)-LexAD (*activation domain*) fusion proteins that are excluded from the nucleus, because they possess an NES, and (2), a LexAD-responsive *leu2* → reporter gene that is only expressed if the hybrid NES-LexAD protein is actively imported into the nucleus (i.e. contains an NLS).

Nuclease: Any enzyme that catalyzes the hydrolysis of → phosphodiester bonds in nucleic acid molecules and leads to their breakdown. Nucleases can be broadly categorized into → exonucleases (releasing nucleotides from the ends of nucleic acid molecules), and → endonucleases (cleaving the nucleic acid molecule at internal sites). There exist nucleases specific for DNA (deoxyribonucleases, DNases) or RNA (ribonucleases, RNases), and for single-stranded or double-stranded polynucleotides. Nucleases generally present problems during the isolation of nucleic acids from animal and plant tissues and are therefore inhibited by the inclusion of various agents (e.g. EDTA, → RNasin; compare also → nuclease-free reagent) in the extraction buffers. See also → micrococcal nuclease, → mung bean nuclease, → nuclease P1, → *Bal* 31 nuclease, → repair nuclease.

Nuclease *Bal* 31: See → *Bal* 31 nuclease.

Nuclease-free reagent: Any chemical that does not contain even traces of RNases and/or DNases. Such chemicals are used to isolate RNA or DNA from cells, tissues, organs, or organisms that are rich in nucleases.

Nuclease P1 (EC 3.1.30.1): A single-strand specific → nuclease (endo- and exonuclease) from *Penicillium citrinum* that catalyzes the degradation of RNA and single-stranded DNA to 5′ phosphomononucleotides. The enzyme also hydrolyzes 3′ mononucleotides (ribo- and deoxyribonucleotides) to nucleosides and inorganic phosphate, and is used for the analysis of the 5′-terminal nucleotide of RNA and DNA.

Nuclease *protection assay* (NPA): A more general term for any technique for the detection, quantitation and characterization of specific → messenger RNA molecules out of complex mixtures of total cellular RNAs. The most frequently used NPAs are → RNase protection assay and → S1 nuclease protection assay.

Nuclease S1: See → S1 nuclease.

Nuclease S1-mapping: See → S1-mapping.

Nuclease S7: See → micrococcal nuclease.

Nucleation: The reannealing of a few complementary bases of two single-stranded DNA or RNA molecules to form a nucleation point for complete renaturation to a duplex molecule.

Nucleic acid: A single- or double-stranded linear polynucleotide containing either deoxyribonucleotides (→ DNA) or ribonucleotides (→ RNA) that are linked by 3'-5'-phophodiester bonds.

Nucleic acid biotool (NAB): A generic name for any synthetic oligonucleotide that binds specifically to a target protein and interferes with its function(s). NABs are used to interfere with physiological or pathological processes, to tag proteins, or to investigate their function.

Nucleic acid chromatography system: See → NACS™.

Nucleic acid hybridization: See → hybridization.

Nucleic acid microarray: A more general term for any → microarray, onto which DNA, RNA or oligonucleotides have been spotted.

Nucleic acid ordered module assembly with directionality (NOMAD): A cloning strategy for the combinatorial arrangement of different DNA fragments in constructs of predetermined structure. NOMAD works with basically two elements, a socalled "assembly vector" with an insertion site, and individual or combined DNA "modules" which are ligated into this site in a sequential or directional mode. In short, specially designed assembly vectors with insertion sites flanked by convergently oriented → recognition sequences for two different type IIS → restriction endonucleases (cutting at a precise distance outside of these sites and producing → sticky ends) are first digested with the appropriate restriction enzyme (e.g. *Bsa*I [recognition sequence: 5'-GAGACC-3'] and *Bsm* BI [recognition sequence: 5'-CGTCTC-3']), then the desired module(s) with compatible cohesive ends are inserted. The second module can be inserted either 5' or 3' to the first one. When the assembly vector is cut by *Bsm* BI, then the second module is inserted at the *Bsm* BI site, whereas a *Bsa* I cut directs the second module to this site. This procedure leaves the restriction site intact, so that the vector can be cut again with the same endonuclease(s), and other modules can be inserted adjacent to the already inserted ones. It is also possible to ligate previously assembled multimodule blocks ("composite modules"). The modules themselves can be sequentially added in any desired order and can also be released as desired and recloned into another modular construct. NOMAD allows the modular construction of → chimeric genes and therefore composite proteins, and creation of new cloning vehicles (e.g. by recombining modules for → origins of replication, → transcription termination signals, → selectable marker genes, and → reporter genes).

Nucleic acid-programmable protein array (NAPPA, self-assembling protein microarray): A glutathione S-transferase (GST)-coated glass slide variant of the conventional → protein array, onto which → expression plasmid DNAs, each containing a distinct gene (or genes) of interest, are spotted and cross-linked to a → psoralen-biotin conjugate via UV light. Then avidin, a polyclonal

GST antibody, and bis (sulfosuccinimidyl) suberate are added to the biotinylated plasmid DNAs, which subsequently are arrayed on glass slides treated with 3-aminopropyltriethoxysilane and dimethyl suberimidate-HCl. The GST antibody serves to capture (immobilize) the protein on the → microarray. Subsequently the microarrays are incubated with → rabbit reticulocyte lysate together with → T7 RNA polymerase. All the different genes are simultaneously transcribed/translated in this cell-free → *in vitro* transcription/translation system (in which the glass slide is immersed). All the resulting proteins (in the femtomol range per spot) contain C-terminal GST tags, are immobilized *in situ* by the polyclonal GST antibody and detected with a → monoclonal antibody raised against GST and concomitant → tyramide signal amplification. NAPPAs allow to detect protein-protein interactions, because both the target protein (on the array) and the test protein (used to probe the array) are transcribed and translated in the same extract. As other high-throughput protein arrays, NAPPAs are also influenced by interfering inhibitors (from the cell-free expression systems), and the peptide tags may sterically block binding domains. Compare → antibody array.

Nucleic acid scanning: The search for distinct sequence motifs (e.g. the → TATA box, → start or → stop codons) in a nucleic acid molecule.

Nucleic *a*cid *s*equence-*b*ased *a*mplification (NASBA, self-sustaining sequence replication, SSSR, 3SR): A technique for the isothermal *in vitro* → amplification of a target nucleic acid sequence that allows to start with the RNA transcribed from the target. In short, the RNA is first reverse-transcribed (by → reverse transcriptase, RTase) into the → first strand of a → cDNA using a → primer complementary to the 3′ end of the template and carrying a → T7 RNA polymerase → promoter sequence. The RNA template is destroyed by *E. coli* → RNase H, the → second strand synthesized by RTase, and the promoter sequence becomes double-stranded and functional. Transcription-competent cDNAs are then used to produce multiple (50–1000) copies of → anti-sense RNAs of the original target with T7 RNA polymerase. The amplified anti-sense transcripts serve as templates and are immediately transcribed into double-stranded cDNA copies, using a second T7 promoter-containing primer. These cDNAs in turn can be used as transcription templates for cDNA synthesis in the cyclic phase of NASBA. This process continues in a self-sustained mode under isothermal conditions (e.g. 42°C), until enzymes are inactivated or compounds in the reaction mixture become limiting.

These continuous cycles of reverse transcription and RNA transcription lead to the production of up to 10^8 copies of the target molecule in only half an hour. It is thus an interesting alternative to the widely used → *p*olymerase *c*hain *r*eaction (PCR).

Nuclein: An outdated synonym for DNA, originally coined by Friedrich Miescher, who isolated DNA (probably) for the first time in 1869.

Nucleobase: A less frequently used term for a → base in an → RNA or → DNA molecule.

Nucleocapsid: The protein coat (→ capsid) of a → virion or → virus together with the enclosed nucleic acid molecule (DNA or RNA).

Nucleocidin: A → nucleoside antibiotic.

Nucleo-*d*elta *p*eptide (NDP): Any artificial peptide that forms the basic unit of biopolymers with importance for chip and nanotechnology.

Nucleofection: A technique for the → direct gene transfer into the nucleus of a cell. Basically, current nucleofection methods rely on → electroporation of the foreign DNA into the target cells and its guidance into the nucleus by cell-type specific solutions ("Nucleofector", composition not disclosed), resulting in high transfection efficiencies.

Nucleoid (karyoid, DNA plasm): The region within a prokaryotic cell that contains the DNA. A nucleoid is analogous to the → nucleus of eukaryotic cells, though not engulfed by a nuclear membrane. Nucleoids are also constituents of mitochondria and plastids (e.g. chloroplasts).

Nucleoid-*a*ssociated *p*rotein (NAP; histone-like protein, *h*istone-like *n*ucleoid structuring protein, H-NS): Any one of about 20 different low-molecular weight DNA-binding bacterial proteins that are involved in → recombination, → replication and → repair of DNA, and change the degree of → supercoiling and thereby influence DNA compaction in the cell. For example, H-NS of *E. coli* is a 15.6 kDa protein highly conserved in Gram-negative bacteria. It is present in the *E.coli* cell in 20,000 copies, and binds unspecifically to DNA with a preference for AT-rich bend regions. Binding is mediated by a flexible loop → domain between an α-helix and two anti-parallel β-sheets within the C-terminal domain of the protein. The N-terminal domain functions in protein-protein interaction(s). The active H-NS is a dimer, but can form higher-order oligomers. H-NS stabilizes → negative superhelicity, and thereby regulates multiple genes via changes in → DNA topology. The protein also directly represses → transcription by bridging of neighboring DNA regions that leads to → promoter trapping (promoter occlusion). In the case of the *hdeAB* promoter, the → RNA polymerase itself bends the promoter and creates a loop, to which single N-HS dimers bind to the two strands, and then recruit other N-HS molecules, thereby forming a DNA-H-NS-DNA bridge. H-NS regulates about 200–300 genes, mostly negatively, and is responsible for the regulation of pathogenicity genes, stress-responsive genes, and control of foreign DNA transferred into the cell by → horizontal gene transfer. Other NAPs are the *i*ntegration *h*ost *f*actor (IHF), the *f*actor for *i*nversion *s*timulation (FIS), and the *l*eucine-*r*esponsive regulatory *p*rotein (LRP), to name few.

Nucleolar organizer: See → nucleolus organizer region.

Nucleolar *r*emodelling *c*omplex (NoRC): A multi-protein nucleolar machine, composed of TIP5 (consists of an MB domain, binds to a subset of → small nucleolar RNAs) and SN72h that silences rRNA genes by recruiting → chromatin-modifying enzymes such as → histone methylases and histone deacetylases. NoRCs therefore catalyze chromatin remodelling, histone modifications and → nucleosome shifting on → promoters.

Nucleolar targeting: The directed transport of specific proteins carrying one (or more) socalled → nucleolar translocation sequences (NTSs) into the → nucleolus.

Nucleolar translocation sequence (*nu*cleolar *t*argeting *s*equence, NTS): The amino acid core consensus sequence H$_2$N-RRQRR-COOH of cellular and viral proteins that targets the protein to the → nucleolus. Examples of such NTSs are (1) H$_2$N-GRKK**RRQRR**AP-COOH for the Tat protein of HIV-1, (2) H$_2$N- HHSRIG II**RQRR**ARNGASRS-COOH for the HIV-1Vpr, and (3) H$_2$N-RQARRNRRRWRE**RQR**-COOH for the HIV-1Rev protein. See → nucleolar targeting.

Nucleolin (C23): A eukaryotic nonribosomal nucleolar phosphoprotein with a tripartite structure. The N-terminal domain interacts with nucleolar → chromatin and is phosphorylated. This phosphorylation, catalyzed by cyclic AMP-independent protein kinase II, modulates chromatin condensation in conjunction with histone H1 and is correlated with nucleolar transcriptional activity. This domain also contains bipartite → nuclear localization sequence motifs. The central domain of nucleolin contains four → RNA recognition motifs. The C-terminal domain consists of glycine- and *a*rginin-rich repeats (socalled GAR repeats). In animals, nucleolin is highly phosphorylated and has a molecular mass of 90–110 kD.

Nucleolin is regulating intranucleolar chromatin organization, → rDNA transcription, and rRNA processing, preribosomal synthesis, ribosomal assembly and maturation. It also is involved in cytoplasmic-nucleolar transport of preribosomal particles from the nucleolus to the cytoplasm.

Nucleolus: The spherical or globular subnuclear organelle associated with the so-called → nucleolus organizer region of chromosomes. It consists mostly of primary → rDNA transcripts, attached ribosomal proteins, and a variety of other proteins such as RNA polymerase I (A) and RNA methylases. In electron microscopic pictures the nucleolus is made up of a → fibrillar zone (pars fibrosa; containing rDNA) and a → granular zone (pars granulosa, containing pre-ribosomal particles). An active nucleolus exports large amounts of ribosomal precursors and exhibits special substructures, such as "pulsing vacuoles", less dense regions within the nucleolus that change their volume rhythmically.

Nucleolus organizer: See → nucleolus organizer region.

Nucleolus organizer region (NOR, nucleolus organizing region, nucleolar organizer, NO, nucleolus organizer): A specific chromosome segment containing the ribosomal RNA genes (→ rDNA) and active in the formation of the → nucleolus.

Nucleolus organizing region: See → nucleolus organizer region.

Nucleome: The microscopical and molecular description of all components of a nucleus of a eukaryotic cell. It encircles the DNA with all its constituents (→ genes, → promoters, repetitive sequences as →

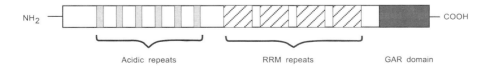

Nucleolin

satellites, → microsatellites, → transposons, → retrotransposons, → telomeres, → centromeres), the RNA (→ ribosomal RNA precursors, all the → small nuclear and → nucleolar RNAs) and the proteins (→ histones, → non-histone proteins, → scaffold proteins, → lamins). See → nucleomics.

Nucleomics: An ill-defined term of the → omics generation for the whole repertoire of technologies applicable to the study of nuclear architecture, → genomes, → transcriptomes, → post-transcriptional modification of transcripts, → post-translational modification of proteins, and nucleo-cytoplasmic interaction(s). See → behavioral genomics, → chemical genomics, → comparative genomics, → environmenral genomics, → epigenomics, → functional genomics, → genomics, → horizontal genomics, → integrative genomics, → kinomics, → medical genomics, → nucleome, → nutritional genomics, → pharmacogenomics, phylogenomics, → proteomics, → recognomics, → structural genomics, → transcriptomics, → transposomics.

Nucleomorph: A remnant gene-rich nucleus of a formerly free-living eukaryotic alga that has been engulfed by another eukaryotic cell and became an endosymbiont in a process called secondary endosymbiosis (where primary endosymbiosis is the acquisition of mitochondria and plastids by a recipient proto-eukaryotic cell). The process of this secondary endosymbiosis certainly occurred frequently in evolution, but nucleomorphs in only two algae groups, the cryptophytes and chlorarachniophytes have been preserved. These nucleomorphs contain three small linear → genomes (chlorarachniophytes: 380 kb; cryptophytes: 600 kb). Nucleomorph DNA encodes a total of 200–300 genes, among them diverse → housekeeping genes (for e.g. → transcription, mRNA → processing, → translation, protein degradation, and signal transduction) and genes for protein subunits needed in multiprotein complexes partly encoded by nuclear DNA of the alga, partly encoded by chloroplast DNA.

Nucleon: See → nuon.

Nucleoplasm (karyoplasm, karyolymph): The non-chromatin fluid phase of a → nucleus.

Nucleoplasmic reticulum: The reticular network of calcium stores within the nucleus of eukaryotic cells that is physically connected to the nuclear envelope and forms an (at least physiological) continuum with the endoplasmic reticulum. The nucleoplasmic reticulum is enriched in *in*ositol 1,4,5-tris*p*hosphate ($InsP_3$) receptors that generate local intra-nuclear calcium signals, thereby stimulating nuclear protein kinase C to translocate to the source of these signals. The nucleoplasmic reticulum is a potential intranuclear compartment involved in time- and space-specific intranuclear signalling.

Nucleoplasmin: The most abundant nuclear protein in some animals (e.g. *Xenopus* oocytes). It interacts as a → chaperone with histones H2A and H2B during the assembly of → nucleosomes, reducing their positive charges. Nucleoplasmin is a pentameric protein with a molecular weight of about 165 kDa, consisting of a highly charged carboxy-terminal tail and a globular amino-ter minal domain.

***Nucleoporin* (nup):** Anyone of a series of about 30 proteins associated with the → nuclear pore complex. For example, the socalled nup 180 (molecular weight: 180kDa) is located close to the annular pore complex at the cytoplasmic side of the pore, whereas the phenylalanine-glycine (FG)-rich nup 153 localizes to the nuclear side, with its C-terminal domain probably involved in the nucleo-cytoplasmic im- and export of RNA molecules.

Nucleoprotein: A complex of nucleic acid(s) and protein(s). For example, basic → histone proteins together with the associated phosphoric acid backbone of DNA form a nucleoprotein complex, the → nucleosome. Compare → ribonucleoprotein.

Nucleoprotein hybridization: A technique to isolate specific genes of an organism as → chromatin. In short, isolated nuclei are digested with appropriate → restriction endonucleases and lysed with → EDTA. Single-stranded termini of the nuclear chromatin fragments are generated by 5'-exonuclease digestion. Then a synthetic, biotinylated oligonucleotide complementary to the sequence adjacent to the restriction site on the targeted gene is hybridized in solution to the chromatin fragments. The oligonucleotide-chromatin hybrids are then immobilized on an → avidin matrix. They may be eluted by cleavage of the disulfide bond in the linker of the biotinylated probe (compare → biotinylation of nucleic acids). This type of → affinity chromatography allows the isolation of specific genes that retain their original chromatin structure.

Nucleoproteome (nuclear proteome): The → proteome of the eukaryotic nucleus, encompassing various classes of (preferentially all) nuclear proteins, as e.g. → chromatin proteins (e.g. → histones, → high mobility group proteins), nuclear matrix and nucleolar proteins, nuclear ribosomal proteins, → heat shock proteins, elongation factors, enzymes involved in DNA and RNA metabolism, RNA binding, nucleo-cytoplasmic trafficking, → nuclear pore transport, nuclear skeleton architecture and nuclear envelope maintenance, to name few.

Nucleosidase: Any enzyme that catalyzes the hydrolysis of → nucleosides to produce free bases and pentoses.

Nucleoside: A → pyrimidine or → purine base covalently linked to ribose (ribonucleoside) or deoxyribose (deoxyribonucleoside) via N-glycosidic bonds. See also → nucleoside antibiotic.

Nucleoside-α-thiotriphosphate (dNTPa-S): A purine or pyrimidine → nucleotide that contains a phosphorothioate diester bond and blocks the 3' → 5' proof-reading activity of → DNA polymerase I. Such nucleotides are used in → DNA sequencing and *in vitro* → mutagenesis procedures.

$$HO-\overset{O}{\underset{OH}{\overset{\|}{P}}}-O-\overset{O}{\underset{OH}{\overset{\|}{P}}}-O-\overset{S}{\underset{OH}{\overset{\|}{P}}}-O-CH_2\overset{O}{\diagup}Base$$
$$OH$$

2'-Deoxynucleoside-5'-O-(α-thio)-triphosphate (NTP)

Nucleoside analogue (NA): Any synthetic or naturally occurring substitute for a → nucleoside that is either incorporated into RNA or DNA and accepted without consequences, or blocks the subsequent synthesis of RNA or DNA. Such nucleoside analogues are used as therapeutic agents to block (or at least interfere) with DNA replication of viruses and tumor cells. The analogue triphosphates (NA-PPPs) are incorporated into the growing DNA chain and lead to an interruption of chain elongation. For example, the triphosphate of the thymidine analogue 3'-acido-3'-deoxythymidine (ACT) is used by the → reverse transcriptase of HIV (AIDS virus) and build into newly synthesized viral DNA. The acido residue at the C3 position of the ribose then blocks chain elongation and interrupts the life cycle of the virus.

3'-azido-3'-deoxythymidine

Ganciclovier (GCV)

Nucleoside antibiotic: Any one of a series of → purine or → pyrimidine nucleosides with → antibiotic activity. These compounds are formed in various bacteria and fungi by modification of → nucleosides, either by derivatization of the sugar (epimerization, isomerization, oxidation, reduction or decarboxylation of D-ribose) or the base moiety (methylation). They are antagonists of their naturally occurring nucleosides, and therefore block the metabolism of purines, pyrimidines, and proteins. Examples for such nucleoside antibiotics are amicetin A and B (*Streptomyces fasciculatus, S. plicatus*), 5-azacytidine (*Streptoverticillius lakadamus*), blasticidin S (*Streptomyces griseo-chromogenes*), cordycepin (*Cordyceps militaris*), nucleocidin (*Streptomyces calvus*), puromycin (*Streptomyces albo-niger*) and tubercidin (*Streptomyces tubercidicus*). Some of them are used in molecular biology (see e.g. → azacytidine, → cordycepin, → puromycin).

Nucleoside bisphosphate (nucleoside diphosphate): Any one of a series of ribose-modified nucleotide analogues that contains phosphate residues at various positions of the ribose moiety, as e.g. the 3' and 5', or the 2' and 5' carbon atoms. Such analogues are used for the mapping of active sites in ribonucleases or other nucleotide-binding enzymes, the inhibition of nucleotide-dependent enzymes, and protein affinity studies.

Adenosine-3', 5'-bisphosphate (pAp), triethylammonium salt

Nucleoside extrusion (base flipping): The opening of base pairs in a DNA double helix, whereby an entire nucleoside is swiveled out of the helix and inserted into the recognition pocket of a DNA-binding protein. Base flipping is induced by the torsional stress imposed onto the double helix by binding the protein.

Nucleoside polyphosphate: Any one of a group of highly phosphorylated bacterial → nucleotides that contain phosphate groups at the 3′-carbon atom of the ribose in addition to the 5′-phosphate (that is normal in nucleotides). Nucleoside polyphosphates resemble so called → alarmones. For example, → guanosine tetraphosphate (responsible for the so called stringent response), guanosine-3′-diphosphate-5′-triphosphate (pppGpp, an intermediate of ppGpp biosynthesis) and guanosine-5′-diphosphate-3′-phosphate (ppGp, a degradation product of ppGpp) are such nucleoside polyphosphates.

Nucleoskeleton: An intranuclear network of fibrils (e.g. of actin and myosin) that is thought to coarsely compartment nuclear reactions. The nucleoskeleton contains for example, anchorage or attachment sites for → looped domains of → chromatin.

Nucleoskeleton theory: A hypothetical attempt to describe the enormous variation in → genome size between related organisms as the result of selection for cell sizes: the bigger the cell size, the bigger the genome. See → C-value paradoxon, → G-value paradox.

Nucleosomal array: A stretch of 10–20 → nucleosomes assembled *in vitro* from → histones (or part of histones) and DNA. Such arrays serve to identify interactions between nucleosomes and various proteins, RNAs, or low molecular weight compounds.

Nucleosome (nu particle, nu body): A disk-shaped structure of eukaryotic chromosomes consisting of a core of eight → histone molecules (two each of H2A, H2B, H3 and H4) complexed with 146 bp of DNA and spaced at roughly 100 Å intervals by "linker" DNA of variable length (8–114 bp; see → nucleosome phasing) to which histone H1 attaches. Nucleosomes mainly serve to package DNA within the nuclei of eukaryotic cells, but play also important roles in gene activation/inactivation. *In vitro* reconstitution of nucleosomes is possible. See → gradient dialysis. See also → lexosome, → nucleosome occupancy, → nucleosome phasing, → nucleosome positioning code.

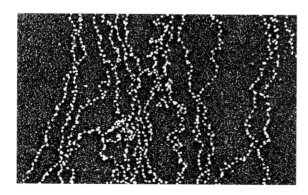

Nucleosomes
of chromatin
of a potato plant

Nucleosome assembly protein (NAP): Any member of a nuclear protein family that participates in → DNA replication, modulates → chromatin formation by assisting in the assembly of → nucleosomes *in vivo* and contributes to the regulation of cell proliferation. Alternative → splicing of the gene encoding NAP-1 produces several transcript variants.

Nucleosome-free region (NFR): A ~150 bp long sequence about 200 bp → upstream of the → initiation codon of many yeast genes that is flanked by → nucleosomes containing the → histone 2A variant H2A.Z, and harbors the → transcription initiation site.

Nucleosome occupancy: The density of → nucleosome positioning along a stretch of DNA, as revealed e.g. by → chromatin immunoprecipitation-chip (ChIP-chip) assays employing protein-specific → antibodies, mostly directed against specific → histones. For example, ChIP-chip experiments in yeast using an antibody against histone H3 or → epitope-tagged histone H2B or H4 revealed that → promoters and → coding regions of transcribed genes generally have fewer (more widely spaced) nucleosomes than non-transcribed genes.

Nucleosome phasing (phasing): The non-random arrangement of → nucleosomes along nuclear DNA of eukaryotic chromosomes. Though the underlying mechanisms of nucleosome phasing are not fully known, it is generally accepted that phasing is a means of gene control. A regulatory sequence within the → promoter of a gene, for example that is supercoiled around a nucleosome generally is not accessible for trans-acting or regulatory proteins. Once such a nucleosome has been partly dissociated (relaxed; → lexosome), the regulatory sequence becomes fully available for binding proteins.

Nucleosome positioning code: A distinctive sequence motif recurring periodically in → genomic DNA of eukaryotes that facilitates the sharp bending of DNA around the → nucleosome and thereby position the formation of a nucleosome from the eight → histone proteins (two each of H2A, H2B, H3 and H4) and the DNA. For example, the 10-bp periodic AA/TT/TA dinucleotide motifs oscillating in phase with each other and out of phase with 10-bp periodic GC dinucleotides represent such positioning codes. This intrinsic organization can explain 50% of the *in vivo* nucleosome positions. The overall nucleosome occupancy of different genomic regions varies. For example, → centromeres have the highest occupancy, probably because centromere function requires enhanced stability of histone–DNA interactions. In contrast, the highly expressed → ribosomal RNA and → transfer RNA genes own low predicted nucleosome occupancy, and nucleosomes are depleted from → transcriptional start sites (TSSs) and → promoters, the depletion being intrinsically encoded in the DNA sequence. Here, the → TATA boxes are virtually free from nucleosomes and localized outside of stably positioned nucleosomes. The location of these stably positioned nucleosomes is conserved across all fungal species (at least yeasts). Eukaryotic genomes direct the transcriptional machinery to functional sites by encoding unstable nucleosomes over the underlying sequence elements, thereby enhancing their accessibility. The nucleosome positioning code facilitates specific chromosome functions including → transcription factor binding, → transcription initiation, and even remodelling of the nucleosomes themselves.

Nucleosome remodeling complex (NuRD): A multiprotein complex capable of changing → chromatin architecture (e.g. the transition from relaxed [active] to tense [inactive] configuration), consisting of at least six different proteins, one of which (called MBD₃) binds to methylated CpG residues and recruits histone deacetylases and histone methylases. Deacetylases convert the acetylated histones such that they are no more inducive for an open chromatin structure. Result: the nearby genes are silenced. See → nucleosome remodelling factor.

Nucleosome remodeling factor (NURF): An ISWI-containing ATP-dependent → chromatin remodeling complex of *Drosophila melanogaster* that regulates → transcription by catalyzing → nucleosome sliding. NURF is a protein complex of four subunits that assists transcription factor-mediated perturbation of nucleosomes in an ATP-dependent manner. Once nucleosome remodeling by NURF is accomplished, a high level of NURF activity is no more required for recruitment of the general transcriptional machinery. See → nucleosome remodelling complex.

Nucleotide (nt): A → pyrimidine or → purine → nucleoside that is esterified with one, two, or three phosphate groups at the 5' carbon atom of the ribose (ribonucleotide) or deoxyribose (deoxyribonucleotide). Ribose-containing nucleotides include ribonucleoside *mono*phosphate (NMP), ribonucleoside *di*phosphate (NDP) and ribonucleoside *tri*phosphate (NTP), deoxyribose-containing nucleotides include *d*eoxy-ribonucleoside *mono-* (dNMP), *di-* (dNDP) and *tri*phosphates (dNTP). DNA contains deoxyadenylate, thymidylate, deoxyguanylate, and deoxycytidylate nucleotides; RNA adenylate, uridylate, guanylate, and cytidylate.

Nucleotide analogue interference mapping: A technique for the detection of amino acids in a protein that interact with a target RNA. In short, a nucleotide is exchanged in a gene by conventional gene technology, the modified gene re-introduced into the target organism and expressed there. The resulting protein is then tested for its ability to bind to the cognate RNA. By using a series of such mutants, each coding for a protein with a specific amino acid replacement, it is possible to map (localize) such amino acids that are involved in the interaction process (or, the nucleotides in the gene that encode these amino acids). This technique has e.g. been used to determine the amino acids in the → RNase P protein from *E. coli* which interact with → transfer RNA during the binding process.

Nucleotide-binding domain ("nucleotide-binding fold"): Any conserved → domain of a protein, containing specific amino acid sequence motifs that form a pocket to bind and accommodate a nucleotide (usually as triphosphate, as e.g ATP or GTP). For example, the ATP-binding motif ("Walker motif") of certain proteins consists of a small stretch of hydrophobic amino acids followed by [gly/ala]-X-X-gly-X-gly-lys-thr/ser (where X is any amino acid). The hydrophobic residues form a buried β-strand, the glycine-rich region a loop ("P-loop") that interacts with the phosphate of the bound nucleotide. For the isolation of the encoding genes, → primers can be designed against these conserved domains and used to amplify parts of the gene from genomic DNA via conventional → polymerase chain reaction techniques. This strategy is employed for the isolation of → resistance gene analogues (RGAs) in plants, where the forward primer could be directed against one of the (three) nucleotide binding sites (NBSs), and the reverse primer against the transmembrane domain or the leucine-rich repeat motifs of the extra-cellular part of such proteins.

Nucleotide diversity: The number of base differences between two (or more) → genomes, divided by the number of base pairs compared.

Nucleotide diversity map: See → diversity map.

Nucleotide diversity per site (π): The frequency with which any two nucleotide sequences differ at a specific site.

Nucleotide excision repair (NER): A prokaryotic DNA repair system, encoded by genes *uvr*A (encoding an ATPase subunit of endonucleases), *uvr*B and *uvr*C (encoding the endonuclease subunits of *E.coli* excinuclease), and *uvr*D (coding for a helicase removing the excised stretch of DNA) that repairs from few to more than several thousands of nucleotides. It is particularly active in the removal of UV photoproducts, alkylated adducts, and oxidized DNA. First the ABC excinuclease recognizes damaged sites ("damage recognition"), cuts at two flanking sites and removes the intervening sequences ("dual incision excision"). Then → DNA polymerase I catalyses repair synthesis, gaps are filled by any of the four DNA polymerases, and the ends ligated. Eukaryotic NER protein machines more or less process DNA damage sites the same way. The initial step is damage recognition by XPC and (in humans) hHR23B (homolog of *Saccharomyces cerevisiae* Rad23) that concertedly recruit other repair proteins. XPB and XPD → helicases mediate strand separation at the lesion site, and XPA identifies the damaged area in an open DNA conformation. The unwound DNA is stabilized by RPA that also positions XPG and ERCCI-XPF endonucleases. These nucleases catalyze the incision around the lesion. Once the lesion is removed, the gap is filled by replication proteins, and the repair process is complete. NER systems are also active in e.g. mammalian organisms (more than 30 different proteins), and their failure causes rare autosomal recessive disorders such as Xeroderma pigmentosum. See → base excision repair, → mismatch repair, → transcription-coupled repair.

Nucleotide heterozygosity (η): The average number of nucleotide differences between two nucleic acid sequences selected at random from a particular population of organisms. h depends on the number of polymorphic sites and their frequency in the nucleic acid region in focus.

Nucleotide mapping: A misleading term for the isolation and characterization of nucleotides resulting from an enzymatic digestion or the chemical hydrolysis of a target DNA or RNA.

Nucleotide pair: See → base pair.

Nucleotide replacement site: Any position in a → codon where a → point mutation has occurred.

Nucleotide sequence: See → DNA sequence.

Nucleotide substitution: The exchange of one → nucleotide in a DNA molecule for another one. Such substitutions are neutral, if the → genetic code is not changed, but have massive consequences, if the genetic code is altered (e.g. result in the synthesis of a non-functional protein).

Nucleotide turnover rate: The maximum number of nucleotides polymerized per molecule of → DNA dependent DNA polymerase per minute. Compare → processivity.

Nucleus: An organelle of eukaryotic cells, surrounded by a double-membrane system (→ nuclear envelope) with pores (→ nuclear pore), and containing the → chromosomes in the form of → chromatin (i.e. associated with a multitude of proteins). Compare → nucleoid.

Nucleus-*encoded* *polymerase* (NEP): Any → DNA-dependent RNA polymerase that is encoded by the nuclear genome. See → plastid-encoded polymerase.

Nucline™: The trademark for a full-length synthetic → messenger RNA (mRNA) of 200–3000 bases, encoding a therapeutic protein of choice and containing the → start codon AUG and a → 5′-untranslated region (5′-UTR) with a short sense sequence of the target mRNA to be detected, to which a 20–200 bases long → antisense oligonucleotide is bound. This antisense oligo forms a specific secondary structure upstream of the AUG start codon and thereby prevents → translation of the corresponding message. Such a secondary structure does not allow the → ribosome to access the start codon. However, if the full-length → pre-mRNA or mRNA of the target gene is present in a cell, the antisense molecule is competed off the nucline molecule through homologous → hybridization ("sense-antisense switch"). In this case, the secondary structure no longer exists, the ribosome translates the nucline-mRNA, and the encoded protein appears in the cell. The nucline is synthesized by ligating an effector gene (to be translated; as e.g. a toxin-encoding or response modifier gene or → oncogen) immediately downstream of an AUG start codon that in turn is located downstream of a 5′-UTR containing a 10–100 bases long sequence ("switch"), and inserting this construct into an → expression plasmid. Then → T7 RNA polymerase is used to amplify the nucline molecule, which is purified and mixed with an antisense oligonucleotide that binds to the sense sequence upstream of the start codon. Nucline RNA can then be introduced into target cells by → biolistics, → lipofection or → electroporation. The potential of nucline can be expanded, if two (or more) sense sequence switches are inserted upstream of the start codon, each switch coding for a different mRNA. For example, if nucline contains both a sense sequence for CD4 and another one for IL-10, then this nucline is only derepressed in cells expressing CD4 and IL-10 (as e.g. in Th2 cells). Also, two separate nucline molecules with different, but multiple switches can operate in the same target cell. For example, nucline A may contain a CD4 and IL-10 sense switch and therefore release the pertussis toxin encoded by the adjacent gene in cells that co-express CD4 and IL-10 proteins (as in Th2 cells). Nucline B, composed of multiple switches of γ-globulin-encoding sequence, targets plasma B cells that overexpress γ-globulin, leading to hypergammaglobulinemia and multiple myeloma (a rare B cell lymphoma). Treatment of this disease is based on the derepression of both nucline A and B in the different cell types and release of the toxin that kills the target cells, but does not affect other cells. The nucline system therefore can be used to introduce a therapeutic RNA into a target cell.

NUE: See → *n*ear-*u*pstream *e*lement.

Null allele: Any → allele whose DNA sequence has been changed by one or more → mutations such that (1) it can no longer be detected by → allele-specific probes in → genomic DNA and (2) the encoded protein is no more functional (i.e. can no longer be detected by e.g. → immunoassays).

Nullisomy: The absence of a complete chromosome pair from the → karyotype of a cell (in a diploid organism: 2n-2). See → disomy.

Null mutation: Any mutation that leads to a complete loss of function of the sequence in which it occurs.

Null promoter: A → promoter that does not contain a → TATA box or the → initiator element (TATA Inr), and therefore allows multiple start sites for → transcription initiation. Some of the null promoters share an intragenic sequence motif (*mu*ltiple start site *e*lement *d*ownstream, MED-1) that is indispensable for null promoter function and replaces the conventional → consensus motifs (as e.g. TATA box).

Num (*nu*clear *m*igration): Any one of a series of proteins of filamentous fungi that is composed of three domains (an NH_2-terminal heptad region, a central region with → direct repeats, and a carboxyterminal PH [*p*leck strin *h*omology] region) and can be translocated from cytoplasm to the nucleus.

Numt: See → nuclear mitochondrial DNA segment.

Nuon (*nu*cleic acid, nucle*on*): Any coding or non-coding DNA or RNA sequence. For example, → genes, → introns, → exons, → retrotransposons, → spacers, → enhancers, → silencers, → microsatellites all are nuons.

n/u orientations: The two orientations possible when a fragment of foreign DNA is inserted into a → cloning vector. N, when both vector and insert have the same orientation; u, when insert and vector are in different orientations.

NUP: See → *nu*clear *p*ore complex protein.

NUPT: See → nuclear plastid DNA.

Nu particle: See → nucleosome.

NURD: See → nucleosome remodeling complex.

Nurse cells: See → feeder cells.

NusA tag (Nus tag): A short peptide sequence from the NusA protein of *E. coli* that can be fused to a target protein and thereby increase the solubility of the fused protein in the bacterial host. The sequence encoding the NusA peptide is cloned into a suitable plasmid vector, fused to the target protein gene, and appropriately flanked by → histidine tag-encoding sequences, a protease cleavage site (for the removal of the tags) and a → T7 RNA polymerase promoter. Expression of the fused protein in the host cell can reach high levels without solubility problems. See → strep tag.

Nutrient broth (NB): A medium rich in mineral salts, vitamins and carbohydrates and otherwise useful compounds that is used for the growth of microorganisms. Contrary to → minimal medium, which contains only basic chemical compounds.

Nutrigenetics: A branch of → genetics that focuses on deciphering the complex interactions between the genetic predisposition and the uptake, processing and utilization of nutrients, as well as their influence on the immune, digestive and metabolic systems of the consumer. In particular, nutrigenetics deals with the genetic variation in the genomes of humans and animals, and how this variation (mostly as → single nucleotide polymorphisms) influences the relationship between diet and disease.

Nutrigenome: The complete set of (still largely unknown) genes that underlies the nutritional qualities of animals and plants (or parts of them) consumed by humans. See → nutrigenomics, → nutritional genomics.

Nutrigenomics: The whole repertoire of techniques designed to decipher the complex interactions between the genetic predisposition and the uptake, processing and utilization of nutrients, as well as their influence on the gene expression, immune, digestive and metabolic systems of the consumer. Do not confuse with → nutritional genomics. See → behavioral genomics, → chemical genomics, → comparative genomics, → environmental genomics, → epigenomics, → functional genomics, → horizontal genomics, → integrative genomics, → medical genomics, → nutritional genomics, → omics, → pharmacogenomics, → phylogenomics, → proteomics, → recognomics, → structural genomics, → trans-criptomics, → transposomics.

Nutritional genomics:

a) An infelicitous term for a series of techniques to improve the nutritional quality of plants through the transfer of foreign, novel, or altered genes encoding enzymes that produce nutritional compounds. These genes are used to increase the levels of essential or desirable micronutrients in crop plants. For example, *Arabidopsis thaliana* converts γ-tocopherol to α-tocopherol (vitamin E) at a very low rate only. The →transformation and → overexpression of the gene encoding a γ-tocopherol methyltransferase (γ-TMT) in this plant allows to increase its α-tocopherol content substantially.

b) A series of techniques for the study of interactions between diet and genome in human (also animal) individuals with the goal to identify genetic predispositions for a certain diet and to propose dietary regimes. An important focus is the → association of certain → single nucleotide polymorphisms (SNPs) and dietary disease risks. For example, diet influences the expression of certain genes, amongst them the gene encoding *methylenetetrahydrofolate reductase* (*MTHFR*). The majority of humans possess an *MTHFR* gene with a cytosine (C) at base pair position 677, whereas about 10% Scandinavians and 15% Mediterranean Europeans carry a thymine$_{677}$ instead. Homozygosity for the $C_{667}T$ polymorphism causes moderate hyperhomocysteinemia (especially in individuals whose folic acid intake is low). Homocysteine is a key intermediate in methionine metabolism, and elevated levels of homocyteine are associated with a higher risk of cardiovascular diseases. The T_{667} → allele is also linked to a lower risk of colorectal adenomas and colon cancer (when folate intake is normal).

See → behavioral genomics, → biological genomics, → cardio-genomics, → chemical genomics, → clinical genomics, → comparative genomics, → deductive genomics, → environmental genomics, → epigenomics, → functional genomics, → horizontal genomics, → integrative genomics, → lipo-proteomics, → medical genomics, → neurogenomics, → neuro-proteomics, → nutrigenetics, → nutritional genomics, → omics, → pathogenomics, → pharmacogenomics, → phylogenomics, → physical genomics, → population genomics, → proteomics, → recognomics, → structural genomics, → transcriptomics, → transposomics.

N-value paradox: See → G-value paradox.

Nylon macroarray: See → macroarray.

Nystatin (mycostatin; fungicidin): A polyene → antibiotic from *Streptomyces nouresii* that affects specifically fungal growth through the formation of complexes with membrane-bound cholesterols. These Complexes generate "pores" in the membrane and lead to uncontrolled leakage of solutes. Since it is not active against bacteria, it is used to keep bacterial cultures free from fungi.

Nytran™: The trade-mark for a nylon membrane that is used to immobilize nucleic acids and proteins. It is positively charged and therefore electrostatically binds the negatively charged nucleic acids or SDS-protein complexes.

O

OATFA: See → oligonucleotide array-based transcription factor assay.

OC: See → open circle.

O^c: See → operator constitutive mutation.

OC-DNA: See → open circle.

OcDNA: See → open circular DNA.

Ochre codon: The triplet UAA in mRNAs which is not recognized by any → tRNA, but signals the termination of → translation, (→ stop codon). See also → ochre mutation, → ochre suppressor.

Ochre mutant: A bacterial mutant that synthesizes mRNA carrying an → ochre mutation.

Ochre mutation: A base substitution which converts an amino acid specifying → codon into the → stop codon UAA (→ ochre codon). Usually such a mutation leads to premature termination of polypeptide synthesis and the formation of abnormally short polypeptides. Its effect can, however, be neutralized by an → ochre suppressor mutation. See also → nonsense mutation.

Ochre suppressor: A mutant gene coding for a mutant → tRNA, which recognizes the → stop codon UAA and causes the insertion of an amino acid into the growing polypeptide chain at the termination site.

Ocs element: Any one of a family of related, bipartite, *cis*-acting 20 bp sequence elements, usually located in between the → TATA box and nucleotide −200 in the promoters of various bacterial, viral and plant genes. Ocs elements, originally identified in the promoter of the → *Agrobacterium tumefaciens* → *oc*topine *s*ynthase (ocs) gene, are also present in other promoters of *Agrobacterium tumefaciens* (e.g. → nopaline synthase gene promoter), promoters of viruses (e.g. 35S promoter of → cauliflower mosaic virus, here the element is called as-1; 19S and 35S promoters of the *figwort mosaic virus*, FMV; also in the badnavirus Commelina Yellow Mottle Virus) and plants (e.g. glutathione S-transferase gene promoter). Plant promoters containing ocs elements are activated by the plant hormone auxin and salicylic acid, which is part of a stress response. Ocs elements are target sites for the highly conserved *b*asic domain-leucine *zip*per (bZIP) transcription factors (socalled *o*cs element *b*inding *f*actors, OBFs). The ocs element contains functionally identical, tandemly arranged nuclear protein-binding sites, each site centered around the consensus core sequence 5′-ACGT-3′ and harboring a binding site for plant transcription factors ("OTFs"). Occupation of both binding sites is required for ocs element function. The 16 bp palindromic consensus sequence of the ocs element is 5′-TGACGTAAGC-GCTTACGTCA-3′ (dashes: variable nucleotides). The ocs element is used for tissue-specific expression of genes in → transgenic plants.

ocs gene: See → octopine synthase gene.

Octamer-based genome scanning (OBGS): A technique for the detection of sequence length differences between over-represented, strand-biased octamer nucleotide stretches in the *E.coli* genome. The technique exploits the presence of about 150 different over-represented oligomers, whose occurrence is skewed to one strand (the leading strand) of the genome. Of these, 23 octamers are represented from 515–867 times, and probably function as priming sites for discontinuous DNA replication. For OBGS, fluorescently labeled octamer-based primers (octamers from the leading strand) are mixed with unlabeled octamer-primers (octamers biased to the lagging strand) and used to amplify the octamer-octamer regions in a → conventional polymerase chain reaction. The size distribution of the fluorescent products is then measured on automated sequencers followed by establishments of binary files from the absence and presence of bands.

Octamer-binding transcription factor (OTF): One of several nuclear proteins that bind to the consensus sequence 5'-ATTTGCAT-3' present in promoters of several protein-coding genes (e.g. histone H2B and immunoglobulin light chain genes) and enhancers (e.g. an enhancer of the → RNA polymerase II-dependent U1 and U2 RNA genes). The octamer sequence in → class II genes interacts with two distinct transcription factors (OTF-1, ubiquitous; OTF-2, B-cell specific).

oct gene: See → octopine synthase gene.

Octopine (N-α-[D-1-carboxyethyl]-L-arginine): An amino acid derivative that is synthesized in plant cells transformed by the octopine strain of the soil bacterium → *Agrobacterium tumefaciens*. This bacterium, after contact with wound-exposed plant cell walls transfers part of a large plasmid (→ Ti plasmid) into the plant cell where it is integrated in the nuclear DNA. The expression of a gene (*ocs* gene) close to the right border of the transforming DNA (→ TL-DNA) leads to the production of the enzyme → octopine synthase that synthesizes octopine from pyruvate and L-arginine. Octopine cannot be used by the host plant cell, but is secreted and serves as a carbon, nitrogen and energy source for agrobacteria possessing *occ* (octopine catabolism) genes on their Ti-plasmid (see → genetic colonization). Octopine is an → opine.

$$\begin{array}{c} HN \\ \diagdown \\ C-NH-(CH_2)_3-CH-COOH \\ H_2N\diagup | \\ NH \\ | \\ H_3C-CH-COOH \end{array}$$

Octopine synthase (octopine synthetase; EC 1.5.1.11): An enzyme present in → crown gall tumor cells and encoded by the *ocs* gene (see → octopine synthase gene) of the → T-DNA originating from the → Ti plasmid of → *Agrobacterium tumefaciens*. Octopine synthase catalyzes the synthesis of the unusual amino acid → octopine from pyruvate and L-arginine.

Octopine synthase gene (ocs gene, oct gene): A gene of the → Ti-plasmid of → *Agrobacterium tumefaciens* that encodes the enzyme → octopine synthase and is only expressed in transformed plant cells (see → crown gall). The *oct* gene is frequently used as a → reporter gene in plant transformation experiments, its → promoter (*Pocs*) and → termination sequences (3' t oct) are incorporated in → plant transformation vectors.

Octopine synthetase: See → octopine synthase.

Oct-protein: Any one of a series of DNA-affine proteins that bind specifically to *oct*amer sequences. See → homeodomain.

ODN: See → oligodeoxynucleotide.

Odorant receptor (OR) gene: Any one of a large → gene family that encodes a seven-trans-membrane → domain protein functioning as odorant receptor. For example, *Drosophila* has a highly diverse family of such *OR* genes, and the individual members are expressed in different subsets of *o*lfactory *r*eceptor *n*eurons (ORNs). ORNs expressing the same odorant receptor project into the same glomerulus (functional processing unit) in the antennal lobes of the fly.

OFAGE: See → *o*rthogonal-*f*ield-*a*lternation *g*el *e*lectrophoresis.

O'Farrell electrophoresis: See → two-dimensional gel electrophoresis.

O'Farrell gel: See → two-dimensional gel electrophoresis.

O'Farrell gel electrophoresis: See → two-dimensional gel electrophoresis.

Off-gel electrophoresis: A free-flow technique for the separation and isoelectric purification of proteins according to their charge that works with a flow chamber of minute dimensions ($4 \times 0.6 \times 40$ mm), where one wall consists of an → *i*mmobilized *p*H *g*radient (IPG) gel such that it buffers a thin layer of solution (without any carrier ampholytes). The protein molecules migrate in this solution rather than in a gel matrix as in conventional electrophoresis methods. An electric field is applied perpendicular to the flow of the solution. Due to the buffering capacity of the IPG gel, proteins with an isoelectric point close to the pH of the gel in contact with the flow chamber stay in solution, because they are neutral. Other proteins are charged, when approaching the IPG gel and are migrating into the gel. The positively charged proteins migrate to the cathode and penetrate the gel, the negatively charged proteins penetrate the gel towards the anode. The proteins are recovered in free flowing solution. Off-gel electrophoresis separates faster and achieves a higher resolution than gel-based separation techniques. The separated proteins are immediately ready for further analyses such as → two-dimensional gel electrophoresis, crystallization, → protein microarrays or → mass spectrometry.

Off-strand: A laboratory slang term for the strand in a double-stranded → microRNA or → siRNA that is not incorporated into the → RISC complex. See → active strand.

Off-target effect ("side effect"): The influence of a chemical substance (e.g. an inhibitor) on a process or processes, which are no actual targets for it. For example, → *s*mall *i*nterfering RNA (siRNA) designed to block the expression of a specific gene, also interferes with the expression of other unrelated genes ("off target"). In this specific case, off-target effects can be reduced by decreasing the intracellular, active siRNA concentration (normal range: 5–200 nM).

Off-target silencing: The undesirable silencing of a gene (or genes) that has a similar sequence to a gene targeted by →

RNA interference. In the normal RNA interference process, the → antisense strand of the → siRNA binds to the cognate → messenger RNA (mRNA) within the so-called → RISC complex. If, by chance, the sense strand of a different mRNA with far-reaching sequence identity is also identified by the siRNA, then this mRNA is destroyed as well, though not intended.

OH⁸dG: See → 8-hydroxy-2′-deoxyguanosine.

Ohnolog: A region of eukaryotic genomes, in which functionally (and phylogenetically) related genes are clustered. These homologues arose by → gene amplification and are frequently duplicated in a genome. Ohnologs are named after Dr. Ohno, who first postulated that much of eukaryotic genomes consists of duplicated sequences.

Ohno's law (Ohno's rule): The far-reaching conservation of the gene order on X chromosomes (also autosomes) of different mammals. For example, genes located on human X chromosomes are also part of mouse X chromosomes and vice versa. However, as a consequence of → genomic rearrangements during evolution over the past 80–90 millions of years (evolutionary separation of both species), the relative order of blocks of conserved sequences is less or not colinear. This means that within such blocks the gene order is well conserved, but the relative order of the syntenic groups between mouse and man is variable. See → macrosynteny, → microsynteny, → synteny.

Ohno's rule: See → Ohno's law.

Okayama-Berg cloning (Okayama-Berg method): An efficient method to construct a → library of full-length → cDNAs, using oligonucleotide-tailed vector fragments that allow cDNA synthesis and cloning in one coordinate experiment. In short, the poly(A)-containing mRNA is first annealed to an oligo(dT)-tailed plasmid primer and → reverse transcriptase used to synthesize a cDNA. The generated vector-mRNA-cDNA hybrid is oligo(dC)-tailed at the 3′ OH-terminus of the full-length cDNA using → terminal transferase (non-full-length cDNAs are not efficiently dC-tailed and therefore eliminated). Then the plasmid primer is trimmed with the → restriction endonuclease Hind III to remove the unnecessary oligo(dC)-tailed segment and a Hind III-linker is annealed to the oligo(dC)-tailed plasmid, and ligated with E. coli → DNA ligase. The original mRNA is selectively removed with E. coli → RNase H and replaced by E. coli → DNA polymerase I. The remaining → nicks are sealed with E. coli DNA ligase before transformation of an E. coli host. See also → Okayama-Berg cloning vector, → Honjo vector. Compare → Heidecker Messing method.

Figure see page 1099

Okayama-Berg cloning vector (Okayama-Berg vector): Any cloning vector (usually derivatives of → pBR 322) that is specially designed for the → Okayama-Berg cloning of → cDNA. In short, a pBR 322 molecule is first cut with the → restriction endonuclease Kpn I and → oligo(dT) tails are attached to the termini using → terminal transferase (see → DNA tailing). A subsequent Hpa I digestion removes one oligo(dT) tail, leaving the other for the annealing of the mRNA → poly(A) tail. See for example → Honjo vector.

Okayama-Berg method: See → Okayama-Berg cloning.

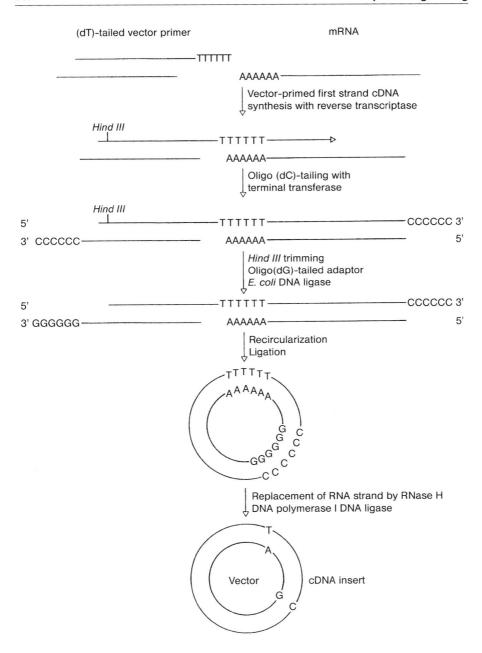

Simplified scheme of Okayama-Berg cloning

Okayama-Berg cloning

Okayama-Berg vector: See → Okayama-Berg cloning vector.

Okazaki fragment: A DNA fragment of several thousands (bacteria) or a few hundred of nucleotides (eukaryotes) that is newly synthesized during DNA replication on the → lagging strand, starting at an RNA → primer (synthesized by an → RNA primase). The Okazaki fragments are covalently linked by ligases to give a continuous strand.

OLA: See → oligonucleotide ligation assay.

Olfactory receptor gene (OR gene): Any one of a superfamily of more than 1000 gene copies in primate genomes that encodes an olfactory G-protein-coupled receptor (binding odorant molecules within the nasal epithelium). The prominent receptor molecules are anchored in the plasmamembrane by seven transmembrane domains (rat), possess a high sequence similarity (60–99%) and are almost exclusively expressed in the nasal epithelium. Whereas in e.g. rodents (with a highly developed sense of smell) almost all *OR* genes are encoding functional receptor molecules, about 70% of all human *OR* genes are non-functional → pseudogenes. This is most probably the cause for a greatly reduced sense of smell relative to other mammals (as e.g. rodents). Some of the olfactory genes (e.g. human olfactory receptor 17-4-encoding gene) are also expressed in sperma cells. The presence of the receptor protein on the surface of sperma allows them to sense gradients of odorants such as bourgeonal or cyclamal, upon which they react with a G-protein-dependent activation of a membrane-bound adenylate cyclase III, a subsequent increase in cAMP concentration and a massive influx of calcium ions into the cells.

Oligo: See → oligonucleotide.

Oligo-capping: A technique for the *in vitro* capping of eukaryotic → messenger RNA (mRNA) to define the 5'-cap site accurately. In short, isolated mRNA is first treated with → alkaline phosphatase to remove the 3'-terminal phosphate, and then with → *t*obacco nucleotide *a*cid *p*yrophosphatase (TAP) to remove the 5'cap of the message. Subsequently, a → T4 RNA ligase is used to ligate a specific 38-mer oligoribonucleotide to the 5'-end of the de-capped message ("re-capping"). The sequence of the 38-mer oligo cap is only rarely represented in mRNA databases. The oligo-capped mRNA is then converted to a stable cDNA by reverse transcriptase employing either a random hexamer or an oligo(dT) primer. The double-stranded cDNA is then purified and used to determine the exact sequence around the original cap site.

Oligoclonics: A somewhat misleading term for a mixture of recombinant human → antibodies highly expressed in clonal cell lines. Oligoclonics are typically directed against multiple epitopes or targets.

Oligodeoxynucleotide: See → oligonucleotide.

Oligo(dT) cellulose: A cellulose matrix, covalently coupled to thymidylic acid oligomers up to 30 nucleotides in length, which is used for the quantitative binding and isolation of poly(A)$^+$-mRNA in oligo(dT) cellulose → affinity chromatography. See also → oligo(U)-sepharose.

Oligo(dT) ladder: A set of single-stranded (dT) oligodeoxynucleotides ranging in size from 4 to 22 nucleotides with 1 bp intervals. This ladder is used as → marker for the precise determination of the size of electrophoretically separated oligodeoxynucleotides (e.g. → linkers, → primers).

Oligo(dT) primer: A synthetic homopolymeric → oligodeoxynucleotide that can be annealed to the → Poly(A) tail of polyadenylated mRNA and used as a → primer to drive → first strand → cDNA synthesis by → reverse transcriptase. See → anchored oligo(dT) primer.

Oligo(dT) priming: The use of a 12–20-mer oligo(dT) deoxynucleotide for the synthesis of the → first strand in → cDNA cloning procedures. See → oligo(dT) primer.

Oligo(dT) tail: A single-stranded tail of deoxythymidine nucleotides added to the termini of linear DNA molecules by the enzyme → terminal transferase.

Oligogene: A vaguely defined term for a gene with a small, but identifiable effect on a → phenotype (e.g. a disease risk), in contrast to e.g. a → major gene (contributing most to a particular phenotype). See → polygene, → polygenic trait.

Oligolabeling: See → random priming.

Oligomer (Greek: olígoi for "some, few" and méros for part): Any macromolecule that is composed of only a limited number of monomeric subunits covalently linked to each other. For example, → oligonucleotides usually consist of up to 100 nucleotides, → oligopeptides contain up to 10–15 amino acids. Compare → polynucleotide, → polypeptide.

Oligomerization: The covalent linkage of identical oligonucleotides to form long DNA molecules (see → concatemer).

Oligomer restriction (OR): A rapid → liquid hybridization method for the detection of a specific → restriction endonuclease → recognition site at any location in genomic DNA. To this end a ^{32}P end-labeled oligonucleotide probe is hybridized under stringent conditions to a specific segment of denatured genomic DNA that spans the target restriction site. Any mismatch within the restriction site prevents subsequent endonucleolytic cleavage of the duplex formed between the probe and its genomic target sequence. This allows the detection of allelic variants.

Oligomer skew: The unequal distribution of short → oligonucleotide stretches (e.g. → microsatellites) at specific regions (e.g. → origins of replication) on both strands of → double-stranded DNA in prokaryotes.

Oligomer walking: See → multiplex walking.

Oligo-mismatch mutagenesis (oligonucleotide-primed mutagenesis, oligonucleotide-directed mutagenesis, oligonucleotide-directed double-strand break repair): The introduction of site-specific → mutations into a target DNA molecule by annealing a specifically designed synthetic → oligodeoxynucleotide (7–20 nt long). The oligo is complementary to the region to be mutated except for one or more "wrong" bases which leads to specific mismatches. After hybridization of the oligonucleotide to the denatured target DNA (usually inserted in a → cloning vector), the →

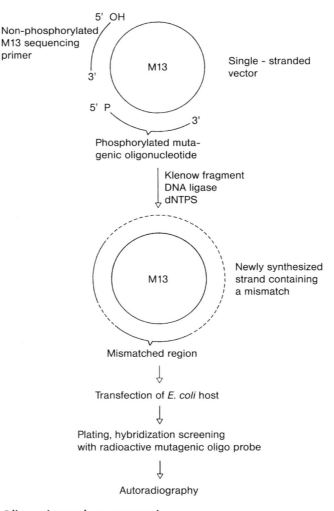

Oligo-mismatch mutagenesis

Klenow fragment of → DNA polymerase I is used to synthesize a complementary strand, where the double-stranded region serves as → primer. Finally DNA ligase seals the → nick. After introduction of this double-stranded molecule into an *E. coli* host, DNA → mismatch repair processes will lead to the occurrence of a mixed population of molecules that consists of 50% wild-type and 50% site-specifically mutagenized clones (because the repair system uses both the original as well as the mutated strand as template). Oligo-mismatch mutagenesis is a way of → site-specific mutagenesis.

Oligonucleotide (oligo): A short nucleic acid molecule of up to 100 nucleotides in length (oligomer), consisting either of deoxynucleotides (oligodeoxynucleotide; general formula: dN_x; for example dAdA-dAdA or $[dA]_4$, a tetramer), or ribonucleotides (oligoribonucleotide; AAAA or [A]$_4$). Oligos may also consist of a mixture of both deoxyribo- and ribonucleotides (in this case deoxyribonucleotides and ribonucleotides are discriminated by the prefix "d" or "r" to avoid confusion).

Oligonucleotide adaptor: See → adaptor.

Oligonucleotide array (oligonucleotide chip, oligonucleotide microarray): A two-dimensional arrangement of thousands, hundreds of thousands, or even millions of short → oligonucleotides, immobilized on a membrane, silicon, or glass support, and used to screen for complementary sequences by hybridization. For example, in → sequencing by hybridization (SBH), the immobilized oligonucleotides have overlapping sequences and are used to reconstruct the sequence of a target molecule by computer analysis of the resulting hybridization signals. See → cDNA array.

Oligonucleotide array-based transcription factor assay (OATFA): A high-throughput technique for the isolation and characterization of DNA sequences encoding → transcription factors (TFs). In short, → double-stranded DNA (dsDNA) sequences are first → biotin-labeled, then incubated with a nuclear extract containing transcription factors, and the complex between specific dsDNAs and their cognate transcription factors separated from unbound dsDNA by → agarose gel electrophoresis, similar to → electrophoretic mobility shift assays (EMSAs). The dsDNA-TF complexes are isolated from the gel, denatured, and the free ssDNA hybridized to a microarray containing ssDNA sequences. Hybridization events identify the sequences, to which TFs bind. The potentially interfering second strand from the dsDNA is excluded from hybridization by adding a 5 bp → overhang to the strand designed to be complementary to the immobilized DNA, which increased the T_M and favored the hybridization of the longer strand to the template.

Oligonucleotide capture: See → DNA capture.

Oligonucleotide carrier: Any organic molecule that is covalently attached to an oligonucleotide (e.g. → antisense oligonucleotide), and functions to introduce the oligonucleotide into a cell. For example, bile acids are such oligonucleotide carriers for e.g. the invasion of liver cells. Constructs of bile acids and an oligonucleotide directed towards the hepatitis C virus are used to inhibit virus replication *in vivo*.

Oligonucleotide

A•G•C•ATGACGTACGT•G•G•T–5'
|
Fluoresceine-3'

• = Phosphorodiester modified as thioate (S$^\ominus$) or methylphosphonate

Oligonucleotide carrier

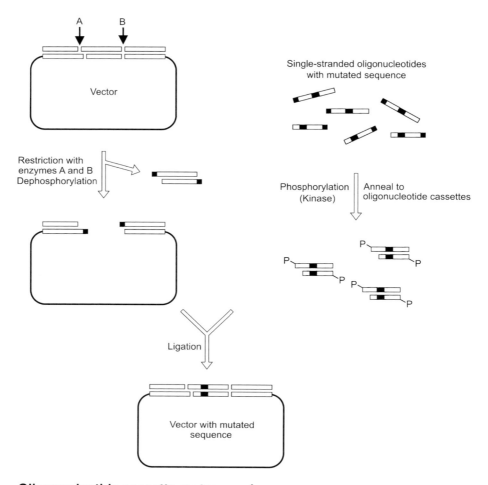

Oligonucleotide cassette mutagenesis

Oligonucleotide cassette mutagenesis: A technique for the introduction of a → mutation into a DNA target sequence. In short, a pair of → oligonucleotides with complementary sequence containing the mutated sequence are first designed and then synthesized, phosphorylated (by e.g. → polynucleotide kinase) and annealed to form a double-stranded DNA fragment. The ends of the fragment are designed to be compatible with the ends generated by → restriction endonuclease digestion of a suitable → vector. This oligonucleotide cassette fragment is then cloned into the compatible site of this vector and creates the desired mutation. See → oligo-mismatch mutagenesis.

Oligonucleotide cataloguing: An outdated technique for the comparison of sequences (originally → ribosomal RNA sequences) from different organisms. The rRNA molecule was first fragmented by cutting it at every → guanosine residue, then each of the resulting fragments ("oligonucle-otides") was again subfragmented by enzymes cutting at different residues, so that a catalogue of fragments characteristic for a specific rRNA molecule was produced.

Oligonucleotide chip: See → oligonucleotide array.

Oligonucleotide-directed double-strand break repair: See → oligo-mismatch mutagenesis.

Oligonucleotide-directed mutagenesis: See → oligo-mismatch mutagenesis.

Oligonucleotide-directed RNase H cleavage: See → RNase H mapping.

Oligonucleotide fingerprinting: A method to screen any genome for the presence of repetitive, highly polymorphic sequences (e.g. short → consensus sequences with varying copy numbers at various loci, such as → variable number of tandem repeats) using synthetic oligonucleotides as probes (see → DNA fingerprint, → DNA fingerprinting).

Oligonucleotide ligation assay (OLA): A method to detect single-base substitutions in a target DNA sequence. In brief, the target DNA is denatured and hybridized to two → oligonucleotide probes in a way that the 3′ end of one oligo is immediately adjacent to the 5′ end of the other ("head-to-tail juxtaposition"). If → DNA ligase is added, it will covalently join the two oligos through the formation of a → phosphodiester bond, provided the nucleotides at the junction are correctly base-paired with the target DNA. This will not be the case if single base substitutions at the junction site have occurred. The ligation of the two oligos may be detected more conveniently if one oligo is labeled with → biotin, and the other one with ^{32}P.

Oligonucleotide microarray: See → oligonucleotide array.

Oligonucleotide mimic: See → DNA mimic.

Oligonucleotide-primed mutagenesis: See → oligo-mismatch mutagenesis.

Oligonucleotide priming: See → oligo(dT) priming.

Oligonucleotide purification-elution cartridge (OPEC): A cartridge packed with a resin (e.g. a polystyrene-divinylbenzene copolymer) that allows the purification of → oligonucleotides in a minimum volume of solvent (e.g. acetonitrile) and a minimum of time.

Oligonucleotide screening: A procedure to identify specific clones in a → cDNA library by using specific synthetic → oligonucleotides of 15–30 nucleotides in length. The cDNA library is plated out, the DNA transferred onto appropriate membranes (e.g. → nitrocellulose) and hybridized to either one specific radiolabeled oligonucleotide or a mixture of similar oligonucleotides with slightly differing base composition, taking into consideration the → degenerate code (see → mixed oligonucleotide probe). The hybridized probe can then easily be detected by autoradiography.

Oligonucleotide-tagged multiplex assay (OTM): A technique for the detection of socalled Src homology 2 (SH2) → domains and consequently phosphorylated tyrosine residues in peptides and proteins (especially signaling proteins) of a cell (see → Src homology 2 profiling (SH2 profiling). In short, the SH2 domains are first expressed in bacteria, monobiotinylated, and different such SH2 domains labeled with biotinylated domain-specific → oligonucleotides. These double-stranded oligonucleotides consist of internal code sequences specific for each SH2 domain that are flanked by invariable primer-binding sites for an amplification in a conventional → polymerase chain reaction (PCR). The various, specifically labeled SH2 domains are then combined in equimolar ratios for a multiplex assay, and incubated with denatured cell extracts immobilized on membranes of e.g. polyvinylidene difluoride (PVDF), an inert thermoplastic fluoropolymer. After binding and washing, the specifically bound SH2 domains are eluted from the membrane by phenylphosphate that displaces the SH2 domains from the phosphotyrosine. The eluates are purified, and the DNA tags PCR amplified via the primer-binding sites. The PCR products are labeled with → digoxigenin via the → primers, then denatured, and differentially hybridized to biotinylated capture oligonucleotides complementary to the internal coding sequences. Finally the capture oligonucleotides are immobilized in → streptavidin-coated wells of → microtiter plates, and the domain-specific PCR products detected with a → peroxidase-conjugated anti-digoxigenin → antibody and a chromogenic substrate. The OTM assay therefore establishes quantitative SH2 profiles in cell extracts, or, since SH2 domains bind to phosphotyrosine residues in peptides and proteins, the → tyrosine phosphoproteome of the cellular extract.

Oligonucleotide therapeutic: Any usually synthetic → oligonucleotide, oligonucleotide analogue or oligoribonucleotide that exerts a therapeutic influence on the symptoms of a disease. For example, → triplex-forming oligonucleotides, oligonucleotides interfering with → transcription of specific genes or the → translation of the resulting messenger RNAs, and oligoribonucleotides for → ribozymes or → aptamers (see also → protein epitope tagging) are such oligonucleotide therapeutics.

Oligopeptide: A short peptide of up to 40–50 amino acids in length (oligomer). Compare → polypeptide.

Oligoribonuclease: A special → exoribonuclease of E.coli, encoded by the chromosomal *orn* gene that specifically degrades small oligoribonucleotides to nucleoside monophosphates. Such small oligoribonucleotides arise by → RNase II and → polynucleotide phosphorylase-catalyzed fragmentation of messenger RNA. Oligoribonuclease is essential for cell viability.

Oligoribonucleotide: See → oligonucleotide.

Oligo(T)-*p*eptide *n*ucleic acid (oligo(T)-PNA): An artificial, negatively charged → peptide nucleic acid (PNA) with high affinity to polyadenylated → messenger RNA that is used to isolate and purify poly(A)$^+$-mRNA, especially in combination with → trans-4-hydroxy-L-proline PNA (HypNA). Oligo(T)-PNA, like conventional PNAs, lacks polarity (i.e. binds to target RNA in parallel and antiparallel orientation) and cannot be degraded enzymatically.

Oligo(U)-sepharose: A → sepharose dextran matrix to which oligouridylic acid (oligo[U]) chains of more than 10 nucleotides in length are covalently bound. Oligo(U)-sepharose is used in → affinity chromatography to isolate polyadenylated RNA (poly[A]⁺-RNA) from complex RNA mixtures. See also → oligo(dT)-cellulose.

Olson-Riles clone: Any one of a set of overlapping → cosmid and → phage → clones covering the entire yeast → genome that are ordered on the basis of *Eco*RI and *Hind*III → restriction mapping.

Ome: A frequently used syllable of the genomics era. See → biome, → bisulfitome, → cellome, → cellular proteome, → chondriome, → complexome, → composite genome, → core proteome, → cybernome, → cytome, → degradome, → enzymome, → epigenome, → expressome, → foldome, → functome, → genome, → glycome, → immunome, → interactome, → kinome, → lipidome, → localisome, → metabolome, → metabolon, → metagenome, → methylome, → methylosome, → microbiome, → morphome, → neurogenome, → neuroproteome, → nucleome, → nutrigenome, → onco-proteome, → operome, → ORFeome, → osmome, → peptidome, → phenome, → phylome, → phosphoproteome, → physiome, → plastome, → protein interactome, → proteome, → pseudo-genome, → ribonome, → secretome, → signalome, → spliceome, → sub-genome, → sublimone, → sub-proteome, → toponome, → transcriptome, → translatome, → transplastome, → unknome.

Omega (Ω): The ratio between → non-synonymous to → synonymous exchange rates (dN/dS) in protein-coding regions of a gene. Values of Ω = 1 are indicative for neutral evolution, Ω < 1 for negative (purifying) selection, and Ω > 1 for adaptive evolution.

Omega (Ω) loop: An irregular secondary structure on the surface of many globular proteins that consists of 6–16 amino acids folded into a rigid, tensely packaged loop, in which the N- and C-termini of the protein are brought into close proximity. The resulting looped structure resembles an Ω.

Omega nuclease: See → meganuclease.

Omega sequence (ω sequence): The sequence H₂N-DGRGG-COOH at the → C-terminus of the → virD2-encoded protein of → *Agrobacterium tumefaciens* that seems to be involved in correct folding of the virD2 protein. This folding may be necessary for the targeting of the → T-strand to the nuclear DNA of the recipient plant cell. See → crown gall.

Omega transposase: See → meganuclease.

o-micron DNA: See → two micron circle.

Omics: A funny abbreviation (coined by J. N. Weinstein) for the various newly generated terms of the genomics and postgenomics era (e.g. → allergenomics, → array-based prote*omics*, → behavioral gen*omics*, → biological gen*omics*, → bi*omics*, → cardiogen*omics*, → cell*omics*, → cellular gen*omics*, → chemical gen*omics*, → chemical prote*omics*, → chemogen*omics*, → chromatin*omics*, → chromos*omics*, → clinical prote*omics*, → comparative gen*omics*, → computational gen*omics*, → crop gen*omics*, → cybern*omics*, → cyt*omics*, → deductive gen*omics*, → degrad*omics*, → econ*omics*, → environmental gen*omics*, → epigen*omics*, → epit*omics*, → expression gen*omics*, → expression pharmacogen*omics*, → express*omics*, → functional gen*omics*, → functional prote*omics*, → gen*omics*, → glyc*omics*, → glycoprote*omics*, → horizontal gen*omics*, → immun*omics*,

→ industrial prote*omics*, → *in silico* prote*omics*, → integrative gen*omics*, → interaction prote*omics*, → interact*omics*, → kin*omics*, → lateral gen*omics*, → lingando*mics*, → lipid*omics*, → lipoprote*omics*, → medical gen*omics*, → metabolic phen*omics*, → metabol*omics*, → metabon*omics*, → metagen*omics*, → methyl*omics*, → microgen*omics*, → neurogen*omics*, → neuropharmacogen*omics*, → neuroprote*omics*, → nucle*omics*, → nutrigen*omics*, → nutritional gen*omics*, → oncopharmacogen*omics*, → one cell prote*omics*, → oper*omics*, → pathogen*omics*, → peptid*omics*, → pharmacogen*omics*, → phen*omics*, → phosphoprote*omics*, → phylogen*omics*, → phyloprote*omics*, → physical gen*omics*, → physi*omics*, → population gen*omics*, → prote*omics*, → quantitative prote*omics*, → recogn*omics*, → reconstruct*omics*, → regul*omics*, resist*omics*, → ribon*omics*, → riboprote*omics*, → RNA gen*omics*, → RN*omics*, signal*omics*, → structural gen*omics*, → subcellular prote*omics*, → tel*omics*, → three-dimensional prote*omics*, → 3D prote*omics*, → tissue prote*omics*, → topological prote*omics*, → toxicogen*omics*, → toxicoprote*omics*, → transcript*omics*, → transpos*omics*, → xenogen*omics*.

OMIM: See → Online Mendelian Inheritance in Man.

Omni Molecular Recognizer Application (OmniMoRA): The Reveo trade name for a still conceptual DNA sequencing device that relies on physical rather than indirect chemical methods, and uses arrays of nano-knife edge probes to directly and non-destructively read the sequence. As the nano-knife edges are stepped with sub-Å resolution over a stretched and immobilized single-stranded DNA, molecular vibrational characteristics are measured and recorded for each of the nucleotides. Since the OmniMoRA uses principles from semi-conductor electronics and photonics, it has the potential to achieve the necessary speed, cost and accuracy improvements over existing sequencing instruments.

On-chip multi-spot *polymerase chain reaction* (on-chip multi-spot PCR): A variant of the conventional → polymerase chain reaction, for which various → primers and the corresponding DNA polymerase (e.g. → *Taq* DNA polymerase) are spotted on a glass surface, where they unspecifically bind. The spots are then dried, and then → template DNA dissolved in buffer containing all necessary PCR reagents added to the → microarray such that a single droplet covers only one spot (or a selected number of spots). The reagents are then mixed and the primers dissolved by e.g. → surface acoustic waves, and the droplets covered with oil. Finally, PCR is carried out separately for each spot, but in parallel.

On-chip PCR: See → on-chip polymerase chain reaction.

On-chip *polymerase chain reaction* (on-chip PCR, "solid-phase PCR"): A variant of the conventional → polymerase chain reaction technique, which allows to perform all the cycling processes on a glass or silicon chip ("PCR chip"). In short, → primers are first spotted onto silanized glass chips, covalently coupled via their 5'-phosphates by e.g. the EDC-methylimidazol method, blocked with succinic anhydride in dimethylformamide, and then hybridised to a mixture of target DNAs. The target then hybridizes to its complementary sequence on the primer, and an added

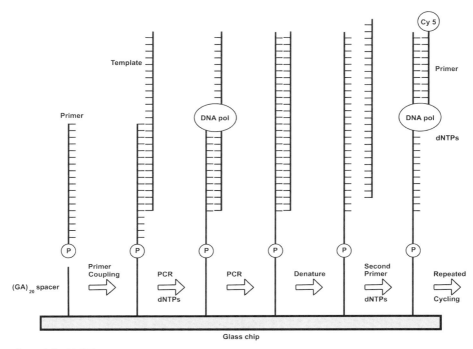

On-chip PCR

DNA polymerase extends the primer and synthesizes a chip-bound double-stranded DNA. Then a second PCR cycling denatures this dsDNA, and now a second fluorophore-conjugated primer anneals to the previously synthesized strand and the DNA polymerase completes the synthesis of a fluorescently labelled dsDNA that can be detected with a laser beam. It is favourable for on-chip PCR to double the amount of primers, dNTPs and DNA polymerase, and to extend the PCR time by 100% (except for the synthesis step). On-chip PCR allows to amplify fragments of up to 7 kb in length.

On-chip single-base primer extension: A variant of the → primer extension technique, in which the extension reaction occurs on a chip-bound target molecule. This configuration allows to use high → stringency for the hybridization reaction between primer and target.

Oncogene (transforming gene): Any one of a class of mutated and/or overexpressed variants of normal genes (→ cellular oncogenes) in animal and human cells that encodes a protein which transforms the normal cell into a tumor cell. Cellular oncogenes (c-*onc*) may become part of → retroviruses and are then designated as v-*onc* (→ viral oncogene). At present, about 40 different oncogenes isolated from acutely transforming viruses of animals and humans are known (see table). See also → oncogenic virus, → *ras* gene.

Oncogene (transforming gene)

Some oncogenes and tumor viruses

Acronym	Virus	Species	Tumor origin
abl	Abelson mouse leukaemia (Ab-MLV)	mouse	Chronic myelogenous leukaemia
erbA	Avian erythroblastosis (AEV)	chicken	
erbB	Avian erythroblastosis (AEV)	chicken	
ets	E26 myeloblastosis	chicken	
fes (fps)	Snyder-Theilen feline sarcoma (SM-FeSV)	cat	Gardner-Arnstein sarcoma
fgr	Gardner-Rasheed sarcoma	cat	
fms	McDonough feline sarcoma (SM-FeSV)	cat	
fps (fes)	Fujinami sarcoma (FuSV)	chicken	
fos	FBJ osteosarcoma	mouse	
hst	Non-retroviral tumor	human	Stomach tumor
int1	Non-retroviral tumor	mouse	MMTV-induced carcinoma
int2	Non-retroviral tumor	mouse	MMTV-induced carcinoma
jun	ASV 17 sarcoma	chicken	
kit	Hardy-Zuckerman 4 sarcoma	cat	
B-lym	Non-retroviral tumor	chicken	Bursal lymphoma
mas	Non-retroviral tumor	human	Epidermoid carcinoma
met	Non-retroviral tumor	mouse	Osteosarcoma
mil (raf)	Mill Hill 2 acute leukaemia	chicken	
mos	Moloney mouse sarcoma (Mo-MSV)	mouse	
myb	Avian myeloblastosis (AMV)	chicken	Leukaemia
myc	MC29 myelocytomatosis	chicken	Lymphomas
N-myc	Non-retroviral tumor	human	Neuroblastomas
neu (ErB2)	Non-retroviral tumor	rat	Neuroblastoma
raf (mil)	3611 sarcoma	mouse	
Ha-ras	Harvey murine sarcoma	rat	Bladder, mammary and skin carcinomas
Ki-ras	Kirsten murine sarcoma (Ki-MSV)	rat	Lung, colon carcinomas
N-ras	Non-retroviral tumor	human	Neuroblastomas, leukaemias
rel	Reticuloendotheliosis (REV-T)	turkey	
ros	UR2	chicken	
sis	Simian sarcoma (SSV)	monkey	
src	Rous sarcoma (RSV)	chicken	
ski	SKV 770	chicken	
trk	Non-retroviral tumor	human	Colon carcinoma
yes	Y73, Esh sarcoma	chicken	

Oncogene amplification: The increase in copy number of one or more → oncogenes in genomes of late-stage cancers of many human organs, probably induced by the inactivation of p53. Oncogene amplification is clearly associated with tumor progression and has prognostic significance. See → gene amplification.

Oncogenesis: The gradual progression of a previously normal cell to a cell with changed genetic, cellular and cytological properties, the most prominent of which are lost contact inhibition and permanent proliferation, altogether leading to the formation of a tumor.

Oncogenic virus (tumor virus): A virus that transforms animal and human cells in culture and induces cancerous growth in animals and humans. Such viruses either contain DNA (e.g. Papovaviridae, Herpetoviridae) or RNA (Retroviridae, see → retrovirus) as genetic material.

Oncogenomics: See → cancer genomics.

Oncolytic virus: Any virus that is genetically modified to selectively target and enter tumor cells, where it replicates and destroys these cells ("oncolysis"). For example, the human neurotropic *herpes simplex virus* (HSV) of the α-herpesvirus subfamily consists of two serotypes, type 1 (HSV-1) and 2 (HSV-2), which are ubiquitous in humans, but only rarely cause severe diseases. Wild-type HSV carries all genes to productively infect normal cells. Oncolytic HSV variants (e.g. G207), in which essential genes for pathogenicity (e.g. the neurovirulence gene γ34.5 and *UL39* gene locus encoding the viral ribonucleotide reductase necessary for viral replication) are mutated, are attenuated and infect, but do not replicate in normal cells (→ abortive infection). Tumor cells carrying → mutations in tumor suppressor genes or → oncogenes complement the deleted genes in G207 (i.e. provide the enzymes required for viral replication). Therefore G207 replicates in tumor cells, and lyses them. Consequence: new attenuated virus progeny is synthesized and released, and infect other tumor cells. Additionally, the engineered HSV induces inflammatory cytokine responses and T-cell-mediated immunity.

OncomicroRNA (oncomiR): Any one of a series of → microRNAs (miRNAs) that are involved in tumor induction or maintenance. Specific oncomiRs can act as oncogenes, others as tumor suppressors.

Onco Mouse™: A transgenic mouse carrying an activated *ras* → oncogene in all germline and somatic cells. Developed as a transgenic *in vivo* model to study oncogenesis, the mouse will predictably undergo carcinogenesis within some months. Compare → knock-out mouse.

Onconase P-30: A 12 kDa basic lectin-like → ribonuclease from oocytes (e.g. *Rana pipiens*) that binds to membrane receptors of a target cell, is channeled actively (i.e. ATP-driven) into the cytoplasm, inhibits ribosomal protein synthesis by the degradation of tRNA, 5S rRNA, 18S and 28S rRNA, and is therefore highly cytotoxic.

Oncopharmacogenomics: A branch of → pharmacogenomics that uses the whole repertoire of → genomics, → transcriptomics, and → proteomics technologies to identify genes and/or mutations in genes involved in cancerogenesis and to design and develop new drugs to control cancerous proliferation of cells and the dissemination of such cells in an individual. See

→ neuropharmacogenomics. Compare → pharmacogenetics.

Onco-proteome: A part of the → proteome that consists of proteins expressed primarily or exclusively in tumor cells. The presently and diagnostically interesting oncoproteins comprise e.g. CA 19-9, CD 44v5, CD 44v6, CEA, c-erb B2, c-myc, kathepsin D, kathepsin L, MDR-1, melanoma, MMP2, MMP9, p53 mutant, PSA, p21 ras, und 3, urokinase, and VEGF. Such oncoproteins are detectable at relatively early stages in tumorigenesis (present level of detection; few cells released from tumors of about 1 cm^3 volume), and serve not only as markers for an early diagnosis, but also for the targeting of the tumor cells and tissues.

One cell proteomics: The whole repertoire of techniques to analyze and characterize the → proteome of a single cell at a given time.

One-chip-for-all: See → universal array.

One gene-one enzyme hypothesis: A hypothesis largely based on the assumption that one single gene codes for a specific enzyme. Since many enzymes are the product of two or more genes, the more precise term would be one gene-one polypeptide chain.

One-hybrid system (yeast one-hybrid system): A technique for the *in vitro* isolation of novel genes encoding proteins that bind to a target DNA (bait) sequence (→ DNA-binding proteins) that is based on the fact that many eukaryotic transcriptional activators are composed of a target-specific → DNA-*binding domain* (DBD) and a target-independent activation domain (AD). The one-hybrid system uses a cDNA candidate encoding a potential DNA-binding protein fused to a sequence encoding an AD. The complex of DBD and AD drives the expression of a → reporter gene (e.g. *His3* gene, β-galactosidase gene, or others). In short, a cassette containing tandem copies of a DNA target element for DNA-binding proteins is first inserted into a → multiple cloning site immediately → upstream of *His3* or *lac Z* reporter gene → promoters in a socalled yeast → integration vector. This linearized vector is then transformed into competent yeast cells at

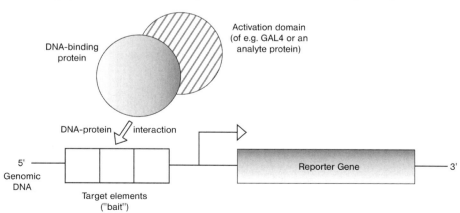

One-hybrid System

high frequency, and integrates at specific sites of the genome. Transformants are selected by their URA3 phenotype. These transformants are called reporter strains. Then an activator domain fusion library containing candidate cDNA clones is transformed into the yeast reporter strain. Whereas *His3* reporter gene expression without any induction is very low in the reporter strain, it is greatly enhanced after an AD-DNA-binding protein hybrid interacted with the target DNA element. *HIS3* expression allows yeast colony growth on minimal medium lacking histidine, so that transformants will be selected. His$^+$ clones contain cDNAs for putative DNA-binding proteins. The cDNAs can easily be isolated, sequenced, the sequence compared to database entries and used in *in vitro* DNA-binding assays. See → dual-bait two-hybrid system, → interaction trap, → LexA two-hybrid system, → reverse two-hybrid system, → RNA-protein hybrid system, → split-hybrid system, → split ubiquitin membrane two-hybrid system, → three-hybrid system, → two-hybrid system.

1-methylguanosine: See → rare base.

1-methylinosine: See → rare base

One-sided PCR: See → one-sided polymerase chain reaction.

One-sided polymerase chain reaction (one-sided PCR): A modification of the conventional → polymerase chain reaction for the direct targeting, amplification, and sequencing of uncharacterized → cDNAs. In short, a specific cDNA sequence ("core region") is selected from a cDNA collection by using two imperfect oligomer → primers that are synthesized *in vitro* based on sequence information derived from homologous cDNAs (or also proteins) from related organisms. These specific primers can be complementary to any region within the message, can be located adjacent to the region to be amplified or may partially overlap it, and prime the amplification of the core sequence in the polymerase chain reaction. Then, based on this core sequence, specific primers are designed that permit the amplification of regions both upstream and downstream of the core, if combined with a second non-specific primer complementary to the 3' → poly(A) tail, or to an *in vitro* enzymatically added d(A)-tail at the 5' end. The pairwise combination of specific and nonspecific primers allows the amplification of the cDNA core with both 3' and 5' flanking regions. The amplified fragments are then inserted into → cloning vectors from which the → insert can be sequenced directly.

One-step gene disruption: The production of a stable, non-reverting gene → mutation in a target genome by (1) transformation of a construct containing the cloned gene interrupted by a → selectable marker gene, and (2) its → homologous recombination with the homologous sequence(s) in the target genome via highly recombigenic termini. This process leads to a replacement of the wild-type gene with the disrupted copy.

One-step protocol: An experimental design that combines two normally independent steps of a procedure such that both reactions occur simultaneously in the same reaction tube. For example, the → reverse transcription of eukaryotic → messenger RNA employing an oligo(dT) primer and → reverse transcriptase leads to the synthesis of a double-stranded → cDNA. This cDNA can be used as →

template for a subsequent amplification using conventional → polymerase chain reaction and → *Thermus aquaticus* DNA polymerase. In a one-step protocol both reactions occur concomitantly in the same reaction tube.

***o*-nitrophenyl-β-D-galactoside (ONPG, o-nitrophenyl-galactoside):** An artificial substrate for β-galactosidase which is cleaved into galactose and the yellowish o-nitrophenol, the concentration of which can be easily measured.

ONPG

***o*-nitrophenyl-galactoside:** See → *o*-nitrophenyl β-D-galactoside.

On-line DNA sequencing: See → automated DNA sequencing.

Online Mendelian Inheritance in Man (OMIM): A directory of human genes and genetic disorders, with links to literature references, sequence records, maps, and related databases.

ONPG: See → *o*-nitrophenyl-β-D-galactoside.

Oocyte translation assay: The translation of foreign mRNA(s) in *Xenopus laevis* oocytes after → microinjection of nanogram amounts of this message.

Opacity gene (*opa* gene): Any one of a family of constitutively transcribed bacterial genes (e.g. in the veneric disease-causing *Neisseria gonorrhoeae*) that harbors → microsatellite motifs (e.g. [5'-CTCTT-3']$_n$) in the → leader peptide encoding sequence. The encoded proteins are effective antigenic determinants, recognized by defense mechanisms of potential hosts (e.g. the phagocytes of the immune system). Bacteria with such proteins are *opa*que in appearance. However, antigenic diversity through → frame-shifts as a consequence of → slipped strand mispairing is exploited by pathogenic bacteria to escape immune detection by the host. If the number of the microsatellite motifs in the leader sequence is changed by slipped strand mispairing, then differ- ent lengths of the leader peptide result, and → out-of-frame mutations lead to a translational switch in favor of an altered protein that is no longer an antigen for the host organism. For example, if a CTCTT unit is lost during strand mispairing, the resulting shorter protein can no longer bind to the host cells (i.e. the bacterium can no longer penetrate the cell). With a new round of replication, this error can be reversed, and the bacterium regains infectivity ("phase shifting", "phase variation").

Opa gene: See → opacity gene.

Opal codon: The → stop codon UGA. See also → opal mutation, → opal suppressor.

Opal mutation: Any → mutation that converts a → codon into the → stop codon UGA (→ opal codon). See also → nonsense mutation.

Opal suppressor: A gene that encodes a mutated transfer RNA (tRNA), whose → anticodon recognizes the → termination codon UGA (→ opal codon) and allows the continuation of polypeptide synthesis.

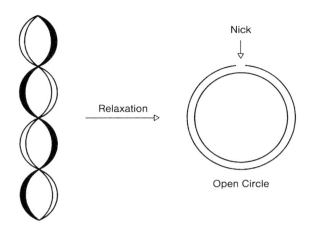

Supercoiled DNA

Open circle

Opaque-2: A mutant of *Zea mays* which contains kernels with lysine-rich proteins.

OPEC: See → *o*ligonucleotide *p*urification-*e*lution *c*artridge.

Open circle (OC, oc-DNA, open circular DNA, relaxed circle, form II-DNA, nicked circular DNA): A non-supercoiled conformation adopted by a circular double-stranded DNA molecule, when one or both polynucleotide strands carry → nicks, so that it cannot form supercoils, but adopts a relaxed conformation.

Open circular DNA: See → open circle.

Open differential gene expression technology (open DGE technology): Any technique for the genome-wide profiling of differentially expressed genes (see → differential gene expression) in two (or more) cells, tissues, organs, or organisms that does not require any *a priori* knowledge of the → transcriptome, so that the field of discovery is "open". However, most open systems exploit existing expressed genome databases to identify known and novel expressed (or suppressed) genes more efficiently. Such open DGE systems are e.g. → cDNA-AFLP, → differential display reverse transcription PCR, → massively parallel signature sequencing, → serial analysis of gene expression, → SuperSAGE, → total gene expression analysis, and others. Compare → closed differential gene expression technology.

Open mitosis: A somewhat misleading term for the mitotic processes in the cytoplasm of lower and higher → eukaryotes after the disassembly of the nuclear envelope, as compared to the → closed mitosis in *Saccharomyces cerevisiae* and *Schizosaccharomyces pombe* that takes place within an intact → nucleus.

Open promoter complex: A → promoter configuration in which the DNA double helix is locally unwound to facilitate the binding of various → transcription factors and → RNA polymerase to form a → pre-initiation complex.

Open reading frame (ORF):

a) A nucleotide sequence in DNA in between two in-frame → stop codons.
b) A nucleotide sequence in DNA that potentially can be translated into a protein or be transcribed into an RNA, and begins with an ATG → start codon and terminates with one of the three → stop codons. A good ORF candidate for coding a *bona fide* cellular protein has a set size requirement: it should have the potential to encode a protein of 100 amino acids or more. An ORF is not necessarily equivalent to a → gene or → locus, unless a → phenotype can be associated with a mutation in the ORF, and/or a → messenger RNA or, generally, a gene product generated from the ORF's DNA is detected. The ORFs of e.g. *Saccharomyces cerevisiae* are designated by a symbol consisting of three uppercase letters followed by a number and then another letter. For example, Y (for "Yeast"), A to P for the chromosome harbouring the ORF (where "A" is chromosome I, and "P" is chromosome XVI), L or R (for left or right arm), a three-digit number corresponding to the order of the open reading frame on the chromosome arm (starting from the → centromere and counting out to the → telomere), and W or C for the location of the open reading frame on either the → Watson or → Crick strand. Compare → closed reading frame.
c) The complete → exon-intron structure of the protein-coding region of a mature → messenger RNA. In this context, a → primary transcript that is alternatively spliced, represents more than one ORF.

Open reading frame (ORF) vector (open reading frame expression vector, fusion vector): A → plasmid → expression vector for the construction of → fusion proteins that carries a bacterial → promoter, a → ribosome binding site and an → initiation codon (ATG) in front of a truncated *lac Z* gene ('*lac Z*). The β-galactosidase encoded by this truncated gene is not active, since it cannot form a tetramer (e.g. because of the deletion of the codons specifying 9 carboxy-terminal amino acids). The first 25 wild-type amino acids at the carboxy terminus may, however, be substituted by up to a few hundred of unrelated amino acids without interfering with tetramerization. Insertion of foreign DNA into cloning sites within this noncritical coding region of *lac Z*, if it is in-frame with a translational start signal, will lead to the formation of a fusion protein with β-galactosidase activity that can be detected even if it is fused to longer proteins (→ insertional activation).

Open reading frame cloning: The use of → open reading frame vectors for expression cloning.

Open reading frame expressed sequence tag (OREST): Any → expressed sequence tag (EST) that is derived from the central part of a → cDNA rather than its 3' or 5' end.

Open reading frame expressed sequence tags (ORESTES, ORF ESTs): A collection of → expressed sequence tags (ESTs) from → cDNAs of various cells, tissues or organs of an organism that contains sequences from the central portion of each transcript rather than from its ends (as in most other related techniques). In short, total RNA is first isolated, treated with → DNAse I to remove → genomic DNA, then poly(A)$^+$ is extracted, reverse transcribed with an 18–25 nucleotide → primer of random sequence (GC content: 50%), and the single-stranded cDNA again amplified with the same primer. The complexity of the preparation is then checked by → polyacrylamide gel electrophoresis, and the amplification pool with multiple bands cloned into a plasmid vector (e.g. → pUC18). Finally the inserts are sequenced. The ORESTES technique generates a better coverage of the → transcriptome of the cell, and facilitates the construction of → contigs of transcript sequences.

Open reading frame expression vector: See → open reading frame vector.

Operational code: A set of rules by which the aminoacyl-tRNA synthetases recognize their cognate → transfer RNA molecule.

Operational gene: Any gene encoding a protein involved in the catalysis of a step in a normal metabolic pathway. Synonym for → house-keeping gene. Operational genes are horizontally transferred between bacterial (and maybe eukaryotic) genomes, but it is not clear whether this → horizontal gene transfer occurred in few massive ancient transfers before diversification of modern prokaryotes ("early massive horizontal gene transfer hypothesis") or is a continuous process ("continual horizontal gene transfer hypothesis").

Operator: A palindromic nucleotide sequence (→ palindrome) with dyad symmetry, localized at the proximal end of an → operon that allows the formation of a → cruciform structure. An operator constitutes the recognition site for a specific → repressor protein and controls the expression of the adjacent → cistrons. See also → operator constitutive mutation, → operator zero mutation.

Operator constitutive mutation (O^c): Any mutation of the → operator leading to increased or constitutive expression of the → cistrons in the adjacent → operon. Compare → operator zero mutation.

Operator zero mutation (O^0): Any mutation of the → operator leading to the loss of function of the → operator. The expression of the → cistrons of the adjacent → operon is rendered impossible. Compare → operator constitutive mutation.

Operome: Another term of the excrescent → omics era, describing the part of the → proteome, which contains proteins with as yet unknown functions. Do not confuse with → operon. See also → biome, → cybernome, → genome, → immunome, → interactome, → metagenome, → microbiome, → morphome, → transcriptome.

Operomics: The whole repertoire of technologies for the study of the complete molecular architecture, composition and functions of a cell, including tools from → genomics, → transcriptomics, → proteomics, and → metabolomics. Another excrescent term of the → omics era.

Operon: A unit of adjacent prokaryotic cistrons → the expression of which is under the control of a common → operator and leads to the synthesis of a single → polycistronic messenger RNA. See also → inducible operon, for example → *lac* operon; → operon fusion, → operon network.

Operon fusion: The head-to-tail ligation of two → operons by recombinant DNA techniques in a way that the coding sequences of the second operon come under the control of the regulatory sequences of the first operon.

Operon head gene: The first gene in an → operon.

Operon network: A series of → operons with their associated → operators interacting in such a way that the proteins encoded by one operon either activate or suppress another operon.

Operon™ primer: The trade mark for a series of 10-mer → primers of arbitrary sequence for the randon amplification of genomic DNA and the generation of → random amplified polymorphic DNA (RAPD) markers.

Opine: One of a series of unusual amino acid or sugar derivatives specifically synthesized by → crown gall tumor cells incited by the soil bacterium → *Agrobacterium tumefaciens*, but not by normal plant tissues. The opine genes reside on the → Ti plasmid close to the right border of → T-DNA. They are not or only weakly transcribed in *Agrobacterium*, but are constitutively expressed once integrated into the plant nuclear genome. Expression results in the appearance of → opine synthases which catalyze the formation of opines. The latter cannot be metabolized by the tumor cells and are therefore secreted. *Agrobacterium* can take up and degrade them because it possesses Ti-plasmid genes for opine catabolism, and thus may use opines as a source of carbon, nitrogen and energy.

Moreover, opines activate the → *tra*-genes of the Ti-plasmid, and thus serve to spread it in a bacterial population. Opines serve as tumor markers in plants. See → agrocinopine, → agropine, → histopine, → leucinopine, → lysopins, → mannopine, → nopaline, → nopalinic acid, → octopine.

Opine synthase (opine synthetase): An enzyme catalyzing the synthesis of → opines (e.g. → octopine, → nopaline).

OPT: See → optical projection tomography.

Optical fiber: A cylindrical dielectric waveguide of glass or plastic material that transmits light along its axis by → total internal reflection. The fiber consists of an inner ring ("core") surrounded by an outer ring ("cladding," "cladding layer"). To confine the optical signal in the core, the refractive index of the core must be slightly higher than that of the cladding. The boundary between the core and cladding may either be abrupt, as in the so called step-index fiber, or gradual, in graded-index fibers. Such fibers are routinely used to transmit light signals in high-speed communication systems (e.g. telephone, Internet, video signals). Individual optical fibers can be converted into → DNA sensors by attaching a → single-stranded DNA probe to the outside of the core. Upon → hybridization of a fluorescently labeled complementary target sequence, labeled → double-stranded DNA is formed that can be excited by a light source. The excited → fluorophore emits light that is captured by the optical fiber, transmitted to the other end of this fiber, where a detection system (e.g. a → CCD camera) separates the excitation signal from the emitted signal. DNA arrays are also made from optical fibers by physically bundling multiple fibers.

Optical fingerprinting: A misleading term for the detection of interactions between thousands of → probe molecules (e.g. DNAs, oligonucleotides, RNAs, peptides, or proteins) immobilized on the surface of a chip (e.g. glass, silicon) and target compounds (e.g. DNAs, oligonucleotides, RNAs, peptides, proteins, also low molecular weight compounds such as metabolites) by RAMAN spectroscopy (analyzing the unique structure of the cross-reacting molecules). Compare → DNA fingerprinting.

Optical mapping (visual mapping): The visualization of → genes, generally DNA sequences, along a chromosome, or a chromosome fiber, or along a → BAC or → YAC clone that are extended (see → DNA combing), by → in situ hybridisation of → fluorochrome-labeled → probes (representing e.g. genes), and detection of fluorescence emission. The threshold of direct visual mapping is about 3.0 kb, so that single genes can be detected. Optical mapping is also used for creating e.g. → restriction maps from a series of single DNA molecules. In short, large DNA molecules are first dropped onto specially prepared glass surfaces, linearized in parallel through a fluid flow across the surface, and then affixed onto the glass. Subsequently → restriction endonucleases are added to produce ordered patterns of restriction fragments, which are stained with → fluorochromes and visualized with a fluorescence microscope. The restriction sites are represented as gaps. The various microscopic images are captured one at a time, processed, and the images of the various restriction fragments aligned to match the restriction sites. Then multiple maps are merged into large → contigs, using map assembly programs. For example, a complete optical restriction map is available for the bacterium *Deinococcus radiodurans*.

Optical noise: An undesirable contribution of reflected light from a → microarray support (e.g. a glass or quartz slide), reflections from any object in the laboratory room, leaking light or even cosmic rays to the readings of the fluorescence detection instrument. See → background subtraction, → dark current, → electronic noise, → microarray noise, → sample noise, → substrate noise.

Optical projection tomography (OPT): A microscopic technique for the production of high-resolution three-dimensional images of fluorescent or also non-fluorescent biological samples of up to 15 mm thickness. The specimen (e.g. a complete mouse embryo) is first stained with a diagnostic fluorescent antibody (e.g. an HNF3β antibody for developing endoderm and the floorplate of the spinal cord, or a neurofilament antibody for developing neurons), then positioned in a cylinder of agarose, and rotated continuously for 360 degrees. Any light emitted by the embryo is focused by lenses onto camera-imaging chips (CICs), and recorded such that a three-dimensional image is generated. OPT allows to map specific messenger RNAs or proteins in intact organs or embryos and can reconstruct gene expression patterns during developmental processes.

Optical trap (optical tweezers): An experimental arrangement, in which the radiation pressure (forces in the picoNewton range) of tightly focussed single-beam infrared lasers trap and hold, or move molecules. Frequently, the molecules are bound with one end to micrometer-sized dielectric beads (e.g. polystyrene beads), which allow better manipulation. The focussed laser light exerts two forces onto the bead with the molecule. The so called gradient force draws the particle towards the beam center, where the light field is strongest. The so called scattering force derives from the radiation pressure exerted on the particle by absorbed or scattered photons, which "blow" the particle down the optical axis. If both forces are balanced, the particle (and the molecule) is held slightly downstream of the laser light focus (is "optically trapped"). Optical traps can measure molecular displacements by only few nanometers (e.g. kinesins moving along microtubules and actin-myosin dynamics), and monitor protein or DNA unfolding. See → dual-beam optical tweezer.

Optical tweezers: See → optical trap.

Optimal codon:

a) Any → codon that is utilized very often in a given organism. In → transgenic organisms, → codon optimization is necessary to achieve → overexpression of the → transgene. See → codon bias, → rare codon.
b) Any → codon that is translated more efficiently than its → synonymous codon.

Optimized stringent random amplified polymorphic DNA technique (OS-RAPD): A variant of the conventional → random amplified polymorphic DNA method that works with optimized amplification reaction mixtures (optimized with regard to concentration of → buffer, Mg^{2+}, dNTPs, → primers, → template DNA, and → Taq DNA polymerase) and DNA amplification at elevated → annealing temperatures, thus increasing → stringency, and avoiding spurious amplification artifacts. OS-RAPD therefore produces reproducible and reliable genomic fingerprint patterns.

OR: See → oligomer restriction.

ORC: See → origin recognition complex.

Ordered array: Any → microarray, onto which regular rows and columns of spots (consisting of oligonucleotides, cDNAs or DNAs) are immobilized.

Ordered clone bank: See → ordered clone library.

Ordered clone library (relational clone library, ordered clone bank): Any → genomic library that contains clones with terminal overlaps which can be arranged so that they represent the complete DNA from which they are derived. See → ordered clone map.

Ordered clone map: A graphical description of the linear arrangement of overlapping DNA fragments, cloned into an appropriate → cloning vector (e.g. → bacterial artificial chromosome, → cosmid, → mammalian artificial chromosome, → yeast artificial chromosome, or even → plasmid). The order of the clones in such a map reflects their original positions on the DNA (or chromosome). See → macro-restriction map.

Ordered fragment ladder far-Western blotting: A technique for the detection of protein-protein interaction(s) and the identification of specific domains involved in such interaction. The method uses a labeled protein → probe that reacts with fragments of a target protein containing the interacting domain. The interaction is then detected by → autoradiography or → phosphorimaging. In short, the isolated and purified target protein, or a whole cell lysate containing this protein is first cleaved chemically (e.g. with 2-nitro-5-thiocyanobenzoic acid, or hydroxylamine) or enzymatically (e.g. with thermolysin or trypsin), the cleavage fragments separated by → SDS polyacrylamide gel electrophoresis, the separated fragments electrophoretically transferred onto a → nitrocellulose membrane and reacted with a ^{32}P-labeled test protein (e.g. labeled with a kinase). The dried blot is then exposed to X-ray film, or analyzed in a phosphorimager. If a binding of a test protein to one (or more) of the target peptides occurs, the interaction can be visualized by → autoradiography, and the interacting domains of the target protein be identified and mapped ("chemical cleavage mapping").

Oregon Green: The → fluorochrome Oregon Green 488-X that is used as a marker for → fluorescent primers in e.g. automated sequencing procedures, or for labeling in → DNA chip technology. The molecule can be excited by light of 492 nm wave-length, and emits fluorescence light at 517 nm. Since the wave-length of the excitation and emission maxima is pH-dependent, the exact values vary.

OREST: See → open reading frame expressed sequence tag.

ORESTES: See → open reading frame expressed sequence tags.

ORF: See → open reading frame.

ORFan: Any hypothetical gene in an organism that has no homologues in other organisms. See → orphan gene, → orphon.

ORF clone: Any → clone (e.g. a → plasmid clone) with an inserted → open reading frame (ORF) sequence.

ORFeome:

a) The complete set of → open reading frames (ORFs) in a particular → genome. Specific ORFeomes are designated according to their organism of origin (e.g. hORFeome for *h*uman ORFeome).

b) A laboratory slang term for a set of full-length → cDNA clones transcribed from a particular genome at a particular developmental stage of the carrier, derived from → cDNA chip analysis. Such ORFeomes circumvent → cDNA libary construction, and ideally contain each transcribed gene sequence in equimolar concentrations.

ORF ESTs: See → open reading frame expressed sequence tags.

ORFmer: A set of two → primers that allow the amplification of an → open reading frame (ORF) from → genomic DNA using conventional → polymerase chain reaction techniques. One ORFmer (A-primer) contains a 13 bp non-variable sequence (→ adaptamer) including the → start codon 5'-ATG-3' (→ amino-terminus of the encoded protein) and a *Sap* I → restriction site at its 5' terminus, followed by a 20–25 bp gene-specific sequence (ORF sequence I). The adaptamer sequence is 5'-TTGCTCTTCC**ATG**-3'. The other ORFmer (C-primer) also carries a 13 bp adaptamer, containing a *Sap* I site and the → stop codon 5'-TAA-3' (sequence: 5'-TTGCTCTTCGTAA-3') at its 5'-terminus, adjacent to a 20–25 bp gene-specific sequence (ORF sequence II). The stop codon signals the → carboxy termi-

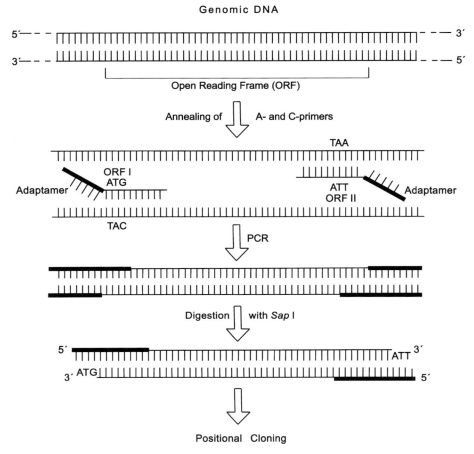

ORFmer

nus of the encoded protein. The length of the gene-specific sequences in each primer is identical to achieve optimal → melting temperature of template-primer duplexes. PCR products generated with both primers contain Sap I sites at both termini. These can be cleaved by Sap I, producing a product with an ATG start and TAA stop codon as → 5' overhangs, which can be positionally cloned into → cloning vectors containing corresponding Sap I sites. Using many different ORFmers it is possible to perform → expression profiling, → expression vector cloning for the characterization of the ORF, and gene → mutagenesis.

ORF sequence tag (OST): Any short (~0.5 kb) sequence from the 5'- and 3'-end, respectively, of a cloned and sequenced protein-coding → open reading frame (ORF), originally derived from amplification of → genomic DNA with two ORF-specific → primers in a → conventional polymerase chain reaction (PCR). OSTs can be aligned to the sequence of a →

genome to verify ORFs predicted by appropriate software (as e.g. GeneFinder (http://ftp.genome.washington.edu/cgi-bin/genefinder_req.pl).

ORF vector: See → *open reading frame vector*.

Organellar gene: Any gene that resides on the genome(s) of an → organelle (as e.g. → chloroplast, → mitochondrium). Distinct from → nuclear gene.

Organellar genome: The → genome of a → mitochondrion (in all eukaryotic organisms) and → plastids (in plants), as different from the → nuclear genome. See → chloroplast DNA, → mitochondrial DNA, → nuclear DNA.

Organellar proteome: The complete set of proteins expressed in an organelle (e.g. → chloroplast or its variants, the nucleus, the endoplasmatic reticulum, the Golgi apparatus, and → mitochondria) at a given time. Since the term organelle is highly fluent, an organellar proteome may also encircle the proteome of → centromeres, → nucleoli, centrosomes, and even → ribosomes and → spliceosomes, to name few. Research into the composition of organellar proteomes starts with the isolation of the target organelle (often by → density gradient centrifugations), the digestion of the proteome into small peptide fragments with an appropriate protease, the fractionation of the peptide mixture by e.g. column chromatography (i.e. by size or charge). The peptides are then ionized as they emerge from the column and their mass analyzed by → mass spectrometry. The resulting mass spectra of the peptides are finally matched against protein sequence databases to identify the proteins, from which the peptides originated. Organellar proteomes are complex. For example, the nucleolus requires more than 700 different proteins for its various functions. Different organelles of the same kind in the same cell, more so in different tissues usually own different proteomes. For example, the mitochondrial proteome in tissues as e.g. brain, heart, kidney and liver share only 85% identical proteins, amongst them a → core proteome set with proteins functioning in the respiratory chain and related pathways. The rest of proteins is quite different from tissue to tissue. See → *cellular proteome*.

Organelle: A membrane-bounded compartment within the cytoplasm of a eukaryotic cell that contains a specific set of proteins catalyzing reactions in one (or more) specific pathway(s). For example, a mitochondrium is such an organelle that is specialized on the β-oxidation of fatty acids, citric acid cycle reactions, electron transport and generation of ATP. Other organelles are the nuclei, plastids, vacuoles, lysosomes, Golgi apparatus.

Organ-specific element (OSE): A *cis*-acting DNA sequence motif of 20–100 bp in → promoters of eukaryotic genes that is responsible for their organ-specific expression. If deleted, transcription from the resulting mutant promoter is no longer organ-specifically regulated. The OSEs are target sites for the binding of specific → transcription factors.

OR gene: See → *olfactory receptor gene*.

Ori: *Ori*gin, see → *origin of replication*, also → oriA, → oriT.

OriA (*origin* of *a*ssembly): A specific sequence of *tobacco mosaic virus* (TMV)

RNA, located within the coding region for P30 (a 30 kDa protein catalyzing the movement of viral RNA from host cell to host cell). The oriA sequence has the potential to form three hairpin loop structures (→ fold-back DNA) and functions in the assembly of coat proteins and viral RNA to new virus particles.

OriC (*origin* of *chromosome replication*): The sequence of a replicon at which chromosome replication is initiated. For example, the *E. coli* oriC region spans 0.245 kb and contains → consensus sequences for replication initiation proteins. Compare → oriV.

Orientation-specific cDNA cloning: See → forced cloning.

Origin: See → origin of replication.

Original synteny: A somewhat misleading term for any genomic sequence in a potential ancestral organism that has been preserved in lineages derived from it. Such original → synteny can be inferred from a comparison of complete sequences of → chromosomes or → genomes of two (or more) related organisms. For example, sequence comparisons of socalled → syntenic anchors between mouse and man can identify such original synteny, and ultimately allow to reconstruct the ancestral mammalian → karyotype.

Origin of *assembly*: See → oriA.

Origin of *chromosome replication*: See → oriC.

Origin of replication (origin, ori; replication origin): A specific sequence of a → replicon at which DNA → replication is initiated. In eukaryotes, the origins of different organisms do not share sequence → homology, but contain similar structural elements (e.g. socalled *base unpairing regions*, BURs, *DNA unwinding elements* (DUEs), → palindromes, and → CpG islands). See also → oriC, → oriV.

Origin of transfer (oriT): The sequence of a → replicon at which an → endonuclease (in → F factors the products of plasmid genes *traY* and *traZ*) introduces a → nick into the → H strand of the replicon, thus generating the substrate for transfer from a donor to an acceptor cell (by e.g. → conjugation).

Origin of *vegetative replication*: See → oriV.

Origin recognition complex (ORC, also *origin replication complex*): A six-subunit → protein complex (see → protein machine) that binds to the socalled → origin of replication (*ori*) and coordinates the assembly of a pre-replication complex (pre-RC) at each *ori* sequence. The ORC first recruits the initiation factors Ctd1 and Cdc6 together with the *mini-chromosome maintenance* (MCM) complex to form the socalled *pre-replication complex* (pre-RC). During the S-phase of the cell cycle, this pre-RC, containing at least six different, but related MCM proteins, is converted to an active replication fork by protein kinases Cdc7 and Cdk2, requiring the binding of at least two additional initiation factors, MCM10 and Cdc45, to the origin, subsequently initiating DNA synthesis.

OriT: See → origin of transfer.

OriV (*origin* of *vegetative replication*): The sequence of a → replicon at which its replication during vegetative growth of the host (vegetative replication) is initiated. Compare → origin of transfer.

Orphan drug: Any drug that has been developed to treat diseases occuring in less than 0.1% of the total population.

Orphan gene (orphan): Any one of a series of → open reading frames discovered in genome sequencing projects, whose function is unknown and whose sequence does not reveal any homology with entries in the sequence databanks. Do not confuse with → orphon. See → fast evolving gene, → orphan gene cluster, → pioneer sequence.

Orphan gene cluster: Any cluster of → open reading frames in a genome, whose functions are not known yet, but suspected to encode proteins for a distinct metabolic pathway (e.g. catalyzing the synthesis of a natural product). See → orphan gene, → orphan protein.

Orphan protein: Any protein, for which no substrate or interaction partner is yet found.

Orphan receptor: Any receptor protein that is known from its encoding genomic sequence, but for which no ligand is yet identified.

Orphon: An isolated → pseudogene that is related to and probably originates from tandemly repeated → multigene families or → gene batteries (for example → histone genes). Orphon genes are not necessarily located close to the gene(s) from which they originate. Compare → orphon gene.

Orthogonal-field-alternation gel electrophoresis (OFAGE): A method to separate DNA molecules in the size range from 50 kb to over 750 kb in → agarose gels by subjecting the molecules alternately to two approximately orthogonal electric fields. Compare → crossed field gel electrophoresis.

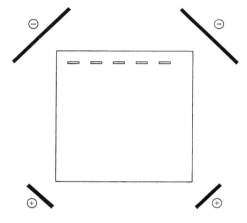

Electrode configuration of OFAGE

Ortholog (orthologous gene): One of two or more genes (generally: DNA sequences) with similar sequence and identical function(s) in two different genomes that are direct descendants of a sequence in a common ancestor (i.e. without having undergone a → gene duplication event). Also called "homology by descent". See → paralogs.

Orthologous domain shuffling: The substitution of → exons, encoding specific → domains of a protein in one species, with equivalent exons from the same gene of a different species. See → orthologous exon shuffling, → paralogous domain shuffling, → paralogous exon shuffling.

Orthologous exon shuffling: A variant of → in vitro exon shuffling that allows to replace → exons from a particular gene of one species with the equivalent exons from the same gene of a different species. See → de novo protein assembly, → functional homolog shuffling, → orthologous domain shuffling, → paralogous domain shuffling, → paralogous exon shuffling.

ortho-nitrophenyl-β-D-galactoside: See → o-nitrophenyl-β-D-galactoside.

Osmome (*osmotic response genome*): The entirety of all genes responding to variations in osmotic pressure within a cell that is sensed as osmotic stress.

Osmotic response genome: See → osmome.

OS-RAPD: See → optimized stringent random amplified polymorphic DNA.

OST: See → ORF sequence tag.

OTF: See → octamer-binding transcription factor.

Ouchterlony technique: See → agarose gel diffusion.

Ouchterlony test: See → agarose gel diffusion.

Outlier: A laboratory slang term for any data point or also cDNA (or, generally, transcript) that differs significantly from the majority of data in repeated sets of → microarray experiments.

Overdigestion: An infelicitous term for the fragmentation of a given DNA substrate by exposing it to an excess of a given → restriction endonuclease and conducting the restriction overnight. Such overdigestion assays allow to test the purity of the enzyme (i.e. to exclude exo- or endonuclease contaminations).

Overdrive:

a) The sequence motif 5'-TAARTYNCTG TRTNTGTTTGTTTG-3' adjacent to the right border of the → T-region in → Ti-plasmids that enhances the efficiency of → T-strand transfer into wounded plant cells. DNA sequences at similar sites with similar → core sequences (5'-TGTTTGTT-3') are present in the → nopaline plasmid pTi T37 and the Ri plasmid pRiA4. The overdrive sequence is also called T-DNA transmission enhancer.

b) A special state of bacterial → DNA-dependent RNA polymerase, in which the enzyme is resistant to → pause, → arrest and → termination signals. Overdrive is e.g. caused by the binding of → bacteriophage l protein Q that stabilizes RNA polymerase in the overdrive conformation, thereby optimising RNA-DNA template contacts and inducing RNA polymerase to ignore the signals. Also, → antiterminator proteins prevent → hairpin formation in RNA that otherwise leads to RNA pausing, or stabilize the elongation complex against disruption by such RNA hairpins. See → antitermination.

Overexpression: The transcription of a gene at an extremely high rate so that its mRNA is more abundant than under normal conditions. Such overexpression usually occurs in host cells that have been transformed with a → cloning vector containing a gene driven by a very strong promoter, allowing the accumulation of its protein product (in some cases this will form up to 40% of the total cellular protein of the host cell). Overexpression may also be due to the presence of a → runaway plasmid in a bacterial cell. In eukaryotes it can be responsible for the transforming activity of → oncogenes. See → overexpressor.

Overexpressor (OEX): A laboratory slang term for any → transgenic or → mutant organism that overexpresses a gene transferred by natural or → direct gene transfer, or that is mutated. Frequently, the gene itself is called overexpressor. See → overexpression.

Overgo hybridisation: A technique to isolate specific sequences from a → bacterial artificial chromosome (BAC) library of the genome of organism A (e.g. human), using genomic sequence information from organism B (e.g. mouse). In short, sequence alignments over the region of interest from both genomes allow to recognize homologous stretches. These in turn are used to design socalled overgo oligonucleotide primer pairs that are each 24 bases long and overlap by 8 bases (i.e. form → Watson-Crick bonds over 8 base pairs) at the 3′-end. Overgo primers must be non-redundant. They are then exploited as primers to fish BAC clones from the organism with the unknown target region (in the example, human). The overgos are then extended by DNA polymerase in the presence of radiolabeled C or G nucleotides, and the homologous region (e.g. a gene) isolated and characterized.

Overhang: See → protruding terminus. Don't confuse with "hang-over".

Overlap: See → contig.

Overlap hybridization: See → chromosome walking.

Overlapping clone: See → contig.

Overlapping code: See → overlapping genes.

Overlapping genes: Genes with overlapping nucleotide sequences (e.g. gene E of phage F X 174 which overlaps with gene D). Overlapping genes produce two different polypeptides (or RNAs), because the corresponding → messenger RNAs are translated in two different → reading frames. For example, in the mitochondrial genome of many metazoans several → transfer RNA (tRNA) genes overlap with neighboring tRNA genes over one (most frequently) to six → nucleotides. Since these overlapping genes are encoded on the same strand (either H- or L-strand) and are part of a single → transcription unit, only one complete tRNA can be released endonucleolytically from one → primary transcript. The other tRNA lacks the overlapping nucleotide(s) and dissociates as a truncated molecule. Regularly, the → downstream located tRNA represents the complete, the → upstream tRNA the truncated product. The truncated form cannot be charged with its cognate amino acid, since it lacks part of the → acceptor stem and position 73 (the socalled "discriminator position") necessary for charging. It is repaired by → insertional editing.

Overlapping reading frames: Any two (or more) → reading frames in the coding part

```
    met  gly  gln  tyr  asn  ala  ile  val  thr  gly  phe           Gene 1
···A U G G G G C A A U A U A A U G C A A U U G U C A C A G G G U U U···
                          met  gln  leu  ser  gln  gly              Gene 2
```

Overlapping genes

of a → gene that are generated by the presence of two (or more) → stop codons. Overlapping reading frames allow to synthesize two (or more) different polypeptides from a single gene.

Overlapping transcript: Any transcript that overlaps for at least 20 nucleotides with another transcript.

Overnight culture: Any liquid bacterial culture that has been grown for more than 12 hours (overnight) and has reached its stationary growth phase.

Overproducer: Any → mutant cell or organism producing large quantities of a chemical compound that occurs in the wild type in minute amounts only. See → overexpression.

Overwinding: See → positive supercoiling.

ox: Laboratory slang term for overexpression of a gene, or also *overex*pressor (an organism overexpressing a gene).

Oxamycin: See → cycloserine.

Oxford grid: A compilation of probe distribution patterns on homologous chromosomes from different organisms that allows a direct comparison of their → genome structure.

O⁰: See → operator zero mutation.

P

p:

a) Symbol for a *p*hosphate group (e.g. ppCpp).
b) Abbreviation for → *p*lasmid (e.g. → pBR 322).

P:

a) Abbreviation for *p*rotein.
b) Abbreviation for the amino acid *p*roline.
c) Abbreviation for *p*arental generation, compare → F1, → F2.
d) Abbreviation for → *p*romoter.

PAA gel: See → *p*oly*a*cryl*a*mide gel.

Pab DNA polymerase: See → *Pyrococcus abyssi* DNA polymerase.

PAC:

a) See → *p*hage *a*rtificial *c*hromosome.
b) See → P1-derived artificial chromosome.
c) See → *p*rotein *a*ssociation *c*loning.

PACA: See → *p*olymerase chain reaction *a*ssisted *c*DNA *a*mplification.

PACE:

a) See → PCR-assisted contig extension.
b) See → *p*olyacrylamide *a*ffinity *c*oel*e*ctrophoresis.
c) See → *p*rogrammable *a*utonomously-*c*ontrolled *e*lectrodes gel electrophoresis.

Packaging (package): The process by which a nucleic acid molecule is encapsulated in a phage (or generally, viral) head particle. This packaging process takes place within the host during normal phage growth. → Lambda phage concatemeric DNA (→ concatemer), produced by a → rolling circle replication mechanism for example, is first cleaved into monomers. One of these monomers is now introduced into the phage head precursor which mainly consists of the major → capsid protein encoded by gene *E*. Then the product of gene *D* is incorporated into the growing capsid, and the products of genes *W* and *FII* (and others) link the capsid to a separately assembled tail to form the complete (mature) phage particle.

***P*ackaging *RNA* (pRNA):** The hexagonal ring of RNA molecules that assists the ϕ29 phage to channel its DNA through the pores of an envelope synthesized by the virus prior to replication.

"Packed array" hybridization: A method to detect specific sequences in up to 2400 different clones simultaneously by transferring 96 clones from one microtiter plate at one time to an agar plate, and repeating the same process with other microtiter plates containing 96 other clones, except that these are then applied at a slightly different position. This packed array of 2400 clones can be transferred to → nitrocellulose filters (up to 20 times) and hybridized to specific → probes in a → Southern blotting experiment.

The Dictionary of Genomics, Transcriptomics and Proteomics, 4th Ed., Günter Kahl
Copyright © 2009 WILEY-VCH Verlag GmbH & Co. KGaA, Weinheim
ISBN: 978-3-527-32073-8

Packing ratio (packaging ratio): The ratio of DNA length to the unit length of the → chromatin fiber (e.g. nucleosome or 10 nm, and solenoid or 30 nm fiber) which it forms.

Pacmid: See → P1 cloning vector.

PACS: See → preferential amplification of coding sequences.

Padlock probe (circulariable probe): A linear single-stranded oligodeoxynucleotide with target-complementary sequences of 20 bp located at both termini, which are separated by a central spacer element of about 50 bp. Upon hybridization of such a padlock → probe to a target sequence, the two ends of the probe are brought into juxtaposition, in which they can be joined by enzymatic ligation (i.e. by → DNA ligase). This leads to a circularization of the oligonucleotide. This intramolecular reaction is highly specific, and discriminates among very similar sequences from two genomes (that differ by only one or few nucleotides). The circles can then be amplified and identified by e.g. → hybridization to → a microarray. Padlock probes are used for the detection of gene variants (→ genetic variation) and → mutations (e.g. determination of copy numbers of specific genomic sequences).

PAGE: Abbreviation for → *poly*acrylamide gel *e*lectrophoresis.

pA gene: See → putative alien gene.

PA-GFP: See → photoactivatable green fluorescent protein.

PAI: See → pathogenicity island.

Paired-**box gene (pax gene):** Anyone of a series of genes that share a socalled *paired* box (pax) sequence element and encode → transcription factors, which regulate the expression of other genes during ontogenesis in a strict spatial and temporal pattern. Therefore, pax genes themselves are transcribed during the development of e.g. the vertebrate embryo in highly specific patterns. The paired box (from the gene "paired" [prd] of *Drosophila melanogaster* encodes a protein → domain of 128 amino acids with DNA-binding specificity. Many pax genes additionally contain a → homeobox and an octapeptide-encoding sequence in between the paired box and the homeobox. Pax genes are known from echinoderms, molluscs, nematodes, insects, fish, birds, and mammals. See → paired box protein.

Paired **box protein (PAX protein):** Any protein encoded by a → paired box gene, and containing a 128 amino acid "paired box" domain, an octapeptide (consensus sequence: NH_2–HSIDGILG-COOH) and a → homeodomain. Paired box and homeodomains in concert function in sequence-specific binding to DNA. Each paired domain consists of two similar globular subdomains, and each subdomain in turn of three → α-helices, of which helices 1 and 2 run antiparallel, and together almost vertical towards helix 3. The aminoterminal subdomain binds to recognition sequences in the small groove DNA (consensus sequence: 5'-[G/T]T[C/T][A/C][T/C] GC-3') through contacts of terminal amino acids. Helix 3 binds within the large groove. The carboxyterminal subdomain of the paired domain contacts the consensus sequence 5'-(C/G)A-T(G/T)-(C/T)-3' in the next turn of the DNA helix. The complex between protein and DNA leads to a → bend in the DNA. The carboxyterminal regions of most paired box proteins frequently contain proline, serine, and threonine, the socalled PST region, which functions as → transcription factor. The target genes for the PAX proteins are not known, but they control different programs of organ differentiation (e.g. PAX 6 regulates the development of eyes and frontal lobe of the brain, PAX 4 the inner ear).

Paired end ditagging (PETting): The ligation of 18bp long sequence signatures from the 5'- and 3'-ends, respectively, of a → cDNA molecule to form a ditag that can be concatenated with other ditags from other → transcripts, be sequenced and mapped to a → physical map of a → genome to localize the corresponding gene, and simultaneously determine the boundaries of the corresponding transcript. PETs are the basis for the socalled → gene identification signature (GIS) technique that allows to isolate tags from both ends of virtually all full-length transcripts of a cell at a given time. The sequence of the PETs can further be exploited to design → primers for the amplification of the intervening transcripts by conventional → polymerase chain reaction (PCR) techniques.

Paired-end mapping (PEM): A technique for the large-scale identification of → structural variants (SVs) of 3 kb or larger than two (or more) genomes. In short, → genomic DNA is first sheared into fragments of 3 kb in size, then biotinylated hairpin-adaptors ligated onto the ends of these fragments, and the fragments circularized. Subsequently, the circularized fragments are randomly sheared and linker-fragments isolated. The resulting library is directly sequenced with one of the next generation sequencing technologies, and the length of the paired ends bioinformatically estimated. A computational approach is then used to map the DNA reads onto a reference genome. Any DNA rearrangements result in significant differences in the DNA fragment lengths of the reference genome as compared to the analyzed genome.

Paired-end sequence (mate pair): The → raw sequence of 500–600 bp in length from both termini of a → double-stranded insert of a clone (e.g. a → plasmid, → bacterial artificial chromosome, or → yeast artificial chromosome clone).

Pairing center (PC): Any region of meiotic chromosomes that is required for accurate segregation of homologous chromosomes during meiosis. The term is largely synonymous to → pairing site. See also → homologue recognition region.

Pairing-sensitive silencing (PSS, homolog recognition region): The increase in silencing efficiency through the physical pairing of two → homologous or partly → homologous genes in a → genome. For example, artificially introduced *Fab-7* transgenes juxtapose to the endogenous *Fab-7* locus in *Drosophila melanogaster* (even when located on different chromosomes), which finally leads to *Fab-7* gene silencing.

Pairing site (pairing element): Any region of a meiotic chromosome that facilitates or regulates the pairing of homologous chromosomes in meiosis. Pairing sites usually map near to, or are comprised of repetitive sequences. For example, a 240 bp repeat sequence in the intergenic spacer between → ribosomal RNA genes clustered on the X and Y chromosomes of *Drosophila melanogaster* is such a pairing element. Multiple copies of this sequence facilitate the pairing and subsequent segregation of both chromosomes during meiosis in

Drosophila males. See also → homologue recognition region, → pairing center.

Palaeogenomics: The whole repertoire of techniques that allow to isolate, purify, characterize and sequence DNA (mostly → mitochondrial DNA, but also residual → genomic DNA in favorable cases) from extinct plants and animals, including prehominids and hominids. For example, *Mammuthus primigenius* (mammoth) or also *Ursus spelaeus* (an extinct cave bear) DNA can be isolated and genotyped from specimen conserved to some extent in permanently frozen soil (e.g. in Siberia), and its sequence similarity with other genomes (e.g. from elephants or recent bears, respectively) detected by → comparative genomics.

Palaeoploidization: See → ancient polyploidization.

Palindrome (Greek: palindromos, running back):

a) Any sequence of letters or words that can be read in either orientation to give the same sence. For example: "Madam, I'm Adam". Or: "A man, a plan, a canal: Panama!"
b) Any sequence in duplex DNA in which identical base sequences run in opposite directions, with the property of rotational (dyad) symmetry, e.g.

↓
5'-GATGCGCATC-3'
3'-CTACGCGTAG-5'
↑

Arrows indicate rotational axis.

Palindromes are target sites for various → DNA-binding proteins (e.g. → restriction endonucleases, → RNA polymerases and → transcription factors), and occur in many → promoters, DNA replication → origins and transcription termination sequences. See also → inverted repeat, → perfect palindrome.

Palindromic unit: See → repetitive extragenic palindromic element.

PALM: See → photoactivated localization microscopy.

Palmitome: See → palmitoyl proteome.

Palmitoylation (S-acylation): The reversible thioesterification of fatty acids, usually palmitic acid, to cysteine thiols of peptides and proteins that tethers the proteins to the cytoplasmic surfaces of cellular membranes. *In vivo*, protein palmitoylation is catalyzed by *p*rotein *a*cyltransferases (PATs), but direct chemical reaction of acceptor thiols with palmitoyl-CoA can occur *in vitro*. Many key proteins of cellular signaling, membrane trafficking, synaptic transmission, and cancer are palmitoylated, as e.g. many G proteins such as Ras- or Rho-like proteins, many non-receptor tyrosine kinases as e.g. Fyn, Lck and Yes, the epithelial nitric oxide synthase eNOS, G-protein-coupled receptors, diverse ion channels, ionotropic neurotransmitter receptors, Golgi-localized mannosyltransferases, plasmamembrane-localized phosphatases, SNARE proteins, the mediators of vesicular fusion, and plasmamembrane-bound amino acid permeases, AAPs. See → palmitoyl proteome.

*Palm*i**toyl proteome (palmitome):** The subproteome consisting entirely of cellular or organellar proteins carrying palmitoyl groups. See → palmitoylation.

PALR: See → promoter-associated long RNA.

PAM: See → prediction analysis of microarrays.

Pancreatic *deoxyribonuclease*: See → pancreatic DNase I.

Pancreatic DNase I (pancreatic *deoxyribonuclease*): An enzyme from bovine pancreas that (in the presence of Mn^{2+}) catalyzes the cleavage of internucleotide bonds in single-stranded and double stranded DNA, preferentially between adjacent purine and pyrimidine residues. The enzyme is used for the limited digestion of DNA, and the removal of DNA from DNA-RNA mixtures. See → DNase I.

Pan-editing: A special type of → RNA editing, in which entire genes are edited, in contrast to partial RNA editing limited to the 5′ termini of editing domains (→ 5′ editing). Pan-editing is probably the more primitive character (e.g. in ancestral trypanosomatid mitochondria pan-editing is prevalent).

Pan-genome: The part of a genome that is present (i.e. conserved) in all organisms of a species or a population. Compare → core genome, → dispensable genome.

Panning (biopanning): A procedure to screen a → random peptide display library for protein-protein interactions and to enrich interacting clones of this library. It starts with the immobilization of bacteria or phages, displaying a target peptide or protein on a solid phase (e.g. → microtiter plate, glass, agarose, or magnetic beads), which is then incubated with the → random peptide display library and left to react. Then the unbound cells are washed off, and the bound cells released by mechanical shearing. These cells (presenting potential protein-binding proteins) are subsequently grown in suitable media, and plasmid DNA from individual clones is isolated and sequenced.

Figure see page 1134

PAP:
a) See → pokeweed antiviral protein.
b) See → pyrophosphorolysis-activated polymerization.

***par*:** See → *par*titioning functions.

Paracentric inversion: Any segment of DNA that is reversed in orientation relative to the rest of the chromosome, but does not involve the → centromere. See → pericentric inversion.

Parafilm™: A paraffin wax film used to seal laboratory glassware (e.g. tubes, Petri dishes).

Paralogon: Any duplicated (or putatively duplicated) DNA sequence blocks characterized by pairs of non-overlapping chromosomal regions. Such paralogons are enriched with → paralogs (paralogous gene pairs).

Paralogous compensation: The functional replacement of a specific gene by one of its → paralogs.

Paralogous domain shuffling: The substitution of → exons, encoding specific → domains of a protein, in a gene from one species by homologous exons from different genes of the same species. See → orthologous exon shuffling, → paralogous exon shuffling.

Paralogous exon shuffling: A variant of the → *in vitro* exon shuffling that allows to replace → exons from a particular gene of one species with the homologous exons

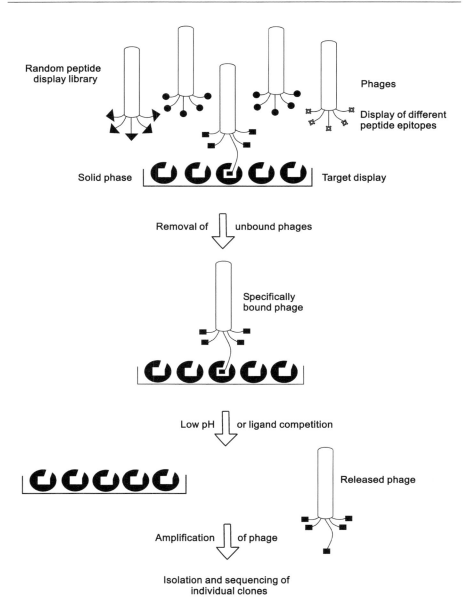

Panning from different genes of the same species. See → de novo protein assembly, → functional homolog shuffling, → orthologous domain shuffling, → orthologous exon shuffling.

Paralogous sequence variant (PSV): Any → polymorphism in → paralogous sequences in a genome, introduced by → mutation or ectopic exchanges (see → ectopic recombination).

Paralogs (paralogous genes): Homologous DNA sequences in two organisms A and B that are descendants of two different copies of a sequence originally created by a → gene duplication event in the genome of a common ancestor. See → orthologs.

Paramagnetic particle: See → magnetic bead.

Paramagnetic particle technology: A term encompassing techniques that involve specially prepared → magnetic beads (paramagnetic particles) as a solid-phase support for the separation of DNA or RNA molecules from complex mixtures of biomolecules. For example, poly(A)$^+$ mRNA can be affinity-purified using this technique. In short, a biotinylated oligo(dT) probe is hybridized in solution to the 3′ → poly(A) tail of eukaryotic mRNA. The biotin-oligo(dT)-mRNA hybrids are then captured with streptavidin covalently coupled to → paramagnetic particles. After removal of nonspecific hybrids by high-stringency washing, the mRNA can be separated from the solid phase by elution with RNase-free water. This procedure yields an enriched poly(A)$^+$-mRNA fraction after one single step of purification. See also → magnetic crosslinking.

Paramutation: Any heritable change in the activity of one → allele induced by the corresponding allele on the homologous chromosome. Such changes are brought about by modifications of → chromatin structure or cytosine methylation (i.e., are epigenetic), and do not result in, or depend on changes of the underlying DNA sequence. Compare → mutation.

Paranemic joint: A region of contact between two complementary DNA → strands that do not form a conventional → double helix.

Paranome: Any collection of → paralogous genes.

Paraspeckle: Any one of about 10–20 nuclear domains, located adjacent to → nuclear speckles and containing paraspeckle proteins1 (PSPα and β), PSP2, p54^{nrb}, and cleavage factor 1m68 (CF1m68), which all contain → RNA recognition motifs (RRMs). Paraspeckles function as depots for sequestration of A-to-I-edited nuclear RNAs.

Parent-ion-scan technique: A method for the identification of modified amino acid residues in a protein that is based on the production and mass analysis of peptide fragments in a triple-quadruple mass spectrometer. The generated peptide ions are separated in the first quadrupole according to their mass/charge ratio (m/z), and fragmented in the second quadrupole (collision cell) by collision with inert gas molecules (e.g. argon), a process called collision induced dissociation (CID). The third quadrupole again functions as mass analyzer, which permits the transmission of a fragment with a defined m/z only (e.g. m/z = 79, corresponding to the mass of a PO_3^-- fragment ion, if phosphopeptides are to be detected). This filter allows to detect peptide ions only that produce this fragment. The parent-ion-scan therefore filters specific peptides (e.g. the phosphopeptides) out of a complex mixture of fragment ions.

PARM-PCR: See → priming authorizing random mismatches polymerase chain reaction.

PARN: See → poly(A)-specific 3′ exoribonuclease.

PARP: See → poly(ADP-ribose) polymerase.

***par* region:** See → partitioning function.

Parsing: The use of algorithms to dissect data into components for an extensive component analysis.

***par* site:** See → partitioning function.

Partial: See → partial digest.

Partial *coding* sequence (partial CDS): Any → coding sequence (i.e. a region of a → genome translated from the → start to the → stop codon) that lacks either the start or the stop codons, or both.

Partial denaturation: An incomplete unwinding of a duplex DNA. See → denaturation.

Partial digest ("partial", incomplete digest):

a) The fragments arising from endonucleolytic cleavage (see → endonuclease) of a DNA molecule, in which not all the potential cleavage sites have been restricted. Partials with e.g. → four base cutter enzymes, as *Sau*3A, are used as a collection of overlapping DNA fragments for the establishment of an → ordered gene library.
b) An incomplete enzymatic proteolysis. For example, the → Klenow fragment of → DNA polymerase I is obtained by partial proteolysis.

Partial editing: See → 5′ editing.

Partial gene bank: See → minilibrary.

Partial intron retention: The inclusion of part of an → intron in a final → messenger RNA. Normally, the introns are spliced out of the pre-mRNA, but in rare cases a → splice site within an intron can be recognized and the corresponding residual intron be left in the message. See → intron retention.

Partially denaturing high performance liquid chromatography (partially denaturing HPLC): A variant of the conventional → high performance liquid chromatography (HPLC) technique, which allows to discriminate DNA → hetero- from → homoduplexes and is therefore employed for → mutation detection (e.g. the discovery of → single nucleotide polymorphisms). In short, a 200–1000 base-pair target fragment is first amplified from genomic DNA of at least two chromosomes, using → primers flanking the target site. Then the amplified fragments are denatured at 95 °C for some minutes and allowed to reanneal by gradually lower the temperature within the separating column from 95 °C to 65 °C over 30 minutes. In the presence of a mutation in one of the chromosomes, not only the original homoduplexes form upon reannealing, but also the sense and anti-sense strands of either homoduplex form two heteroduplexes. These heteroduplexes are thermally less stable than the homoduplexes. Therefore all these different duplex molecules can be separated from each other by their different retention time in an alkylated non-porous poly(styrene-divinylbenzene) column during their elution with acetonitrile. The more extensive, but still partial denaturation of the heteroduplexes in a temperature range between 50 and 70 °C (which depends on the GC content and size of the fragments, the influence of the nearest neighbour base of both the matched and the mismatched base pairs and column temperature) typically leads to a reduced retention time of the heteroduplexes and their separation from the homoduplexes. As a consequence, one or more additional peaks appear in the chromatogram. See → denaturing high performance liquid chromatography.

Partially intronic noncoding EST contig (PIN): Any → expressed sequence tag (EST) derived from a → contig that spans an → exon and parts of both flanking → introns. PINs are transcribed into → noncoding RNAs (→ PIN RNA). See → antisense PIN, → antisense TIN, → PIN RNA, → TIN RNA, → totally intronic noncoding EST contig.

Partially inverted repetitive DNA (PIR-DNA): A tandemly repeated sequence family on chicken chromosome 8, whose basic repeat units are 1.43 kb in length and consist of a central core of about 0.6–1.0 kb (in different animals). An 86 bp flanking sequence forms a → palindrome with the core (therefore partially inverted repetitive DNA).

Particle acceleration technique: See → particle gun technique.

Particle bombardment: See → particle gun technique.

Particle gun technique (biolistics, microprojectile bombardment, particle acceleration technique, particle bombardment): A method for → direct gene transfer into cells, tissues, organs or whole plants. Tungsten or gold particles are coated with DNA and shot through target cells. On their way through the cells the DNA on the particle surface is stripped off and may then be inserted into the nuclear genome. The particle gun technique has certain advantages over other direct gene transfer methods, e.g. the transfer does not require → protoplasts, but is possible with intact tissues.

Partition (plasmid partitioning): The → segregation of → plasmids to daughter cells during bacterial cell division. Segregation may depend on random distribution to the daughter cells (as e.g. for → Col E1, or other → multicopy plasmids), or may involve → partitioning functions (as e.g. for → F-factors, or other → low copy number plasmids).

Partitioning function (partitioning region, *par* site, *par* region, *par* locus, *par*): A particular nucleotide sequence of → plasmids responsible for their precise → segregation at each cell division. The partitioning activity normally ensures that each daughter cell receives about the same number of plasmids. Not all plasmids, however, have a *par* site. This region has been deleted in e.g. → pBR 322 which is consequently segregated at random during cell division. See → partition.

Partitioning region: See → partitioning function.

Partially intronic non-coding EST contig (PIN)

PIN RNA

PAS: See → primosome.

PASA: See → allele-specific polymerase chain reaction.

PAS gene: See → peroxisome assembly gene.

PASR: See → promoter-associated small RNA.

Passage:

a) The serial infection of different hosts by one and the same parasite.
b) The repeated sub-culture of cells from a cell culture.

Passenger DNA: A synonym for → insert DNA.

Passenger domain: A misleading term for any region of a peptide or protein that is fused to a bacterial membrane protein (e.g. a porin such as OmpA, OmpC, or LamB, the adhesin-like intimin, or the autotransporter EstA of *P. aeruginosa*, to name few), localized to the outer membrane of the bacterial host and exposed to the environment. Libraries with hundreds of millions of combinatorially generated passenger domains are screened with ligands to identify and select potential interaction partners. Passenger domains are applied in various screening processes, e.g. as → antigens (for the production of specific → antibodies for vaccines), as adsorbents of toxic substances (e.g. for detoxification), or as catalytic centers for novel enzymes with higher substrate affinity and other improved catalytic properties. See → display library, → microbial cell-surface display, → panning, → phage display, → random peptide display, → ribosome display.

Passenger protein (target protein): Any protein of interest expressed in appropriate host cells (e.g. *E.coli, Bacillus* or *Staphylococcus* strains, *Saccharomyces cerevisiae*) as a → fusion with a so called carrier protein that transports, anchors and exposes the passenger on the cell's surface. See → microbial cell-surface display, → peptide display, → phage display, → ribosome display.

Passenger strand (anti-guide strand): One of the two RNA strands in double-stranded → small interfering RNA (siRNA) that is not recognized by the double-strand RNA-binding protein R2D2 and therefore not incorporated into the → RISC-loading complex (RLC), but instead destroyed by the argonaute protein Ago2. However, its complementary strand, the → guide strand, is recognized by R2D2 and finally incorporated into the → RNA-induced silencing complex (RISC), and guides the destruction of complementary → messenger RNA. The passenger strand is excluded and destroyed.

Passive sliding ("translocational equilibrium", "positional equilibrium", "Brownian ratchet"): The movement of → DNA-dependent RNA polymerase II along the DNA that consists of sliding back and forth of the enzyme on the DNA in response to molecular collisions. Passive sliding presupposes a relaxed structure of the socalled clamps of the RNA polymerase molecule. See → power stroke mechanism.

Pass rate: The efficiency with which viable information is derived from each well in a 384-well microtiter plate.

Pasteur pipet: An open-end glass tube with one end pulled out to a capillary. Such Pasteur pipets allow the transfer of small volumes of liquids with the aid of a rubber bulb.

Patched circle polymerase chain reaction (PC-PCR): A variant of the → polymerase chain reaction technique that can be used for → site-directed mutagenesis. In short, the target DNA is cloned into a specific plasmid → cloning vector between opposing → T3 and → T7 RNA polymerase promoters. Then one amplification primer (→ amplimer) is annealed to sequences within the T7 promoter, and a second amplimer – in opposite direction as compared to the first – is annealed to a sequence flanking the region to be deleted. The supercoiled plasmid then serves as a template for PCR. During the amplification process linear DNA molecules accumulate which lack the region to be deleted. A third oligodeoxynucleotide primer, base-pairing with the two ends of the linear molecules is then used to form patched circles for direct transformation of *E. coli*. Appropriate and rapid screening procedures allow the isolation of clones that lack the deleted fragment. See also → polymerase chain reaction mutagenesis.

Paternal leakage: The relatively rare inheritance of paternal → mitochondrial DNA (mDNA) in primates (including humans) and rodents. For example, human mDNA is almost exclusively transmitted by maternal mitochondria, which leads to the presence of only maternal mDNA in all mitochondria of the PRogeny. However, sometimes maternal and paternal mDNAs are both present in progenial mitochondria, and also recombine to form mixed maternal-paternal hybrid molecules. The frequency of paternal leakage is very low (mice: 10^{-3} to 10^{-4} per fertilization).

Paternal X chromosome (X^P): One of the two X chromosomes of female diploid organisms that originates from the male parent. Compare → maternal X chromosome.

Paternity test: The proof or disproof of paternity in questionable cases, which is based on → DNA fingerprinting. Usually, at least 12 and maximally 20 independently segregating → short tandem repeat (STR) or → microsatellite loci are tested. For example, the human *TH01* locus is present as nine different → alleles (corresponding to nine different lengths of sequences,

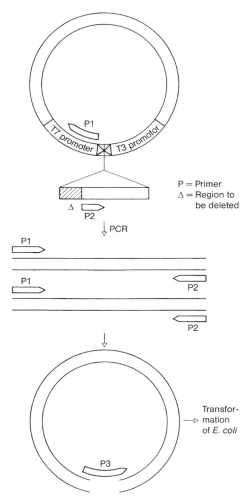

Patched circle PCR

depending on the → variable number of tandem repeats) in a population. The various alleles occur at different frequencies in any given population. In Berlin (Germany) about 25% of tested individuals possess the *TH01* allele 6, 16% allele 7, 11% allele 8, 18% allele 9, 30% allele 9.3, and 3% allele 10. The latter allele 10 (if it occurs in putative father and child) is more discriminatory than e.g. allele 9.3. The combination of 12 (or more) such loci in a paternity test results in probabilities of more than 99.99%.

PAT gene: See → *p*hosphinotricin *a*cetyl-*t*ransferase gene.

Pathochip: A laboratory slang term for a → DNA chip or → protein chip, onto which either specific genes, gene-specific oligonucleotides, cDNAs, peptides or proteins are spotted that are diagnostic for a particular pathogen (as e.g. *Staphylococcus aureus*), or even a specific strain or isolate of the pathogen. Such pathochips are used to screen samples from e.g. hospitals for particular pathogens.

Pathogen chip: See → pathogen detection array.

Pathogen-derived resistance (PDR): The resistance of a plant towards a virus, bacterium, or fungus that is engineered by the stable transformation of the plant with a → transgene derived from the virus, bacterium or fungus. For example, the gene for the coat protein or the movement protein of a virus can be transferred to target plants, stable integrated into their genome, and expressed. The resulting proteins then either coat the viral genome (preventing its replication in the host cell) or inhibit the dispersal of the virus throughout the plant, especially if the movement protein gene is genetically engineered such that its protein product still lines the plasmodesmata (channels between two plant cells, through which the viruses are transported for a systemic infection), but does not support viral transport anymore. At least one variant of PDR works with → RNA interference.

Pathogen detection array ("pathogen chip"): A specially designed → microarray, onto which diagnostic sequences of various pathogens (virus, bacteria, protozoa, fungi, parasites) are immobilized, against which test samples containing RNA or DNA are hybridized to detect the presence and (tentatively) the concentration of a pathogen. For example, 60-mer → oligonucleotides, designed from databases (e.g. Entrez Genome and Nucleotide Database, www.ncbi.nlm.nih.gov/entrez/query.fcgi) and representing → *o*pen *r*eading *f*rames (ORFs) of many viral (e.g. endogenous and exogenous retroviruses, → dsDNA, → dsRNA, → ssRNA and delta viruses), bacterial and fungal pathogens are spotted. Additionally, → genic sequences identifying immediate early, early and late genes of a virus are included that allow to define different stages of pathogen infection. Total RNA from infected and non-infected human cell cultures is then isolated, amplified, and reverse transcribed into → cDNA with simultaneous incorporation of → cyanin3 dUTP (green, control sample) or → cyanin 5 (red, test sample), respectively. After mixing, the samples are hybridized to the pathogen detection array, and hybridization events monitored by a laser detection device. Pathogen detection arrays allow to detect a broad spectrum of pathogens in a single experiment and are efficient tools in the fight against → bioterrorism. See → pathochip.

Pathogen-inducible promoter: Any → promoter that carries sequence elements mediating the activation of the adjacent gene upon a pathogen attack. For example, the promoters of plant defense genes contain three groups of → cis-acting elements, the GCC-like boxes (5'-AGCCGCC-5'), the D-boxes (5'-GGAACC-3'), and the W-boxes (5'-[T]TGAC[C/T]). GCC-like elements are active in a series of promoters driving the expression of genes involved in jasmonate and elicitor-responsive expression (JERE, 5'-AGACCGCC-3'), in cold-, salt stress- and dehydration-responsive expression (DRE, 5'-TACCGACAT-3'), and fungus-induced expression ("Box S", 5'-AGCCACC-3'). The D-box responds to wounding and some pathogenic fungi. The W-box is the address site for the WRKY family of → transcription factors. Different sequence motifs are also combined in certain promoters. The socalled Gst1 box of the potato gst1 promoter is composed of a W box and an S box, and therefore the adjacent gst1 gene is activated by both WRKY and AP2/ERF transcription factors. Or, the parsley WRKY1 gene promoter harbors a W box and a GCC box.

Pathogenesis-related (PR) proteins: An operational term encircling a characteristic group of proteins accumulating in pathogen-infected or elicitor-induced plant cells. These proteins have mostly low molecular weight and acidic isoelectric points (e.g. phenylalanine ammonia lyase, PAL; 4-coumarate ligase, 4CL; β-1,3-glucanases, chitinases, thaumatin-like inhibitors, proteinase inhibitors, hydroxyproline-rich glycoproteins, peroxidases, and others).

Pathogenicity: See → virulence.

Pathogenicity island (PAI): A distinct, instable region in a bacterial genome that contains two or more virulence-associated genes. For example, the toxin complex (Tc) genes of the bacterial symbiont and pathogen *Photorhabdus luminescens* are organized in four PAIs (loci *tca, tcb, tcc,* and *tcd*). *P. luminescens* lives symbiotically in the gut of nematodes that invade larvae of the tobacco hawkmoth. Once inside the larvae, the bacteria synthesize Tc proteins that kill the insects. Both nematodes and bacteria then feed on the dead larvae.

Pathogenomics: The whole repertoire of techniques for the sequencing and characterization of genomes of pathogens, for the identification of genes involved in pathogenicity and virulence (e.g. pathogenicity islands) or other functions relevant for the efficiency and fitness of the pathogen, and for the detection of pathogen genes with high homology to host genes (able to mimic host gene function).

Pathway mapping: The estimation of – preferentially all – possible interactions between all proteins of a biochemical pathway (e.g. glycolysis, steroid biosynthesis, protein degradation).

Pathway slide: A laboratory slang term for a → microarray, onto which → cDNAs or → oligonucleotides are spotted that represent the transcripts of genes encoding enzymes of a particular metabolic pathway (e.g. glyolysis, pentose phosphate shunt, phenyl-propanoid pathway). Pathway slides allow activity profiling of all genes encoding all proteins of such a pathway.

Pattern filtering: The detection of conserved functional genomic elements ("signatures") in one genome by aligning the sequences of two (or more) other genomes and filtering out regions of extensive → synteny. See → macrosynteny, → microsynteny, → synteny mapping. Compare → evolutionary footprinting.

PATTY: See → *PCR-aided transcript titration assay*.

Pauling-like DNA (P-DNA): A specific conformation of DNA, experimentally produced by stretching and overwinding, in which the sugarphosphate backbone is oriented towards the center, and the unpaired bases turned outside. The extreme stretching allows only 2.6 base pairs per turn (→ B-DNA: 10–11 bp/turn). Hypothetically, P-DNA could occur *in vivo* in front of a moving → RNA polymerase molecule, where a positive torsional stress leads to overwound DNA. A similar configuration was proposed by Linus Pauling before the discovery of B-DNA by Watson and Crick, therefore the somewhat misleading name "Pauling-like DNA". See → A-DNA, → B-DNA, → C-DNA, → D-DNA, → E-DNA, → ε-DNA, → G-DNA, → H-DNA, → M-DNA, → V-DNA, → Z-DNA.

Pause signal (pause site): Any (usually short) specific sequence element in a gene that adopts a specific secondary structure, where the movement of → DNA-dependent RNA polymerase during the → elongation phase of → transcription temporarily slows down or comes to a halt ("pausing"). Pausing signals reduce → nucleotide addition to the growing → messenger RNA up to thousandfold, and the pause is the initial step in arrest and → termination. Pause sites frequently are multipartite (i.e. allow multiple protein-DNA contacts) and fall into two basic classes. Class I sites, unique to bacteria and related to ρ-independent terminators, form RNA structures in the socalled exit channel (see transcription) of the → transcription elongation complex (TEC) that cooperate with sequences in the downstream DNA and the hybrid (see transcription) to open the exit channel and cleft such that 3′-nucleotide misalignment occurs. At class II pause sites, a weak hybrid (rich in rU·dA) triggers RNA polymerase to reverse translocate (backtrack, see → translocation) on the DNA, thereby shifting the hybrid to a more stable upstream register and threading 3′ single-stranded RNA into the NTP entry tunnel. This translocation may cause transcriptional arrest. See → arrest signal, → termination signal.

Pax gene: See → paired-box gene.

Pax protein: See → paired box protein.

PB: See → *piggyback*.

pBeloBac 11: A 7.507 kb single-copy → bacterial artificial chromosome (BAC) → cloning vector for the cloning of large DNA fragments (up to 1 Mb) in *E. coli* that contains an oriS replicon of the fertility (F) factor of *E. coli*, Sop AB (Par AB) functions for active → partitioning (acting at Sop C [IncD, Par C] such that each daughter cell receives a plasmid copy during cell division), replication initiation factor Rep E (Rep A) sequences (the encoded protein mediates the assembly of a replication complex at ori 2), a truncated copy of a site-specific recombinase (red F), and a → chloramphenicol acetyltransferase gene from → transposon 9 as a → selectable marker. The cloning region encompasses a → lambda phage → cos site (representing a unique cleavage site and enabling the → packaging into phage particles), a → lox P site, two (in variants of the orginal pBeloBac vector more) cloning sites (e.g. *Bam*HI and *Hin*dIII sequences), a series of → rare cutter sites (e.g. *Sfi*I), → SP6 RNA polymerase and → T7 RNA polymerase promoters flanking the cloning site (for the generation of RNA → probes from the insert). The large DNA inserts in

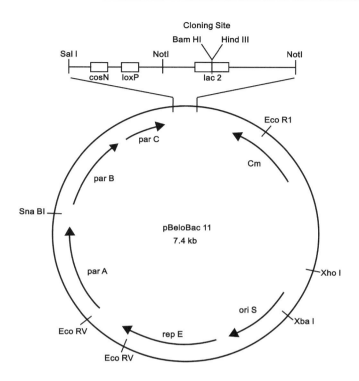

pBeloBac11

the single-copy pBeloBac 11 vectors are stable and do not interfere with the viability of the host cell. BAC cloning has superseded the traditional → yeast artificial chromosome (YAC) cloning, because BAC cloning procedures are easier, and → chimerism of inserts a rare event. See → BIBAC, → mammalian artificial chromosome, → P1 cloning vector, → *Schizosaccharomyces pombe* artificial chromosome.

P body: See → processing body.

pBR 322 (and derivatives): A series of comparatively small, → multicopy (15–20 copies/cell), → non-conjugative plasmid cloning vectors containing → ampicillin and → tetracycline resistance genes and several unique cloning sites (or, in derivatives, → polylinkers). The latter are located within one or the other resistance gene, so that the insertion of foreign DNA can be detected by → insertional inactivation of the antibiotic resistance function. The notation "BR" is derived from *Bolivar* and *Rodriguez*, two Mexican molecular biologists who synthesized the plasmid using the tetracycline resistance gene from pSC 101, the origin of replication (ori) and rop gene from the Col E1 derivative pMB1, and the ampicillin resistance gene from → transposon Tn3. The plasmid replicates in *E. coli* under → relaxed control, but is slightly unstable and has a relatively narrow host range (*E. coli, Serratia marcescens*). Therefore, more advanced derivatives have been designed (see → pUC).

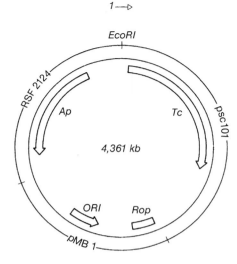

Simplified map of pBR 322

PBS: See → primer binding site.

Pc:

a) See → pseudogene.
b) See → pseudouridine.

PCA:

a) See → principal component analysis.
b) See → protein fragment complementation assay.

Pc box: An approximately 15 amino acids long sequence motif within the → C-terminus of the Polycomb (Pc) protein of the Polycomb repressive complex 1 (PRC 1) of Drosophila melanogaster and many other animals and also plants. The Pc box mediates transcriptional → repression of target genes of the → Polycomb group proteins.

P call: A laboratory slang term for the presence ("P") of a particular molecule (e.g. a → messenger RNA) in a particular population (e.g. the → transcriptome).

PcG: See → polycomb group protein.

PC-PCR: See → patched circle polymerase chain reaction.

PCR: See → polymerase chain reaction.

PCR add-on primer (restriction site add-on): A synthetic → oligonucleotide that carries a → recognition site for a → restriction endonuclease and still serves as a → primer for → Thermus aquaticus DNA polymerase. Such add-on primers with 5′ overhanging termini are annealed to the target DNA and the DNA amplified by the conventional → polymerase chain reaction. The amplified product then contains the desired restriction site(s), and can easily be cloned into appropriately cut → cloning vectors. See also → add-on sequence.

Figure see page 1145

```
       5'
        C
         G  ╲  EcoRI
          G  ╲  add-on
           A  ╲  sequence        PCR primer
            A  ╲
             T  ╲
              T  ╲
               C  ╲
                  C G C A C T C G A G C T T G G C A G 3'
                  ║ ║║ ║║ ║ ║║ ║ ║║ ║║ ║ ║ ║║ ║ ║║ ║║ ║ ║
3'···T C G A G T A G C G C G T G A G C T C G A A C C G T C T A C C G···5'
```

Template DNA

PCR add-on primer

PCR-aided transcript titration assay (PATTY): A technique for the quantification of a specific → messenger RNA, which capitalizes on the co-amplification of a mutated, and therefore different form of the target messenger RNA. In short, first the mutated cDNA is generated by → site-directed mutagenesis such that a single base exchange occurred and a new → restriction recognition site is generated. Then identical amounts of total RNA (containing an unknown amount of wild-type mRNA) are mixed with decreasing, but known amounts of mutated mRNA. After → reverse transcription to cDNA and → polymerase chain reaction amplification with → primers complementary to sites within the target cDNA, the amplified fragments are restricted (only mutated cDNA is cut, and therefore differentiated from the wild-type cDNA). The cDNA (tar get) or cDNA fragments (mutant) are then separated by → agarose gel electrophoresis and hybridized to a radioactively labeled subfragment of the target cDNA. The hybridization signals then allow to identify one particular sample, which contains equal or nearly equal amounts of both types of cDNAs, reflecting equal starting concentrations of the original mRNAs.

PCR amplification of specific alleles: See → allele-specific polymerase chain reaction.

PCR array: A laboratory slang term for a → microarray, onto which DNA is spotted that is generated by → polymerase chain reaction amplification of target DNA.

PCR-assisted contig extension (PACE): A technique for the closure of → gaps remaining in unfinished bacterial genome sequences that involves the generation of stepwise extensions from the ends of → contigs by a conventional → polymerase chain reaction (PCR), until the closure of the individual gaps is achieved. In short, specific internal and → nested primers are first derived from the sequenced contigs. In a first step, the specific internal forward primer is used in combination with a reverse primer of arbitrary sequence (→ "arbitrary primer") to amplify → genomic DNA outwards of the contig. The nested primers are designed approximately 150 bp from the contig ends and 40 bp apart from each other. In a subsequent second step, the amplification products of the first step are diluted and again amplified with a nested primer and a perfectly matching primer derived from the sequence of the first amplicon under higher → stringency. The products are then electrophoresed in agarose gels, stained with → ethidium bromide, and single bands isolated and sequenced directly with the same specific primer used for amplification. The contig →

extensions have to be verified by specific PCR and sequencing.

PCR carry-over prevention: See → *polymerase chain reaction carry-over prevention*.

PCR clamping: A technique for the detection of → deletions, → insertions, → mutant alleles, or → point mutations in a target DNA that is based on the increased affinity and specificity of → *peptide nucleic acids* (PNAs) for their complementary target sequences and the inability of → DNA polymerase to recognize and extend a PNA primer. In short, a 15–18-mer peptide nucleic acid complementary to the wild-type sequence is synthesized. The PNA oligomer is then mixed with two DNA primers, one of which is complementary to the mutant allele sequence (forward primer), whereas the other one serves as a → reverse primer to amplify the target sequence. In the subsequent → polymerase chain reaction, the wild-type PNA competes with the mutant DNA primer for the same target priming site. Hybridization of the DNA primer and subsequent amplification will only occur, if the target is a mutant allele (amplification product can be visualized by e.g. → ethidium bromide staining). In the absence of a mutant allele, the PNA will bind to the target and prevent amplification (no amplification product can be visualized). Two PCR clamping configurations are possible. First primer exclusion, where a PNA oligomer competes with a DNA primer for binding at the target site, as described above. The DNA outcompetes the PNA, binds to the target, and allows its extension only when it is fully complementary to the mutant site. Point mutations at various positions in the target can be identified by altering the sequence of the primer. Second, elongation arrest is a result of the stronger binding of PNAs to their targets (PNA/DNA duplexes at physiological ion strength are about 1 °C/base more stable than the corresponding DNA/DNA duplexes), which prevents the elongation of a primer that binds outside the target DNA.

PCR fingerprinting: The amplification of distinct highly polymorphic target DNA sequences (e.g. → simple repetitive DNA sequences), using → polymerase chain reaction techniques to establish a → DNA fingerprint of the target. See for example → arbitrarily primed polymerase chain reaction.

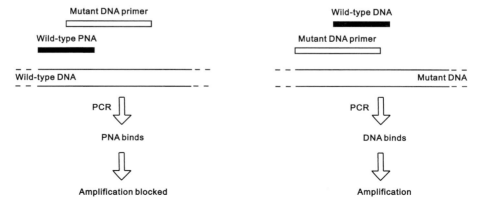

PCR clamping

PCR *in situ* hybridization: A variant of the → polymerase chain reaction, in which DNA is amplified and detected within morphologically intact cells or tissues. In short, cells or tissue specimens are fixed, mounted on a silane-coated microscope slide and digested with → protease. All PCR reagents are added and diffuse into the cells. Then the slide, on an aluminum foil, is placed directly on the thermoblock of a → thermocycler, → *Thermus aquaticus* DNA polymerase is added and the slide overlaid with mineral oil. The amplified product can be detected by → *in situ* hybridization or by direct incorporation of → biotin- or → digoxygenin-labeled nucleotides into the PCR product. Since the diffusion of the product away from its original location is a problem, either → multiple overlapping primer PCR or → concatemer PCR are used to generate large PCR products that do not freely diffuse. See → *in situ* hybridization.

PCR-ligation-PCR mutagenesis (PLP mutagenesis): A technique for the generation of → fused genes, site-directed mutagenesis, or introduction of specific → deletions, → insertions or → point mutations into target DNA. For example, the fusion of two (or more) genes starts with the amplification of each gene in a separate → polymerase chain reaction. The amplification products are then phosphorylated using → T4 polynucleotide kinase and ligated with → T4 DNA ligase, creating different combinations of joined fragments. The fused gene is then specifically PCR-amplified out of this heterogeneous mixture with a primer directed to the 5′ end of the upstream gene and a primer complementary to the 3′ end of the downstream gene. The resulting amplification product is then subcloned into the original target sequence, creating a type of insertion mutation. PLP mutagenesis relies on a DNA polymerase with exonuclease (i.e.

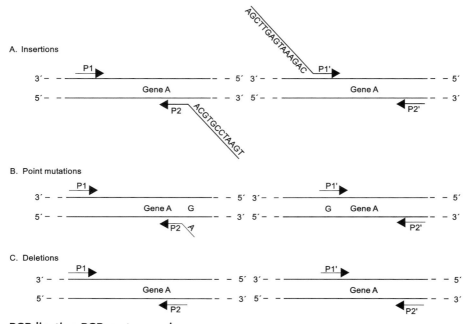

PCR-ligation-PCR mutagenesis

proof-reading) activity, so that the blunt-ended fragments match exactly with the primer sequence. Compare → splice overlap extension polymerase chain reaction.

PCR-mediated chromosome splitting (PCS): A technique for the fragmentation of a eukaryotic → chromosome into (preferably) stable and (preferably) defined → minichromosomes. In short, gene cassettes with *loxP*-marker-*loxP* and CEN4 sequences are first generated by conventional → polymerase chain reaction (PCR) using two → oligonucleotides, one harbouring a short region of → homology (45 or 80 bp) to the chromosomal target site and a → Cre/*loxP* system (5'-*loxP*-marker gene-*loxP*-3', for repeated chromosome splitting by the successive, targeted deletion of several different → marker genes), and one single marker gene (e.g. a → kanamycin resistance gene), and the other one carrying several copies of the → telomeric seed motif 5'-CCCCAA-3' repeat (e.g. 5'-$[C_4A_2]_6$-3'). These constructs are then transformed into recipient cells together with a third vector encoding and expressing → Cre recombinase that removes the corresponding marker gene. PCS is based on the → integration of the chromosome-splitting vector containing the cassettes with the target-complementary sequence, a → selectable marker gene, → centromeric repeat sequences (e.g. for yeast chromomosome splitting: *CEN*4), and inverted repeats of termini of the *Tetrahymena* → ribosomal DNA sequence (*Tr*) into the chromosomal target site, which induces the splitting of the recipient chromosome at the target site into two monocentric chromosomes. Splitting is a result of the resolution of the *Tr* ends into two new → telomeres. PCS is employed for the generation of minichromosomes, the replacement, → fusion, or transfer of chromosome segments, or the test for minimal chromosome constitutions. See → chromosome fragmentation, → chromosome splitting.

PCR mimic: See → heterologous competitive standard.

PCR mutagenesis: See → polymerase chain reaction mutagenesis.

PCR-RFLP: See → polymerase chain reaction restriction fragment length polymorphism.

PCR-SAGE: See → polymerase chain reaction serial analysis of gene expression.

PCR-SSCP: See → single-strand conformation analysis.

PCR technology: A myriad of techniques to amplify a specific DNA segment and to modify the amplified sequence simultaneously (e.g. by the application of → PCR add-on primers, the introduction of → mutations in → PCR mutagenesis, or the in-vitro recombination of two specific DNA fragments in → recombinant PCR). See → polymerase chain reaction.

P-DNA: See → Pauling-like DNA.

P$_E$: See → proportion of essential genes.

Pectinase ("driselase"; EC 3.2.1.15): An enzyme that catalyzes the degradation of plant pectin and is therefore used to degrade the cell walls in → protoplast isolation procedures.

PEER: See → primer extension enrichment reaction.

PEG: See → *polyethylene glycol*.

Pegylation: The attachment of *polyethylene glycol* (PEG) chains to peptides or proteins to increase their size and to protect them from degradation. For example, pegylation of interferon α-2a increases its half-life time from 9 (unpegylated) to 77 hours and at the same time reduces its renal clearance 100-fold, such that the compound is less susceptible to destruction in the digestive tract and remains in the body for a longer time.

P element: A member of a family of transposable elements (→ transposons) in *Drosophila* species that is randomly distributed in the genomic DNA of so-called P strains (*paternally contributing*) in 30–50 copies. The P element prototype is 2.9 kb in length, and the other members of the family have evolved to 0.5–2.5 kb elements by different internal deletions. Each P element is flanked by perfect → inverted repeats of 31 bp at which excision takes place. The insertion of the P element at another locus is accompanied by a duplication of a short 8 bp target sequence that can be found on either side of the integrated P sequence. The internal fragment of the P element prototype carries four → open reading frames (ORF 0, 1, 2 and 4), one of which encodes an 87 kDa → transposase and another one a 66 kDa transposition repressor protein. → Transposition requires the activity of this transposase, which is active in so-called M (*maternally contributing*) cytotype cytoplasm, but mostly inactive in so-called P-cytotype cytoplasm because of the presence of the transposition repressor protein in the cytoplasm of P strains. Transposition activity of P elements is strictly limited to germ line cells.

P elements may insert into control or coding sequences of genes, which are thereby inactivated. Together with concomitantly occurring chromosome breakages these insertions lead to the disease syndrome of "hybrid dysgenesis" (P-M hybrid dysgenesis, i.e. genetic abnormalities such as chromosomal aberrations, high frequencies of lethal mutations and high rates of sterility).

P elements can be exploited as → gene transfer vectors. Any foreign DNA can be cloned into a P element which in turn can be inserted into a → plasmid. After → microinjection of this plasmid into *Drosophila* embryos the P element together with the foreign DNA can transpose into germline chromosomal DNA. P elements can also be used to search for specific genes of *Drosophila* via → transposon tagging. In this case, the P elements function as → mutagens which lead to a loss of gene function through their → insertion (→ insertion mutation).

P element transformation: The integration of specific DNA fragments into germ line chromosomes of *Drosophila* using the transposable → P elements as transposing sequences.

Pellet: Any packed material sedimented by centrifugation.

PEM: See → *paired-end mapping*.

pEMBL: A family of single-stranded 4 kb → plasmid cloning vectors derived from → pUC, containing the → *bla* (ampicillin resistance) gene as → selectable marker, a short DNA segment coding for the α-peptide of → β-galactosidase that carries a → polylinker, and the intragenic region of the → fl phage. Upon superinfection with phage fl, these plasmids may be encapsidated as single-stranded DNA, and the virions are excreted into the culture medium. pEMBL vectors can be used for DNA → Sanger sequencing, for → site-directed mutagenesis, → S1-mapping and hybridization to mRNA and cDNA. These vectors are smaller than the → M13 vectors and are relatively stable even with large inserts. Without superinfection, the replication of the double-stranded pEMBL plasmids is initiated at the → col E1 → origin of replication. EMBL stands for European Molecular Biology Laboratory. Do not confuse with the lambda phage derived → EMBL vectors.

Penelope retrotransposon: Any one of a class of autonomous vertebrate → retroelements that contains genes for a → reverse transcriptase and a UvrC and intron-encoded → endonuclease (URI) and disappeared in the human lineage in the course of evolution.

Penetrance: The frequency (also probability) of the expression of an → allele or → gene. See → complete penetrance, → expressivity, → incomplete penetrance, → nonpenetrance.

Penicillin: Any of a series of → antibiotics synthesized by *Penicillium notatum* and related molds (e.g. *Aspergillus, Trichophyton, Epidermophyton*). Penicillins are derivatives of 6-amino-3,3-dimethyl-7-oxo-4-thia-1-azabicycloheptan-2-carboxylic acid

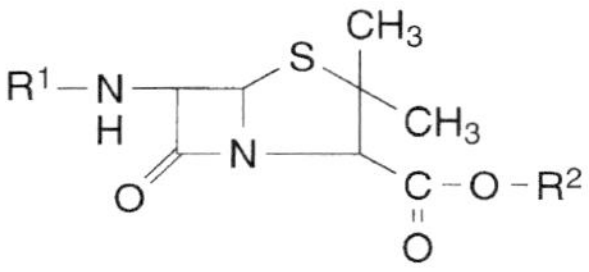

6-Aminopenicillanic acid

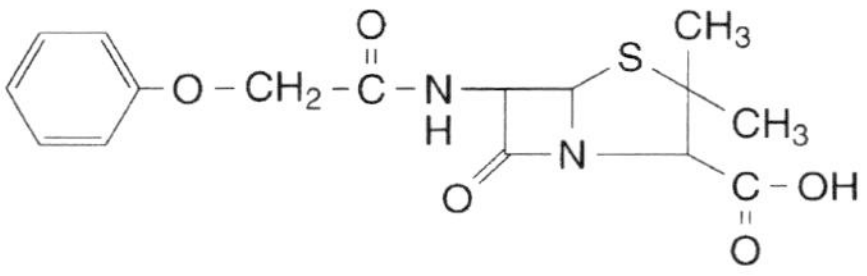

Penicillin V

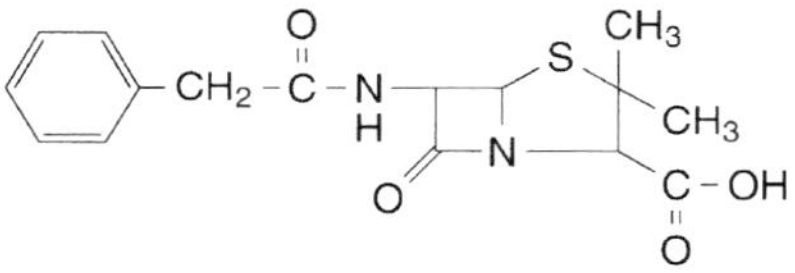

Penicillin G

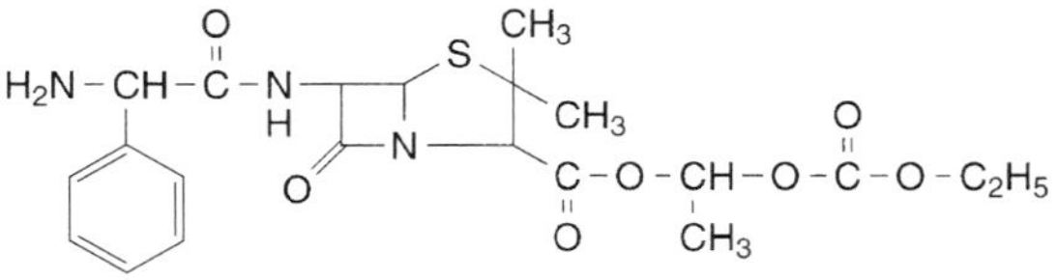

Bacampicillin

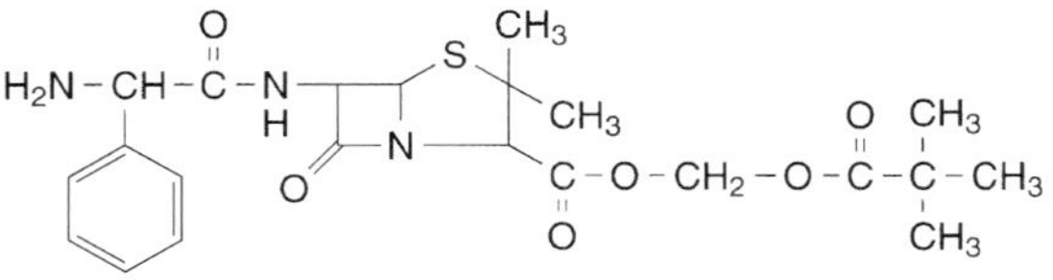

Pivampicillin

Penicillin

(6-aminopenicillanic acid). Different penicillins differ from each other in the structure and number of the side chains, (one in e.g. penicillin G, penicillin V). Penicillins block the cross-linkage between parallel peptidoglycan chains, and thus prevent the completion of the synthesis of bacterial cell walls.

Penicillinase: See → (b)-lactamase.

PENT: See → primer extension-nick translation.

Pentaplex DNA: A self-assembled higher-order structure of DNA, in which the naturally occurring base 2′-deoxy-iso-guanosine (iG) is assembled around a caesium ion in a quintet geometry (iG-quintet/Cs+/iG-quintet), and the caesium ion is positioned between two quintet layers.

Penta-snRNP: A pre-formed complex of all five → small nuclear RNAs (snRNAs) U_1, U_2, U_4, U_5 and U_6 and about 13 different proteins that associates with the → messenger RNA (mRNA) as a single discrete particle. The experimental proof of this particle is somehow conflicting with the socalled stepwise assembly model of the → spliceosome, which predicts that the U_1snRNP first recognizes its substrate mRNA and binds to the 5′splice site. After this interaction, the U_2snRNP contacts the branch point region of the message. And only then a complex of $U_4/U_5/U_6$ j → with the → intron removed and the two adjacent → exons joined.

Pentatricopeptide repeat protein (PPR): Any one of a large family of proteins encoded by → nuclear genes that are synthesized on cytoplasmic → ribosomes, in their majority transported into either → mitochondria or → chloroplasts of plants, where they function in maturation, stability or → editing of individual → messenger RNAs, in RNA cleavage, → splicing, and → translation initiation. PPR proteins are composed of a series of pentatricopeptide repeat motifs (where the number of repeats vary in individual proteins) and associated with mitochondrial → RNA polymerase and ribosomes. In *Arabidopsis thaliana*, the 451 PPRs comprise about 15% of all soluble mitochondrial proteins, and fall into two categories: the P family of PPRs are constituents of all eukaryotes with mitochondrial genomes, and the PLS family proteins are confined to land plants.

PEP:

a) See → plastid-encoded polymerase.
b) See → primer-extension preamplification.

PEPSI: See → polyester plug spin insert.

Peptibody: Any one of thousands of peptides generated by → phage display that owns promising therapeutic poperties. For example, a specific peptibody targeting and binding the *t*hrombo*p*oietin (TPO) receptor protein exerts its stimulating influence on precursoe platelets (megakaryocytes) to mature into platelets.

Peptide: A molecule consisting of two or more amino acids. Peptides range in size from 400 to 9,000 Da, and are mostly secreted. Biologically active peptides are e.g. adrenomedullin, glucagon, ghrelin, and orexin-A, to name few. See → peptide bond, → peptide map, → polypeptide.

Peptide amphiphile: Any one of a series of engineered nanomolecules consisting of a hydrocarbon tail attached to a peptide that additionally contains amino acid sequences with a cellular function. For example, a specific peptide amphiphile carries sequences that stimulate neurons to connect to neighboring neurons, and is therefore a candidate for nanomedicine. This particular amphiphile self-assembles into fibers ("nanofibers") that form networks with neurostimulatory properties.

Peptide array (peptide microarray, peptide microchip): An inert membrane or glass slide (or other solid support), onto which thousands of short, 24 amino acids long peptides are spotted in an ordered array to allow the visualization of interactions with labeled ligands (e.g. peptides, proteins, antibodies, low molecular weight effectors). For a specific variant of a peptide array the peptides are synthesized on modified cellulose disks. The cellulose with the covalently bound peptides is subsequently dissolved, and the peptides are spotted onto a solid support (e.g. a glass slide), on which a three-dimensional layer of peptide-cellulose conjugates form. Then target proteins labeled with e.g. → biotin are added, and an interaction between an immobilized peptide and one of these proteins can occur. A subsequent reaction with → streptavidin-alkaline phosphatase (AP) conjugate leads to the capture of biotin by streptavidin. If a colorless substrate for AP is then added and catalytically processed, the formed coloured product can easily be detected (see → alkaline phosphatase). Peptide arrays are used for e.g. → epitope mapping, definition of protein binding domains, immunogen selection, vaccine design, and drug discovery. Compare → protein chip.

Peptide biomarker: Any native peptide that (alone or in combination with other peptides) is characteristic for a specific condition of a cell, tissue or organ, for a normal or pathological process, or for pharmacological reactions upon therapeutic interventions, and can therefore be used as diagnostic marker. Peptide biomarkers, usually in the size range up to 150 amino acids or 20 kDa, can be identified by → differential peptide display and fall into several broad categories: the socalled disease markers (diagnostic for a disease), the staging markers (allow the division of the course of a disease into different stages), the stratification markers (allow to group individuals into collectives), the bridging or translational markers (can be used in both the pre-clinical and clinical phases), the efficacy markers (indicate the benefit of a specific treatment), toxicity markers (indicate undesirable effects of a drug), the predictive markers (permit conclusions about the course of a disease or treatment), the screening markers (used for early diagnosis of a disease), and the prognostic markers (prognosticate the outcome of a disease or treatment). See → peptidome, → peptidomics.

Peptide bond: Any covalent bond between two amino acids arising from linkage of the α-aminogroup of one to the α-carboxyl group of the second molecule with concomitant elimination of water.

Peptide chip: See → protein chip.

Peptide computer (protein computer): A special variant of a → biocomputer that performs computational tasks with peptides or proteins. One of the major advantages of peptide computers over → DNA computers is that every position of a peptide can be occupied by 20 (or more, for example, synthetic or artificial) amino acids as compared to only four bases in DNA). Instead of a hybridisation reaction of two nucleic acid molecules in DNA computers, peptide computers exploit the (usually stereo-specific) interaction(s) of e.g. → antibodies with peptide → antigens.

Peptide display: See → phage display.

Peptide fingerprint: The specific pattern of peptide fragments generated by proteolytic cleavage of a protein and displayed on e.g. stained gels after their electrophoretic separation. Peptide fingerprints are the products of → peptide or → protein fingerprinting. See → peptide fragmentation fingerprint, → peptide mass fingerprint.

Peptide fingerprinting: See → protein fingerprinting.

Peptide fragmentation fingerprint (PFF): The specific pattern of fragments arising from a singl e peptide of a → peptide fingerprint. The target peptide is first isolated from the peptide mixture in the mass spectrometer and subsequently fragmented. The molecular weights of these fragments can be determined precisely and altogether represent a fingerprint of the peptide. See → peptide fingerprint, → peptide mass fingerprint.

Peptide map: A characteristic → peptide fragment pattern, generated by → protein fingerprinting. The comparison of such peptide maps from two (or more) proteins allows the detection of similarities or dissimilarities between the corresponding proteins on a large scale.

Peptide mapping: A procedure for the establishment of a → peptide map of a protein. In short, peptide mapping starts with the unfolding of the isolated and purified protein, its reduction, and alkylation to prevent re-formation of disulfide bridges. After extensive dialysis to remove excess reagents, the protein is proteolytically digested, and the resulting peptide fragments separated by reversed-phase chromatography. Peptide mapping provides informations about protein structure and reveals substitutions of amino acids and post-translational modifications. See → protein fingerprinting.

Bromocyan

Val-Val-Arg-Asn-Lys↑-Ile-Tyr-Thr-Ser-Met↓
-Ser↑-Asp-Leu-Phe

Endoproteinase Lys-C Endoproteinase X-Asp

Enzymes/reagents	Specificity
Chymotrypsin	Aromatic amino acid-X
Endoproteinase Arg-C	Arg-X
Endoproteinase Asp-N	X-Asp
Endoproteinase Lys-C	Lys-X
Factor Xa	Ile-Glu-Gly-Arg-X
Pepsin	Leu_X, Phe-X, Met-X, Trp-X
V8 proteinase	Glu-X, Asp-X
Trypsin	Lys-X, Arg-X
Bromocyan	Met-X
Iodobenzoic acid	Trp-X

X: any amino acid

Peptide mass fingerprint (PMF): The specific peptide fragment pattern arising from e.g. cleavage of a protein by proteolytic enzymes and analysed by → matrix-assisted laser desorption ionization mass spectrometry (MALDI-MS). PMFs are specific for the target proteins and can be used as search query against a database of PMF-like entries (e.g. produced by theoretical digests of protein sequences). See → peptide fingerprint, → peptide fragmentation fingerprint.

Peptide-mediated non-covalent protein delivery: See → non-covalent protein delivery.

Peptide microarray: Any solid support (e.g. a glass slide) that is coated with → avidin or → streptavidin and onto which (usually synthetic) biotinylated 13–15-mer → peptides are coupled via biotin-avidin (streptavidin) interactions. Peptide microarrays allow to detect and characterize → antibodies by e.g. → immunofluorescence sandwich assays.

Peptide mimicry: The synthesis of a biologically active protein or peptide that retains all or most of the structural features in addition to all functional domains. Compare → peptide morphing.

Peptide morphing: The design and synthesis of a derivative of a naturally occuring protein or peptide that eliminates all non-functional amino acids, but retains its functional domains (for e.g. binding of ligands, protein-protein interactions, catalysis, protein-DNA- or protein-RNA interactions). Peptide morphing aims at increasing the chemical and biological stability of the "morphed" peptide and lowering its polarity by reducing its amide bonds such that it can be used as therapeutically active compound. Compare → peptide mimicry.

Peptide MS/MS-fragmentome: See → fragmentome.

Peptide nanowire: A short → oligopeptide engineered to bind one or more cobalt ions to its surface such that it has conductive properties. For example, a 33 amino acid long → domain of a → transcription factor is a such a peptide nanowire, which is used to join a carbon nanotube element of a nanoelectrode array and a redox protein such that it spaces the protein from the nanotube, minimizing interference from surface effects, and at the same time connects the active site of the protein to the electrode.

[Peptide]n⁺-fragmentome: See → fragmentome.

Peptide nucleic acid (PNA; polyamide nucleic acid): A relatively simple, synthetic chimeric polymer with a neutral achiral polyamide (peptide-like) backbone composed of N-(2-aminoethyl) glycine units, to which nucleic acid bases are covalently bound via carbonyl methylene ($-CH_2-CO-$) linkers. Such PNAs are increasingly used as substitutes of normal DNA. PNAs also form duplexes with complementary PNA strands via Watson-Crick base pairing. The resulting PNA-PNA hybrids are more stable than PNA-DNA duplexes. But also PNA-DNA and PNA-RNA hybrids are more stable than the corresponding DNA-DNA and DNA-RNA complexes, because no repulsion occurs between the charged phosphodiester backbone of DNA (or RNA) and the neutral PNA backbone. In contrast to DNA-DNA hybrids, PNA-DNA duplex stability is little affected by changes in salt concentration. Since single base → mismatches in PNA-DNA duplexes are more unstable than corresponding single base mismatches in DNA-DNA hybrids, a higher specificity results. Therefore, PNAs easily allow discrimination between perfect matches and mismatches of bases. As an artificial molecule, a PNA is no substrate for proteases or nucleases.

Homopyrimidine PNAs invade intact double-stranded DNA. For example, a PNA complementary to CAG repeats binds to its target DNA even in intact chromatin. PNA strand invasion results in stable PNA-DNA complexes, especially within transcriptionally active regions. Following a digestion of chromatin with a mixture of → restriction endonucleases (that do not cleave within the CAG repeats), the CAG-containing fragments are then first bound to mercurated → paramagnetic beads via thiol-reactive → nucleosomes (→ lexosomes), then released from the beads, hybridized to the biotinylated PNA → probe and the resulting PNA-DNA hybrids captured on → streptavidine-coated beads. DNA is then released and tested for CAG triplet content that allows to diagnose → triplet expansion-based diseases.

PNA oligomers serve as hybridization probes in → in situ hybridization, → Northern and → Southern analyses, but cannot be used as → primers in conventional → polymerase chain reaction techniques, since they lack 3′ OH groups and therefore cannot be recognised by → DNA polymerases.

PNAs can be labeled by → biotin, → digoxygenin, → Cy 5, → Cy 3, → fluoresceine, → rhodamine, or ^{32}P and ^{125}I. They can also be attached to a solid phase (e.g. a glass or quartz chip, see → DNA chip) and used as probes to screen for complementary → cDNA or DNA sequences. Compare → pyranosyl-RNA.

Peptide nucleic acid inhibitor probe (PNA inhibitor probe): A short → peptide nucleic acid (PNA) sequence that prevents undesired product formation on → genomic DNA templates (contaminants) during → reverse transcription polymerase chain reaction (RT-PCR) amplification of specific RNA sequences. The inhibitor probe is designed to bind to a genomic sequence overlapping one of the PCR primer-binding sites within the sequence of interest. Hybridization of the blocking probe precludes PCR → primer attachment to DNA without affecting attachment of the same primer to → cDNA. A specific pre-primer annealing step in the RT-PCR protocol proceeds at a temperature high enough to allow the PNA inhibitor probe to anneal to genomic DNA, but not cDNA. The inhibitor PNA therefore blocks any PCR amplification of genomic DNA contaminants, but does not affect the amplification of cDNA. The probe can either be designed to target the 3'-end of the cDNA (as the → forward PCR primer does) corresponding to the 3'-end of the → messenger RNA ("terminal inhibition"), or target the region amplified by the → reverse primer used in the RT-PCR ("internal inhibition").

Peptide nucleic acid-phosphono peptide nucleic acid chimera (PNA-pPNA chimera): A synthetic hybrid polymer composed of alternating stretches of → peptide nucleic acids and → phosphono peptide nucleic acids. Phosphono peptide nucleic acid monomers of basically two types containing N-(2-hydroxyethyl)phosphono-glycine or N-(2-aminoethyl)phosphono-glycine are linked through amide or phosphonate monoester bonds to PNA derivatives N-(2-hydroxyethyl)glycine or N-(2-aminoethyl) glycine. The resulting chimeric oligomers form stable complexes with complementary single-stranded DNA or RNA molecules, are resistant to nucleases and possess good water solubility. PNA-pPNA chimeras are used for nucleic acid hybridisations.

PNA-pPNA chimeras

Peptide sequencer: An instrument for the automated estimation of the order of amino acids in → peptides and → proteins that is based on the → Edman degradation (cleavage of the N-terminal amino acid of peptides and its subsequent identification). A full sequencing cycle consists of the exposure of the target peptide to phenylisothiocyanate, the acid hydrolysis of the carbamylpeptide and the separation of the thiohydantoin derivatives of the terminal amino acid. Various types of peptide sequencers are available, as e.g. solid phase sequencers (where the peptide is covalently bound to a solid carrier that allows the removal of the cleaved amino acid by simple washing), gas phase and liquid phase sequencers.

Peptide topography: The three-dimensional arrangement of the side chains of the amino acids in a peptide.

Peptide vaccine: Any peptide that induces a strong immune response and therefore can be used for vaccination. Such peptides must carry an appropriate allele-specific T-cell epitope for the recipient species, and need to be attached to a carrier protein to enhance immunogenicity (longer peptides elicit a strong humoral response). For example, lipopeptide vaccines trigger humoral and cellular immune responses very effectively. These vaccines are heat-stable, non-toxic, completely biodegradable and are synthesized on the basis of minimized epitopes. They activate the antigen-presenting macrophages and B lymphocytes.

Peptidome: The complete set of (specifically biologically active) peptides and small proteins (molecular weight up to 20 kDa) in an organelle, a cell or a tissue at a given time. For example, all the peptides secreted by neuroendocrine cells or glands represent the neuropeptidome. The peptidome is very dynamic, i.e. changes during development, cell differentiation, generally with the stage of a cell, and as a consequence of many endogenous and environmental influences. See → fragmentome, → peptidomics, → proteome.

Peptidomics (*peptide*-gen*omics*): The whole repertoire of techniques to detect, analyze and characterize the → peptidome (the low-molecular weight proteome with peptides up to 20 kDa) of an organelle or a cell, encircling peptide isolation, chromatographic or electrophoretic fractionation and separation, analysis by → MALDI mass spectrometry, se-quencing, including determination of modifications (such as e.g. acetylation, glycosylation, methylation, phosphorylation), immunocytochemical detection and quantification, and storage and analysis of the resulting informations. See → functional genomics, → genomics, → proteomics, → recognomics.

Peptidyl *transfer* RNA (P-tRNA): The → transfer RNA molecule that is bound to the peptidyl chain during protein synthesis on the → ribosome. See → A-tRNA, → E-tRNA.

Peptoid tag: Any → peptoid sequence with a unique mass, covalently attached to a synthetic → oligonucleotide that is used as a → probe for the detection of complementary RNA. After → hybridization of the tagged oligonucleotide to the target RNA, the tag is chemically separated from the oligonucleotide and its mass determined by e.g. → mass spectrometry. Since different oligonucleotides are tagged with peptoid tags of different masses, each oligonucleotide can be discriminated from an other one unequivocally.

Peptone: A misleading term for incompletely degraded (partially hydrolyzed) proteins. The term is still in use for an incomplete enzymatic hydrolysate of proteins by pepsin or trypsin that consists of free amino acids (~30%), and di-, tri- and oligopeptides. Depending on the origen of the proteins, casein, meat, soybean or milk peptones can be distinguished that are all used as additives to nutrient media for bacteria.

Percent *i*dentity *p*lot (PIP): A graphical depiction of a comparison of two (or more) related nucleotide or amino acid sequences from two (or more) different organisms that allows to infer the extent of sequence identity. Computer programs such as PipMaker (http://bio.cse.psu.edu/) and VISTA (http://www-gsd.lbl.gov/vista/) assist to establish a PIP.

Percoll: An inert colloidal silica coated with *poly*v*inyl*p*yrrolidone (PVP) that is used for generating gradients which allow the separation of subcellular organelles (e.g. nuclei, mitochondria, plastids), viruses, and cells.

Perfect *m*atch (PM): The complete correspondence of two (or more) bases in two (or more) strands of a DNA molecule. Perfect matches are only possible by → Watson-Crick base pairing of A=T and G≡C pairs, respectively. Any other combination inevitably leads to a → mismatch.

Perfect palindrome: Any sequence in duplex DNA in which completely identical base sequences run in opposite directions (e.g. 5'GAATTC 3'). Such perfect palindromes frequently are recognition sites for → restriction endonucleases. Compare → palindrome.

Perfect repeat: Any stretch of → repeated sequences that consists of elements with identical sequence (e.g. 5'-CATCATCATCAT-3'). See → compound microsatellite, → imperfect repeat.

Pericentric inversion: Any segment of DNA that is reversed in orientation relative to the rest of the chromosome, and does involve the → centromere. See → paracentric inversion.

Perinatal genetics: A branch of → genetics that focusses on the detection of chromosomal and DNA abnormalities in newborn human beings, using the whole repertoire of classical → cytogenetics and → molecular genetics from chromosome banding to → DNA chip technology.

Periodicity: The number of base pairs per turn of the DNA double helix.

Permanent cell line: Any cell line (→ cell strain) with an unlimited life time.

Permanganate oxidation of DNA: An outdated technique for the detection of methylated cytosines in a DNA molecule that uses potassium permanganate at pH 4.3 to degrade 5-methylcytosine to barbituric acid derivatives, but does not attack cytosine itself. Since this treatment is not specific for 5-methylcytosine, but also degrades thymine, it was changed to include a combination of hydrazine degradation of C and T (but not 5-methylcytosine) and permanganate oxidation with little further improvement. The hydrazine and permanganate-modified nucleotides can be removed with piperidine and detected by sequencing techniques. See → combined bisulfilte restriction analysis, → methylation assay,

→ methylation-sensitive amplification polymorphism, → methylation-sensitive single nucleotide primer extension, → methylation-specific polymerase chain reaction.

Permissive cell (permissive host): Any cell in which a particular virus may cause a production of progeny viruses (productive infection).

Permissive condition: A condition that allows the survival of a → conditional lethal mutant.

Permissive host: See → permissive cell.

Permissive temperature: The temperature at which a → temperature-sensitive mutant is able to grow.

Permissivity: The ability of cells to support the growth of phages (or plasmids).

Permutation: Any permanent mutation in a gene without phenotypic consequences. Permutations predispose the carrier for further mutation(s).

Peroxidase-conjugated antibody (POD-conjugated antibody; immunoperoxidase): An → antibody to which a horseradish peroxidase (HRP) molecule is covalently attached. Such conjugates are used to detect a specific protein or nucleic acid sequence in e.g. biotinylation- and digoxygeninbased detection systems (see → biotinylation of nucleic acids and → digoxigenin labeling), where the antibody binds to its antigen (e.g. a biotin-avidin complex), and the complex is detected by the H_2O_2-dependent conversion of e.g. luminol (5-amino-2,3-dihydro-1,4-phthalazinedion) with concomitant emission of light. This reaction can be enhanced by the presence of an enhancer, see → enhanced chemiluminescence detection. Compare →immunophosphatase, see also→enzyme-conjugated antibody.

Peroxin: Any one of a series of *peroxi*somal proteins that are synthesized on cytoplasmic → ribosomes and imported into peroxisomes. All peroxins carry one or two targeting signals (peroxisomal targeting signal, PTS) that allow their specific transport to and into peroxisomes. PTS1 is localized at the → carboxy terminus of the peroxins and is composed of the tripeptide serine-lysine-leucine (SKL), or variants of this motif. PTS2 is part of the → amino terminus and contains up to 30 amino acids (consensus sequence: [R/K][L/I/V] X_5 [H/Q][L/A]. In contrast to PTS1, some of the PTS2 signal sequences are processed after the import of the corresponding peroxins. Most peroxins are membrane-bound, some contain → zinc finger domains. See → peroxisome assembly gene.

Peroxisome assembly gene (PEX gene, PAS gene): Any one of a series of genes encoding proteins (socalled PEX proteins) for the biogenesis of socalled peroxisomes, organelles of eukaryotic cells that contain catalases, peroxidases, a β-oxidation system, and enzymes of the glyoxylate cycle (plants), or glycolysis (glycosomes of the trypanosomes). PEX genes encode → peroxins.

Peroxisome proliferator-activated receptor **(PPAR):** Any one of a superfamily of nuclear hormone receptors that bind agonists (e.g. eicosanoids or unsaturated fatty acids) and then form heterodimers with the 9-*cis* retinoic *acid* receptor (RXR). The resulting complex in turn binds to socalled *PPAR response elements* (PPREs) composed of direct 5'-AGGTCA-3' repeats in specific → promoters, and modulate the → transcription of the adjacent gene(s). For example, PPARγ, predominantly expressed in adipose tissues, represents the receptor for *trans*-resveratrol (3,4',5-trihydroxy-trans-stilben, a secondary metabolite of plants, especially grapes) and regulates the differentiation of fat cells, and the release of cytokines that are involved in the insulin sensitivity.

Peroxysomal targeting signal (PTS): The conserved tripeptide sequence motif H_2N-SKL-COOH (exception: H_2N-SRL-COOH in soybean Hsp16.2) at the extreme C-terminus of proteins that are synthesized on cytoplasmic → ribosomes and subsequently imported into peroxisomes. The import process starts with the binding of the PTS to the cytoplasmic receptor Pex5 (PTS1 proteins) or Pex7 (PTS2 proteins), upon which the complex Pex5-PTS1-protein or Pex7-PTS2-protein, respectively, translocates through the peroxisomal membrane into the matrix of the organelle. Recognition and binding of the PTS to the receptor requires a C-terminal domain with seven tetratricopeptide repeats (TPR1-7), whereas the N-terminus of the receptor mediates binding to the docking receptor Pex14 (in a complex consisting of Pex14, Pex13 and Pex17). After docking, the translocation of the cargo-receptor complex is initiated. After completed translocation, the import receptor is recycled into the cytoplasm. Oligomeric proteins can also be transported into the organelle via Pex5, and in some cases, oligomerization is a prerequisite for transport (e.g. isocitrate lyase: tertramer; thiolase: dimer; acyl-CoA oxidase: pentamer). Proteins without a PTS motif can bind to proteins containing a PTS, and are cargoed into peroxisomes. Mutations in the gene encoding Pex5 cause the peroxisomal disorder neonatal adrenoleukodystrophy. See → nuclear localization signal.

Perpendicular denaturing gradient gel electrophoresis (perpendicular DGGE): A method to determine the → melting behavior of a DNA duplex molecule in an → agarose gel containing a gradient of denaturants perpendicular to the direction of electrophoresis. The DNA is applied to the gel in a single large slot. In the gel region with low denaturant concentration the DNA fragments run far into the gel (i.e. do not melt), in the gel region with high denaturant concentration they do hardly migrate (i.e. melt extensively). In between these extreme positions intermediate mobilities of the DNA fragments may be observed. After → ethidium bromide staining the fragment pattern in the gel resembles a → C_0t curve, and therefore allows the calculation of the number of melting domains in a DNA fragment as well as the estimation of the → T_m for each individual fragment.

Perpendicular DGGE: See → perpendicular *d*enaturing *g*radient *g*el *e*lectrophoresis.

Persistence length (p): The number of → base pairs between two bends in → double-stranded DNA.

PERT: See → *p*henol *e*mulsion *r*eassociation *t*echnique.

PEST protein: Any protein rich in *p*roline (P), *g*lutamate (E), *s*erine (S) and *t*hreonine (T). Hypothetically, PEST proteins are more rapidly turning over than non-PEST proteins.

PET: Any short sequence containing both the 5'- and 3'-ends of a → transcript.

Petite: A mutant strain of *Saccharomyces cerevisiae* that suffered mutations in either one or more mitochondrial genes (called vegetative petits), or on nuclear genes (called segregational petites). Petites grow only slowly and as small colonies, a consequence of respiratory deficiency.

PETRA: See → primer extension telomere repeat amplification.

Petri dish (Petri plate): A disposable, round and flat plastic culture dish with a lid that is used for the culture of bacteria or fungi on solid media.

Petrifilm™ plate: Any ready-to-use, water-thin substitute for a conventional agar plate for the culture of bacteria. The Petrifilm plate contains a dehydrated nutrient medium (as e.g. → LB medium), a gelling agent and indicators for → blue-white screening. The bacteria are simply added onto the medium, and the top film used to seal the plate, which can then be incubated. All routine plating procedures (e.g. library screening) can be done with the Petrifilm plates.

PETting: See → paired end ditagging.

PEV: See → position effect variegation.

PEX gene: See → peroxisome assembly gene.

pEX vector: Any one of a series of → plasmid → expression vectors that is designed for the expression screening of → cDNA libraries in *E. coli*, and for the expression of β-galactosidase → fusion proteins. Each pEX vector contains a *cro* – *E. coli lacZ* gene fusion driven by the strong P_R promoter. A → polylinker at the 3' end of the *lacZ* gene allows the insertion of a foreign sequence in such a way that it is placed in all three → open reading frames alternatively. In one of these constructs the insert will thus be in frame with the vector, allowing its expression as a hybrid β-galactosidase protein. Downstream of the polylinker site → fd phage transcription terminators and a synthetic translation stop signal are inserted.

PFF: See → peptide fragmentation fingerprint.

PFGE: See → pulsed-field gel electrophoresis.

p53 (TP53): A conserved 393 amino acids human → transcription factor, encoded by → exons 2–11 of the 8 kb *p53* tumor suppressor gene on chromosome 17, and folded into four structurally and functionally different domains: an acidic N-terminal region harbouring the 42 amino acid long transactivation domain, a hydrophobic proline-rich region comprising amino acids 64–92, a central sequence-specific DNA-binding domain (amino acids 102–292), a tertramerization domain (amino acids 324–355), and a highly basic C-terminal regulatory region (amino acids 363–393). P53 as tetramer binds to defined DNA consensus motifs, represses transcription of specific genes encoding proteins functioning in multiple cellular pathways (e.g. cell proliferation, cell survival [→ apoptosis], → translation, redox regulation and maintenance of genomic integrity), and activates a different set of genes involved in cell cycle control, causing growth arrest prior to DNA replication in the G1 phase of cell cycle, or mitosis in the G2 phase. For example, cell proliferation of damaged cells or cells with damaged DNA is prevented by p53. In normal cells, this suppressive effect of p53 is inhibited by a continuous poly-ubiquitinylation catalyzed by E3 ligase Mdm2, which leads to the degradation of p53 in the proteasome. In damaged cells, p53 is multiply phosphorylated (humans: phosphorylation at about 23 different sites by stress-activated DNA protein kinase casein-kinase I and II, and cyclin-dependent kinases), which interferes with its ubiquitinylation and subsequent degradation. As a consequence, the damaged cell accumulates p53 that in turn arrests the cell cycle. P53 is also acetylated at multiple lysine residues by CBP/p300 and pCAF, which is supposed to prevent p53 degradation. Any → mutation in the *p53* gene by e.g. genotoxic stresses (UV light, X-rays, γ-irradiation, carcinogens, chemotherapeutic drugs), oncogenic stresses (activated oncogenes), or non-genotoxic stresses (oxygen radicals) therefore may promote genomic instability, checkpoint defects (e.g. suppressing G1 arrest) and non-programmed cell survival, which altogether lead to uncontrolled proliferation of damaged cells. Actually *p53* mutations are most frequent in human cancers, where between 30 and 70% of tumors of almost every organ contain at least one → point mutation in one of the two gene copies. Both nature and distribution of more than 22,000 *p53* mutations vary between different cancer types. More than 75% of all *p53* mutations are → missense (substitution), → nonsense (stop) or → splice site mutations located in the central DNA-binding region encoded by exons 5–8 (amino acids 102–292), which destroy the transcription factor function of p53. The DNA-binding domain of p53 is by far the most frequently mutated region. Moreover, → deletions, → insertions and → frame-shift mutations comprise 12% of all mutations in the gene. A series of p53 isoforms are known: p53β (46 kDa, wild-type p53: 53 kDa), p53γ (46 kDA), Δ40p53 (48 kDa), Δ40p53β (41 kDa), Δ40p53γ (41 kDa), Δ133p53 (35 kDa), Δ133p53β (25 kDa), Δ133p53γ (25 kDa), and IntΔp53 (~46 kDa). All mutations in the 1.3 kb region of exons 2–11, including the flanking → intron sequences of → splice junctions, can be detected by → hybridization of labeled target RNA or → cDNA to a high-density → oligonucleotide microarray ("p53 gene chip") and subsequent laser excitation and scanning. Differences in patterns and fluorescence intensities between a reference DNA microarray and the target DNA array are computed with a mixture detection algorithm. Generally, mutations in the p53 gene are associated with poor prognosis in many human cancers and are also negative predictors for a tumor's response to chemo- or radiotherapy. The R11 release of http://www-p53.iarc.fr/ contains 23,544 somatic mutations, 376 germline mutations, functional data on 2314 mutant proteins and TP53 gene status of 1569 cell-lines. See → p73, → p63.

PFM: See → physical functional marker.

PFP: See → *protein fusion and purification technique*.

pfu: See → *plaque forming unit*.

***Pfu* DNA polymerase:** See → *Pyrococcus furiosus* DNA polymerase.

PGRS: See → *polymorphic GC-rich repetitive sequence*.

pGV 3850: A → cointegrate vector for the transfer of foreign genes (generally, DNA) into target plants via → *Agrobacterium-*

mediated gene transfer. pGV 3850 is a derivative of the → *Agrobacterium tumefaciens* → Ti-plasmid, in which the T-region has been substituted for a modified → pBR 322. The latter is flanked by the two T-DNA borders in this construct. The pBR 322 portion allows the insertion of foreign DNA into pGV 3850 by homologous recombination with a conventionally constructed recombinant pBR 322 As pGV 3850 does not contain an → ampicillin resistance gene, *Agrobacterium* cells containing such a → cointegrate structure can be selected on ampicillin-containing medium. The foreign sequence can then be transferred to compatible plant cells via *Agrobacterium*-mediated gene transfer because it is flanked by T-DNA border regions. See also → coculture or → leaf disk transformation.

Phage: See → bacteriophage.

Phage bank: See → phage library.

Phage cloning vector (phage vector): A → cloning vector derived from a → bacteriophage. See for example → autocloning vector, → broad host range vector, → expression vector, → lambda phage-derived cloning vector, → P1 cloning vector, → SP6 vector.

Phage conversion (lysogenic conversion; prophage-mediated conversion): The acquisition of new properties by bacterial cells harboring a → prophage (for example the property of immunity against phage superinfection, see → phage exclusion). If the prophage is lost, the new characters disappear.

Phage cross: The exchange of genetic material between phages. Occurs during multiplication of → bacteriophages after their entry into the host cell. If a single bacterium is infected with several phages differing at one (or more) genetic loci, then recombinant progeny phages can be recovered upon → lysis of the host cell. These recombinants carry genes derived from two parental phages.

Phage display (slang: Ph.D., peptide display): A technique for the presentation of distinct peptides or proteins on bacterial surfaces that uses → bacteriophages (e.g. → M13, fd, f1) as carriers for these display molecules and allows to identify peptides or proteins with desirable binding properties. Genes for the display peptides are integrated in the single-stranded DNA genome of the phage, and the corresponding peptides expressed as → fusion proteins with a viral coat protein. The fusion proteins are then exposed to the surrounding medium. For example, the M13 phage carries a single-stranded circular DNA genome of 6408 bp that is packaged by various viral DNA-encoded proteins (e.g. g3p, g6p, g7p, g8p, g9p), of which g8p is the major coat protein (about 2700 copies per phage). The phage particle itself is a flexible, 900 nm long filament (diameter: 6 nm), and on its surface the coat proteins (especially g3p) are exposed. If the coat protein is fused to a foreign protein, the latter is also presented. Phage display is used for the establishment of libraries of peptide or protein (e.g. enzyme) variants, or oligopeptide inhibitors for various target molecules, for the isolation of enzyme variants with a better or modified binding affinity for their substrates and changed catalytic properties, or for the detection of enzyme variants with increased stability. See → display library, → panning, → random peptide display. Compare → *Bacillus* spore display, → *Baculovirus* expression system, → bifunctional phage display, →

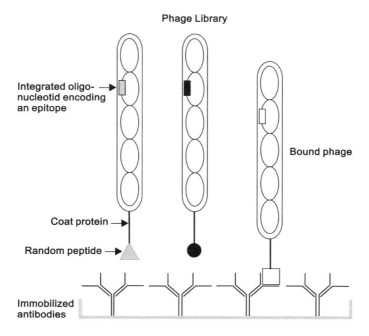

Phage display

CIS-display, → differential genome display, → differential peptide display (DPD), → dual-bait two-hybrid system, → interaction trap, → LexA two-hybrid system, → mammalian cell display, → messenger RNA display, → one-hybrid system, → reverse two-hybrid system, → phagemid display, → ribosome display, → RNA-protein hybrid system, → split-hybrid system, → three-hybrid system, → two-hybrid system, → yeast/bacterial two hybrid system (YBTH), → yeast display.

Phage display library: See → display library.

Phage display peptide library: A DNA library, established in → phages that contains the → insert fused to the gene of the coat protein gene of the phage (→ fused gene), and allows to detect the insert-encoded peptide on the surface of the phage. See → display library, → panning, → phage display, → phagemid display.

Phage exclusion (phage immunity, prophage interference, superinfection immunity): The immunity of a host bacterium that contains a → prophage in its genome (→ lysogenic bacterium), against a secondary infection by the same or a related phage. The inserted prophage codes for the synthesis of → repressor proteins that bind to the → operator sequences of the superinfecting phage and interfere with its transcription. In the case of the → lambda phage the repressor protein is encoded by the gene → *cI* and acts in concert with the products of the genes *rex A* and *rex B* (immunity loci, *imm* loci) to suppress the genes necessary for the lytic cycle of the superinfecting phage.

Phage fd: See → fd phage.

Phage f1: See → f1 phage.

Phage immunity: See → phage exclusion.

Phage induction: The stimulation of a → prophage to enter the productive, i.e. → lytic cycle, usually by exposure of lysogenic cells to UV light, X-rays or → mutagens (e.g. nitrogen mustard, hydrogen peroxide). Phage induction allows the initiation of transcription of phage genes, the excision of the prophage from the host chromosome, and the synthesis of phage DNA and capsid proteins.

Phage lambda (λ): See → lambda phage.

Phage library (phage bank): A collection of random DNA fragments, cloned into a phage → cloning vector (e.g. → M13 or → lambda phage-derived vector) and ideally encompassing the entire genome of a given species. See also → gene library.

Phage lifting: See → plaque hybridization.

Phagemid ("*phage*-plas*mid*"): A chimeric → plasmid vector (→ hybrid vector) that contains an → origin of ssDNA replication such as the f1 or M13 intergenic region (IG). Phagemids replicate as normal plasmids in *E. coli*. If the host cells are infected with a helper bacteriophage (→ helper virus, e.g. M13 KO7) that supplies the functions necessary for ssDNA replication and packaging, phage-like particles are synthesized and released through the bacterial cell walls in a non-lytic process. The ssDNA can then easily be recovered from the culture medium. See for example → Bluescript[R], → expression phagemid, → lambda ZAP, → multi-functional phagemid.

Phagemid display: A technique for the presentation of distinct peptides or proteins on bacterial surfaces that uses → phagemids as carriers for the display molecules. In conventional → phage display, the size of displayed peptides is limited, because the fusion product of target peptide and viral coat protein should not exceed a certain threshold, otherwise the function of the coat protein is inhibited. This size limitation is relaxed in phagemid display. Compare → *Bacillus* spore display, → *Baculovirus* expression system, → CIS-display, → mammalian cell display, → messenger RNA display, → ribosome display, → yeast display.

Phage M13: See → M13.

Phage Mu: See → Mu phage.

Phage φ X 174: See → φ X 174.

Phage Q-beta: See → Q-beta.

Phage therapy: The treatment of a bacterial infection of humans with a preparation of a → bacteriophage specific for the causative bacterium. Phage therapy is specific such that non-host bacteria are not attacked. Moreover, intravenously applied phages cross the blood-brain barrier, are self-replicating, and are simply excreted, if no host bacterium is encountered. Since pathogenic bacteria exist in a series of different serotypes, socalled phage cocktails consisting of three to five phages with different host spectrum are employed for an effective treatment and prophylaxis. Phage therapy, reported to be effective against human pathogenic bacteria with antibiotic resistance (e.g. methicillin-resistant *Staphylococcus aureus*, vancomycin-resistant enterococci, pathogenic strains of *E.coli, Pseudomonas* a*eruginosa,*

Streptococcus pyogenes, Proteus vulgaris) still meets reservations, because the temperate phages could adopt host genes for virulence or resistance and spread them in a bacterial population. Bacteriophages can also be exploited as delivery vehicles for antimicrobial peptides (only active inside bacterial cells) or agents for food processing and food safety. For example, phage preparations can be sprayed on chicken eggs or cut fruits and vegetables to reduce *Salmonella* contamination (at least active against the five to six serogroups most commonly associated with human illness).

Phage typing: The classification of bacteria on the basis of their susceptibility towards infection by various → bacteriophages.

Phage vector: See → phage cloning vector.

Pharmaceutically tractable genome (PTC): A subset of genes from a genome that represents (preferably) all drug targets (as e.g. genes encoding cell surface proteins such as receptors, circulating proteins, or proteins modulated by small molecules as e.g. drugs). The human PTC probably consists of 6,000–8,000 genes.

Pharmacogenetic marker: Any → splice variant of a → pre-messenger RNA that is either directly or indirectly responsible for, or at least linked to a specific subtype of a disease.

Pharmacogenetic single nucleotide polymorphism (pharmacogenetic SNP): Any → single nucleotide polymorphism that is located within a → gene encoding a drug target protein and confers drug resistance, reduced drug sensitivity or drug hypersensitivity onto the encoded protein (and the carrier). Pharmacogenetic SNPs are frequently responsible for the differential efficacy of a distinct drug in different patients. So, individuals with pharmacogenetic SNPs do either not at all respond to the administration of a pharmacon ("non-responders"), or suffer from adverse effects ("toxic responders"). Therefore the detection of pharmacogenetic SNPs in a patient's genome helps to individually adjust the level of a pharmacon or to substitute it for a more efficient or tolerated drug with less side-effects. See → individualized medicine.

Pharmacogenetics: The detection, isolation and characterization of → genes and the encoded proteins as potential targets for pharmaceutically active compounds. Moreover, pharmacogenetics aims at establishing individual gene profiles, i.e. to detect sequence polymorphisms at strategic sites of a particular gene between e.g. patients. For example, → single nucleotide polymorphisms (SNPs) – the human genome probably contains 3 million SNPs, of which the majority is already mapped – in specific genes may determine the capacity of the encoded proteins such that e.g. a wild-type protein transports a certain drug, the mutated protein does not. The patient with the SNP mutation in the transporter gene does not respond to the drug. Another example is the *multi-drug resistance* (MDR)-1 gene that encodes the socalled P glycoprotein (a membrane-bound protein, eliminating compounds recognized as xenobiotics). This gene harbors at least 35 polymorphisms, of which the socalled TT variant occurs in about 25% of humans. This mutation leads to a highly reduced production of P glycoprotein in the intestines, so that the uptake of drugs from the intestinal tract to blood proceeds uncontrolled. Therefore the drugs are present in very high concentrations in the blood, increasing the incidence of side effects. Patients with the TT variant can be advised to reduce the drug dosis. Still another important example capitalizes on the genes encoding cytochrome P450 enzymes (CYP, in this case CYP3A enzymes), which metabolise about 50% of all common therapeutics as well as natural compounds such as estrogene, testosterone, and bile acids. Specific SNPs in the CYP3A genes reduce or abolish the individual's capacity to metabolize a drug (i.e. they determine, how the individual is susceptible to the drug and its side effects). See → functional genomics, → genomics, → medical genomics, → pharmacogenomics, → proteomics, → recognomics. Compare → comparative genetics, → cytogenetics, → developmental genetics, → forward genetics, → interphase genetics, → molecular genetics, → reverse genetics.

Pharmacogenomics: The whole repertoire of techniques to explore the effects of drugs on the structure of → genomes and → genes and the expression of these genes, as well as the implication(s) of → mutations in specific genes and, as a consequence, amino acid replacements in the encoded protein(s) for the effectiveness of pharmaca. For example, the genomes (or particular genes) of socalled responders (individuals responding positively to a specific drug) may be different from the genomes or genes of socalled non-responders (individuals not responding to the drug). By profiling the potential users of such a pharmacon, a prediction can be made about the effectivity of a specific drug application ("right drug for the right patient"). Pharmacogenomics aims at the identification of previously unknown target molecules for the development of fitting drugs ("drug targets"; e.g. the design of cyclooxigenase inhibitors for the treatment of arthritis, based on gene expression analysis), at the recognition of genetic polymorphisms in genes encoding drug-metabolizing enzymes (e.g. the phase I drug metabolizing P-450 superfamily of monooxygenases), or the definition of all genes contributing to a specific disease phenotype coupled to a better, more effective drug application ("personalized drug therapy"; "individualized medicine"). Compare → behavioral genomics, → biological genomics, → cardio-genomics, → chemical genomics, → clinical genomics, → comparative genomics, → deductive genomics, → environmental genomics, → epigenomics, → functional genomics, → horizontal genomics, → integrative genomics, → lipo-proteomics, → medical genomics, → neurogenomics, → neuroproteomics, → nutritional genomics, → omics, → pathogenomics, → pharmacogenomics, → phylogenomics, → physical genomics, → population genomics, → proteomics, → recognomics, → structural genomics, → transcriptomics, → transposomics.

Pharmacological profiling: The monitoring of the effects of pathway-specific drugs and peptide or protein hormones on protein-protein interactions and determination of the cellular localization of these protein-protein interactions by e.g. protein-fragment complementation assays (PCAs).

Pharmacoperone (*pharmaco*logical chap*erone*, chemical chaperone): Any synthetic → chaperone that binds to proteins and corrects their incorrect folding. For example, the gene encoding the receptor protein for gonadotropin-releasing hormone (GnRH) is frequently mutated, and consequently the encoded receptor can be non-functional. This is the cause for e.g. hypogonadotropic hypogonadism (phenotype: faulty sexual development). Thirteen of the 14 most frequent mutations change a single amino acid in the receptor, each of which leads to an incorrect folding of the receptor protein. If the faulty proteins are exposed to a synthetic GnRH antagonist, the pharmacoperone, then the receptor can be folded normally and regains function. Even unspecific chaperones such as 4-*phenylb*utyric *a*cid (PBA) can be employed for correct folding of target proteins *in vitro*. Pharmacoperones are potential agents for the causative treatment of many diseases caused by defective folding of proteins, as e.g. → prion diseases, Alzheimer disease, and Chorea Huntington.

Pharmacophore map: A list of descriptors defining the physical, chemical, and structural properties of pharmaceutically potent compounds.

Pharmacoproteomics: The whole repertoire of techniques for the identification and characterization of peptides and proteins in a specific cell, tissue, organ or individual that are expressed as a consequence of drug administration. Such → protein expression data can be used to predict drug toxicity and efficacy, and to understand the mechanism of action of a drug.

Pharmanome: The pharmaceutically relevant portion of a genome, including genes encoding receptor proteins, proteins for drug metabolism (e.g. the various cytochrome P450 proteins), proteins for signal transduction, and cell surface antigens, to name few.

Phase:

a) The arrangement of → codons downstream of the → start codon AUG (or GUG) in → messenger RNA. These codons are *in phase*, if they can be read in → triplets starting from the AUG

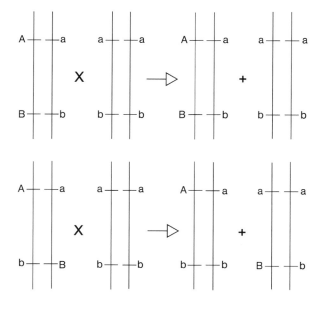

Phase (b)

(i.e. function as codons for amino acids during → translation). They are *out of phase*, if the → reading frame has shifted by one or two nucleotides (i.e. the new arrangement does no longer code for the same amino acids as before the change). See → reading frame shift mutation.

b) The distribution of specific → alleles on → homologous chromosomes. For example, alleles A and a occupy one, alleles B and b another locus. An individual with the phenotype AaBb may have two possible genotypes, the *cis* and *trans* phases (in analogy to chemical isomeres). The phase can be determined in a particular individual by a genetic cross with an appropriate (and informative) partner.

Phase lock gel (PLG): A gel block that serves to trap the organic phase and interphase material in phenol or phenol/chloroform extraction of nucleic acids. The PLG forms a seal between the organic and aqueous phases after centrifugation and either allows to decant the aqueous phase or to pipet it without contamination.

Phase lock procedure: A technique to separate the organic phase and interphase (containing denatured proteins) from the aqueous phase (containing nucleic acids) during phenol-chloroform extractions. A chemically inert and hydrophobic gel ("phase lock gel") is included that forms a solid barrier between the aqueous and organic phases during centrifugation of the phenol-chloroform mixture, thereby trapping the organic and interphase, leading to complete separation of the phases and easy recovery of nucleic acids from the aqueous epiphase.

Phase shift: See → reading frame shift.

Phase shift mutation: See → reading frame shift mutation.

Phase variation: The reversible loss or gain of → intragenic → microsatellite repeats in certain bacteria, leading to a loss or gain of specific function(s). For example, the socalled Opa genes of *Neisseria gonorrhoe* (encoding 12 outer-membrane proteins that make the bacterial colonies appear *opa*que and allow the bacteria to adhere to and invade epithelial cells, e.g. respiratory tract epithelia) contain a microsatellite composed of multiple copies of 5'-CTCTT-3'. As a con-sequence of → slipped strand mispairing during → replication, one such repeat can be lost, which leads to a shorter protein. Cells with such a truncated protein can no longer enter epithelial cells. This deficiency turns into a selective advantage, if the bacterium is unable to invade e.g. phagocytotic cells, which would destroy them. See → contingency gene.

Phasing: See → nucleosome phasing.

Phasmid: A → hybrid vector consisting of a → plasmid with a functional → origin of replication and → lambda phage sequences (in particular, the l origin of replication and one or more → attachment site(s). Foreign DNA may be conventionally cloned into the plasmid vector. The recombinant plasmid can then insert into a phage genome ("lifting"), exploiting the function of the l *att* sites. Phasmids may be propagated in appropriate *E. coli* strains either as a plasmid (non-lytic route), or as a phage (lytic route). Reversal of the lifting process releases the plasmid vector.

PHD finger: A zinc-finger-like amino acid motif with the unique pattern Cys_4-His-Cys_3 that occurs in → transcription factors (e.g. the plant → homeodomain proteins AtHAT3.1 and MzHOX1A) and other proteins (e.g. Trithorax, Polycomblike, Peregrin, and Neuro-D4 protein, to name few). Such motifs either stand alone, or are duplicated or triplicated in regions of 50–80 amino acid residues. PHD fingers bind Zn^{2+} ions and interact with DNA and RNA.

Phenol emulsion reassociation technique (PERT): A variant of the genomic subtraction technique that employs phenol to increase the rate of → hybridization. PERT is a form of competitive hybridization between two related, but slightly differing genomes (e.g. genomes of male and female plants) that preserves only the unique sequences of one genome (e.g. the female one) in a clonable form. See → normalization.

Phenol extraction: A procedure for the denaturation and removal of proteins from solutions containing nucleic acids and proteins, using buffer-saturated phenol.

Phenome: See → phenotype.

Phenomic fingerprint: See → molecular phenotype.

Phenomics: The whole repertoire of techniques to decipher all molecular processes leading to a → phenotype (phenome). Phenomics encompasses → transcriptomics, → proteomics, and → metabolomics.

Phenotype (phenome): The entirety of observable structural and functional properties of an organism, which results both from its → genotype and the environment.

Phenotype array (*p*henotype *m*icroarray, PM): A solid support, onto which single cells or cell colonies are arrayed, and used to detect their interactions with small molecular weight compounds (e.g. metabolites, drugs) in solution by monitoring cell or colony growth (i.e. the phenotype). See → microarray.

Phenotype microarray: See → phenotype array.

Phenotype mixing: The packaging of the genome of one virus into the protein → capsid of a second, unrelated virus.

Phenotypic enhancement: The enhancement of any → phenotype by a → mutation or → overexpression of one gene associated with mutation or overexpression of another gene. See → phenotypic suppression.

Phenotypic suppression: The suppression of any → phenotype by a → mutation or → overexpression of one gene associated with mutation or overexpression of another gene. See → phenotypic enhancement.

Phenyl*m*ethysulfonyl *f*luoride: See → PMSF.

φ80: An *E. coli* strain that carries the lambdoid phage φ80.

φ29 DNA polymerase: A highly processive enzyme of the *Bacillus subtilis* phage φ29 that catalyzes the replication of template DNA, possesses strand displacement and inherent 3' → 5' exonuclease → proofreading activities.

φ X 174: A small icosahedral → bacteriophage infecting *E. coli* (→ coliphage) with a circular single-stranded DNA genome of 5.386 kb. Its replication proceeds through a double-stranded circular → replicative

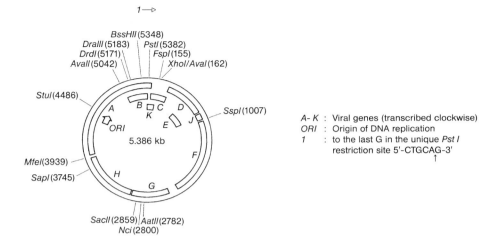

Simplified map of Phi X174 (with unique restriction sites)

form. Some of its genes have been used for the construction of → cloning vectors.

Phleomycin: One of a series of glycopeptide → antibiotics of *Streptomyces verticillus* that binds and intercalates DNA and destroys the integrity of the double helix by its metal-chelating domain. Phleomycin is an effective selective drug for mammalian cells, but can also be used for prokaryotes, fungi, plants, and generally animal cells.

pho **A promoter:** See → alkaline phosphatase promoter.

pho-box (*pho*sphate uptake box): A regulatory sequence element of the → promoter of bacterial genes involved in phosphate uptake and metabolism. The pho-box is the address site for PhoB, a positive regulator protein inducing or enhancing the transcription of these genes.

Phosphatase: An enzyme catalyzing the removal of phosphate residues from substrates (including polymers such as nucleic acids). See → alkaline phosphatase.

Phosphatase and *ten*sin homolog (PTEN): A tumor suppressor protein that dephosphorylates phosphatidylinositol-3,4,5-phosphate (PIP3) at the plasmamembrane and thereby negatively regulates (i.e. inhibits) phosphatidylinositol-3-kinase (PI3K) signal transduction cascade necessary for growth, proliferation, cell migration and survival, invasion, angiogenesis, genomic instability and stem cell self-renewal. The class I family of PI3Ks is activated → downstream of receptor tyrosine kinases (RTKs) or G protein-coupled receptors (GPCR), catalyzing the conversion of phosphatidylinositol-4,5-phosphate (PIP2) to phosphatidylinositol-3,4,5-phosphate (PIP3) leading to the activation of AKT kinase and other downstream effectors. PTEN contains two PEST motifs that are frequent constituents in proteins targeted for degradation by the → ubiquitin pathway. Consequently, PTEN is poly-ubiquitinylated and thereby degraded in the cytoplasm, whereas its mono-ubiquitinylation increases its nuclear localization. The nuclear pool of PTEN maintains chromosomal stability. The gene encoding PTEN, *pten*, is one of the most frequently mutated genes in human cancer. → Somatic mutations in *pten* occur in multiple sporadic tumors and result in PTEN inactivation. → Germline mutations of *pten* cause the inherited hamartoma and cancer predisposition syndrome called Cowden disease.

Phosphatidylinositol (PI) kinase: Any one of a family of enzymes catalyzing the transfer of phosphate groups onto one (or several) of the five hydroxyl goups of the inositol moiety of membrane lipid phosphatidylinositol. The products, phosphatidylinositol-4-phosphate (PI-4-P), phosphatidylinositol-4,5-bisphosphate (PI-4,5-P2), and phosphatidylinositol-3-phosphate (PI-3-P) are precursors of second messengers. For example, PI-4,5-P2 leads to the synthesis of diacylglycerol and inositol-1,4,5-trisphosphate in response to growth factors, and PI-3,4,5-P3, a membrane-embedded second messenger, regulates growth, and, when overproduced, results in cell transformation.

Phosphatome: Another term of the "ome" era describing the complete pattern of phosphorylated amino acid side chains in a protein. For example, the socalled tyrosine phosphatome, a result of the action of protein tyrosine kinases (PTKs) and phosphatases (PTPs), and therefore changing continuously, is an important element in signal transduction pathways underlying tumorigenesis in mammals.

Phosphinotricin *a*cetyl*t*ransferase gene (PAT gene): A gene (*bar*) from *Streptomyces hygroscopicus* encoding the enzyme phosphinotricin acetyltransferase that catalyzes the inactivation of the herbicide phosphinotricin (PPT). PPT is an analogue of glutamic acid and inhibits plant glutamine synthase. The PAT gene is used as a → selectable marker gene in plant transformation experiments.

Phosphodiester: An imprecise term for a molecule containing the group depicted below, where R^1 and R^2 are carbon-containing groups. For example, in RNA or DNA the 5' carbon of a pentose (ribose or deoxyribose) and the 3' carbon of an adjacent sugar moiety are linked by a phosphodiester type bond. See → phosphodiester bond.

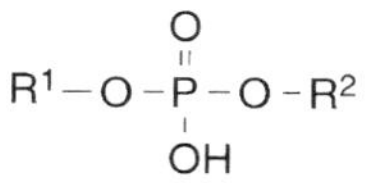

$$R^1-O-\overset{\overset{O}{\|}}{\underset{OH}{P}}-O-R^2$$

Phosphodiesterase: An enzyme that catalyzes the hydrolysis of phosphodiesters into a phosphomonoester and a free hydroxyl group. See for example → phosphodiesterase I.

Phosphodiesterase I (5' exonuclease, snake venom phosphodiesterase; EC 3.1.4.1): An enzyme from *Crotalus adamanteus* that catalyzes the hydrolysis of both DNA and RNA by processive exonucleolytic attack of the free 3'hydroxy terminus to produce 5'-mononucleoside phosphates.

Phosphodiester bond: The covalent linkage between the phosphate group of the → 5' position of one pentose with the hydroxyl group of the → 3' position of the next pentose in a → nucleotide polymer (e.g. DNA, RNA).

Phosphodiester method: See → chemical DNA synthesis.

Phosphono peptide nucleic acid (pPNA): A negatively charged → peptide nucleic acid, in which the monomer units are attached to an N-(2-aminoethyl)phosphono glycine backbone and connected by phosphonester bonds. These → DNA mimics recognize complementary target DNA or RNA by → Watson-Crick base pairing. If composed of homo-T stretches (e.g. containing a chain of 14 thymine pPNA momomers), pPNA binds strongly to complementary poly(A)⁺-strands and can be used to isolate polyadenylated → messenger RNAs with e.g. short poly(A)-tails or complex secondary structures (e.g. → stem-loops) around the poly(A)-tail. pPNA oligomers do not possess a → polarity, and therefore bind in both parallel and antiparallel orientation to RNA. They also bind double-stranded RNA by invading the RNA:RNA duplex and displacing one strand, forming a stable → displacement loop (D-loop). Also, pPNAs are excellently water-soluble, but not enzymatically degraded by nucleases and therefore stable *in vivo*. See → peptide nucleic acid-phosphono peptide nucleic acid chimera, → trans-4-hydroxy-L-proline PNA, → oligo(T)-PNA.

Phosphonyl-*methoxy*propyl-*a*denine (PMPA): An → adenine derivative that inhibits the activity of → reverse transcriptase of retroviruses and is therefore employed in antiretroviral treatment strategies.

Phosphopantetheinyl *transferase* (PPTase): An enzyme catalyzing the transfer of a 4′-phospho*pant*etheinyl (Ppant) group of coenzyme A (CoA) onto a conserved serine residue in *p*eptidyl *c*arrier *p*roteins (PCPs) in → non-ribosomal peptide synthetases (NRPSs) or acyl carrier proteins (ACPs) in polyketide synthetases (PKSs). Ppantheneylation converts both NRPSs and PKSs from their inactive apo forms into the functional holo forms, inciting the synthesis of polyketides and nonribosomal peptides, respectively. Many of the polyketides and non-ribosomal peptides promise therapeutic use, as in anticancer (bleomycin, epothilone) and antibiotic (erythromycin, vancomycin) treatment.

Phosphoproteome: A → sub-proteome, consisting of phosphorylated peptides and proteins of a cell.

Phosphoproteomics: The whole repertoire of techniques to study phosphorylated peptides and proteins, the corresponding phosphokinases and phosphatases and the consequence of one or more phosphorylations of amino acid residues of a protein onto its function(s). See → peptidomics, → phosphoproteome, → proteomics.

Phosphorimaging: A technique for the sensitive detection of radioisotopes that employs a polyester plate coated with fine crystals of photostimulatable phosphor (BaFBr:Eu^{2+}) as an imaging plate. This plate accumulates and stores the energy emitted by the respective isotope. The sample (e.g. a nylon membrane) is simply covered by e.g. Saran wrap and exposed on the imaging plate (IP) inside a cassette. After exposure, the IP is scanned with a laser beam and emits → luminescence (proportional to the recorded radiation intensity), which is collected into a photomultiplier tube and converted to electrical signals. The IP is reusable, after the image data are erased (e.g. by exposure to light).

Phosphoroamidite technique: See → chemical DNA synthesis.

Phosphoro*dithio*ate oligodeoxyribonucleotide (PS₂): Any deoxyribonucleotide in which both oxygen atoms of the nucleotides are substituted by sulfur atoms. PS$_2$s are chemically very stable, achiral, resistant towards exonucleases, moderately resistant towards endonucleases, and hybridize with the normal oligodeoxyribonucleotides, though with a decreased stability of the duplex. Also, the → antisense properties of PS$_2$s are inferior to the normal oligodeoxyribonucleotides (as measured by → *in vitro* translation inhibition of specific → messenger RNAs), and the capacity to bind proteins is reduced. See → phosphorothioate oligonucleotide.

Phosphorodithioate oligodeoxyribonucleotide

Phosphorolysis: The cleavage of a covalent bond by orthophosphate.

Phosphorothioate antisense oligonucleotide (PS antisense oligo): Any → antisense oligodeoxynucleotide, in which some or all phosphate groups are replaced by → phosphorothioate groups. Such PS antisense oligonucleotides are used to block e.g. pharmacologically interesting DNA sequences or proteins. For example, the PS antisense oligonucleotide EPI 2010 (sequence: 5'-GATGGAGGG**C**GGCATG G**C**GGG-3') targets the AA1R protein, and is in clinical trials as an athma drug.

Phosphorothioate bond (PS): Any chemical bond in the phosphate backbone of an → oligonucleotide or DNA molecule, in which a non-bridging oxygen is substituted for a sulfur atom. This modification protects the internucleotide linkage from nuclease degradation. Phosphorothioate bonds can be introduced at either the 5'- or 3'-end of an oligonucleotide to prevent → exonuclease attack, or also internally to limit → endonuclease action. See → phosphorodithioate oligodeoxyribonucleotide, → phosphorothioate group, → phosphorothioate oligonucleotide.

Phosphorothioate group: A modified phosphate group, in which one of the oxygen atoms is replaced by a sulfur atom.

3'-Phosphorothioate

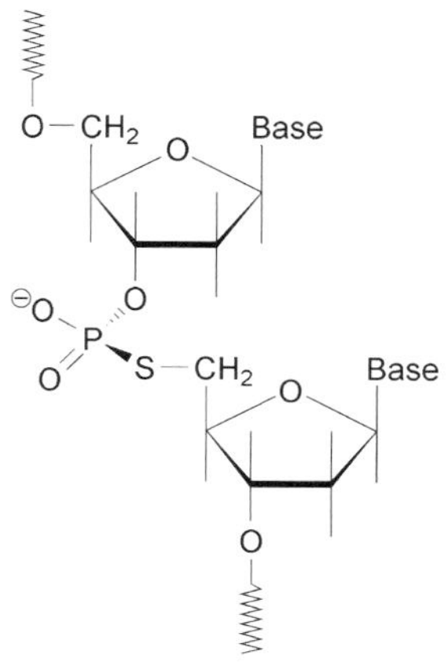

5'-Phosphorothioate

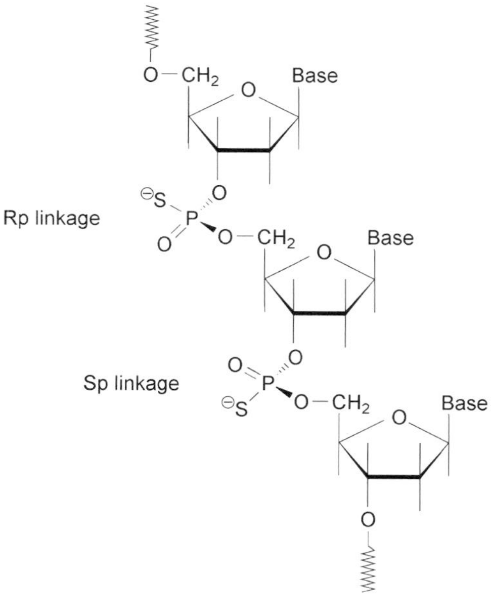

Rp linkage

Sp linkage

Phosphorothioate group

Phosphorothioate interference: The enzymatic replacement of a nonbridging oxygen atom at a 5'-phosphate group of an oligonucleotide (generally, DNA) molecule with sulfur (see → phosphorothioate group) and the use of this modification to detect the function of the substituted oxygen or, more precisely, of the specific phosphor atom to which it is covalently linked. For example, binding of a metal ion to a specific phoshate group is changed, if the latter is exchanged for a sulfur atom. This change can be measured, and the interactive phosphate be defined. For interference studies with oligonucleotides, phosphorothioates are generally incorporated by transcription.

Phosphorothioate oligonucleotide ("S-oligo"; phosphorothioate): Any → oligodeoxynucleotide in which some or all of the internucleotide phosphate groups are replaced by → phosphorothioate groups. Such modified oligonucleotides are resistant towards attack of most exo- and endonucleases, and could therefore be useful as intracellular → antisense oligonucleotides. See → phosphorodithioate oligodeoxy-ribonucleotide.

Phosphorothioate sequencing: A method for the → sequencing of DNA that uses Sanger techniques (→ Sanger sequencing) in combination with chemical cleavage reactions (→ chemical sequencing). In short, a synthetic oligodeoxynucleotide (→ primer) is annealed to the single-stranded target DNA. The reaction mixture is then aliquoted into four separate tubes that contain all four deoxynucleoside triphosphates, and additionally a → nucleoside-α-thiotriphosphate (dNTPaS) that also serves as substrate for the polymerization reaction and is incorporated at random. Then 2-iodoethanol or 2,3-epoxy-1-propanol is used to form a phosphorothioate triester with the incorporated dNTPaS. These esters are more easily hydrolyzed than → phosphodiesters. A careful hydrolysis can therefore lead to DNA fragments that can be used directly in the Sanger sequencing procedure.

Phosphorothioated DNA: See → S-DNA.

Phosphorylation: A frequent → posttranslational modification of proteins, mediated by specific phosphotransferases.

Phosphorylation site-specific antibody (PSSA): An → antibody raised against specific phosphorylated amino acid residues that is used for the detection and quantitation of the phosphorylation status of these amino acids in target peptides or proteins.

Phosphorylome: Another term of the omics era, describing the complete set of protein substrates for all cellular kinases.

Phosphotriester technique: See → chemical DNA synthesis.

Photoactivatable fluorescent protein (PAFP): Any one of a series of → autofluorescent proteins, whose emitted → fluorescence can be increased by additional irradiation with light of a specific wavelength and intensity. The PAFPs fall into several broad categories. First, the socalled irreversibly photoconverted, → photoactivatable → green fluorescent protein (GFP), the photoswitchable → cyan fluorescent protein (PS-CFP) and its enhanced version PS-CFP2 represent mutant variants of the natural green fluorescent proteins from *Aequorea victoria* and *Aequorea coerulescens*. These proteins contain a chromophore that initially exists in a neutral state with an absorption maximum at 400 nm. They all can be excited, emitting at 515 nm (PA-GFP), or 468 nm (PS-CFP, PS-CFP2), respectively. Irradiation with more intense UV or violet light (350–420 nm) induces irreversible chromophore transition from a neutral to an anionic state, resulting from light-driven decarboxylation of glutamate residue 222. This transition is accompanied by a 100- to 400-fold increase in excitation at 500 nm, with green emission at 515 nm. Second, the Anthozoa-derived green-to-red convertible proteins fold and form the chromophore to the green fluorescent state, and irradiation with UV-light irrversibly transform them into a red fluorescent state. Examples are Kaede, EosFP, mEosFP, KikGR, Dendra and → Dendra2. Third, and in contrast, a series of reversibly convertible PAFPs exist that allow repeated excitation and quenching. For example, chromoprotein as e.g. FP595 and its mutants called kindling fluorescent proteins (KFPs) or → Dronpa belong to this class. They can be transformed from non-fluorescent to red fluorescent states by irradiation with intense green or blue light. Within seconds or minutes after excitation these proteins spontaneously relax into the inactive state. Dronpa fluoresces green upon blue light excitation. After intense blue light irradiation, Dronpa is quenched to the non-fluorescent state. Dronpa can be re-activated to the green fluorescent state by a short pulse of UV. PAFPs allow to photolabel living cells, organelles or intracellular molecules (e.g. proteins, → fusion proteins), to visualize their spatial and temporal movement, and to monitor their half-life time and localization.

Photoactivatable green fluorescent protein (PA-GFP): An engineered variant of the → green fluorescent protein from *Aequorea victoria* that is extremely stable under aerobic conditions (more than a week), and – after excitation with light at 488 nm wavelength – increases fluorescence emission by a factor of 100 (compared to wild-type GFP). PA-GFP allows to explore temporal and spatial intracellular protein trafficking *in vivo*. See → photoactivatable fluorescent protein.

Photoactivated cross-linking: A technique to locate the sites of effective contacts between a nucleic acid sequence (e.g. a → promoter) and its cognate protein (e.g. one or more → transcription factors) by UV irradiation which leads to a complex formation between both partners.

Photoactivated localization microscopy (PALM): A technique for the intracellular visualization of proteins in tissue sections, fixed cells, or thin sections of organelles (mitochondria, lysosomes) at spatial resolution in the nanometer range. For example, cultured mammalian cells, transformed to express a *photoactivatable fluorescent protein* (PA-FP) are fixed, and processed on cover slides that are placed in a microscopic chamber and continuously excited by a laser at a wave-length close to the excitation maximum of the expressed PA-FP (e.g. λ_{exc} = 561 nm). The cells are imaged by → *total internal reflection fluorescence* (TIRF) microscopy onto an electron-multiplying charge-coupled device (EMCCD) camera (that detects single photons). Thereby the proteins can precisely be localized to intracellular compartments to a few nanometers. See → photoactivatable green fluorescent protein.

Photoactivation: The rapid conversion of light-activatable molecules to a fluorescent state by intense irradiation.

Photoaptamer: Any synthetic single-stranded → oligonucleotide (→ aptamer), into which → BrdU is incorporated instead of thymidine. This BrdU can be covalently cross-linked to a target protein by UV light, if it fits into the three-dimensional structure of its target region on the protein. This extremely specific interaction is exploited with the design of → photoaptamer arrays. Photoaptamers are selected *in vitro* by combinatorial chemistry (see → systematic evolution of ligands by exponential enrichment).

Photoaptamer array: Any glass slide, onto which thousands of → photoaptamers are spotted in an odered array that allows to detect many proteins of a protein mixture simultaneously on the basis of their specific interactions with the immobilized aptamers. In short, the protein mixture to be analyzed is first incubated with the photoaptamer array, specific interactions take place between the photoaptamers and some cognate proteins, and the bound proteins are cross-linked to their target aptamers by UV irradiation. Then non-bound proteins are removed by washing. After detection of the bound proteins, socalled protein profiles can be established.

Photobiotin: A → biotin molecule (vitamin H) attached to a photo-activable azido group via a spacer arm and used for the → non-radioactive labeling of single-stranded RNA and DNA, and double-stranded DNA. The labeling reaction involves exposure of the compound to strong visible light. This converts the azido group into a highly reactive nitrene that forms stable complexes with nucleic acids (single- or double-stranded DNA and RNA). Compare → biotinylation of nucleic acids; similarly used is → photodigoxigenin.

Figure see page 1178

Photobleaching: The irreversible light-catalyzed degradation of any → fluorochrome. Photobleaching determines the half-life time and thereby the utility of a particular fluorochrome for e.g. → fluorescent *in situ* hybridization.

Photocrosslink: Any covalent bond formed between a → photoaptamer and its target protein by *u*ltra*v*iolet light (UV).

Reaction of photobiotin with DNA (or RNA)

Photodigoxigenin: A → digoxigenin molecule linked to an azido phenyl residue via a hydrophilic spacer, and used to introduce digoxigenin into nucleic acids or proteins by simply exposing the reactants to UV irradiation of 260–300 nm wavelength. Incorporation of digoxigenin by photoactivation is less efficient than enzymatic labeling. See → digoxigenin-labeling; similarly used is → photobiotin.

Photodynamic protein: Any protein that is capable of transforming light energy into either a change of color ("photochromic protein"), an electromotive force ("photovoltaic protein"), or a change in absorbance ("nonlinear optical protein"), or to use a photon to energize a process.

Photo-footprinting:

a) Psoralene footprinting. A method for the detection of specific contacts between one or several proteins and a DNA duplex molecule, using → psoralene and UV light. In the presence of UV light, psoralene reacts with DNA through the formation of a photo adduct and pyrimidine monoadducts. These complexes influence or prevent the binding of → DNA-binding proteins.
b) Photo-footprinting technique. A method to detect contacts between specific sequences of DNA and regulatory proteins *in vivo*. After UV irradiation of intact cells, DNA is isolated before cellular → repair of the DNA damage begins. After purification it is subjected to a series of chemical reactions that break its sugar-phosphate → backbone only at the sites of UV damage. The DNA is then denatured, and labeled using direct or indirect methods. After electrophoresis on a polyacrylamide → sequencing gel the resulting fragment pattern is visualized by → autoradiography. Because protein-DNA contacts can inhibit or enhance UV photoproduct formation, differences in the strand-breakage patterns of protein-free and protein-associated DNA can be used to detect protein-DNA contacts at the basepair level.

Photohydrate: Any → pyrimidine base to which a hydroxyl group has been added onto C5 or C6 as a result of ultraviolet light radiation.

Photo-leucine: A derivative of the naturally occurring amino acid L-leucine that contains a diazirine moiety for the cross-linking of proteins *in vivo*. Photo-activation of diazirine with UV light leads to a reactive carbene intermediate that irreversibly cross-links proteins within protein-protein interaction domains at almost zero distance. Photo-leucine is added to growing cells in a leucine-free medium, substitutes L-leucine, and is incorporated into newly synthesized proteins. Then the cells are exposed to UV light at $\lambda = 365$ nm, which cross-links the proteins carrying photo-leucine. Such cross-linked protein complexes can be identified by a reduced mobility in → SDS-polyacrylamide gels or → Western blotting. See → photo-methionine.

Photolithography: A technique for the light-dependent engraving of a specific pattern on a solid support, used in printing processes. The solid support ("plate") is coated with a light-sensitive emulsion and overlaid by a photographic film. Then the coated plate is illuminated, and the image of the film is reproduced on the plate. Photolithography is employed in → DNA chip technology, where modifications of the usual phosphoramidite reagents are used (i.e. the *di*methoxytrityl [DMT] group that protects the 5'hydroxyl, is replaced by a photolabile protective group). The synthesis of the oligonucleotides on the chip proceeds by photolithographically deprotecting all the areas that will receive a common nucleoside, and coupling this nucleoside by exposing the entire chip to the appropriate phosphoramidite. This is achieved by socalled masks made from chromium/glass that contain holes at positions, where

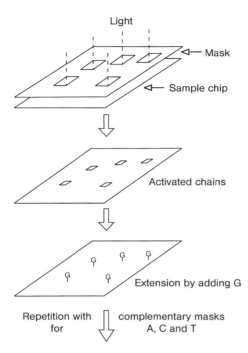

Oligonucleotide chip construction using photolithography

Photolitography

deprotection is desired. A more advanced procedure exploits a socalled virtual mask. Up to 480,000 (or even more) digitally controlled micro-mirrors allow the illumination of only defined spots on a DNA chip depending on their precise angular position (→ "mask-less photo-lithography"). After the oxidation and washing steps the procedure has to be repeated for the next nucleoside.

Photolyase: A 54 kDa repair enzyme from *E. coli*, catalyzing the removal of → pyrimidine dimers.

Photo-methionine: A derivative of the naturally occurring amino acid L-methionine that contains a diazirine moiety for the cross-linking of proteins *in vivo*. Photo-activation of diazirine with UV light leads to a reactive carbene intermediate that irreversibly cross-links proteins within protein-protein interaction domains at almost zero distance. Photo-methionine is added to growing cells in a methionine-free medium, substitutes L-methionine, and is incorporated into newly synthesized proteins. Then the cells are exposed to UV light at λ = 365 nm, which cross-links the proteins carrying photo-methionine. Such cross-linked protein complexes can be identified by a reduced mobility in → SDS-polyacrylamide gels or → Western blotting. See → photo-leucine.

Photon (Greek: phos for "light"): The quantum of light. Its energy is proportional to its frequency: $E = h \cdot \upsilon$ (where E is energy, h is Planck's constant [6.62×10^{-27} erg-second], and υ is frequency).

Photoprotective group: Any chemical compound covalently bound to the 5'-deoxyribose of a nucleic acid base that prevents any reaction of this deoxyribose with the 3'-OH of another deoxyribose. This protective group can, however, be removed by e.g. UV light. For example, for the production of a special kind of → microarray, the oligonucleotides are synthesized directly on the glass support of the array. Each base that is coupled onto the glass surface, carries a protective group, frequently *m*ethyl*n*itro*p*iperonyl*o*xy*c*arbonyl (MeNPOC) on the 5'-OH position. MeNPOC can be removed by half a minute UV irradiation, and a second base can be coupled that in turn carries a MeNPOC group at its 5'-OH position. See → photolithography.

Figure see page 1181

Photoprotective group

Photo-reactivation: The breakage of carbon-carbon bonds in the cyclobutane ring of → thymidine dimers generated in DNA by ultraviolet radiation, to restore the normal base sequence. This process is catalyzed by → DNA photolyases that bind to pyrimidine dimers in the dark, but utilize light energy (365–405 nm, 435–445 nm) to break the cross-links (photoreactivation).

Figure see page 1182

Photoreactive crosslinker: Any → crosslinker, whose chemical reactivity is induced by UV illumination. Photoreactive crosslinkers are used for crosslinking molecules at defined sites (e.g. specific cells or organs). For example, 4,4'-*dia*zido*dip*henyl ethane (DADPethane) or 4,4'- *dia*zido*dip*henyl ether (DADPether) that react non-specifically with two different biomolecules, sulfosuccinimidyl-6 (4'-*azi*do-2'-*ni*tro*p*henyl*am*ino)hexanoate (sulfo-SANPAH), sulfosuccinimidyl-4(p-*a*zido*p*henyl)butyrate (sulfo-SAPB), N-*h*ydroxysuccinimidyl-4-*a*zido benzoate (HSAB), N-*h*ydroxysulfosuccinimidyl-4-*a*zido benzoate (sulfo-HSAB) or N-*h*ydroxysuccinimidyl-4-*a*zido salicylic *a*cid

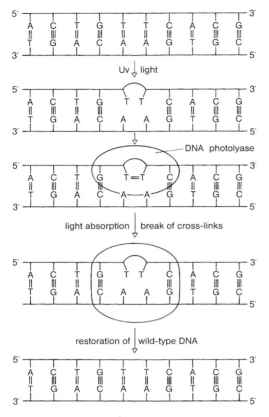

Photo-reactivation

(NHS-ASA) that react with amines, are such photoreactive crosslinkers. See → heterobifunctional crosslinker, → homobifunctional crosslinker.

PhotoSELEX: The trademark of an *in vitro* combinatorial chemistry technique for the generation of a large pool of potential → aptamer → oligonucleotides and the identification of some oligonucleotides specifically binding to a DNA target. In short, a starting pool of about 10^{15} different oligonucleotides (i.e. a million billions) is exposed to the target DNA of interest. The oligonucleotides contain → bromodeoxyuridine (BrdU) instead of → thymidine, which allows the best binding ones to be cross-linked to the target by ultraviolet light (→ "photoaptamers"). Oligonucleotides that are tenaciously bound to the target sequence, are therefore covalently cross-linked by UV, and the oligonucleotides with lower or no affinity to the target are washed off. Then the cross-linked oligonucleotides are amplified by conventional → polymerase chain reaction. See → photoaptamer array, → systematic evolution of ligands by exponential enrichment.

Photoswitchable autofluorescent protein: Any → autofluorescent protein, whose emission light wave length can be changed upon irradiation. For example, a → cyan fluorescent protein 2 from *Phialidium sp.* (PS-CFP2) is such a dual-color monomeric photoconvertible protein. If it absorbs light of 405 nm, it undergoes an irreversible photoconversion from the normally occuring cyan to a → green fluorescent form with a concomitant dramatic (about 2000fold) increase in green-to-cyan fluorescence intensity. PS-CFP2 is used for *in vivo* labeling of proteins, organelles and cells to monitor their real-time movement *in situ*.

Phototoxicity: The toxic effects of proteins (or their aberrant aggregates) on various cellular structures and processes. For example, late-onset human neurodegenerative diseases such as Alzheimer's disease (AD), Huntington's and Parkinson's disorders are genetically and pathologically linked to aberrant protein aggregation. In AD, aggregation of *a*myloid *p*recursor *p*rotein (APP)-derived peptides aggregate into $A\beta_{1-42}$ oligomers that are deposited within and between neurons in the brain of the afflicted individuals and interfere with neuronal function(s).

PHRAP: A software program that allows to assemble → raw sequence data into sequence → contigs and to assign a specific quality score to each position of the DNA sequence, based on → PHRED scores of the raw sequence reads. A PHRAP quality score of X corresponds to an error probability of approximately $10^{-X/10}$. Thus, a PHRAP quality score of 30 corresponds to 99.9% accuracy for a base in the assembled sequence.

PHRED: A software program that allows to analyze → raw sequence data, to generate a base call and a linked quality score for each position in the sequence. See → PHRAD.

pHyg: A mammalian → expression vector containing the *E. coli* gene for → hygromycin B phosphotransferase that can be used as a dominant → selectable marker in transfection experiments.

Phylogenetic footprint: The conservation of certain sequence motifs in → orthologous genes of many different species.

Phylogenetic footprinting: A bioinformatics approach to the *in silico* identification of gene-regulatory elements in moderately to highly conserved regions of genomes that is based on the alignment of → orthologous sequences and the definition of non-coding regions (e.g. → *t*ranscription *f*actor *b*inding *s*ites, TFBSs) with unexpectedly high evolutionary conservation. Such regions are protected from random drift by selection (i.e. are under selective pressure), implying a slower evolution than surrounding sequences. Several software packages assist phylogenetic footprinting. For example, FootPrinter (http://bio.cs.washington.edu/software/html) exploits available sequences and uses an algorithm for *de novo* discovery of short conserved motifs, or ConSite (www.phylofoot.org/consite) identifies conserved regions and binding-site characteristics to select active transcription factor binding motifs. Multiple alignment of sequences from various genes together with their upstream sequences identifies binding motifs for specific transcription factors. Example: transcription factor → Sp1.

Human gene	Putative binding site for Sp1 (highly conserved regions in bold face)
Chorionic somatomammotropin	5'-**ATGTGTGGGAGGAGC**TTCT-3'
Growth hormone 1	5'-**ATGTGTGGGAGGAGC**TTCT-3'
Growth hormone 2	5'-**ATGTGTGGGAGGAGC**TTCC-3'
AMPK gamma-2	5'-CTCTGGGA**ATCTGTGGGAGGAGC**CGAGA-3'
PPP1R1B (also *DARPP-32*)	5'-TG**TGTGTGGGAGGA**CACGTG-3'

Phylogenetic footprinting

Once potential gene-regulatory elements are known, techniques as e.g. → ChIP-on-chip (ChIP-chip), a combination between → chromatin immunoprecipitation (proteins are reversibly crosslinked to fragmented DNA and the resulting complex precipitated by a transcription factor-specific antibody. After precipitation, the protein-bound DNA is released and fluorescently labeled) and → DNA chips (→ genomic microarrays), to which the fluorescent DNA probe is hybridized to map its genomic position, are used to validate their function(s).

Phylogenetic microarray ("phylochip"): Any → microarray, onto which → oligonucleotides are immobilized that are derived from highly variable as well as highly conserved sequence motifs in → ribosomal DNA (rDNA) detected by sequence → alignment of 100,000s of entries in databanks. For each rDNA of each organism at least three different oligonucleotides are designed and placed onto the chip. Then rRNA from environmental samples is isolated, labeled with a → fluorochrome, and hybridized to the phylochip. Hybrids are then detected by a laser scanner. Phylogenetic microarrays allow to identify (preferably) all organism (e.g. bacteria) from a sample and to establish socalled phylogenetic fingerprints of a community of organisms.

Phylogenetic profiling: A computational screen for proteins that always occur together in many, if not all organisms. Phylogenetic profiling aims at inferring functional linkage between proteins from their simultaneous presence in a multitude of cells, tissues, organs, or organisms.

Phylogenetic shadowing: A variant of the → phylogenetic footprinting technique that is based on the comparison of → orthologous DNA sequences from a set of related species, and allows to detect conserved (and therefore functional) regions of a → genome. For example, in a set of 15 primate species (Old World and New World monkeys and hominoids including man), fast-versus-slow mutation rates for each aligned nucleotide site of a selected genomic region can be identified. The slow-mutation sites usually encompass → coding regions, in particular → exons, sequences for the binding of proteins, and other functional elements shared between humans and evolutionarily distant mammals.

Figure see page 1185

Phylogenetic tree: A graphical representation of the genealogical or evolutionary relationship(s) among individuals of a group of molecules or organisms.

Phylogenomic map: A topographical depiction of the similarity structure of a phylogenomic matrix, in turn based on sequence data from partial, single or multiple genomes that are used to produce protein predictions, which are aligned to a database of → proteomes from hundreds of completely sequenced genomes. This matrix detects genes, whose encoded proteins are consistently co-inherited. In a phylogenomic map, products from genes with similar evolutionary histories cluster together.

Phylogenomic mapping: The process to establish a → phylogenomic map.

Phylogenomics (*phylogenetics/genomics*): A branch of → genomics that exploits existing sequence information from various organisms ("evolutionary information") in the databases to assign a specific function to a particular sequence, and

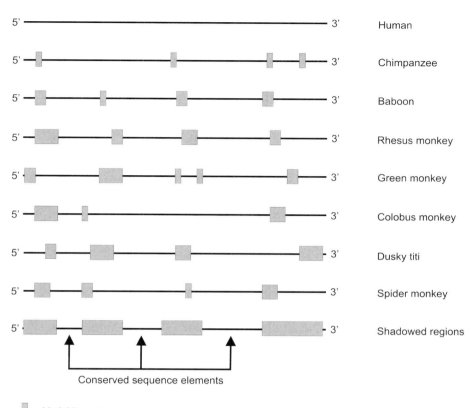

Phylogenetic shadowing

links genome analysis to phylogenetics. Integrating evolutionary analysis improves the accuracy and specificity of functional annotation. Functional predictions are improved by concentrating on questions as e.g. *how* genes became similar in sequence during evolution rather than focusing on sequence similarity *itself*. The term "phylogenomics" also describes the integration of genomic and comparative genomic data in e.g. species tree reconstruction. See → behavioral genomics, → biological genomics, → cardio-genomics, → chemical genomics, → clinical genomics, → comparative genomics, → deductive genomics, → environmental genomics, → epigenomics, → evolutionary developmental genetics, → functional genomics, → horizontal genomics, → integrative genomics, → lipo-proteomics, → medical genomics, → neurogenomics, → neuroproteomics, → nutritional genomics, → omics, → pathogenomics, → pharmacogenomics, → physical genomics, → population genomics, → proteomics, → recognomics, → structural genomics, → transcriptomics, → transposomics.

Phylome: Another term of the → ome era, describing the complete set of phylogenetic trees for the genes of a given genome.

Phyloproteomics: A branch of → proteomics that aims at deciphering phylogenetic relationships between organisms on the basis of peptide and protein sequence and structure.

Physical containment: A package of physical-technical security measures to prevent the escape of living organisms containing → recombinant or otherwise dangerous (e.g. pathogenic) DNA from a laboratory or an industrial production plant. Generally, four levels of biosafety (BL 1–4) of various stringencies are characteristic for the guidelines of most countries:

BL 1: The lowest level does not require a separate laboratory, nor any specific containment equipment or specially trained personnel (that should be familiar with microbiological techniques).

BL 2: This level demands a limited access to the laboratory, biological safety benches and autoclaves in addition to the requirements of BL 1.

BL 3: This more stringent level requires additionally that the laboratory is only accessible to authorized and specially trained personnel (i.e. personnel familiar with handling of pathogenic or potentially lethal agents) who wears protective clothing. Protected laboratory bench surfaces, biological safety benches, airlocks, and negative pressure within the BL 3 area are obligatory.

BL 4: The most stringent level requires additionally a separate, window-less building, air- and liquid-decontamination, airtight doors and positive pressure protective clothing. See also → biological containment, → containment.

Physical functional marker (PFM): Any → molecular marker generated with → genomic DNA from organism A by the amplification of a gene with a gene-specific primer from organism B (using a conventional → polymerase chain reaction) that can be mapped on the → physical map of organism B.

Physical genomics: The whole repertoire of techniques for the analysis of an organism at the → genome level, encompassing large insert libraries (→ bacterial artificial chromosome libraries, → yeast artificial chromosome libraries), genome sequencing, the establishment of expressed sequence tag databases, the (preferably) complete inventory of the → transcriptome, → proteome and → metabolome, and the relevant → bioinformatics tools. Compare → biological genomics.

Physical map: The linear arrangement of genes or other markers on a chromosome as determined by techniques other than genetic recombination (e.g. → heteroduplex analysis, → DNA sequencing). Usually, map distances are expressed in numbers of nucleotide pairs between identifiable genomic sites (e.g. → contigs, → sequence tagged sites, or → restriction sites). See → contig mapping, → macro-restriction map, → ordered clone map, → restriction mapping. Compare → map, → mapping.

Physiologic epigenome: A misleading term for the entirety of → histone modifications varying between different cell types and the changes in binding patterns of chromosomal and other proteins to → chromatin under various regimes (as detected by e.g. → chromatin immunoprecipitation).

Physiological quantitative trait locus (pQTL): A genomic region, or two or more separate genetic → loci that cooperatively contribute to the establishment of a specific physiological phenotype. See → Expression Quantitative Trait Locus (eQTL), → Quantitative Trait Locus (QTL).

Physiome: An additional term of the → omics era for the description of the complete physiological condition of a cell, a tissue, an organ, or an organism. See → genome, → metabolome, → physiomics, → proteome, → transcriptome.

Physiomics: The whole repertoire of techniques for the comprehensive and quantitative description of the → physiome of a cell, a tissue, an organ, or a complete organism.

Physisorption: The functionalization of a cantilever tip of an *atomic force microscope* (AFM) by the strong physical adsorption of a biotinylated protein (e.g. bovine serum albumin, BSA). This physisorbed protein coat may be reacted with → avidin or → streptavidine and serves as a matrix for modification with biotinylated ligands. See → chemisorption.

Phytochelatin (PC): A member of a class of small, cysteine-rich peptides with high heavy metal ion-binding capacity, which is mediated by thiolate coordination. These plant peptides function as traps for cadmium, copper, lead, mercury and zinc. The synthesis of phytochelatins proceeds without → translation and is catalyzed by phytochelatin synthase.

π: See → nucleotide diversity per site.

P$_i$: Symbol for *i*norganic *p*hosphate.

pI: Abbreviation for → *i*soelectric *p*oint.

Pi: See → protein interference.

PIC:

a) See → pre-initiation complex.
b) See → polymorphism information content.
c) See → preintegration complex.

***Pichia* expression system (*Pichia pastoris* expression system):** An *in vivo* system for the high-level expression of heterologous recombinant proteins, based on the methylotrophic yeast *Pichia pastoris* (or *Pichia methanolica*). This yeast can metabolize methanol as sole carbon source, if the preferred substrate glucose is absent. The first step in methanol utilization is the alcohol oxidase-driven oxidation of methanol to formaldehyde. Expression of this enzyme, which cannot be detected in the absence of methanol and which is encoded by the AOX1 gene, is therefore tightly regulated and induced by methanol to very high levels (e.g. >30% of the total suluble cellular proteins represent alcohol oxidase). Expression of the AOX1 gene is controlled by the strong AOX1 promoter, which has therefore been cloned into *Pichia* expression vectors. In short, the gene of interest is first cloned into such a *Pichia* vector, designed for intracellular expression, or intracellular expression and secretion, the linearized construct transformed into appropriate competent *Pichia* cells or → spheroplasts, the transformants selected by their resistance phenotype (e.g. if a *HIS4*-selectable marker is used, a histidine-deficient medium is employed; in case of → zeocin-selection, this antibiotic is added to the medium), and analyzed for the integration of the gene of interest at the correct locus and in the correct orientation. Then a small-scale pilot expression by some 10–20 colonies is tested (to verify the presence of the recombinant protein, using → SDS polyacrylamide gel electrophoresis and → Western blot analysis), before an upscaled production in fermenters is started. Since the AOX1 gene promoter is very strong, expression of the foreign gene leads to extraordinarily high levels of the recombinant protein (e.g. grams per liter on average). For high-level methanol-independent expression, vectors are equipped with the constitutive promoter of the glyceraldehyde-3-phosphate dehydrogenase gene. Also, expression vectors are available that allow to detect multiple insertion events (occurring spontaneously at a frequency of 1–10%). These multi-copy *Pichia* expression vectors carry the → kanamycin resistance gene, conferring resistance to → geneticin. Multiple insertions can therefore be identified by increased levels of resistance to this antibiotic.

The *Pichia* expression system combines the advantages of *E. coli* (inexpensive and easy handling, high-level expression) and the eukaryotic *Pichia* (protein folding, post-translational modifications, protein processing, secretion), and allows production of nearly all proteins in high quantity (e.g. enzymes, enzyme inhibitors, membrane proteins, regulatory proteins, antigens and antibodies). Compare → Baculovirus expression system.

Picking robot: An automated machine for the transfer of bacterial colonies onto → microtiter plates. Usually the robot station uses a camera, which generates a digital image of the colonies on the petri dishes. A suitable image analyzing software then transforms the positions of the colonies into robot coordinates. An xyz system moves a picking tool which allows to individually guide picking pins to the colonies. After 96 such pins have taken up different bacterial colonies, they dive into the wells of microtiter plates which are filled with nutrient medium. In between two picking processes, the pins are sterilized in ethanol and dried in a hot air stream. The capacity of picking robots ranges from 5,000–10,000 clones picked per hour.

Picoliter reactor sequencing: See → fiber-optic reactor sequencing.

Picotiter plate: A variant of the → microtiter plastic plate that contains up to 300,000 of 50–75 picoliter wells. Such picotiter plates are used for highly multiplexed amplification and sequencing reactions for whole genome analysis.

PICS: See → 7-*propynyl iso*carbostyril.

PID: See → pre-implantation diagnostics.

Piezoarray: A laboratory slang term for a → microarray, onto which solutes in the nano- and picoliter range are spotted contact-free by piezoelectric forces. In this volume range, the surface tension of the solute is greater than its kinetic energy, so that satellite drops or a backward movement of the drop is prevented. Piezoarrays are preferentially loaded with proteins.

Pif1p: A highly conserved 5' → 3' → DNA helicase encoded by the *PIF1* gene in *Saccharomyces cerevisiae*, associated with → telomeres, and controlling telomere length by inhibiting → telomerase activity.

***piggyback* (PB):** A 2.472 kb DNA → transposon from the cabbage looper moth *Trichoplusia ni* that carries 13 bp → inverted terminal repeats (ITRs) and a gene encoding a 594 amino acids long → transposase, and inserts into the tetranucleotide 5'-TTAA-3' site of a target DNA flanked by stretches of A and T nucleotides. Upon → insertion, the TTAA sequence is duplicated. *PB* can carry multiple genes (of up to about 15 kb) and efficiently and preferentially transpose into → transcriptional units (97% in → introns, 3% in → exons) in mouse and human cell lines and mice as well. *PB* (or other members of the *piggyback* family) are components of the → genomes of phylogenetically diverse species from fungi to mammals, and are used to transform the germline of more than a dozen species spanning four orders of insects, and to generate → transgenic animals (e.g. mice).

Pilot protein: Any protein that mediates the transfer of DNA from a donor to a receptor CEll during bacterial → conjugation.

Pilus (sex pilus, conjugative pilus): Extracellular filamentous organelle of Gram-negative bacteria containing a → conjugative plasmid. Pili serve to form mating pairs between donor and recipient cells and are the site of adsorption for certain bacteriophages (see e.g. → fd phage, → f1 phage). The F-pilus for example is a hollow cylinder 80 Å in diameter with a 20 Å axial hole and is composed of a single subunit protein (pilin) arranged in four parallel helices with a 128 Å repeat.

PIM: See → protein interaction mapping.

PIN: See → partially intronic noncoding EST contig.

Pin and ring spotter (PARS): An instrument for the → spotting of → probes onto → microarray supports that works with a circular metal loop ("ring") to load sample liquid by capillary action, and a solid pin moving up and down through the liquid in the loop to deposit the probes onto the microarray by contact printing (i.e. by direct contact with the support).

Ping: Any one of a class of → miniature inverted repeat transposable elements (MITEs) that spans about 5,500 bp with a central region of 4,900 bp flanked by a 252 bp left and a 178 bp right part, and is present in low copy numbers in the genome of e.g. rice (*Oryza sativa* spp. *japonica*: 60–80; *O. sativa* spp *indica*: 14). A Ping sequence ends in TTA duplications. The central part contains two putative → open reading frames (ORFs), and can be excised, giving rise to the highly conserved 430 bp socalled *mini*ature Ping element (miniPing or mPing) that in turn is flanked by 15 bp terminal inverted repeats (TIRs), but does not contain any ORF. MPing excision is activated by stress (e.g. culture stress, γ-rays), its reinsertion occurs at new loci, also within exons of genes (e.g. exon 2 of the *Waxy* (*Wx*) gene of maize. Pings seem to be active, giving rise to mutable seed phenotypes. Compare → *pong*.

PIN RNA: Any → RNA transcript that is complementary to a → partially intronic noncoding EST contig. See → TIN RNA.

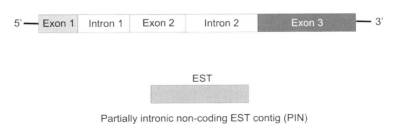

Partially intronic non-coding EST contig (PIN)

PIN RNA

Pioneer sequence: Any novel DNA sequence, for which no related sequence exists in the databases. Since the sequence databases contain immense amounts of sequences, in particular gene sequence information, pioneer sequences most probably have species-, family- or kingdom-specific functions, evolved to meet special demands of the particular organism or group of organisms, from which they originate. See → orphan gene.

PIP: See → percent identity plot.

PIRC: See → Piwi-interacting RNA complex.

piRNA: See → Piwi-interacting RNA.

piRNA cluster: Any one of several gene clusters of *Drosophila melanogaster*, where each gene encodes a specific → Piwi-interacting RNA. Transcription occurs from both strands within the cluster (exception: the *flamenco* locus, where only one single strand is expressed). The piRNAs transcribed from such clusters function in the control of mobile genetic elements. For example, the piRNAs from the *flamenco* locus (about 180 kb, maps to the pericentromeric heterochromatin on the X chromosome) repress transposition of → retrotransposons *gypsy*, *ZAM*, and *Idefix*.

PIS: See → proviral insertion site.

PISA: See → protein *in situ* array.

PITC: Phenyl *iso*thio*c*yanate, a compound used for → protein sequencing.

Pitch: The length of one complete turn of a DNA double helix along its vertical axis (as measured e.g. in °).

Piwi domain (*P*-element-*i*nduced *wi*mpy testis domain): A highly conserved → motif at the carboxy terminus of → Argonaute proteins that adopts an → RNase H fold essential for the → endonuclease activity of → RISC. The piwi domain contains two sequence motifs, a GxDV and an RDG motif highly conserved in eukaryotes, and specifically associates with so called → piwi-interacting RNAs.

Piwi gene: A *Drosophila* gene that encodes a nuclear protein of the → Argonaute family, and is essential for germ stem cell self-renewal (stem cell maintenance) and the silencing of → LTR retrotransposons in testes (for example, *piwi* mutations lead to a repression of the endogenous → retrotransposon → copia), thereby controlling their mobilization in the male germline. See → piwi domain, → piwi-interacting RNA.

Piwi-*i*nteracting RNA (piRNA): Any one of a series of highly abundant, small, 26–31 nucleotides long, poorly conserved RNA molecules of *Drosophila* germline cells that interact with the → piwi domain of → Argonaute proteins, carry a 5'-monophosphate group and a 2'-O-methyl modification at their 3'-ends (added by a Hen-1 family RNA methyltransferase) and are essential for spermatogenesis. Additionally, piRNAs prevent the spreading of selfish genetic elements within a genome. piRNAs match repetitive elements throughout the *Drosophila* genome, but piRNA-encoding genes are frequently clustered in socalled → piRNA clusters ranging from several to hundreds of kilobases that are enriched in → transposons and other repeats. piRNAs associate with multiple Piwi proteins. The number of distinct mammalian pachytene piRNAs (appear around the pachytenic stage of meiosis, become abundant, and persist up to the haploid round spermatid stage, then gradually disappear during sperm differentiatition) alone amounts to >500,000.

Piwi-interacting RNA complex (PIRC): The complex between a → piwi-interacting RNA and the → piwi domain of → Argonaute proteins.

Pixel plot: Any two-dimensional representation of a → microarray spot, in which all pixel intensities are plotted at two fluorescence emission wavelengths (e.g → cyanin 3 versus → cyanin 5). A pixel plot allows the experimentor to verify the quality of the spot (i.e. whether or not the distribution of pixel intensities is uniform and normal, as should be expected from a successfully printed spot).

PKR: See → double-stranded RNA-activated protein kinase.

Plab: See → plant antibody.

PLAC: See → plant artificial chromosome.

Planar array: Any → microarray, whose elements (individual spots) are immobilized on a planar surface, in contrast to any microchip based on microfabricated channels (see → electrophoresis chip, → lab-on-a-chip, → microfluidic chip, → suspension array).

Planar waveguide chip: Any → chip support that consists of the chip material itself (e.g. glass, silicon), a $SiSiO_2$ porous layer machined onto it and target molecules (e.g. DNA, → oligonucleotides, or proteins) bound to this layer. These molecules are targets for other molecules that bind to them (e.g. DNA fragments with complementary sequence, → antibodies). Interaction between → probe and target molecules is achieved by → hybridization (DNA-DNA interaction) or protein-protein interaction (antigen-antibody interaction). For a detection of this interaction the target molecules are labeled (with e.g. a → fluorochrome), and this fluorochrome can be excited by light. This is guided and mostly contained within by the $SiSiO_2$-layer. Some of the light quanta escape from the waveguide layer and excite the fluorophors, whose emittance light can then be detected. In a variant of the conventional planar waveguide chip, a thin film of e.g. tantalium pentoxide /Ta_2O_5) covers the chip. This layer with a high refractive index exclusively guides the laser light on the chip's surface, permitting the selective detection of labeled and captured (i.e. surface-bound molecules only, but not free label in solution.

Plant antibiotic resistance marker (plant ARM): Any plant gene that confers resistance towards an → antibiotic. For example, the *Atwbc19* gene from *Arabidopsis thaliana* encodes an ATP-binding cassette (ABC) transporter protein. If this gene is overexpressed in a → transgenic plant, this plant is resistant to the antibiotic → kanamycin. Such plant ARMs avoid the problem of → horizontal gene transfer from a transgenic plant to soil bacteria and a (hypothetical or real) increase in antibiotic resistance of bacteria generally. Plant ARMs are therefore better suited as → selectable markers in → genetic engineering experiments than bacterial genes.

Plant antibody (plab; plantibody): Any → monoclonal antibody that is synthesized by → transgenic plants. For example, the genes for the heavy and light chain peptide of → IgG antibodies have been transferred into separate tobacco mesophyll cells by → *Agrobacterium*-mediated gene transfer, which were regenerated to complete plants expressing the foreign gene. Conventional crossing of these two transgenic plants leads to a plant harboring both the gene encoding the k → light chain and the gene for the g → heavy chain. This plant is able to synthesize a complete IgG antibody.

Plant ARM: See → plant antibiotic resistance marker.

Plant artificial chromosome (PLAC): A → cloning vector containing plant → centromere DNA and → telomere repeats that can be introduced and maintained in both yeast and a target plant as a stable autonomous → minichromosome. PLACs additionally are equipped with → selectable marker genes and are designed to optimally function in diverse plant cells and to adopt → genomic DNA in the megabase range. See → bacterial artificial chromosome, → human artificial chromosome, → mammalian artificial chromosome, → pBeloBac11, → P1 cloning vector, → transformation-competent articificial chromosome vector, → yeast artificial chromosome.

Plant cloning vector (plant vector; plant cloning vehicle): Any → cloning vector that is designed to introduce foreign DNA into a plant's genome. Such vectors may be based on the → Ti-plasmid of → *Agrobacterium tumefaciens*, or DNA plant viruses. See also → plant expression vector.

Plant expression vector: A → plasmid cloning vehicle, specifically constructed so as to achieve efficient transcription of the cloned DNA fragment(s) and translation

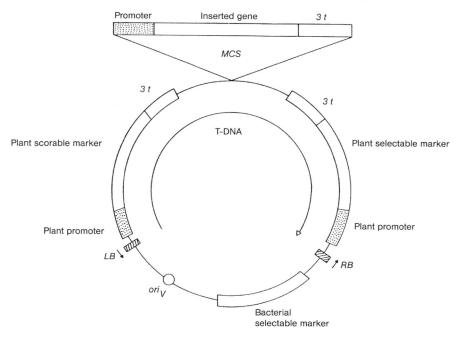

Plant expression vector

of the corresponding transcript(s) within a target plant cell. Such cloning vectors contain either a constitutive and highly active ("strong") → promoter sequence (e.g. CaMV35-promoter, → nopaline synthase promoter) or an inducible (e.g. hormone- or light-inducible) or regulated promoter. Immediately downstream of the promoter appropriate cloning site(s) and a plant transcriptional → termination sequence have been inserted. Any promoter-less foreign gene, cloned into such a vector will be expressed at a high level in the transgenic plant.

Plant gene therapy: The use of → chimeric oligonucleotide-directed gene targeting to correct or introduce single-nucleotide mutations in plant genomic DNA.

Plantibody: See → plant antibody.

Plant-made pharmaceutical (PMP): Any one of a series of pharmaceutically active substances produced by e.g. → transgenic plants. For example, transgenic socalled bioreactor plants (as e.g. potatoes) synthesize cationic antimicrobial peptides transcribed from human β-defensin genes isolated from human keratinocytes. These defensins attack and permeabilize the cell membranes of invading Gram-positive and -negative bacteria, fungi, encapsulated viruses and other parasites and are potential pharmaceuticals effective against e.g. *Candida albicans*.

Plant vector: See → plant cloning vector.

Plaque: A clear or turbid area in a bacterial lawn on a culture dish caused by → phage growth and subsequent death and → lysis of the bacteria. Each plaque contains from 10^6–10^7 infectious phage particles. The plaques of virulent phages are generally clear, the plaques of temperate phages are turbid. The term plaque is also used for cell-free areas in cell culture lawns, caused by → viruses. See also → plaque count, → plaque-forming unit, → plaque hybridization, → turbid plaque.

Plaque assay: See → plaque count.

Plaque blotting: See → plaque hybridization.

Plaque count (plaque assay): The determination of the number of complete, infective → bacterio phage particles or infected bacterial cells in a particular suspension. For plaque count, the sus pension is spread onto the surface of an agar plate that is covered with a thin layer of susceptible bacteria ("lawn"). Appropriate dilutions ensure that not more than one phage can infect one host cell, and the number of → plaques are counted that develop on the host cell lawn.

Plaque forming unit (pfu): Usually defined as the number of infectious → virus particles per unit volume, or alternatively, any single infectious particle that generates a single → plaque under defined conditions.

Plaque hybridization (Benton-Davis technique, Benton-Davis procedure, Benton-Davis hybridization; lifting, plaque lifting, plaque blotting, plaque screening, phage lifting): An *in situ* → gene screening procedure for the direct detection of a particular DNA sequence within a population of transformed → bacteriophages harboring vast amounts of different cloned sequences (→ phage libraries). Detection is made possible by hybridization *in situ* with radioactively labeled RNA or DNA → probes with complementarity to the sequence sought. In short, a → nitrocellulose filter is placed on an → agar plate containing bacteriophage → plaques. Unpackaged recombinant phage DNA is bound to the filter (plaque lift, phage lift), denatured and fixed to the filter by baking. The radioactive probe is then hybridized to the filter-bound DNA and the position of the plaque containing the complementary sequence is located by → autoradiography. Interesting plaques can then easily be recovered from the master plate.

Plaque lifting: See → plaque hybridization.

Plaque screening: See → plaque hybridization.

Plasmaprinting: A technique for the functionalization of glass, silicium or polymer/copolymer surfaces that is based on a socalled *d*ielectric *b*arrier *d*ischarge (DBD, also corona discharge) under atmospheric pressure. Such discharges form between two electrodes after aplication of an alternating current (AC), if the current flow is impeded by at least one dielectric barrier (the socalled isolator, e.g. the glass slide of a → microarray). In such a configuration gases are activated by 5–10 nsec microdischarges ("filaments"), and chemically very reactive compounds (e.g. radicals) are formed. These in turn allow to either polymerize suitable monomers or modify the surface of the support with a series of functional groups, as e.g. amino, epoxy, hydroxy, carbonyl or carboxyl groups. Amino and epoxy groups can further be used to couple → biotin or → protein A onto the surfaces. The impeded discharge raises the mean gas temparature between the electrodes by only few degrees Kelvin, so that the discharge remains "cold" (and does not damage temperature-sensitive substrates). Plasmaprinting can also be localized, if a specific configuration of the dielectricum is used that creates small voids (from 10 to 100 µm dimensions), in which the plasma develops. Gas can enter the dielectricum through the fine meshwork of one of the electrodes. The plasma then allows to spot functional groups in an array format of desirable geometry. Plasmaprinting therefore is used to functionalize chips for DNA and protein analysis.

Plasmid: A closed circular, autonomously replicating, extra-chromosomal DNA duplex molecule ranging in size from 1 to more than 200 kb and in copy number from one to several hundred per bacterial cell. The copy number of plasmids may depend upon environmental factors. The average number of e.g. → pBR 322 plasmids per cell on rich (→ LB) medium is 55, immediately before cell divisions it increases to 80. Plasmids generally confer some selective advantage to the host cell (e.g. → antibiotic resistance). → Conjugative plasmids harbor a set of genes capable of transferring the plasmid to other, plasmid-less cells. → Cryptic plasmids are naturally occurring plasmids with unknown genotype and biological function. Different plasmids may interfere with the replication and inheritance of each other, see → plasmid incompatibility. Plasmids are also constituents of mitochondria and plastids in eukaryotic organisms. Bacterial plasmids have been extensively used for the construction of → cloning vectors, see → plasmid cloning vector. See also → chimeric plasmid, → helper plasmid, → multicopy plasmid, → natural plasmid, → non-conjugative plasmid, → plasmid promiscuity, → plasmid rescue, → plasmid sequencing, → plasmid stability.

Plasmid chip: Any glass or plastic chip, onto which plasmids are immobilized. Such plasmid chips are used for e.g. → reverse transfection.

Plasmid cloning vector: Any → plasmid designed to allow the → cloning of foreign DNA with recombinant DNA techniques. Plasmid vectors are preferentially small in size, replicate under → relaxed control, contain → selectable marker genes (coding for example for → antibiotic resistance), scorable marker genes (coding for enzymes which can easily be monitored), and unique → restriction sites or → polylinkers at locations not necessary for plasmid function (e.g. not in regions needed for replication). Plasmid vectors for a great number of specific experimental needs and different host systems have been developed, see for example → artificial chromosome, → ARS plasmid, → broad host range vector, → expression vector (→ open reading frame vector, → pEX vector), → intermediate vector, → low copy number plasmid vector, → multifunctional plasmid, → mini-Ti, → Okayama-Berg cloning vector, → pBR 322, → pEMBL, → promoter plasmid, → pUC, → restriction site conversion plasmid, → ribozyme auto-cleavage vector, → shuttle vector, → yeast cloning vector, → cosmid vector; also → helper plasmid.

Plasmid conjugation: See → conjugation.

Plasmid curing: See → curing.

Plasmid DNA (pDNA): The covalently closed circular (ccc) double-stranded DNA molecule that represents a → plasmid.

Plasmid end pair: The sequence reads from both ends of a → plasmid clone. See → BAC end pair.

Plasmid-enhanced PCR-mediated (PEP) mutagenesis: A variant of the → splice overlap extension PCR (SOE-PCR) that allows to introduce mutations (e.g. → deletions or → insertions) into a target DNA. In short, the target DNA is first amplified as two parts using two primer pairs, designed to introduce the mutation and two → restriction sites (which are incorporated into the most distal primers). The internal 5'-phosphorylated primers permit an efficient → blunt-end ligation of the two parts. For a deletion mutation, the targeted sequence is simply omitted, for an insertion it is incorporated into one of the primers. The two parts together with the cloning plasmid are then digested with the two → restriction endonucleases, ligated and used to transform bacterial host cells. The efficiency and orientation of the blunt-end ligation process is controlled by sequence-specific overlapping interactions with the plasmid.

Plasmid incompatibility: The inhibition of replication and thus inheritance of a given → plasmid by the presence of another coresident plasmid in the absence of external selection pressure. Incompatibility is based on several mechanisms. First, in the competition of both plasmids for common membrane binding sites one of them, but not the other may be successful. Usually such binding occurs at the → origin of vegetative replication (oriV) of the plasmid and induces → replication. Second, an inc-gene at the origin of → replication of a resident plasmid may encode an RNA (RNA 2) that functions as → primer for DNA replication. If the complementary RNA (RNA 1) is also synthesized, RNA 1 and 2 will anneal and the primer is masked. In this way, the RNA 1 of the resident plasmid may inhibit the replication of the incoming plasmid. See also → plasmid incompatibility group.

Plasmid incompatibility group: A class of closely related → plasmids that are mutually exclusive (i.e. cannot be stably maintained in the progeny of a particular host cell). Since incompatibility is based on the action of *inc* (incompatibility) genes, incompatibility groups are designated as incA, incB, incC and so on. See → plasmid incompatibility.

Plasmid instability: The relatively short existence of → plasmid in a host cell, before it is eliminated. Elimination is a consequence of the burden on the cellular metabolism to maintain the plasmid. For example, the growth rate of a plasmid-containing cell is significantly reduced relative to that of a plasmid-free cell, because plasmid → replication and → transcription of its genes as well as → protein production from the resulting → messenger RNAs requires energy that is withdrawn from the normal energy metabolism of the host. Therefore, cells losing a plasmid in e.g. a fermentation process are appreciably more fit than plasmid-free cells, so that the former outcompete the latter in the bacterial population. Plasmid instability is a major concern in protein production of genetically engineered cells. Several measures can be taken to avoid plasmid loss, as e.g. the → complementation of an essential mutated chromosomal gene by a wild-type → allele inserted into a plasmid. The mutatnt host is then unable to synthesize an essential amino acid without a plasmid carrying the gene that provides this function. Another strategy is the use of the → separate-*c*omponent stabilization (SCS) system. See → plasmid stability.

Plasmid-*l*ike DNA (plDNA): A circular → plasmid of filamentous fungi (e.g. *Podospora anserina*) that is derived from the first → intron of the cytochrome oxidase subunit I gene. The plDNA is involved in age-related rearrangements of mitochondrial DNA of this fungus.

Plasmid maxiprep: The isolation and purification of large amounts (>100 µg) of → plasmid DNA from comparably large volumes of bacterial cultures (>10 ml). Compare → plasmid miniprep.

Plasmid miniprep: The isolation and purification of minute amounts (<20 µg) of → plasmid DNA from comparably small volumes of bacterial cultures (<1 ml). Compare → plasmid maxiprep.

Plasmid partitioning: See → partition, → partitioning function.

Plasmid promiscuity: The ability of a → plasmid to promote its own transfer to and replication in a wide range of host cells. A promiscuous plasmid is for example → RP4.

Plasmid rescue (homologous assist): The → recombination of a donor → plasmid with a homologous resident plasmid to form a stable → cointegrate preserving the function(s) of the donor DNA that would otherwise be destroyed by → restriction. This rescue is observed predominantly in *Bacillus subtilis* and several other Gram-positive bacteria. Usually the donor plasmid will be linearized upon entry into the recipient cell and degraded by restriction. It may be rescued, however, by a → homologous recombination with a resident plasmid (homologous helper plasmid), a process requiring the *recE* gene product. Plasmid rescue may be used for → shotgun cloning of heterologous DNA in *B. subtilis*. Monomeric vector DNA is ligated to a foreign sequence and the construct is linearized. After its uptake it pairs with homologous sequences of a resident plasmid, and both the non-homologous vector DNA and insert DNA are rescued.

Plasmid sequencing (double-strand sequencing, "supercoil sequencing"): The → sequencing of linearized → plasmid DNA that has been heat-denatured (i.e. made single-stranded). The single strands can either be separated and each annealed to a synthetic, strand-specific → oligonucleotide → primer, or both strands can remain in the same reaction mixture, if only one strand-specific primer is used. The primer-annealed plasmid strands can then be sequenced following the → Sanger sequencing procedure.

Plasmid shuffling: A technique for the identification of conditional lethal mutations in an essential cloned gene on a → plasmid, preferentially developed for yeast. In short, the chromosomal copy of the essential genes is first deleted (or disrupted), but functionally replaced by an intact copy on a → yeast episomal plasmid, YEp (that additionally carries the URA3 gene). Then a second plasmid (e.g. → centromere plasmid) with a temperature-sensitive copy of the gene of interest is introduced, relieving the selection pressure on the YEp at permissive temperatures (normally 23–25 °C), thereby generating URA⁻ derivatives (loss of the YEp). At non-permissive temperatures (usually 36 °C), the YEp is absolutely essential (no URA⁻ segregants), and loss of this plasmid can be monitored by → replica plating single colonies on 5-fluoroorotic acid (5-FOA) plates at permissive and non-permissive temperatures. URA⁻ segregants will then appear as 5-FOA-resistant papillae.

Plasmid stability: The persistence of a → plasmid through many generations of → host cells. Plasmid stability is a function of the number of plasmid copies per cell, and the → partitioning function (par sequence) of the plasmid. The par sequences direct an equal distribution of plasmid molecules into each daughter cell after cell division. See → plasmid instability.

Plasmid vaccine: See → DNA vaccine.

Plasmid vector: See → plasmid cloning vector.

Plasmone: The entire DNA of both → mitochondria and → plastids of a cell. See → chondriome, → plastome.

Plastid-encoded polymerase (PEP): Any → DNA-dependent RNA polymerase that is encoded by the plastid (e.g. chloroplast) → genome. The enzyme requires additional nuclear-encoded sigma-like factors (SLFs) for initiation of transcription. See → nucleus-encoded polymerase.

Plastome: The genetic information of plastids. See → chloroplast DNA, → mitochondrial DNA. Do not confuse with → plasmone.

Plastome mutation: Any mutation in plastid DNA (→ chloroplast DNA). See → mitochondrial DNA.

Plate: Any Petri dish that contains a solid medium (mostly → agar) for the growth of microorganisms. See for example → gradient plate.

Plateau effect: The decrease of the exponential amplification of a target sequence in the later cycles of a conventional → polymerase chain reaction such that the amplification rate reaches a plateau (the specific product does no more accumulate). In the plateau phase, a preferential amplification of unspecific side products ensues.

Plate lysate: A solution of mature → bacteriophage particles which are released from bacterial → host cells that grow on an → agar plate. Usually a special medium (e.g. SM buffer containing NaCl, $MgSO_4$ 7 H_2O, Tris-HCl, pH 7.5, and gelatin) is layered onto the surface of a plate showing confluent lysis, and the phages diffuse into that medium.

Plating: The process of inoculation of a solid growth medium in a Petri dish with microorganisms (in particular bacteria), so that the inoculum is either uniformly distributed or present as stripes.

Plating efficiency (*efficiency of plating*, EOP): The efficiency with which → bacteriophages infect bacteria. A plating efficiency of 1.0 means that each phage particle causes a productive infection (visualized as → plaque). A plating efficiency of 1.0 10^{-4} indicates that only one out of 10^4 phage particles infects the host cell productively.

Plectonemic winding: The winding of two DNA strands around each other to produce the normal DNA → double helix.

Pleiotropic gene: Any gene that influences two (or more) phenotypic characters. See → pleiotropy.

Pleiotropy: The effect of one particular gene on other genes to produce apparently unrelated, multiple phenotypic traits.

Plexisome: Any linear DNA molecule with a known sequence at both ends. Plexisomes can be generated by fragmenting → genomic DNA with a → restriction endonuclease, the → ligation of an → adapter A to one end, subsequent → nick-translation of the fragment such that it stops half way to the 5′end, and by adaptor B ligation to the 5′end. Both ends of this plexisome carry known sequences (adapter A and adapter B).

PLG: See → *p*hase *l*ock *g*el.

P-loop (also motif A, Walker A motif, kinase-1a, G-1): A highly conserved sequence motif (consensus sequence: GVGKTT) in both ATP- and GTP-binding proteins that interacts directly with the phosphate of the nucleotide triphosphate bound to the socalled nucleotide-binding site of these proteins.

PLP mutagenesis: See → PCR-ligation-PCR mutagenesis.

Plus: Located → downstream of the → cap site. See → minus.

Plus-minus hybridization: See → subtractive hybridization.

Plus-minus screening: A nucleic acid → hybridization method for the isolation of tissue- or organ specific or developmentally regulated (generally, inducible) → cDNA sequences. In short, plus minus screening of e.g. hormone-inducible sequences first requires the establishment of a → cDNA library from mRNAs of hormone-treated tissues. Replica filters carrying identical sets of recombinant clones are then prepared. One of these filters is then probed with radiolabeled → messenger RNA (or cDNA) from control cells, and one with radiolabeled mRNA (or cDNA) from hormone-treated cells. Some colonies will give a signal with both probes (i.e. carry cDNA sequences from mRNAs

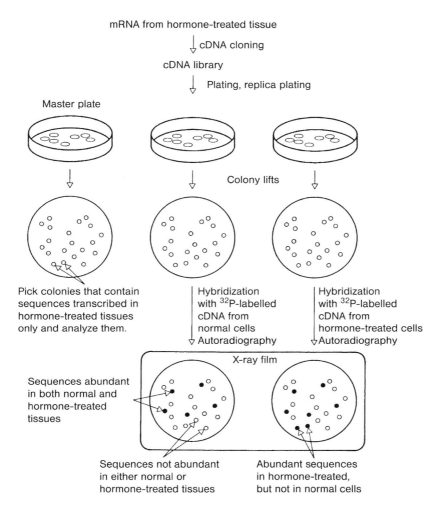

Plus-minus screening

present in both tissues). Some colonies will give a signal with the cDNA from hormone-treated cells only. These represent sequences which are induced by the hormone. The corresponding clones can easily be recovered from the master plate. A similar technique, which allows faster isolation of cell-specific clones is the method of → subtractive hybridization.

Plus strand (plus viral strand, positive strand, + strand):

a) The DNA strand contained in a single-stranded DNA virus (or any strand with identical sequence).

b) The RNA strand of single-stranded RNA viruses with the same polarity as viral mRNA and encoding viral proteins, compare → minus strand, definition b. See for example → Q-beta, → positive-strand RNA virus.

PM: See → perfect match.

PMAGE: See → polony multiple analysis of gene expression.

pMB9: A small multicopy → plasmid carrying a → tetracycline resistance gene and a → Col E1 type → origin of replication that was used for the construction of the universal → cloning vector → pBR 322. See also → yeast episomal plasmid.

PMF: See → peptide mass fingerprint.

pMON: A series of plant transformation vectors that allow the transfer of foreign genes (generally, DNA) into target plants via a → cointegrate with → Ti plasmid vectors. MON stands for MONSANTO company (USA). See also → *Agrobacterium*-mediated gene transfer.

PMP:

a) See → paramagnetic particle.
b) See → plant-made pharmaceutical.

PMSF (phenylmethylsulfonyl fluoride): An effective inhibitor of trypsin- and chymotrypsin-type proteases that is used to block serine protease activity during protein isolation procedures.

$$\text{C}_6\text{H}_5-\text{CH}_2-\overset{\overset{\displaystyle O}{\|}}{\underset{\underset{\displaystyle O}{\|}}{\text{S}}}-\text{F}$$

PNA: See → peptide nucleic acid.

PNA array (PNA microarray, PNA chip): Any → microarray, onto which → peptide nucleic acids (PNAs) instead of conventional nucleic acids with deoxyribose-phosphate backbones are bound via N-terminal groups. PNA arrays do not require any labelling of hybridisation → probes with radioisotopes, stable isotopes or fluorochromes. Another advantage of PNA arrays is the neutral backbone of PNAs and the increased strength of PNA-DNA pairing. The lack of inter-strand charge repulsion improves the hybridisation properties in DNA-PNA duplexes as compared to DNA-DNA duplexes (e.g. the higher binding strength leads to better sequence discrimination in PNA-DNA hybrids than in DNA-DNA duplexes). PNA arrays are used for genome diagnostics, sequencing of DNA or RNA, detection of sequence polymorphisms and identification of expressed genes.

PNA chip: See → PNA array.

PNA-directed PCR clamping: A technique for the detection of rare sequence variants that is based on the competition between a → peptide nucleic acid (PNA) oligomer and a DNA → primer in a → polymerase chain reaction (PCR) leading to the suppression of the amplification. In short, a PNA oligomer, complementary to the sequence of the original → wild-type allele is added to a mixture consisting of the → template DNA, and two primers (one → forward primer complementary to the → mutant allele, and one → reverse primer complementary to both the wild-type and mutant alleles, the socalled common primer). During the ensuing PCR the wild-type PNA oligomer competes with the mutant DNA primer for the target sequence. If this sequence is completely complementary to the wild-type PNA oligomer (i.e. no mutation occurred in it), then the amplification is blocked (the DNA polymerase cannot extend the PNA oligomer). If, however, the mutant primer binds, an amplification of the target allele is possible.

PNA opener: A laboratory slang term for a special type of → *p*eptide *n*ucleic *a*cid (PNA) that invades into double-stranded DNA in a sequence-specific way and forms P-loops. The open (looped) region is accessible for hybridization with DNA or RNA → probes and binding of a → sequencing primer or primer for a → polymerase chain reaction. On the other hand, the PNA opener interferes with the binding of enzymes that recognize double-stranded DNA (as e.g. → restriction endonucleases).

PNA-pPNA chimera: See → peptide nucleic acid-phosphono peptide nucleic acid chimera.

PNA2-DNA2 hybrid quadruplex: An artificially produced four-stranded hybrid molecule of two G-rich → peptide nucleic acid (PNA) strands and two homologous G-rich DNA strands that is extremely stable. Such quadruplexes are used as samples to study the suitability of PNAs as a → probe to target homologous regions in a → genome for e.g. → silencing.

PNK: See → *p*oly*n*ucleotide *k*inase.

PNPase: See → *p*oly*n*ucleotide *p*hosphoryl*ase*.

PNPG (*p*-*n*itro*p*henyl-*g*lucuronide): A synthetic substrate for → β-glucuronidase.

POD-conjugated antibody: See → *p*er*o*xi*d*ase-conjugated antibody.

Point mutation (microlesion, micromutation): A → mutation involving a chemical change in only one single → nucleotide. See → single nucleotide polymorphism, → transition, → transversion.

Pokeweed *a*ntiviral *p*rotein (PAP): A protein synthesized in pokeweed in response to a viral infection that inactivates ribosomes by depurinating a specific adenosine residue in the highly conserved α-sarcin loop of the large → ribosomal RNA. Thereby → translation of → messenger RNAs (mRNAs) is inhibited. PAP additionally binds to the m7GpppG → cap of cellular mRNA and depurinates the mRNA, so that it cannot be translated. Both functions of PAP are probably part of a defense mechanism of the attacked plant.

***pol*:** The genetic notation for the retroviral gene encoding → reverse transcriptase.

Polar effect: A laboratory slang term for the positive influence of an → upstream → alternative exon in a → pre-messenger RNA on the inclusion of a → downstream alternative exon within the same → transcript.

Polarity: The phenomenon that a → nonsense mutation introduced into a gene transcribed early in an → operon has the secondary effect of repressing expression of genes downstream. See → polar mutation.

Polar mutation (dual effect mutation): Any gene → mutation of the → nonsense or → frame shift type in an → operon producing two effects, namely the repression of non-mutated genes located farther downstream and the failure to express the gene subsequent to the mutation site.

Polished ends (polished termini): The double-stranded termini of duplex DNA molecules generated by the → polishing of → sticky ends with DNA polymerase. Such sticky ends have been generated by restriction with specific type II → restriction endonucleases.

Polished termini: See → polished ends.

Polishing: The synthesis of short sequences complementary to single-stranded protrusions ("tails") generated in DNA molecules by either mechanical shearing or digestion with → restriction endonucleases. Polishing is accomplished by → Klenow polymerase or → T4 DNA polymerase, and leads to complete DNA duplexes.

Pollen electro-transformation: A technique for the introduction of foreign → genes into pollen, which is then used to pollinate flowers. The developping seeds contain the new genes, and can easily be grown to complete plants. Pollen transformation circumvents the tedious transformation of single somatic cells (→ protoplast transformation) and their multi-step regeneration to plants.

Polo-like kinase (Plk): Any one of a family of highly conserved serine/threonine phosphokinases (named after the kinase Polo of *Drosophila melanogaster*) that is involved in various steps of the cell cycle. Four Polo-like kinases are present in humans, denoted as Plk1-4. For example, Plk1, a 68 kDa protein contains two highly conserved sequence motifs, the C-terminal polo box and the N-terminal kinase domain. The polo domain consists either of only one (Plk4), or two polo boxes (Plk1–3), each of which is 60–70 amino acids long. The expression of the corresponding genes is dependent on the cell cycle phase. For example, Plk1 expression is low at the onset of the G1 phase, but increases during the cycle, reaching a maximum in G2/M, mostly parallel to cyclin B1. During prometaphase, Plk1 is attached to centrosomes or kinetochores, but during the early and late anaphase the enzyme moves to the socalled mid-body, a region located between the newly forming daughter cells. In the telophase or cytokinesis it localizes to the socalled mid-zone, the last junction between the daughter cells. Plk1 to 4 are overexpressed in malign tissues, and this overexpression is a negative prognostic marker.

Polony:

a) See → polymerase colony technology.
b) Any tiny colony of DNA molecules about 1 µ in diameter that is grown on a microscopic glass slide. In short, a solution of DNA fragments is poured onto the glass slide, and a → DNA polymerase added to repeatedly copy each fragment, thereby creating millions of polonies (*pol*ymerase col*onies*). Each single dot on the slide then contains only copies of one original DNA fragment.

Polony multiple analysis of gene expression (polony multiplex analysis of gene expression, PMAGE): A technique for the genome-wide quantitative detection and profiling of → messenger RNAs (mRNAs), including → low-abundance mRNAs encoding signaling molecules and → transcription factors participating in disease, from any cell or tissue that incorporates a → ligation-based step in the sequencing of 14 nucleotide long → tags from individual mRNAs. In short, total RNA is converted to double-stranded → complementary DNAs (cDNAs), which are bound to →

oligo(dT) ferromagnetic beads, cleaved with a socalled anchor endonuclease, here *Nla* III, leaving a 4–base pair (bp) 3′ overhang, which is ligated to a forward → adapter containing an *Acu* I type IIs → endonuclease recognition site. Digestion with the tagging enzyme *Acu* I cleaves 16 bases from the nonpalindromic recognition sequence and releases the adapter with its 10-bp cDNA sequence tag (plus a 4-base CATG anchor sequence). The tag is subsequently ligated to a reverse adapter containing a 2-bp 3′ degenerate overhang, and amplified onto 1 µm beads carrying → primers with sequences that match the forward adapters. Each bead is then called a *pol*ymerase col*ony* or polony (also polymerase chain reaction of colony), a small colony of DNA. Individual cDNA template tags are then clonally amplified in a water-in-oil emulsion – each polony bead in an emulsion oil droplet among millions of other compartmentalized droplets formed within the emulsion serve as tiny → po*l*ymerase *c*hain *r*eaction (PCR) chambers (see → *e*mulsion polymerase chain reaction, emPCR). Amplification generates millions of tag copies per polony. Polony beads carry DNA templates with terminal 3′ OH groups that are capped by → ligation, using annealed bridging primers, to → oligonucleotides containing 3′ amines. These amines are cross-linked to aminosilylated glass cover slips with amino-ester bridges, providing a gel-less monolayer milieu for sequence chemistry. About 5 million polonies are arrayed in a flow cell for parallel polony sequencing-*by*-ligation (SBL), exploiting the high discriminatory power of → DNA ligase to iteratively label each bead with a → fluorophore encoding the identity of a base within the template. A computer program then matches the tags to known genes. The more tags associated with a gene, the higher the expression of that gene. PMAGE integrates the advantages of → SAGE over hybridisation-based techniques (such as → DNA microarrays), and does not require antecedent library amplification, → concatenation, or → subcloning. See also → See → cap analysis of gene expression (CAGE), → 5′-SAGE, → LongSAGE, → SuperSAGE, → 3′-SAGE.

Poly(A): A homopolymer consisting of adenine nucleotide residues. See → poly(A) tail.

Poly(A) addition signal (poly[A] signal; poly[A] site, PAS; poly[A] addition site; poly[A] signal sequence; polyadenylation site):

a) A hexanucleotide → consensus sequence (animals: 5′-AATAAA-3′, 5′-ATTAAA-3′, 5′-AATTAA-3′, 5′-AATAAT-3′, 5′-CATAAA-3′ or 5′-AGTAAA-3′; plants: 5′-AATAAN-3′, generally 5′-AATAA-3′ sequence) close (within the last 50 bp) to the 3′-end of most eukaryotic genes transcribed by → RNA polymerase II.

b) The consensus sequence 5′-AAUAAA-3′ in an → mRNA molecule that directs the cleavage of the message 10–30 bases 3′ of the element. The cleaved mRNA then serves as a substrate for processive poly- adenylation. First, the socalled *p*olyadenylation *s*pecificity *f*actor (CPSF), a tetrameric protein with subunits of 33, 73, 100, and 160 kDa binds to the 5′-AAUAAA-3′ signal, then the trimeric *c*leavage-*st*imulating *f*actor (CstF; 50, 64, and 77 kDA) binds to a GU-rich sequence element further downstream of the RNA. The → poly(A)polymerase (PAP) binds in between the two elements. This complex is joined by two (or more)

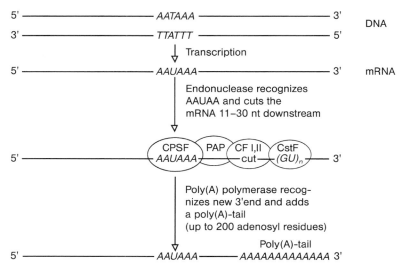

Poly(A) addition signal

other proteins, of which the cleavage factors CF I and CF II are positioned upstream of the GU-rich box and terminate the mRNA.

Poly(A)-assisted transcript degradation (polyadenylation-assisted degradation): A nuclear post-transcriptional quality control mechanism for → transcripts generated by → DNA-dependent RNA polymerases I, II, and III in yeast that is based on the targeting of the transcripts by a nuclear → poly(A)polymerase for exosomal degradation. This mechanism limits the accumulation of transcripts resulting from inappropriate → transcription of intergenic regions of the genome. The faulty intergenic transcript is first polyadenylated by non-conventional proteins as e.g. Trf4p, associated with Air1p and Air2p and the → RNA helicase Mtr4p that activates the nuclear → exosome. In the absence of this polyadenylation complex (e.g. in → mutants), the intergenic transcripts are nearly completely stable.

Polyacrylamide affinity coelectrophoresis (PACE): A variant of the conventional → affinity coelectrophoresis, in which agarose is replaced by polyacrylamide (PAA) as electrophoretic matrix, allowing to assay the binding interaction(s) of RNA-peptide or RNA-protein complexes. In short, the gel tray is initially turned by 90 °C relative to its orientation during electrophoresis, and a series of PAA plugs with increasing peptide (protein) concentrations sequentially poured such that a discrete step gradient of peptide concentration from the left side of the final gel to the right side is created. Then the gel is turned to its electrophoresis orientation, a PAA mixture without peptide poured onto its top, a → comb inserted, and the radioactively labeled RNA (generally: ligand) applied. If any interaction(s) between the peptide and the RNA occurs, then the electrophoretic mobility of the complex is retarded, which can be detected by → autoradiography. PACE allows to follow a saturation kinetics, because increasing peptide concentrations will cause increased mobility shifts, unless the RNA is completely bound (i.e. no change in shift occurs anymore). See → affinity electrophoresis. Compare → mobility-shift DNA-binding assay.

Polyacrylamide gel (PAA gel): An insoluble three-dimensional matrix of acrylamide monomers cross-linked with N,N'-methylene-bisacrylamide in the presence of polymerization catalysts (e.g. TEMED and *a*mmonium *pe*rsulfate, APS). The crosslinking process starts with the interaction of TEMED and APS, producing a TEMED radical cation, a sulfate radical, and a sulfate anion. The sulfate radical transfers its unpaired electron onto an acrylamide monomer, thereby converting the acrylamide to a free radical, which then reacts with other acrylamide monomers to polymers. These polymeric chains are then statistically crosslinked by N,N'-methylene-bisacrylamide, forming the effective gel matrix that is used for → polyacrylamide gel electrophoresis. The gel pore sizes can be varied by varying the relative proportions of the ingredients and thus can be adapted to the electrophoretic separation of molecules such as proteins and nucleic acid fragments of various size classes. See also → stacking gel.

Polyacrylamide gel electrophoresis (PAGE): A technique to separate macromolecules (e.g. nucleic acids, proteins) on the basis of their size and structure by electrophoresing them through an inert matrix consisting of cross-linked acrylamide. The separation of proteins is usually carried out in the presence of → sodium dodecyl sulfate (see → SDS polyacrylamide gel electrophoresis).

$$CH_2=CH-\underset{O}{\overset{\|}{C}}-NH_2$$

Acrylamide

$$CH_2=CH-\underset{O}{\overset{\|}{C}}-\overset{H}{\underset{|}{N}}-CH_2-\overset{H}{\underset{|}{N}}-\underset{O}{\overset{\|}{C}}-CH=CH_2$$

N,N'- methylenebisacrylamide

```
    |
-CH₂-CH-[CH₂-CH-]ₙCH₂-CH-[CH₂-CH-]ₙCH₂-
        |            |            |
        CO           CO           CO
        |            |            |
        NH₂          NH           NH₂
                     |
                     CH₂
                     |
                     NH
                     |
                     CO
                     |
-CH₂-CH-[CH₂-CH-]ₙCH₂-CH-[CH₂-CH-]ₙCH₂-
        |            |            |
        CO           CO           CO
        |            |            |
        NH           NH₂          NH₂
        |
        CH₂
        |
        NH            NH₂
        |             |
        CO            CO
        |             |
-CH₂-CH-[CH₂-CH-]ₙCH₂-
```

Cross-linked polyacrylamide

Polyacrylamide gel

Complete denaturation of proteins before PAGE is achieved by heating, especially in the presence of SDS. Nucleic acids, or the products of DNA-sequencing reactions can also be denatured by heating but furthermore by formamide, urea or methyl mercuric hydroxide before being electrophoresed in strongly → denaturing gels to separate fragments differing by only one single nucleotide. Optimal polyacrylamide concentrations for the separation of e.g. differently sized linear double-stranded (ds) DNA molecules are different. See also → horizontal polyacrylamide gel electrophoresis, → Laemmli gel.

Non-denaturing acrylamide concentrations for resolution of linear dsDNA

Acrylamide (%)	Size of linear dsDNA fragments separated (bp)	Dye migration (bp)	
		Xylene cyanol	Bromophenol blue
3.5	100–1000	460	100
5.0	80–500	260	65
8.0	60–400	160	45
12.0	40–200	70	20
15.0	25–150	60	15
20.0	6–100	45	12

Dye migration in denaturing polyacrylamide gels

Polyacrylamide/urea (%)	Dye migration (bp)	
	Xylene cyanol	Bromophenol blue
5.0	140	35
6.0	106	26
8.0	75	19
10.0	55	12
20.0	28	8

Polyacrylamide gel percentages for resolution of proteins

Gel percentage (%)	Protein size range (kDa)
8	40–200
10	21–100
12	10–40

Polyacrylamide-oligonucleotide conjugate: A copolymer of → polyacrylamide and an → oligonucleotide, a → polynucleotide (e.g. DNA) or a → DNA mimic (as e.g. → peptide nucleic acid [acrylamide PNA], → phosphono peptide nucleic acid [acrylamide-pPNA], → trans-4-hydroxy-L-proline peptide nucleic acid [acrylamide-HypNA], or chimeras of alternating PNA and pPNA residues or pPNA and HypNA monomers). Such conjugates are produced by (1) derivatizing oligonucleotides or DNA mimics with acrylamide groups at their 5′- or 3′-termini, (2) co-polymerizing these acrylamide-modified oligonucleotides (or mimics) with polyacrylamide, and (3) the covalent attachment of the conjugates onto the surface of non-derivatized microscope slides treated with either monoethoxydimethylsilylbutanal, 3-mercaptopropyl-trimethoxysilane or 3-aminopropyltrimethoxysilane, and activated with phenylisothiocyanate. The PAA support owns high chemical and thermal stability and low non-specific adsorption of biological macromolecules. The polyacrylamide-oligonucleotide conjugates are used for the capture of complementary probe molecules.

Polyadenylated RNA: An RNA molecule that contains a homopolymeric tail of adenyl residues at its 3′-terminus (e.g. poly[A]$^+$-mRNA). See → polyadenylation.

Polyadenylation: The post-transcriptional addition of → poly(A) tails of up to 200 adenine residues to the 3′-termini of → heterogeneous nuclear RNA and → messenger RNA in eukaryotes. Compare → in vitro polyadenylation, → poly(A) addition signal, see → post-transcriptional modification.

Polyadenylation signal: See → poly(A) addition signal.

Polyadenylation site: See → poly(A) addition signal.

Poly(ADP-ribose) polymerase (PARP): A chromatin-associated 113 kDa nuclear → zinc finger protein that binds to breaks in → single-stranded and → double-stranded DNA and catalyzes the formation of linear or branched → homopolymers of poly(ADP-ribose) bound to a series of different nuclear proteins. The enzyme has at least three different → domains (the zinc finger → DNA-binding domain, DBD, an automodification, and a catalytic domain). The 42 kDa N-terminal DBD extends from the initiator methionine to threonine^{373} (human PARP), contains two zinc fingers and two → helix-turn-helix motifs mediating interaction between the enzyme and DNA. Zinc finger 1 (F1) is located between cysteine^{21} and cys^{56}, zinc finger 2 (F2) between cys^{125} and cys^{162}. The automodification domain occupies the central region, resides between ala^{374} and leu^{525} in human PARP, and harbors a BRCA1 C-terminus (BRCT) domain between ala^{384} and ser^{479}, regulating cell cycle checkpoints and DNA repair. The 55 kDa catalytic domain in the C-terminal region of the enzyme spans from thr^{526} to trp^{1014} (human PARP). The ADPr transferase activity is confined to a 40 kDa region at the extreme → C-terminus ("minimal catalytic domain") that catalyzes → initiation, → elongation and branching of ADPr polymers. PARP, upon binding to DNA breaks introduced by e.g. oxidative stress as O2$^-$, NO$^-$, or OH$^-$, or ionizing radiation, is activated and cleaves its nuclear substrate nicotinamide dinucleotide (NAD$^+$) into nicotinamide and ADP-ribose. The ADP-ribose then polymerizes onto nuclear acceptor proteins (e.g. → histones, → transcription factors, and PARP itself). PARP is activated proportionally to the number of strand breaks. The polymerized poly(ADP)-ribose can then be hydrolyzed by poly(ADP-ribose) glycohydrolase, and a new cycle of automodification in response to DNA damage can take place. Three classes of PARPs exist: Class I PARP is represented by the classical 113 kDa enzyme (responsible for about 90% of total PARP activity), class II enzymes are smaller and involved in DNA repair during → apoptosis, and class III PARPs are large proteins containing socalled ankyrin repeats, the only representative being tankyrase involved in → telomere function. PARP seems to play a pleiotropic role in various cellular processes, but its major function is most likely → DNA repair and maintenance of genome integrity. At the initial processes leading to programmed cell death (apoptosis), a transient activation of PARP with concomitant poly(ADP)-ribosylation of nuclear proteins (e.g. Ca^{2+}/Mg^{2+}-dependent nuclease) is followed by its specific proteolysis by → caspase (caspase 3 and 7). The degradation of PARP leads to a contact of free poly(ADP)-ribose with mitochondria inducing the release of apoptosis-inducing factor (AIF) that is translocated into the nucleus and fragments the nuclear → chromatin (leading to → nucleosomal ladders). Moreover, its action finally leads to the disruption of the nucleus and and the formation of apoptotic bodies. PARP-1 also interacts with diverse → transcription factors and poly(ADP) ribosylates them, mostly triggering transcriptional activity of target genes. See → poly(ADP)-ribosylation, → poly(ADP-ribose) synthetase.

Poly(ADP-ribose) synthetase (PARS): An enzyme catalyzing the covalent bonding of ADP-ribose to nuclear proteins (e.g. → histones). See → poly(ADP-ribose) polymerase.

Poly(ADP)-ribosylation (pADPr): The covalent post-translational modification of proteins (preferably nuclear proteins) catalyzed by → poly(ADP-ribose) polymerase (PARP) that transfers ADP-ribose (ADPr) units from NAD^+ onto acceptor proteins producing linear and/or branched ADPr polymers.

Polyamide dimer: An eight-ring hairpin polyamide molecule consisting of three types of aromatic rings (pyrroles, imidazols, and hydroxypyrroles) that folds back and forms pairs of the three rings. Now the various combinations of each two compounds can discriminate specific → Watson-Crick base pairs in the → minor groove of the DNA helix. For example, through direct hydrogen bonding between pairs of pyrroles on the polyamide and the double hydrogen bond acceptor potential of the O2 of thymine and the N3 of adenine, A-T can be distinguished from T-A base pairs. Or an imidazole (Im)/pyrrole (Py) pair discriminates G-C from C-G, and both of these from A-T and T-A base pairs. Likewise, a hydroxypyrrole(Hp)/Py pair discriminates T-A from A-T, and both of these from G-C and C-G. The sequence-specific recognition of DNA sequences by polyamide dimers is therefore determined by the side-by-side pairing of the residues in these polyamide dimers.

Poly(A)⁻-RNA: Any → RNA that does not carry a → poly(A) tail at its 3′-end (e.g. → ribosomal RNA, → transfer RNA, but also → histone messenger RNA). See → polyadenylated RNA.

Poly(A)-PCR: See → poly(A) polymerase chain reaction.

Poly(A)⁺-mRNA: See → polyadenylated RNA.

Poly(A) polymerase (PAP; Bollum enzyme): A primer-dependent enzyme that catalyzes the polymerization of AMP from ATP onto free 3' hydroxyl groups of mRNA. The enzyme is used for the addition of → poly(A) tails to RNA and for 3'-endlabeling of RNA.

Poly(A) polymerase chain reaction (poly[A]-PCR, sequence-independent primer reverse transcriptase PCR, SIP-RT-PCR): A variant of the conventional → reverse transcriptase polymerase chain reaction technique for the global amplification of → messenger RNA (see → global mRNA amplification) from samples containing only little RNA or → low-abundance messenger RNAs at a lower detection limit (e.g. single cells, biopsies, needle aspirates) that capitalizes on limiting the size of the → first-strand cDNA to 300–700 bases, and therefore avoids a bias against long transcripts during the amplification process. In short, whole cells (or isolated total RNA) are first lysed by heating (which also denatures secondary structures in RNA), cooled, and reverse transcribed into first-strand cDNA by → reverse transcriptase using an oligo(dT) primer. After heat inactivation of the enzyme, the first-strand cDNA is poly(A)-tailed using a → terminal transferase to generate a 5'-oligo(dT) and 3'-poly(A)tailed cDNA. Aliquots of the reaction are then directly suspended into a buffer containing sequence-independent 5'-(T)$_{24}$-X-3' (X: A,C, or G, but not T) primer, and the poly(A)-tailed cDNA is amplified by conventional → polymerase chain reaction. For subsequent gene expression profiling, → Northern blot analysis, → quantitative PCR, or cDNA expression arrays (see → microarray) can be employed. SIP-RT-PCR preserves the relative abundances of specific transcripts present in the original mRNA population (i.e. no distortion occurs).

Poly(A) sequence: See → poly(A) tail.

Poly(A) signal (poly[A] signal sequence): See → poly(A) addition signal.

Poly(A) signal selection: The scanning of an → mRNA molecule for alternative → poly(A) addition signals and the cleavage of one preferred site by an endonuclease recognizing the target sequence 5'-AAUAAA-3'. If several such signals are located on one → primary transcript molecule, then poly(A) signal selection may (1) result in multiple messenger RNAs derived from one transcription unit (if all sites are cut), or (2) the generation of one specific messenger RNA that may be different from cell to cell, or tissue to tissue (if signal selection is cell- or tissue-specific).

Poly(A) site: See → poly(A) addition signal.

Poly(A)-specific 3'-exoribonuclease (poly(A) removing nuclease, PARN): An enzyme catalyzing the 3'-exonucleolytic removal of adenosyl residues from the → poly(A) tail of eukaryotic → polyadenylated messenger RNAs as a first step towards their complete degradation. The enzyme requires Mg^{2+} and a free 3'-OH group, releases 5'-adenosyl monophosphate (5'-AMP) and thereby initiates the decay of many mRNAs. PARN simultaneously binds to the 3' poly(A) tail and the → cap at the 5'end of → messenger RNA. Absence of the cap reduces PARN activity. This exonuclease belongs to the class of → deadenylating nucleases. See → deadenylation, → decapping enzyme, → destabilizing downstream element.

Poly (A) tail (poly[A] sequence): A sequence of 60–200 adenine nucleotides at the 3′ end of most eukaryotic mRNAs, added to the molecule after its transcription by a template-independent → poly(A) polymerase and functioning in the stability of mRNA.

Polybrene transformation: A method for → direct gene transfer into animal cells (e.g. CHO cells), using the polycation polybrene. See → DEAE dextran precipitation.

Poly(C): A homopolymer consisting of cytidylic acid residues.

Polycistronic: See → polycistronic mRNA.

Polycistronic message: See → polycistronic mRNA.

Polycistronic mRNA (polycistronic message): A transcript of an → operon which is transcribed as a single message and codes for all the individual enzymes specified by the operon (e.g. *lac* operon of *E. coli*). Polycistronic messages also occur in eukaryotic organisms, see for example → histone genes. They are occasionally transcribed into → polyproteins.

Polyclonal: The property of molecules or cells to originate from more than one → clone, i.e. from more than one single cell. See for example → polyclonal antibody.

Polyclonal antibody: Any → antibody (immunoglobulin) synthesized by different → clones of B lymphocytes. After challenge with an → antigen, an organism usually produces a heterogeneous mixture of specific antibodies with different binding specificities and affinities. Thus every naturally induced antiserum is polyclonal. Compare → monoclonal antibody.

Polyclonal aptamer: Any one of a pool of synthetic → aptamers generated by e.g. → systematic *e*volution of *l*igands by *ex*ponential enrichment (SELEX) that all specifically react with a particular target molecule. These polyclonal aptamers can be enzymatically amplified or modified by primers at the 3′- or 5′-end, and are used for applications like → polyclonal antibodies. The individual structure of an aptamer of the pool can be determined by → cloning and → sequencing with standard molecular techniques. See → monoclonal aptamer.

Polyclonal bead: Any bead in → emulsion polymerase chain reaction that harbors two (or more) target DNAs. Since these (usually) different target DNAs are simultaneously amplified, the bead contains different clones of DNA, which obsoletes the advantage of emPCR to amplify only one single DNA clone. Such polyclonal beads are therefore un-desirable in e.g. → fiber-optic reactor sequencing.

Polycloning site: See → polylinker.

***Poly*comb group protein (PcG, PcG protein):** Any one of a large family of conserved structurally and functionally diverse, → chromatin-associated proteins conserved from plants to humans that were originally identified in *Drosophila melanogaster* as necessary to maintain cell-fate decisions throughout embryogenesis (by repressing *Hox* genes in a body segment-specific arrangement). PcG proteins bind to the socalled *P*cG *r*esponse *e*lements (PREs) in → promoters of hundreds of transcriptionally silent genes (encoding receptors, signalling proteins and → transcription factors), and are involved in many cellular processes such as body patterning, → X chromosome inactivation in female animals, vernalization in plants, epigenetic cellular memory, pluripotency and stem cell self-renewal, to name few. They form two types of large multimeric complexes, *Polyc*omb *r*epressive *c*omplex 1 (PRC1) and PRC2 that post-translationally modify → histone tails and thereby local chromatin substructure, and cooperate in transcriptional → repression of target genes. PRC1 is composed of proteins dRING, *polyc*omb (Pc), *polyh*omeotic (Ph), and *p*osterior *s*ex *c*omb (Psc). Pc carries an N-terminal *c*hromo*d*omain (CD) and a C-terminal → Pc box, and binds strongly to the trimethylated lysine 27 of histone H3 (H3K27me3). The RING-type zinc-finger protein dRING is an E3 → ubiquitin ligase that mono-ubiquitinylates histone H2A at lysine119 (H2AK119ub), which – in cooperation with H3K27me3 – is involved in PcG-mediated gene → repression. The core of PRC2 consists of five proteins, enhancer of zeste E(z), *e*xtra *s*ex *c*ombs (Esc), → *nu*cleosome *r*emodelling *f*actor 55kDa subunit (Nurf55), the histone-binding protein NURF-55, and → *su*ppressor of zeste *12* (Suz12). E(z) is a histone methyltransferase that catalyzes the trimethylation of histone H3 lysine 27 (H3K27me3) via its SET domain. The various genes encoding PcG proteins underwent multiple → duplication events in evolution. For example, vertebrates own between three and five Pc → homologs (coined *c*hro*mobox*, Cbx) that altogether have highly conserved chromodomains and Pc boxes. Paralogs, however, greatly differ in length and the presence (or absence) of motifs or domains, and are also differentially expressed, suggesting different functions. PRC1 genes were lost in *Caenorhabditis elegans*. In embryonic cells, PcG proteins reside in about 50–100 nuclear foci called PcG bodies.

Polycore probe: Any nucleic acid → probe that consists of a tandem arrangement of several to many so-called → core sequences and serves as a → hybridization probe to detect polymorphic loci in eukaryotic genomes.

Poly*d*imensional single *n*ucleotide *p*oly*morphism microarray (polydimensional SNP microarray): Any → microarray that allows the high-throughput identification of → single *n*ucleotide *p*olymorphisms (SNPs) in → genomes. In short, SNPs are first identified by multiple → alignments of e.g. → expressed sequence tags (ESTs), verified for the genome (or the species) under investigation, and → oligonucleotides designed with their 3'-ends one base adjacent to the identified SNP. These oligonucleotides contain 5'-amino modifications and are covalently and irreversibly immobilized onto epoxy-silanized glass slides ("solid phase detection primers") by a → split pin microarrayer. Then the target SNP loci are amplified in a conventional → polymerase chain reaction (PCR), the amplicons single-stranded, and on-chip hybridized to the detection primers. Finally a → primer extension with fluorophor-labeled → dideoxynucleotides (ddNTPs) is performed and the SNPs identified by the incorporation of one of the four ddNTPs. A polydimensional SNP microarray can either probe 20,000 individuals at a single SNP locus, or 20,000 SNP loci in a single organism. This type of microarray therefore is used for cultivar identification and protection in plants, high-throughput → genetic mapping, mutation analyses and the detection of biodiversity.

***Poly*ester *p*lug *s*pin *i*nsert (PEPSI):** A simple device for the separation of → agarose and DNA after → agarose gel electrophoresis. The fragment of interest is cut out of the gel, placed on top of a polyester fiber plug in a plastic pipette tip inserted in a microcentrifuge tube, and spun in a small volume of elution buffer. The DNA is recovered in the microcentrifuge tube, and the agarose trapped in the PEPSI. Do not confuse with Pepsi Cola.

Polyethylene glycol (PEG; carbowax): A polymeric hydrophilic chemical compound with the general formula $HOCH_2(CH_2OCH_2)_nCH_2OH$, used to destabilize cellular membranes, for example during → bacteriophage purification, to induce liposome-cell and cell-cell-fusions (see → lipofection and → cell fusion, → protoplast fusion), and to increase the → hybridization efficiency.

Polyethyleneglycol-based brush surface (PEG-based brush surface): A special modification of chip surfaces that prevents the unspecific binding of non-ligand material (i.e. contaminants) in solution to surface-immobilized → probes (e.g. oligonucleotides, DNA, antibodies, proteins). The charged surfaces of chips are simply coated with densely packed polyethyleneglycol chains, which adsorb spontaneously and form nanometer-thin layers ("brushes"). PEG can also be functionalised by the covalent attachment of e.g. → biotin.

Poly(G): A homopolymer consisting of guanidylic acid residues.

Polygene: A misleading term for any gene that is responsible for only a small effect on a → polygenic trait. The full expression of the trait is then controlled by two (or many) such polygenes.

Polygenic trait (multigenic trait, genetically complex trait): A (usually phenotypic) feature of an organism that is controlled by more than one, in extreme cases up to ten different genes. See → polygene. Compare → multifactorial trait.

Polyhistidine tag: See → histidine tag.

Polyketide: Any one of a series of structurally diverse and complex organic polymers, composed of acyl-*co*enzyme *A* (CoA) monomers that are constituents of actinomycetes, bacilli, and filamentous fungi. Their functions for the organism, in which they are synthesized, is not in all cases very clear. However, some compounds own pharmaceutical potential. For example, → rifamycin (inhibitor of RNA polymerase), FK 5506 (immunosuppressant, binds calcineurin) and lovastatin (inhibits hydroxymethylglutaryl-CoA reductase) are such compounds. Synthesis of polyketides circumvents ribosomes. Socalled modular *polyk*etide *s*ynthases (PKSs), exceptionally large multi-functional proteins ("megasynthases"), organized as coordinate groups of active sites ("modules"), where each module is responsible for catalysis of one cycle of polyketide chain elongation, replace ribosomal functions. Within each module resides a 75–90 amino acid *a*cyl *c*arrier *p*rotein (ACP) domain, to which the growing polyketide chain is covalently tethered. The size and complexity of the synthesized polyketide are controlled by the number of repeated acyl chain extension steps. Four synthetic phases can be discriminated: priming, chain initiation and elongation, and termination, reminiscent of the → non-ribosomal peptide synthesis.

Polylinker (*m*ultiple *c*loning *s*ite, MCS; polycloning site): Any synthetic → oligonucleotide containing multiple → restriction sites (multiple cloning sites). In a symmetrical arrangement two identical polylinkers flank a DNA fragment (e.g. in

→ M13 mp 7), in a non-symmetrical arrangement the polylinkers at both ends of a duplex DNA are not identical (e.g. in M13 mp 8, mp 10, mp 11, mp 18 and mp 19). Polylinkers allow the cloning of foreign DNA into any of their restriction sites.

Polymer: Any molecule that is composed of many identical subunits linked to each other by the process of polymerisation.

Polymerase: An enzyme that catalyzes the assembly of nucleotides into RNA (→ RNA polymerase), or of deoxynucleotides into DNA (→ DNA polymerase).

Polymerase chain reaction (PCR): An → *in vitro* amplification procedure by which DNA fragments of up to 15 kilobases in length can be amplified about 10^8-fold. In brief, two 10–30 nucleotides long → oligonucleotides complementary to nucleotide sequences at the two ends of the target DNA and designed to hybridize to opposite strands, are synthesized. Excessive amounts of these two oligonucleotide → primers (→ amplimers) are mixed with → genomic DNA, and the mixture is heated for → denaturation of the duplexes. During subsequent decrease of temperature the primers will anneal to their genomic homologs and can be extended by → DNA polymerase. This sequence of denaturation, annealing of primers and extension is repeated 20–40 times. During the second cycle, the target DNA fragment bracketed by the two primers is among the reaction products, and serves as template for subsequent reactions. Thus repeated cycles of heat denaturation, annealing, and elongation result in an exponential increase in copy number of the target DNA. The use of thermostable DNA polymerases (e.g. → *Thermus aquaticus* DNA polymerase; → *Pfu* DNA polymerase; → Vent™ DNA polymerase) obviates the necessity of adding new polymerase for each cycle. About 25 amplification cycles increase the amount of the target sequence selectively and exponentially by approximately 10^6-fold. In later phases of the amplification cycle undesirable, incompletely elongated products may accumulate (→ shuffle clones).

Since its introduction the polymerase chain reaction has multiply been modified and its potential been expanded for a great number of sophisticated applications.

See → ACP-PCR, → adaptor PCR, → adapter-tagged competitive PCR, → allele-specific PCR, → *Alu* PCR, → anchored microsatellite-primed PCR, → anchored PCR, → *a*nnealing *c*ontrol *p*rimer *p*olymerase *c*hain *r*eaction, → antiprimer-based quantitative real-time polymerase chain reaction, → arbitrarily primed PCR, → assembly PCR, → asymmetric PCR, → balanced PCR, → booster PCR, → bubble PCR, → capture PCR, → cascade PCR, → cDNA PCR, → colony PCR, → colony-direct PCR, → comparative reverse transcription PCR, → competitive oligonucleotide priming PCR, → competitive PCR, → competitve quantitative PCR, → competitive reverse transcriptase PCR, → concatemer PCR, → continuous flow PCR, → convection PCR, → degenerate oligonucleotide primed PCR, → degradative polymerase chain reaction, → differential cDNA PCR, → differential display reverse transcription PCR, → differential PCR, → digital PCR, → direct PCR amplification, → duplex PCR, → electronic PCR, → emPCR, → emulsion PCR, → ePCR, → error-prone PCR, → epPCR, → exclusive PCR, → expression PCR, → extender PCR, → fast cycling PCR, → 5′ nuclease PCR, → fluorophore-enhanced repetitive sequence-based PCR, → gd-PCR, → gene dosage polymerase chain reaction, → genomic amplification with transcript sequencing, → gradient PCR, → hemicompetitive PCR, → high-efficiency thermal asymmetric interlaced PCR, → hot PCR, → immuno PCR, → immuno-bead RT-PCR, → in-cell reverse transcriptase PCR, → *in situ* PCR, → *in situ* reverse transcription PCR, → inter-interspersed repetitive element PCR, → inter-retrotransposon amplified polymorphism, → inter-simple sequence repeat amplification, → interspersed repetitive sequence PCR, → interspersed repetitive sequences long range PCR, → inverse PCR, → IP-PCR, → island rescue PCR, → isolated probe PCR, → kinetic PCR, → LAM-PCR, → LATE-PCR, → ligation-anchored PCR, → ligation-independent cloning of PCR products, → limiting dilution PCR, → linear-after-the-exponential polymerase chain reaction, → linear amplification-mediated PCR, → linker-adaptor PCR, → long and accurate PCR, → long distance PCR, → long-extension PCR, → long fragment PCR, → long range PCR, → long reverse transcriptase PCR, → low copy PCR, → low stringency single specific primer PCR, → MB-PCR, → message amplification phenotyping, → methylation-independent PCR, → methylation specific PCR, → methyl-*b*inding *p*olymerase *c*hain *r*eaction, → micro PCR, → microdissection PCR, → microplate-based PCR, → microsatellite-primed PCR, → microwell PCR, → minisatellite-primed amplification of polymorphic sequences, → mixed oligonucleotide-primed amplification of cDNA, → mixed target PCR, → module-shuffling primer PCR, → motif-primed PCR, → multicolor PCR, → multiple overlapping primer PCR, → multiplex PCR, → multiple fluorescence-based PCR-SSCP, → mutagenically separated PCR, → nano-polymerase chain reaction, → nested on chip PCR, → nested primer PCR, → nested RAP-PCR, → non-selective PCR, → nucleic acid-programmable protein array, → on-chip PCR, → one-sided PCR, → patched circle PCR, → PCR mutagenesis, → PCR RFLP, → PCR single-strand conformation polymorphism analysis, → poly(A)-PCR, → priming authorizing random mismatches PCR, → protein PCR, → quantitative PCR, → quantitative reverse transcriptase PCR, → random PCR, → rapid cycle DNA amplification, → Rayleigh-Bénard PCR, → RB-PCR, → real competitive PCR, → real-time PCR, → recombinant PCR, → relative quantitative reverse transcriptase PCR, → relative reverse transcriptase polymerase chain reaction, → repetitive sequence-based PCR, → restriction endonuclease-mediated selective PCR, → restriction fragment differential display PCR, → restriction site PCR, → reverse transcriptase multiplexed PCR, → reverse transcriptase PCR *in situ* hybridization, → reverse transcription PCR, → RNA amplification with transcript sequencing, → RNA arbitrarily primed PCR, → RNA PCR, → rotary microfluidic PCR, → sequence-independent primer reverse transcriptase PCR, → sexual PCR, → simplex PCR, → single cell PCR, → single molecule PCR, → single-sided PCR, → single specific primer PCR, → single-tube PCR, → solid-phase PCR, → solution PCR, → splice overlap extension PCR, → splinkBlunt PCR, → splinkTA polymerase chain reaction, → standardized reverse transcriptase PCR, → staRT-PCR, → step-out PCR, → suicide PCR, → suppression PCR, → TAIL-PCR, → tagged PCR, → tagged random primer PCR, → *Taq*Man PCR, → targeted display PCR, → targeted gene walking PCR, → TD-PCR, → telomerase PCR ELISA, → terminal deoxynucleotidyl transferase-dependent PCR, → thermally asymmetric interlaced PCR, → thermally asymmetric PCR, → touchdown

PCR, → T-PCR, → transcription chain-reaction, → transcription-mediated amplification, → triplex PCR, → two-stage PCR; → two-step gradient PCR, → two-step PCR, → uncoupled RT-PCR, → universally primed PCR, → unpredictably primed PCR, → vectorette PCR, → vector-insert PCR, → virtual PCR, → VPCR, → whole genome PCR. Compare → *h*ybridization *c*hain *r*eaction (HCR).

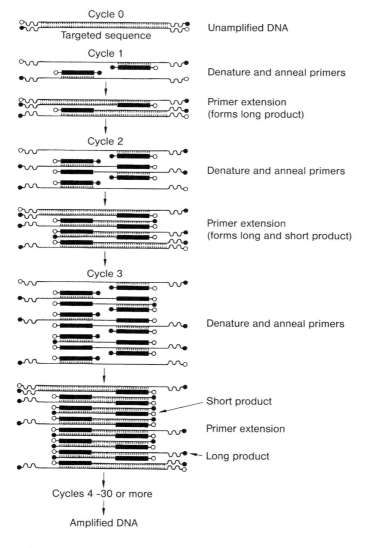

Polymerase chain reaction (PCR) technique

An alternative method for the *in vitro* amplification of nucleic acids which is based on continuous cDNA synthesis and transcription is the → self-sustained sequence replication procedure.

Polymerase chain reaction *a*ssisted *c*DNA *a*mplification (PACA): A variant of the conventional → polymerase chain reaction (PCR) that uses two gene-specific degenerate or non-degenerate primers (amplimers) to amplify specific cDNA fragments ("internal PACA"). A modification works with a gene-specific primer and the unspecific oligo(dT)-primer that hybridizes to a stretch of adenyl residues added to the target cDNA by → terminal transferase. See → anchored polymerase chain reaction.

Polymerase chain reaction carry-over prevention: Any measure to avoid the re-amplification of previously amplified → polymerase chain reaction products that are carried over accidentally to new amplification mixtures. In addition to physical procedures excluding such exogenous templates there exists an enzymatic preventive technique. In short, in the PCR mixture dTTP is substituted by dUTP,

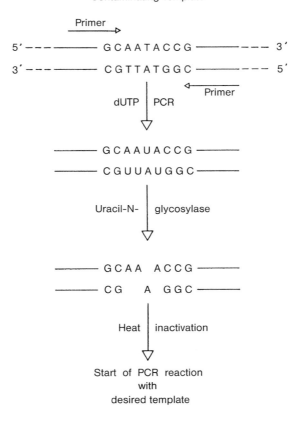

Polymerase chain reaction carry-over prevention

which does not interfere with most subsequent procedures (e.g. → Sanger sequencing). Before starting a new reaction, the potentially left-over template is destroyed with → uracil-DNA-glycosylase that removes uracil bases from the amplified DNA. Then the enzyme is heat-inactivated and the new reaction started.

Polymerase chain reaction mutagenesis (PCR mutagenesis): A variant of the conventional → oligonucleotide-directed mutagenesis, which allows to introduce → insertions, → deletions, or → point mutations in a target DNA with its concomitant amplification using → polymerase chain reaction techniques. See for example → patched circle polymerase chain reaction.

Polymerase chain reaction restriction fragment length polymorphism (PCR-RFLP): Any variation(s) in the length of DNA fragments amplified from → genomic DNA of two or more individuals of a species using conventional → polymerase chain reaction techniques, and subsequently cut with a specific → restriction endonuclease. PCR-RFLP therefore allows to detect restriction endonuclease site variation in two (or more) genomes. The procedure starts with the amplification of a fragment of genomic DNA from organism A and B, respectively, using → primers matching highly conserved → gene sequences (e.g. derived from the chloroplast or mitochondrial genomes). The PCR products are then restricted with a → four-base cutter (e.g. *Rsa* I, or *Dpn* II), and the restriction digests electrophoresed in → agarose gels. The separated fragments are then visualized by staining with → ethidium bromide, and a potential RFLP detected by fluorescence. See → amplification fragment length polymorphism, → restriction fragment length polymorphism.

Polymerase chain reaction serial analysis of gene expression (PCR-SAGE): A variant of the conventional → serial analysis of gene expression (SAGE) technique that is adapted to very low concentrations of input → messenger RNA (mRNA) and works with an initial amplification of the transcriptome by → polymerase chain reaction (PCR) to increase the amount of mRNA from pictogram to microgram levels. In short, → total RNA is first isolated, mRNA extracted, and first strand → cDNA synthesis started from the → poly(A)$^+$-tail of mRNA with a modified → oligo(dT) primer (sequence: 5'-AAGCAGTGGTAACA-ACG CAGGCTAC-T$_{(29)}$VN-3' (V = adenine, cytosine, or guanine; N = any nucleotide). A specific → reverse transcriptase adds a short tract of cytosines to the 5'-end of the mRNA that can hybridize to a synthetic → oligonucleotide containing a short tract of guanines. After hybridization, the reverse transcriptase switches strands and extends the cDNA using the hybridized oligonucleotide sequence as → template ("strand-switching technology"). cDNA synthesis at the 5'-end of the mRNA is completed with the incorporation of an oligonucleotide containing a *Sap*I recognition site(5'-AAGCAGTGGTAACAACGCAG-**GCTCTTC** GGG-3'). *Sap*I recognizes the bold face seven-base sequence, and therefore cuts DNA on average only once per 16,384 bp. The second strand cDNA synthesis occurs during the first PCR cycle. Simce both cDNA strands carry identical 3'-termini, PCR amplification can proceed with a single primer (5'-AAGCAGTGGTAACA ACGCAGGCT-3'), and maintains a *Sap*I site at the 5'-end of the mRNA, but does not introduce such a site at its 3'-end. With the addition of a 5'-biotinylated primer, PCR then generates cDNA with → biotin tags at both ends. Digestion of amplified cDNA with *Sap*I removes the biotin tag only from the 5'-end of the mRNA. The cleaved 31 bp 5'-ends are removed, such that cDNAs biotinylated only at the poly(A)$^+$-end of the original mRNA are produced. The transcript profiling then follows the conventional SAGE protocol. See → SAGELite.

Polymerase chain reaction single-strand conformation polymorphism (PCR-SSCP): The difference in the conformation of two slightly differing single strands of a DNA target sequence isolated from two (or more) different individuals. Such polymorphisms are detected by a → polymerase chain reaction single-strand conformation polymorphism analysis.

Polymerase colony (polony) technology: A technique for the → genotyping and → haplotyping of multiple individualized DNA molecules (e.g. single chromosomes) within a thin → polyacrylamide gel attached to a microscope slide. In short, a small amount of DNA (e.g. from a buccal swap) is first mixed with polyacrylamide solution containing all the reagents for a conventional → polymerase chain reaction (PCR), then degassed and poured on a glass slide and polymerised as a 40 μm thick layer. The slide is partially covered with a Teflon coating that represents a spacer between the glass and a coverslip. Mineral oil is then layered onto the gel (to prevent evaporation of material), the coverslip applied, and the slide as such subjected to a PCR reaction. Two primer pairs (two forward and two reverse primers) flanking the two target SNPs are also included, each amplifying one of the SNPs. This is the simplest arrangement, but several to many different SNPs can be detected in one single experiment by multiplexing. Since the DNA concentration in the gel is low, the chromosome fragments are well separated from each other on the surface of the slide and can be amplified separately side by side. Then the two SNP loci on the chromosomal DNAs are serially amplified in-gel. Since the polyacrylamide matrix restricts the free diffusion of the amplification products, an immobilized colony of double-stranded DNA accumulates around each chromosome ("polony"). Each polony is amplified from a different region of the same chromosome. After amplification, one strand of the amplified DNA remains covalently attached to the acrylamide matrix via the specially designed primers. Therefore it is possible to remove the second strand from all polonies by simple heating to 70°C (in 70% formamide) and washing the slide. At the end, single-stranded templates for subsequent → single base extension (SBE) reactions are produced. For this reaction, the slide is covered with an appropriate chamber, an annealing mixture with the specific primers for SNP 1 added, the unannealed primers washed away, and the templates extended by → Klenow exopolymerase in the presence of a single-strand binding protein from *E. coli* and → cyanin 3-labeled deoxynucleotides (alternatively, also fluorochrome-labeled → dideoxynucleotides can be employed). The slides are scanned on a scanning confocal microscope, and a software used to subtract background. Then the same reaction is run for SNP 2 and → cyanin 5-labeled deoxynucleotides. The images of both reactions are then merged to detect overlapping polonies. The individual polonies can either be genotyped by performing single base extensions with dye-labeled nucleotides, and allow the accurate quantitation of two → allelic variants. The polony technology can also determine the → phase or → haplotype, and two (or more) single-nucleotide polymorphisms

(SNPs) by co-amplifying distally located targets on a single chromosomal fragment. Moreover, the "colonies" of amplified DNA molecules can also be used for fluorescent *in situ* sequencing, a → sequencing-by-synthesis method based on reversibly labeled nucleotides. Almost 10 millions of polonies can be accommodated on a single microscope slide, which is the basis for a high-throughput sequencing platform. See → nanopore sequencing, → single molecule sequencing.

Polymerase infidelity: A laboratory slang term for the mispairing of → nucleotides during → DNA replication, catalyzed by → DNA-dependent DNA polymerase.

Polymerase loading assay (haplotype-specific chromatin immunoprecipitation, "haplo-ChIP"): A technique for the determination of the transcriptional activity of the two → alleles of a gene. In short, first transcriptionally active DNA fragments are isolated by → immunoprecipitation of → RNA polymerase II bound to target → promoters. Then the isolated fragments are assessed for sequence → polymorphisms (e.g. → single *n*ucleotide *p*olymorphisms, SNPs) in heterozygous samples to determine the relative allelic expression.

Polymerase I: See → DNA polymerase I.

Polymer-based biosensor: Any conjugated polyanionic polymer (as e.g. poly[*p*henylene *v*inylene], PPV) that normally emits fluorescence light, but can be quenched by a cationic electron acceptor molecule (e.g. methyl viologen, MV^{2+}). If the quencher is covalently linked to e.g. → biotin, and this biotin together with the quencher removed by the addition of → avidin, then the unquenched polymer will emit fluorescence light. Likewise, the quencher can be linked to an → antibody specifically removed by its specific → antigen, or an oligonucleotide, specifically removed by hybridizing to a complementary sequence. Compare polymer-based DNA chip.

Polymer-based DNA chip: A variant of the conventional → DNA chip, onto which a network consisting of a polymer and multiple DNA molecules bound to this polymer are immobilized per spot. In short, a hydrophilic polymer carrying photoreactive benzophenone goups is first deposited on a plastic or glass slide and illuminated. During illumination, the photoreactive groups react with neighboring polymer chains to form crosslinks and covalently attach the crosslinked polymer to the substrate (e.g. PMMA). Then the DNA is added and crosslinked to the polymer under UV light. Polymer-based chips represent three-dimensional networks with an increased reactivity as compared to two-dimensional chips, which is a result of the higher concentration of DNA on each spot. Moreover, a better signal-to-noise ratio allows to detect even weak signals.

Polymerization fidelity: The accuracy with which a → DNA-dependent DNA polymerase replicates a → template, expressed as average number of nucleotides incorporated, before the enzyme introduces an error.

Polymorph: Laboratory slang term for → DNA polymorphism.

Polymorphic GC-rich repetitive sequence (PGRS): Any one of a series of repetitive sequences in the *Mycobacterium tuberculosis* genome that contains >80% G + C. PGRS underly the socalled PE (name derived from the motif Pro-Glu at the N-terminus) and PPE (name derived from the motif pro-pro-glu at the N-terminus) multigene families encoding acidic, glycine-rich proteins.

Polymorphism:

a) See → DNA polymorphism.
b) The existence of several forms of a phenotypic or genetic character in a population.
c) A localized change in a specific DNA sequence within a genome, generated by → deletions, → inversions, → insertions, or generally → rearrangements. These mutations lead to the existence of different → alleles for a specific locus in a given population. In the case of → repetitive DNA, variations in the number of repeats may lead to → restriction fragment length polymorphisms, see for example → variable number of tandem repeats. Polymorphisms may be detected by → DNA fingerprinting techniques.

Polynucleotide: A linear sequence of deoxyribonucleotides (in DNA) or ribonucleotides (in RNA) in which the 3′ carbon of the pentose sugar of one → nucleotide is linked to the 5′ carbon of the pentose sugar of the adjacent nucleotide via a phosphate group (→ phosphodiester bond). Compare → oligonucleotide.

Polynucleotide kinase **(PNK; T4 polynucleotide kinase):** An enzyme from → T4 phage-infected *E. coli* cells which catalyzes the transfer of the γ-phosphate group of ATP onto the 5′ OH termini of RNA or DNA chains. Used to label the 5′-termini of DNA or RNA prior to → sequencing.

Polynucleotide phosphorylase **(PNPase; EC 2.7.7.8):** An enzyme widely distributed among bacteria that catalyzes the covalent linking of ribonucleotides at random and is used to synthesize artificial RNA (e.g. poly[U], poly[A], or poly[AU] molecules).

Polypeptide: A linear polymer of amino acids that are linked by peptide bonds. Compare → oligopeptide.

Polypeptide tag: Any polypeptide that is conjugated post-translationally (rarely translationally) to target proteins, thereby changing their structure, activity, location, assembly, trafficking or turnover. Conjugation links the C-terminal caboxyl group of the polypeptide tag via a covalent isopeptide bond to ε-lysyl amino group(s) of the target. This process can be reversed by unique proteases (cleaving specifically the isopeptide bond). As examples, → ubiquitin (signal for selective protein degradation by the 26S → proteasome), *s*mall *u*biquitin-like *mo*difier, SUMO, also sentrin, UBL1 or PIC1 in animals, SMT3 in yeast (potential role in protein trafficking from nucleus to cytoplasm and vice versa), *r*elated to *ub*iquitin (RUB), *auto*phagy-defective-12 (APG12), *u*biquitin *c*ross-*r*eacting *p*rotein (UCRP) and Finkel-Biskis-Reilly murine sarcoma virus-*a*ssociated *u*biquitously expressed protein (FAU) are such polypeptide tags, whose genes occur in small → gene families. Usually the tags are short (ubiquitin and RUB: 76 amino acids; SUMO: 93–115 amino acids; APG12: 96–186 amino acids).

Polyphenism: The occurrence of two (or more) different → phenotypes within one species. For example, female social insects exhibit various defined phenotypes such as queens, soldiers, and workers). Polyphenism is not based on a difference in the genomes of the variants, but rather on an epigenetic developmental switch of differential gene expression patterns triggered by worker-controlled nutritional and microenvironmental differences within the nest.

Polyphosphate kinase (*ppk*) reporter gene: A bacterial gene of e.g. *E. coli* that encodes the enzyme polyphosphate kinase (PPK) that serves as → reporter protein to monitor → gene expression in mammalian cells. In short, the *ppk* gene is fused to an inducible → promoter, the construct inserted into an appropriate → expression vector plasmid, the → vector transfected into target cells, and expressed. As a consequence, the PKK catalyzes the synthesis of polyphosphate from ATP, a linear polymer of orthophosphate residues linked by high-energy phosphoanhydride bonds, which can be readily quantified by e.g. ^{31}P magnetic resonance spectroscopy (MRS) or ^{31}P magnetic resonance imaging (MRI). Since the endogenous levels of polyphosphate in mammalian cells is extremely low (below the detection level of MRS), the PKK can be used as a reporter. The PKK system is non-invasive, does not require exogenous substrate or cofactors, and can be applied to internal tissues of multicellular organisms.

Polyploidy (Greek: polis for "many"; ploid for "fold"): The occurrence of more than two complete sets of chromosomes within a cell, a tissue, an organ, or an organism, resulting from chromosome replication without nuclear division or the recombination of two gametes with differing chromosome sets. The normal set is then diploid (diploidy), a triple set is triploid (triploidy), a quadruple set is tetraploid (tetraploidy), and so on.

Polyprotein: Any protein that is produced by the uninterrupted translation of a → polycistronic mRNA transcribed from two or more adjacent genes.

Polyribosome (polysome): The linear array of → ribosomes attached to a molecule of mRNA. Such polysomes may also contain small → translational control RNA.

Polysaccharide sequencing: The estimation of the linear arrangement of individual sugars (or their modified forms) in complex polysaccharides. For example, *h*eparin-*l*ike *g*lycos*a*mino*g*lycans (HLGAGs) that are components of the cell surface and extracellular matrix, usually vary in the number of disaccharide repeat units, and their chemical modifications (at four potential sites). The basic disaccharide unit of HLGAGs is either α-L-*i*duronic acid (I) or β-D-glucuronic acid (G) linked 1,4 to α-D-hexosamine. Together, the four different modifications for an I or G uronic acid isomer containing disaccharide produce $2^4 = 16 \times 2 = 32$ different disaccharide units for HLGAGs. The sequencing of such complex polysaccharides starts with the chemical or enzymatic fragmentation (with e.g. heparinase I and III, or iduronate 2-O sulfatase, iduronidase, and glucosamine 6-O sulfatase), and the determination of the mass of each fragment by → *m*atrix-*a*ssisted *l*aser *d*esorption *i*onization *m*ass *s*pectrometry (MALDI-MS) and the number of sulphate and acetate groups. Mass-identity relationships are then computed.

Polysome: See → polyribosome.

Polysome display: See → ribosome display.

Polysome selection: See → ribosome display.

Poly(T): A homopolymer consisting of thymidylic acid residues.

Polytene chromosome (giant chromosome): A → chromosome consisting of homologous → chromatids that remain attached to each other (synapsed) after repeated chromosomal → replications without nuclear division. Such polytene chromosomes are characteristic for some Ciliatae, the suspensor cells of some plants, and the salivary gland cells of insect larvae (e.g. *Drosophila*) but also occur in other organisms. The DNA of the original chromosomes in such cells replicates in 10 cycles without separation of the daughter chromosomes so that 2^{10} (1024) chromatids may exist in parallel orientation and in strict register. Polytene chromosomes are visible throughout the interphase and are composed of a series of dark condensed bands that are separated by so-called interbands. The pattern of the bands is specific for each chromosome. For example, the *Drosophila* genome contains about 5000 bands and corresponding interbands, each one consisting of a total of 1024 homologous → looped domains. Each band can be assigned a specific number so that a → chromosome map can be established. If a gene has to be transcribed, the corresponding band is decondensed and forms a so-called → puff. The pattern of puffs is characteristic for the physiological and/or developmental stage of the cell, tissue or organ. Specific genes may be localized on polytene chromosomes by → in situ hybridization.

Poly(U): A homopolymer consisting of uridylic acid residues.

Poly(U) sepharose: A sepharose matrix to which poly-uridylic acid residues (poly[U]) are covalently bound and which is used for the binding, isolation and purification of poly (A)-mRNAs in → affinity chromatography. Compare also → messenger affinity paper.

P1 cloning vector (pacmid): A → cloning vector, derived from the phage P1 of *E. coli* that allows the packaging of foreign DNA of up to 100 kb without interference with the phage functions and thus has a much higher → cloning capacity than → lambda-phage derived or → cosmid vectors. P1 plasmid vectors contain a P1 packaging

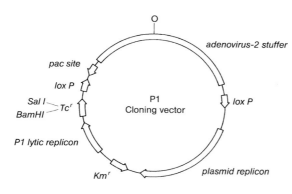

Simplified map of a P1 cloning vector

P1 cloning vector

site (*pac* site) to initiate packaging of vector and cloned DNA into phage P1 particles, two directly repeated P1 recombination sites (*lox P*) flanking the cloned insert and necessary for the circularization of the packaged DNA after its entry into the host cell, a → selectable marker gene (e.g. → ampicillin or → kanamycin resistance), a P1 plasmid → replicon (stabilizing the vector in the host cell at one copy per chromosome), and a *lac* promoter-regulated P1 lytic replicon (allowing the → IPTG-induced amplification of the DNA, see → *lac* operon). Plasmids are propagated in a special *E. coli* strain containing the P1 *cre* recombinase, which mediates recombination between the two *lox P* sites, and thus the circularization of the infecting linear DNA. Packaging of the DNA is initiated when P1-encoded pacase proteins recognize and cleave the *pac* site in the P1 phage DNA. The DNA on one side of the cleavage point is then packaged into an empty phage pro-head. Once this pro-head is filled, a second, non-specific cleavage occurs that separates packaged from non-packaged DNA. Tails are attached to the filled heads to complete the assembly process.

P1-derived *a*rtificial *c*hromosome (PAC): A cloning vector, derived from the bacteriophage P1 and used to clone DNA fragments of 100 to 300 kb insert size (average at 150 kb) in *Escherichia coli*. A variant of → P1 cloning vector.

P1 nuclease: See → nuclease P1.

Pong: A variant of the → miniature inverted repeat transposable element *Ping*, characterized by identical 15 bp → terminal inverted repeats and a generally similar organization as *Ping* (i.e. two → open reading frames and 5,166 bp length).

Pool: The amount of a defined molecule that is available for distinct metabolic reactions (e.g. the ATP molecules that can serve as phosphate donors in phosphorylation reactions, as opposed to the AMP molecules fixed in nucleic acids).

Poolwise directed recombination (poolwise recombination, poolwise shuffling): A variant of the conventional → directed molecular evolution that aims at recombining many related, but mutated parental genes into a single progeny, thereby increasing the number of positive → mutations accumulating between two selection events. See → DNA shuffling, → incremental truncation for the creation of hybrid enzymes, → L-shuffling, → protein complementation assay, → ribosome display, → staggered extension process.

Poor man's cloning: See → *in vivo* cloning.

Population genetics: A branch of → genetics that focuses on the genetic composition (e.g. frequencies of polymorphisms) of whole populations of organisms as influenced by various intrinsic (e.g. mutations, genetic drift, population size and expansion) and environmental factors (as e.g. natural and/or sexual selection, migration, socalled bottle necks) and therefore aims at developing models for evolution. See → population genomics.

Population genomics: The complete repertoire of techniques to develop → molecular markers for genetic variants in whole populations, to associate diagnostic markers with e.g. disease phenotypes by → linkage analysis, and to isolate the underlying genes via → positional cloning. For example, population genomics was applied to different human populations

```
            EcoR I
            recognition  SD site  Poly(A)    Start
            site                  tract      codon
        5'- GAATTCGGAGGAAAAAATTATG              - 3'
        3'-       GCCTCCTTTTTTAATACCTAGG - 5'
                                       Bam H 1
                                       recognition
                                       site
```

Portable SD sequence

(e.g. Estonians and Icelanders), and markers (i.e. distinct variants of *m*ajor *h*istocompatibility *c*omplex [MHC] alleles) were defined for e.g. rheumatoid arthritis.

Population-specific single *n*ucleotide polymorphism (population-specific SNP): Any → single nucleotide polymorphism that is present in one, and absent in another population. For example, the colonization of Polynesia or the Americas led to the development of single base pair exchanges that did not occur in ancestral groups of hominids in Asia. These SNPs can therefore be considered as specific for Polynesian or American Indian populations, respectively.

Porosome: See → nuclear pore.

Portable biosensor: Any → biosensor that allows the monitoring of pesticides, chemical warfare agents or, generally, pollutants directly on the spot. For example, *b*iochemical *o*xygen *d*emand (BOD) biosensors can determine the amounts of metabolizable organic material in e.g. waste water by measuring oxygen consumption by immobilized bacteria or yeast cells. The bacteria can also be genetically engineered to express the → lux gene in response to pollutants, which encodes → luciferase generating light during substrate decomposition. See → affinity biosensor, → biomimetic sensor, → electrode biosensor, → enzyme biosensor, → immunosensor, → synthetic receptor.

Portable promoter: Any isolated and fully characterized → promoter or promoter fragment that contains all regulatory sequence elements for function and can be inserted into any → expression vector and be transformed into any target genome.

Portable SD sequence (portable Shine-Dalgarno sequence): A short synthetic oligodeoxynucleotide sequence that contains the → Shine-Dalgarno sequence 5'-AGGAGGU-3', flanked by appropriate → restriction endonuclease → recognition sites, so that it can be easily cloned into prokaryotic → expression vectors.

Portable Shine-Dalgarno sequence: See → portable SD sequence.

Portable terminator: A sequence containing the 3'-terminus of a eukaryotic gene including the transcription → terminator sequence and → poly(A) addition signal(s) flanked by → polylinkers. Such terminators can be ligated to the coding sequence of any gene to construct a → fused gene whose transcript can be correctly terminated and polyadenylated.

Portable translation *i*nitiation *s*ite (PTIS): A double-stranded DNA sequence that contains a five-base → Shine-Dalgarno sequence with an adjacent 3' poly(A) tract flanked by the translation initiation codon ATG eight bases downstream of the SD tract. This configuration is optimal for correct and efficient initiation of translation. Such "portable" sites are usually flanked by specific → restriction endonuclease → recognition sites to allow the → ligation of a PTIS into a → cloning vector.

Positional candidate gene (positional candidate): Any gene linked to a DNA marker co-segregating with a phenotype of interest, and meeting the criteria for a gene, which could be responsible for the trait. It is usually isolated by → positional cloning.

Positional cloning (map-based cloning; *map-assisted* *c*loning, MAC): The → cloning of a specific gene in the absence of a transcript or a protein product, using → genetic markers tightly linked to the target gene and a directed or random → chromosome walk by linking overlapping clones from a → genomic library.

Position effect: Any change in the expression of one or more genes accompanying a change in its or their position with respect to neighboring (or also distant) genes. Position effects may be brought about by → cross-over or chromosome mutation, and can be seen in → transgenic organisms, where the → chromatin configuration at the integration site determines the expression potential of the foreign gene.

Position *e*ffect *v*ariegation (PEV): The influence of the chromosomal position on the activity of a → gene. For example, if a normally active gene is introduced into a chromosome at a position adjacent to a → heterochromatin domain, its transcription is repressed. In yeast, genes integrated close to the silent mating-type loci or the → telomeres are silenced. This silencing influence of heterochromatin can spread from 5–10 kb (*Drosophila*) to 20–30 kb (yeast). Position effect variegation is a common observation in → transgenic animals or plants: → reporter genes randomly introduced into the genome have highly variable transcription rates. See → enhancer variegation, → position effect, → suppression variegation.

Position-Specific *I*terative BLAST (PSI-BLAST): A specific iterative search program using the → BLAST → algorithm. A profile is built after an initial search, which is then used in subsequent searches. This process is repeated, with new sequences found in each cycle used to refine the profile.

Positive *c*ofactor (PC): Any nuclear protein that facilitates and accelerates the cooperation between the different proteins of the → transcription initiation machinery and thereby leads to the start of transcription initiation. For example, PC4 of *Homo sapiens* supports the function of the general → transcription factors → TFIID and → TFIIH. Result: initiation of transcription of the adjacent gene. See → negative cofactor.

Positive control: Any experimental control element that provides a signal or result, irrespective of the results obtained from the actual experimental components. For example, a positive control on an → expression microarray consists of e.g. the → cDNAs of socalled → house-keeping genes that are active throughout the life cycle of an organism (e.g. β-actin- or polyubiquitin genes). If total and fluorescence-labeled cDNAs of a test organism are hybridised to the array with these positive control cDNAs, they will always give a constant signal (e.g. a → fluorescence signal) notwithstanding the reaction of the other cDNAs spotted onto the array. Positive controls are necessary for a test of the function of the array. See → negative control.

Positive cooperation: See → zippering.

Positive feedback activation: The binding of a transcriptional activator protein to its own → promoter and the subsequent activation of the expression of this activator as well as the induction of its target genes. For example, in *Aspergillus nidulans* the activity of the socalled ALCR transcriptional activator is induced by ethanol, and during this induction process the ALCR also activates its own expression by binding to its own promoter. The ALCR promoter contains two ALCR binding sites, and both are essential for its auto-activation and the activation of downstream genes encoding enzymes for ethanol metabolism. See → feedforward loop, → feedforward loop activation.

Positive interference: The suppression of → recombinations in the vicinity of → heterochromatin (as e.g. the → centromeres).

Positive regulator protein: A protein that activates the transcription of a → gene.

Positive selection: See → direct selection.

Positive selection vector: Any → cloning vector that contains a marker gene or genes, whose mutation can positively and directly be detected. For example, one type of positive selection vectors carries dominant → selectable marker genes. If an → insertion occurs (by e.g. an → insertion sequence element), it may lead to drug resistance or enable the host to grow on a medium containing sucrose. These mutation events generate positively selectable phenotypes in the host cells. Therefore, positive selection vectors are used to isolate mobile genetic elements, as e.g. insertion sequences.

Positive strand RNA virus: A virus containing a single-stranded RNA genome that functions as mRNA template (→ plus strand). The viral RNA is itself infectious. These viruses belong to the Coronaviridae, Flaviviridae, Picornaviridae, Polioviridae, Retroviridae, and Togaviridae.

Positive supercoiling (overwinding): The coiling of a → covalently closed circular DNA duplex molecule in the same direction as that of the turns of its → double helix. Compare → negative supercoiling, → supercoil.

Post-genomic era: The time after the seqencing of the human genome (year: 2001), which is considered to be the time for genome-wide → transcriptomics, the in-depth → proteomics, detailed → metabolomics, and the deciphering of the molecular mechanisms of development, evolution, and disease, to name only few. Generally, → functional genomics is regarded as the central topic of the post-genomic era. Of course, during this era more and more genomes of both prokaryotes and eukaryotes will be fully sequenced, and the handling of the immense quantities of data will be a challenge for → bioinformatics. The vague starting point for the post-genomic era is arbitrary. It could have been the publication date for the sequence of the genome of the first bacterium (i.e. *Haemophilus influenzae*; year 1995).

Postreplication repair: See → mismatch repair.

Post source decay (PSD): The spontaneous fragmentation ("metastable fragmentation") of ionized molecules during their acceleration in the electrical field or after passage of the accelleration electrode in the field-free drift section of a → mass spectrometer. The analysis of PSD-ions in → reflector time-o-flight mass spectrometers allows to extract informations about the structure of the original ionized molecule.

Posttranscriptional gene silencing (PTGS): A more general term for several protective mechanisms of eukaryotic cells against invading viroids, viruses or moving → retrotransposons (generally RNAs) that can be converted to double-stranded RNAs (dsRNAs) within the cell. These dsRNAs are recognized by → Dicer RNase III and cut into → small RNAs that incite the silencing of genes encoding homologous RNAs. PTGS was originally described for plants (e.g. *Arabidopsis thaliana*, *Petunia hybrida*), but its variants are components of defense systems in all eukaryotic organisms. In fungi, → quelling suppresses → transgenes, in invertebrates → RNA interference (RNAi) and → co-suppression are incited by dsRNA, → transgenes and → short hairpin RNA, in vertebrates dsRNA is the prime trigger for RNAi. PTGS can also be transmitted systemically from silenced to non-silenced plant tissues by a → degradation-resistant signal RNA. Invading viruses have evolved more or less effective counter-measures. For example, plant potyviruses encode a protein, HC-Pro that inhibits maintenance of PTGS, cucumoviruses (e.g. cucumber mosaic virus) encode a 2b protein that interferes with DNA methylation in the plant nucleus and prevents signal RNA-mediated intercellular spread of PTGS (see → degradation-resistant signal RNA), and the socalled movement protein of potexviruses (e.g. potato virus X) suppresses the release of degradation-resistant signal RNA from infected to non-infected plant cells. PTGS events are not meiotically transmitted and need to be re-established in each sexual generation. See → virus-induced gene silencing.

Post-transcriptional modification (PTM, posttranscriptional RNA processing, nuclear processing of RNA, RNA-processing, RNA-maturation): Any one of a series of structural modification(s) of → primary transcripts prior to or during their transport into the cytoplasm. Modifications include → splicing (removal of → introns) → capping of 5'-ends, → polyadenylation of the 3'-end, or → methylation of cytidylic (or adenylic) residues within the RNA molecule, thiolation, isopentenylation, → pseudouridine formation, and association with various proteins. See also → RNA editing, the post-transcriptional modification of mitochondrial RNA, and → post-translational modification.

Posttranscriptional RNA processing: See → post-transcriptional modification.

Post-translational cleavage: The enzymatic cleavage of a large protein molecule or a polyprotein at specific sites to produce smaller functional proteins. A → post-translational modification reaction.

Post-translational modification (PTM, protein maturation, protein processing, post-translational processing): Any alteration of polypeptide chains after their synthesis (e.g. acetylation, ADP-ribosylation, biotinylation, glutathionylation, glycosyl-

Post-translational Modifications

Modification	Target site	Reaction
Acetylation	Lys, NH$_2$-terminus	Transfer
ADP-ribosylation	Arg, Cys, Asn, Glu, Lys	Transfer
Amidation	Glycin, C-terminus	Lysis
Deimination	Arg	Hydrolysis
Glycosylation	Asp	Transfer
Hydoxylation	Pro, Lys	Oxidoreduction
Methylation	Arg, Lys, His, Glu, iso-Asp	Transfer
Myristoylation	Gly, NH2-terminus	Transfer
Palmitoylation	Cys	Transfer
Phosphorylation	Ser, Thr, Tyr	Transfer
Prenylation	Cys	Transfer
Sulfation	Tyr	Transfer
Ubiquitinylation Lys		Ligation

ation, hydroxylation, lipidation, lipoylation, oxidation, phosphopantetheinylation, phosphorylation, sulfation, transglutamination, or also conversion of proenzymes into enzymes by specific proteolytic cleavage and epimerization). Compare → post-translational cleavage. See → intein-mediated protein ligation.

Post-translatomics: Another term of the omics era describing the whole repertoire of techniques to detect and characterize the various → post-translational modifications of proteins, especially in connection with their impact on protein function(s).

Potonuon (*potential nuon*): Any → nuon that arose by amplification, duplication, recombination or → retroposition and may acquire a new functional role as a new gene (or part of a gene, e.g. an → exon or an → intron) or new regulatory element (as e.g. an → enhancer, → silencer). If it has acquired the new function, it is called → xaptonuon.

Power stroke mechanism: The movement of → DNA-dependent RNA polymerase II along the DNA that is supported by storing the energy of → dNTP hydrolysis in a transient polymerase conformation. This conformation relaxes at the end of bond formation to propel the → elongation complex by one nucleotide along the template DNA. See → passive sliding.

pp: Abbreviation for *p*hospho*p*rotein.

PPAR: See → peroxisome proliferator-activated receptor.

PPD: See → AMPPD.

PPi based sequencing: See → pyrosequencing.

pPNA: See → phosphono peptide nucleic acid.

PPP: See → promoter prediction program.

P primer: An → oligonucleotide of arbitrary sequence that serves as a forward → primer in → differential display reverse transcription polymerase chain reaction in combination with a reverse oligo(dT) primer targeting at the poly(A) tail of eukaryotic → messenger RNAs ("T primer"). In advanced differential display techniques, these primers of arbitrary sequence are replaced by primers complementary to common sequence motifs found in a comprehensive collection of messenger RNAs.

PPTase: See → phosphopantetheinyl transferase.

PQL: See → protein quantitative locus.

pQTL: See → physiological quantitative trait locus.

Precipitation: The sequestration of an insoluble compound or mixture of compounds in a solution. For example, DNA or RNA can be precipitated from aqueous solutions by extensive dehydration with absolute ethanol. The precipitated material is called precipitate.

Precursor messenger RNA: See → pre-messenger RNA.

Precursor protein: The primary product of the → translation of a → messenger RNA, containing all → exteins and → inteins. See → mature protein.

Precursor RNA: Any → ribonucleic acid synthesized from a gene as a long precursor that is not yet mature, but still contains many different regions cut out or modified in later processing steps. Such modifications include → capping, → polyadenylation, and → splicing, which altogether lead to its final functional form.

Predicted gene: Any DNA sequence that has significant homology to → genic sequences deposited in the databanks (e.g. GenBank), and can therefore be considered a gene candidate. Compare → putative gene.

Prediction analysis of microarrays (PAM): A statistical classifier method for → microarray analyses that identifies a subgroup of genes characteristic for a predefined class. This gene subset is predicted from gene expression data using the PAM software (working under Windows and Unix/Linux) and a modification of the nearest shrunken centroid method, which computes a standardized centroid for each class in the training set. This is the average gene expression for each gene in each class divided by the within-class standard expression deviation for that gene. Nearest centroid classification takes the gene expression profile of a new sample, and compares it to each of these class centroids. The class, whose centroid is closest, is the predicted class for that new sample. PAM can be downloaded: http://www-stat.stanford.edu/~tibs/PAM.

Predictive gene test: The identification of abnormalities in a gene that make an individual susceptible to certain diseases.

Preference gene: Any one of a series of (still hypothetical) genes that determine the choice of a mating partner.

Preferential amplification of coding sequences (PACS): A technique for the detection of differentially expressed genes (or better → cDNAs) that specifically targets at the → coding regions of a → messenger RNA (rather than its 5'- or 3'-non-coding parts). In short, total RNA is first isolated, contaminating DNA removed by RNase-free → DNAseI, and single-stranded cDNA produced by → reverse transcriptase polymerase chain reaction with → primers of random sequence. The double-stranded cDNA is synthesized with an ATG-containing → forward primer and a → double restricton site primer (DRSP) as a → reverse primer in a conventional → polymerase chain reaction. The ATG-complementary primer specifically selects coding parts of messenger RNAs (mRNAs), since ATG is the → initiation codon of almost all organisms (exceptions: some viral [e.g. human T-cell lymphotropic virus type I], chloroplast [mRNA encoded by the infA gene], plant mitochondrial [atp9-rp116 cotranscript], and bacterial mRNAs [encoding ribosomal proteins]).

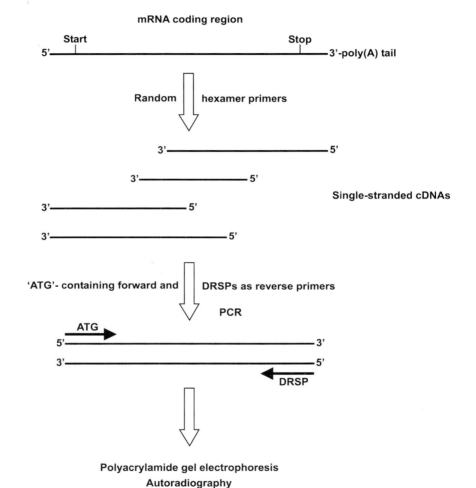

DRSP: Double restriction site primer

Preferential amplification of coding sequences (PACS)

Moreover, ATG codons occur only rarely → downstream of the proper initiation codon. Therefore, generation of multiple → amplicons from a single mRNA by PACS is unlikely. The ATG primer also contains a → restriction site (e.g. *Bam* H1) at its 5′-end for cloning of the amplification product. These amplification products are then separated electrophoretically in 6% → polyacrylamide/urea sequencing gels, the gels dried and subsequently processed by → autoradiography. The resulting pattern represents a differential mRNA fingerprint of the cell, tissue or organ, from which the RNA was extracted. See → adapter-tagged competitive PCR, → enzymatic degrading subtraction, → gene expression fingerprinting, → gene expression screen, → linker capture subtraction, → module-shuffling primer PCR, → quantitative PCR, → targeted display, → two-dimensional gene expression fingerprinting. Compare → cDNA expression microarray, → massively parallel signature sequencing, → microarray, → serial analysis of gene expression.

Prefoldin (PFD, Gim complex): A 90 kDa protein complex in the cytoplasm of Archaea and eukaryotes, consisting of two α- and four β-subunits that emanate from a → β-barrel core as six α-helical coiled-coil protrusions. The tips of these 65Å long coiled coils are partially unwound and expose hydrophobic amino acid residues for the binding of non-native proteins. PFD binds to nascent peptide chains and cooperates with eukaryotic chaperonins in the folding of e.g. actin and tubulin, thereby functionally replacing the → heat shock protein (Hsp) 70 system in some Archaea.

Preformed adaptor (ready-made adaptor; conversion adaptor): A short synthetic, single-stranded → oligonucleotide with a → restriction endonuclease → recognition site that allows complete base-pairing with the → cohesive ends of a DNA duplex molecule and a regeneration of a second endonuclease recognition site. For example, the *Eco* RI → *Sma* I preformed adaptor with the sequence 5′-GAATTCCC

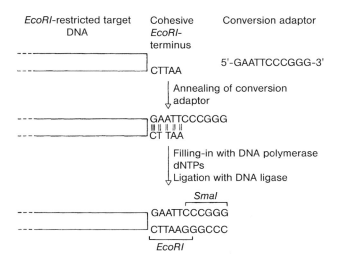

Preformed adaptor

GGG-3' anneals to cohesive *Eco* RI termini of the target duplex. After → filling-in and ligation, a circular DNA molecule is generated that contains an additional *Sma* I recognition site.

Pre-gel hybridization: A technique for the detection of → mutations (e.g. → deletions, → insertions, → single nucleotide polymorphisms) that circumvents the various steps of → Southern blotting (i.e. pre-hybridization, hybridization, and stringency washes). In short, the double-stranded target DNA is denatured at low salt concentrations and in the presence of a short (i.e. 18-mer) labeled → *p*eptide *n*ucleic *a*cid (PNA). At these low salt concentrations, the DNA strands cannot reanneal, and the PNA can bind its complementary DNA target. The resulting PNA/DNA hybrids are then separated by → agarose gel electrophoresis or capillary electrophoresis, and detected after blotting (by e.g. → autoradiography, → luminography, or → fluorescence detection).

Prehybridization: The preparation of → nitrocellulose and nylon-based filters carrying denatured DNA, fixed by → baking or → cross-linking, for hybridization with radioactively labeled → probes. The prehybridization procedure serves to block unspecific binding of the probe to the membrane, and is therefore carried out in solutions containing high concentrations of proteins (→ Denhardt's reagent) and/or detergents (e.g. SDS). Compare → blocking reagent.

***Pre-i*mplantation *d*iagnostics (PID, *pre*implantation *g*enetic *d*iagnosis, PGD, *pre*implantation *g*enetic *s*creening, PGS):** A series of techniques to detect genetic abnormalities (e.g. chromosomal aberrations) in (usually) a single cell of an e.g. eight-cell embryo that is a product of an artificial insemination. For that purpose, the target cell is mechanically removed from the morula (by e.g. capillary forces), which in turn will be implanted into the womb after PID. In case of a positive test (i.e. abnormalities found) and no further implantation, a possible abortion is avoided. Moreover, genetically caused repeated miscarriages can be circumvented. PID is also used to detect chromosomal abnormalities in embryos from women of advanced maternal age undergoing fertility treatment, and to allow patients with chromosomal rearrangements and mutations associated with single-gene disorders to produce phenotypically normal children and to avoid the risk of transmitting the genetic abnormalities to their offspring. PGS usually works with a laser biopsy of oocytes and the isolation of the socalled *p*olar *b*ody (PB) that contains a haploid set of chromosomes. The → karyotype of this PB is then determined by → *f*luorescent *in s*itu *h*ybridization. If a chromonema is missing, it is supposed to be retained within the oocyte, which then is aneuploid, and can therefore not be used for *in vitro* fertilization.

***Pre-i*nitiation *c*omplex (PIC):** An aggregate of general and specific → transcription factors and → DNA-dependent RNA polymerase II subunits that assembles on → promoter elements following the sequence-specific binding of → transcription factor IID (TFIID) to the → TATA-box. The TFIID-promoter complex is recognized by → transcription factor IIB (TFIIB), the RNA polymerase II-TFIIF complex is recruited (see → transcription factor IIF), and the PIC is completed with the binding of → transcription factor IIE (TFIIE) and → transcription factor IIH (TFIIH). The pre-initiation complex converts to an initiation complex after → promoter melting (local disruption of hydrogen bonds within the DNA helix). Then the first → phosphodiester bond of the nascent → messenger RNA is synthesized, the promoter "cleared" (i.e. the RNA polymerase moves out of the PIC), and the → transcript is elongated. TFIIA (see → transcription factor IIA) interacts and stabilizes the PIC any time after binding of TFIID. Within the PIC, the carboxyterminal domain (CTD) of the large subunit of RNA polymerase II is dephosphorylated, but is phosphorylated by CTD kinases before entering the elongation phase, and leaves the PIC. After termination, the CTD is again dephosphorylated by CTD phosphatases and re-initiates transcription on the → core promoter, if the TFIID complex is still assembled.

Preintegration complex (PIC): A large nucleoprotein complex consisting of retroviral DNA and viral proteins that is able to integrate into the host cell's → nuclear DNA. Upon → reverse transcription of the viral RNA genome, the PIC is formed. The two ends of the viral DNA are bridged by proteins, probably assisted by the viral integrase (IN). This enzyme removes two nucleotides from the → blunt-ended viral genome, which triggers transport of the PIC into the nucleus. The 3′ OH recessed viral DNA termini in the PIC are inserted into the host genome by IN, producing proviral DNA. Do not confuse with → polymorphism information content.

Premature polyadenylation: The faulty addition of adenosyl residues within the coding region of a pre-messenger RNA molecule (pre-mRNA) that is recognized by the socalled → non-stop messenger RNA decay system as an aberrant mRNA. Prematurely polyadenylated mRNAs are degraded.

Premature termination codon (PTC): Any → codon in a → messenger RNA that causes termination of → translation within the message. If such a truncated message would be translated, the resulting protein most likely would be non-functional. However, when a PTC is recognized by a ribosome, → nonsense-mediated mRNA decay (NMD) is activated, reducing the levels of PTC-containing messages to 5–30% of the normal levels.

Pre-messenger RNA (pre-mRNA): Any complete → primary transcript from a → structural gene before its → post-transcriptional modification. Pre-mRNA is packaged with proteins into → messenger ribonucleoprotein complexes (mRNPs), also called → heterogeneous nuclear ribonucleoprotein complexes (hnRNPs) that contain e.g. proteins of the hnRNP A family and specific → splicing/mRNA export-associated factors like THO/TREX complexes. Compare → pre-ribosomal RNA.

Pre-messenger RNA splicing: See → splicing (definition b).

Pre-miRNA/intron: See → mirtron.

Premutation: Any mutation in a gene that does not lead to phenotypic consequences, but a predisposition for a disease in the next generation. For example, a normal transmitting male carries a premutation in the FMR1 gene on the distal long arm of the X chromosome. This premutation consists of an increased number of CGG repeats in the 5′-untranslated region of the

FMR1 gene (repeat numbers in normal individuals: 5–44; premutation: 55–200). CGG alleles with intermediate numbers of repeats are considered intermediate alleles (also called "gray zone" alleles). A further expansion of the CGG repeat leads to an inhibition of the transport of the 40S ribosomal subunit from the nucleus and therefore to the suppression of translation. Repeat numbers beyond 200, accompanied by aberrant methylation of cytidyl residues (full mutation), generally cause clinical symptoms of the full-blown fragile X syndrome in males, whereas females with the full mutation are less affected. See → microsatellite expansion.

Prenylation: The covalent addition of either farnesyl (15 carbon atoms) or geranyl-geranyl (20 carbon atoms) isoprenoids to conserved cysteine residues at or near the → C-terminus of proteins via a thioether linkage. Prenylation takes place at the consensus sequence CAAX (C = cysteine; A = any aliphatic amino acid, except alanine; X = carboxyterminal amino acid). First, the three amino acids AAX are removed, and cysteine is activated by methylation (methyl donor: S-adenosylmethionine). Many membrane-associated proteins are prenylated, and therefore prenylation is probably important for trafficking. Also, prenylation promotes interaction(s) of proteins and cellular membranes, and facilitates protein-protein contacts. Prenylation occurs in e.g. nuclear → lamins, fungal mating proteins, Ras and Ras-related GTP-binding proteins (G proteins), protein kinases and viral proteins.

Preparative comb: A special slot former (comb) for horizontal → agarose gels that allows to apply large volume samples. It contains one tooth spanning most of the length of the comb usually flanked by two small teeth for the electrophoresis of → molecular weight markers.

Preparative *i*soelectric *m*embrane *e*lectrophoresis (PrIME): A technique for the isolation of isoelectrically pure proteins from complex protein mixtures, using → isoelectric focusing in immobilized pH gradient gels on a preparative scale. A variant of this technique works with a series of chambers separated by single-pH → polyacrylamide-immobiline membranes, which act as → isoelectric point-selective barriers. Proteins introduced between the membranes migrate through the membranes to focus in the chamber bounded by one membrane with a pH greater than the pI of the protein, and by another membrane with a pH less than the pI of the protein. PrIME allows to separate proteins differing in pI by only 0.005 pH units.

Pre-replication complex (pre-RC): A → protein machine that is assembled at each → origin of replication during the G1 phase of mitosis in eukaryotic cells. The pre-RC consists of various proteins, of which the ORC proteins (*o*rigin of *r*ecognition *c*omplex) bind to the origin in an ATP-requiring reaction, where they remain throughout the cell cycle. The ORC recruits Cdt1 (*c*ell *d*ivision *t*arget), Cdc6 (*c*ell *div*ision *c*ycle), and Mcm (*m*inichromosome *m*aintenance) proteins, in higher eukaryotes assisted by geminin that binds to Cdt1 in the S phase (preventing the formation of a new pre-RC). Cdc6 is rather unstable, and present on the pre-RC only during G1. It becomes phosphorylated at the onset of the S phase and is thereby labeled for degradation. Cdt1 and Cdc6 assemble the Mcm2-7 complex onto the pre-RC. Mcm proteins possess helicase activity. The pre-RC is complete at the end of G1, and converted to the initiation complex (IC) by the action of cyclin-dependent kinases (CDKs) and a helicase that denatures the RNA at the origin such that the socalled *r*eplication *p*rotein *A* (RPA) can bind and stabilize the single-stranded DNA. The socalled Cdc45 (*c*ell *d*ivision *c*ycle 45) can then interact with the IC and bind to a subunit of the DNA polymerase α-primase. Thereby the enzyme is activated which leads to the synthesis of a 10 nucleotide RNA primer. This primer is subsequently extended to a 40 nucleotide RNA-DNA primer by a DNA-dependent RNA polymerase, which is displaced by the *r*eplication *f*actor C (RF-C) that loads the socalled *p*roliferating *c*ell *n*uclear *a*ntigen (PCNA) onto the RNA-DNA primer (ATP-dependent). PCNA recruits the DNA polymerase d or e, which extend the primer by several thousand nucleotides (→ replication).

Pre-ribosomal RNA: The complete → primary transcript from a ribosomal RNA → gene battery (→ rDNA). Its size varies from organism to organism (*Drosophila*: 38S; *Xenopus*: 40S; HeLa cells: 45S). The primary transcript is cleaved in a series of steps to form the → ribosomal RNAs (5.8S, 18S, and 28S rRNA). See → posttranslational modification. Compare → pre-messenger RNA.

Pre-RISC: An intermediate structure of the → RISC assembly pathway, formed by the → *R*ISC-*l*oading *c*omplex (RLC) that recruits the Ago2 protein, and therefore contains the double-stranded → *s*mall *i*nterfering RNA (siRNA) firmly bound to the Ago2 protein. Ago2 cleaves the socalled passenger strand of the siRNA, and initiates the dissociation of this strand from the socalled guide strand, thereby forming the → holo-RISC. Therefore, the holo-RISC contains only the guide strand of the siRNA (fully competent to down-regulate transcript targets), whereas the passenger strand is dissociated from the holo-RISC.

Presentation: The display of small → antigen fragments bound to specialized proteins on the surface of antigen-presenting or virus-infected cells. T lymphocytes only respond to presented antigens.

Pre-spliceosome: A cage-like structure that assembles on the GU splice site of → pre-messenger RNA after ATP-dependent binding of U1 → small nuclear (sn) RNA as a prelude for the formation of a → spliceosome. First a socalled A complex is organized that harbors U2snRNA (binding to the A of the socalled branch site), U1snRNA, and additionally about 70 different proteins. Then U4/U5/U6 are recruited in an ATP-dependent reaction to form the pre-catalytic socalled B complex.

Press-blot: A simple technique for the detection of nucleic acids or proteins in plant organs (e.g. leaves). The tissues are shock-frozen and then fixed onto hybridization membranes by high pressure. The membrane can then be processed for → Southern (detection of DNA), → Northern (detection of RNA), or Western blotting (detection of proteins).

PrEST: See → protein epitope signature tag.

Pretermination cleavage (PTC): The co-transcriptional cleavage of a nascent → messenger RNA (mRNA) mediated by sequence tracts downstream of the → poly(A) site that effectively releases the mRNA from the elongating → DNA-dependent RNA polymerase II (RNAP II). The disengagement of RNAP II from the DNA template downstream of the transcribed gene requires previous transcription of the PTC sequence and a functional poly(A) site. PTC precedes and is required for → transcription termination. See → intrinsic transcription termination.

Prey (P): A part of a hybrid protein component of yeast → two-hybrid systems, encoded by a → hybrid gene consisting of a fusion of a → cDNA or a genomic DNA fragment (the prey *per se*, whose interaction with the socalled → bait has to be tested) and a fused → transcriptional activation domain. If prey and bait interact, the activation domain of the prey construct comes into close proximity with the DNA-binding site, which induces transcription of a → reporter gene.

Prey vector: Any → cloning vector that contains a cDNA-derived sequence encoding a specific protein ("prey") cloned into a → multiple cloning site fused to a sequence encoding a transcription → activation *domain* AD (e.g. B42) upstream. The AD in turn is linked to a → nuclear localization signal and the expression of the prey protein driven by a → promoter. The vector additionally carries replication origins (e.g. the Col E1 origin for replication in *E. coli* and the 2m origin for replication in yeast) and one (or more) → selectable marker genes. Prey vectors are co-transformed with → bait vectors into socalled yeast reporter strains. Simultaneous expression of the genes on both vectors produces the prey cDNA-derived protein ("prey protein") and the corresponding "bait protein", whose potential interaction can then be detected with the → two-hybrid system.

PRF: See → ribosomal frameshifting.

Pribnow box (Pribnow-Schaller box, −10 box): The 6 bp DNA → consensus sequence 5′-TATAATG-3′, located about 10 bp upstream of the → transcription initiation site of prokaryotic → structural genes and functioning as the binding site of the → sigma factor of *E. coli* → RNA polymerase. The Pribnow box facilitates correct initiation and is the equivalent of the eukaryotic → TATA box.

Pribnow-Schaller box: See → Pribnow box.

Primary amplicon: A DNA fragment that is preferentially amplified during → polymerase chain reaction, because the used → primer possesses either complete, or far-reaching → homology to potential target sequences and therefore allows vigorous amplification by → DNA polymerase. Such primary amplicons appear as strong bands in → ethidium bromide-stained → agarose gels, or → silver-stained → polyacrylamide gels, respectively. See → amplicon, → secondary amplicon.

Primary microRNA (pri-miRNA): Any long (up to 1 kb) primary transcript containing a → hairpin of 60–120 nucleotides that encodes a mature → microRNA in one of the two strands. The hairpin is cleaved from the pri-miRNA molecule *in nucleo* by the double-strand-specific ribonuclease Drosha. The resulting precursor miRNA ("pre-miRNA") is transported to the cytoplasm by exportin-5, and then further processed by → Dicer to generate a short, partially double-stranded RNA, in which one strand represents the mature microRNA. The latter associates with a protein complex similar or identical to the → *RNA-i*nduced *S*ilencing *C*omplex (RISC).

Primary response gene: Any gene that is immediately and directly activated by an external or intrinsic signal. Frequently, such primary response genes encode proteins involved in signal cascades.

Primary small interfering RNA (primary siRNA): Any → small interfering RNA (siRNA) that is derived from the original "trigger" → double-stranded (ds) RNA, in contrast to the → secondary siRNA derived from regions → upstream of the double-stranded trigger sequence.

Primary structure: The one-dimensional representation of the covalently linked monomer units of a → nucleic acid (RNA, DNA), socalled → nucleotides, or a protein (socalled → amino acids). For example, the monomers of RNA are composed of an aromatic heterocyclic → base (A,C,G,U) covalently bound to a → ribose molecule. The 5'-carbon of the sugar is covalently linked to a phosphate group. An RNA polymer is formed by covalently joining the 3'-carbon of one → nucleotide with the 5'-phosphate of another one, and so forth. The sequence of these covalently linked bases represents the primary structure of the RNA, written from its 5' end (left) to the 3' end (right). See → secondary structure, → tertiary structure, → quarternary structure.

Primary structure of an RNA molecule:

5'-CGCAAUCUUGACUUUCGGAUGGC
UACAUCUUCAGGUCUCCGAUGAGUUCA-3'

Primary transcript: An RNA molecule immediately after its transcription from DNA (i.e. before any → post-transcriptional modifications take place). The primary transcript corresponds to a → transcription unit. See → pre-messenger RNA and → pre-ribosomal RNA.

Primase: See → RNA primase.

PrIME: See → preparative isoelectric membrane electrophoresis.

Primed in situ labeling (PRINS, DNA-PRINS): A sensitive variant of the → in situ hybridization technique to detect specific DNA sequences in metaphase chromosomes. In short, metaphase spreads are prepared and denatured, synthetic oligodeoxynucleotides or short DNA fragments (e.g. specific for specific chromosomes) are hybridized to the chromosomes in situ, and used as → primers for → DNA polymerase (e.g. → Thermus aquaticus DNA polymerase)-catalyzed extension in the presence of biotinylated or digoxigenin-labeled nucleotides (e.g. digoxygenin-11-dUTP), using the chromosomal DNA as a template. The newly synthesized strand is visualized with fluorescence (e.g. FITC)-labeled → avidin or anti-digoxygenin → Fab fragments, respectively, and the labelled chromosomes visualized under a fluorescence microscope.

PRINS can be used for the detection of DNA sequences (DNA-PRINS) as well as for the visualization of RNA in situ (RNA-PRINS). RNA-PRINS employs oligodeoxynucleotides as primers for → reverse transcriptase (RTase)-catalyzed extension with labeled nucleoside triphosphates, using the RNA (e.g. mRNA) as a template.

Primed synthesis technique: The enzymatically controlled extension of a primer DNA strand in DNA sequencing. See → Sanger sequencing.

Primer: A short RNA or DNA → oligonucleotide which is complementary to a stretch of a larger DNA or RNA molecule and provides the 3'-OH-end of a substrate to which any → DNA polymerase can add the nucleotides of a growing DNA chain in the → 5' to 3' direction. In prokaryotes a specific → RNA polymerase (→ RNA primase) catalyzes the synthesis of such → primer RNAs for DNA → replication (especially of the → Okazaki fragments of the → lagging strand). Primers are also needed by RNA-dependent DNA-polymerases (→ reverse transcriptase).

In vitro, synthetic primers, usually about 10 bp in length, are needed for any DNA polymerization reaction using DNA polymerases or reverse transcriptase. Thus they are necessary for → cDNA synthesis, → Sanger sequencing, the → polymerase chain reaction (see → amplifier), → primer extension and similar techniques. See also → primer adaptor, primer-directed sequencing, primer DNA, → primer hopping, → primed *in situ* labeling, → primer RNA, → primosome; → random priming, → sequencing primer, → unidirectional primer, → universal primer.

Primer-adaptor (adaptor-primer): Any synthetic oligodeoxynucleotide that serves the dual function of a → primer (e.g. for the reverse transcription of a poly(A)⁺-mRNA by → reverse transcriptase) and an → adaptor (e.g. carrying a → recognition site for a specific → restriction endonuclease). An example for a primer-adaptor is the oligo(dT)-*Xba* I primer-adaptor used in → forced cloning of cDNA.

Primer binding site (PBS): A sequence adjacent to the 5′ → long terminal repeat of → retroviruses or → retrotransposons, which is complementary to the 3′ end of a → transfer RNA. Annealing of the tRNA to the PBS produces a priming site for → reverse transcriptase. In plant retrotransposons the PBS is complementary to the initiator methionine tRNA:

Primer dimer: An artifact, representing a non-target amplification product in conventional → polymerase chain reaction techniques that is caused by sequence homologies within → primers and hence partly double-stranded primer-primer ("primer dimer") adducts. Usually such primer-dimer artefacts also occur in controls (e.g. without any → template) and are of low molecular weight. See → primer-primer artifact.

Primer-directed sequencing (primer-directed walking; primer hopping; primer walking; primer jumping): A technique to sequence DNA fragments of more than 1 kilobase in length. In short, the target fragment is first cloned into an appropriate → cloning vector, and a → forward and → reverse → sequencing → primer complementary to flanking vector sequences used to sequence the → insert from both ends by → Sanger sequencing techniques. This procedure leads to sequence information of about 600–800 bp on both ends of the insert. Now primers are designed from the outermost 100 bp at both ends (walking primers) and used for the second sequencing step, and so on. Primer walking thus allows to sequence long stretches of DNA that cannot be sequenced by classical sequencing strategies. See → multiplex walking, → uniplex DNA sequencing.

Primer-directed walking: See → primer-directed sequencing.

Primer DNA (DNA primer): A single-stranded DNA fragment required by →

```
         LTR              PBS
  5'....CA AGTGGTATCAGAGCCTCGTTT....3'
          || ||| ||| || || ||| || ||| || ||| |||    ||| || || ||
           ACCAUAGUCUCGGUCCAAA
           3'part of tRNA_i^met
```

Pimer binding site

DNA polymerase III for DNA replication. → extended. Primer exclusion allows to detect → point mutations.

Primer exclusion: A variant of the conventional → polymerase chain reaction that exploits the competition between a → primer oligonucleotide with a specific sequence and a second oligonucleotide with a slightly different sequence for a common primer binding site. Under conditions of → high stringency the primer with the higher sequence → homology to the target sequence will bind, outcompete (exclude) the competing primer, and gets

Primer extension:

a) A method to precisely map the 5′-terminus of mRNAs and to detect precursors and intermediates of → processing of → messenger RNA. The mRNA is hybridized to a synthetic, 5′ radiolabeled, complementary oligodeoxynucleotide 30–40 nucleotides in length which is then used by retroviral → reverse transcriptase as a → primer. The enzyme completes synthesis of the

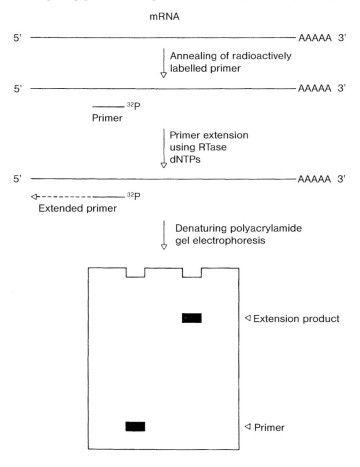

Primer extension

complementary strand (→ cDNA) at the 5'-terminus of the mRNA template. The length of the extended primer, and consequently the 5'-terminus of the → transcript, can be precisely determined by → polyacrylamide gel electrophoresis and → autoradiography. See → *in silico* primer extension.

b) Any DNA polymerization reaction using a single-stranded template and starting with an → oligonucleotide primer. See for example → Sanger sequencing.

c) A technique to detect socalled → single *n*ucleotide *p*olymorphisms (SNPs) in target DNA. In short, the target (e.g. a gene) is first amplified with specific → primers in a conventional → polymerase chain reaction and subsequently denatured. Then a single-stranded oligonucleotide primer is annealed to the single-stranded target DNA such that the primer ends exactly at the SNP site. After annealing, the duplex exposes a 3' OH-group for an extension catalysed by DNA polymerase in the presence of all four → dideoynucleoside triphosphates (ddNTPs; instead of → deoxyribonucleoside triphosphates, dNTPs), each labeled with a specific → fluorochrome. The matching ddNTP will then be incorporated and stops extension. The incorporated ddNTP is then identified by the specific fluorescence emission of its fluorochrome. A comparison with the wild-type sequence at the SNP site allows to identify the type of SNP.

Primer extension inhibition: See → translation toeprinting.

Primer extension-*n*ick *t*ranslation (PENT): A technique for the detection of G-overhangs at → telomeres that starts with the annealing of a C-strand telomeric → primer to the telomeric G-overhang. This primer is then extended with e.g. → DNA polymerase I or → *Taq* DNA polymerase (to stop the → extension at the telomere-subtelomere junction). The newly synthesized strand (Cs) replaces the original telomeric strand (Co) due to the → 3'-exonuclease activity of DNA polymerases. A → nick is therefore left between the Cs and Co. Subsequently, Cs strands are separated from bulk DNA by alkaline electrophoresis, and their lengths measured after their → hybridization with a labeled telomeric probe (5'-TTTAGGG-3'). See → primer extension telomere repeat amplification, → single telomere length analysis.

Primer extension preamplification (PEP): A technique for the sampling of an entire genome that uses a mixture of all possible 4^n → primers (excluding primers composed of only one type of nucleotide, e.g. A_n, or G_n). Each primer has a length of about 15 nucleotides. This extremely complex mixture (4^{15} compounds) is then employed in a conventional → polymerase chain reaction to amplify the majority of sequences in a complex genome present in a single haploid cell (e.g. sperm or oocyte). PEP suffers from multiple template-independent primer-primer artifacts. See → tagged random primer PCR.

Primer extension telomere repeat amplification (PETRA): A technique for the determination of → the length of telomeres at individual chromosome ends in an organism with relatively short telomeres and known sub-telomeric sequences. PETRA starts with the annealing of an → adaptor-primer ("PETRA-T") to the telomeric 3'G-rich → overhang. PETRA-T consists of 12 nucleotides complementary to the 3'-telomeric region, and a unique sequence at its 5'-end ("non-telomeric tag sequence"). This → primer is then extended by → DNA polymerase I. Subsequently, a → primer specific for a known sub-telomeric sequence of a specific → chromosome arm is used in combination with a primer identical to the 5'-non-telomeric tag of the adaptor-primer to amplify the region in a conventional → *p*olymerase *c*hain *r*eaction (PCR). The amplified products are then electrophoretically separated in 0.8–1.0% → agarose gels, transferred to a nylon membrane, and finally detected by → Southern hybridization with a ^{32}P end-labeled telomeric repeat probe (5'-TTTAGGG-3'). This procedure is repeated for each arm of the chromosomes of an individual, and allows to estimate their telomere lengths. See → primer-extension-nick translation, → single telomere length analysis.

Primer hopping: See → primer-directed sequencing.

Primer jumping: See → primer-directed sequencing.

Primer-primer artifact: The appearance of amplified products in a → polymerase

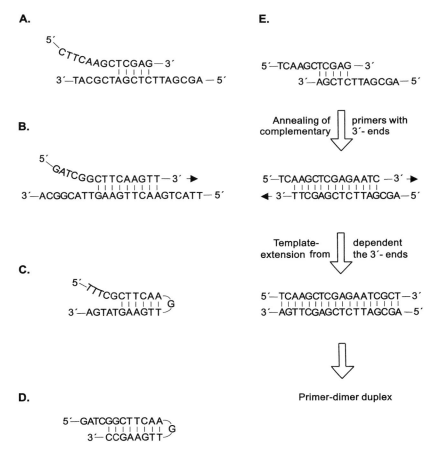

Primer-primer artifact

chain reaction that arise by primer-primer rather than template-primer interactions. Such artifactual interactions are undesirable, since they give rise to spurious bands on → ethidium bromide-stained gels and withdraw primers from the desired primer-template interactions. Primer-primer artifacts are based on different mechanisms. If the primers partially anneal to each other or to template DNA, the DNA polymerase with its 5'→3' exonuclease activity may remove bases from the 5' end (A). If the primers only partially anneal to the template DNA, the DNA polymerase may add bases onto the 3' end, producing non-specific amplification products (B). If a primer forms a → hairpin structure with a 3' overhang, the DNA polymerase with its 5'→3' exonuclease may remove bases from the 5' end (C). If a primer forms a hairpin structure with a 5' overhang, the DNA polymerase may add bases to the 3' end (D). If two primers with complementary 3' ends partially anneal, the DNA polymerase may add bases to the 3' ends, resulting in primer-dimer duplexes (E). See → primer dimer.

Primer RNA: See → RNA primer.

Primer-specific and mispair extension analysis (PSMEA): A technique for the detection of single nucleotide variations (e.g. → deletion, → insertion, → transition, → transversion) between two DNA → templates. The method exploits the highly efficient 3' → 5' → exonuclease proofreading activity of → *Pyrococcus furiosus* DNA polymerase that prevents the → extension of a → primer when (1) an incomplete set of → deoxynucleotide triphosphates is present and (2) a → mismatch occurs at the initiation site of DNA synthesis (i.e. the 3'-end of the primer). For example, in the presence of only dCTP and dGTP, primer 3'-CTCTG·····5' can easily be extended on template A (5'-GAGAC·····3'), because the crucial nucleotide (in bold face) matches. The same primer cannot be extended on template B (5'-AAGAC·····3'), so that genome A can be discriminated from genome B (presence/absence of an extension product). In contrast, the use of dTTP and dGTP allowed the extension of the primer on template B, not on template A. PSMEA therefore allows genotyping of organisms that differ in only one (or few) nucleotide pairs at the 5' end of the primer-binding site.

Primer walking: See → primer-directed sequencing.

Priming: The initiation of the synthesis of a DNA strand by the formation of an → RNA primer or by → self-priming.

Priming *a*uthorizing *r*andom *m*ismatches polymerase *c*hain reaction (PARM-PCR): A rarely used → polymerase chain reaction technique, in which specific → primers are used in combination with highly unspecific annealing conditions (low → stringency) to allow random annealing and therefore universal amplification of target sequences.

pri-miRNA: See → primary microRNA.

Primosome (primosome complex, replisome): A mobile multi-protein DNA replication-priming complex of *E. coli*, consisting of the proteins encoded by genes *dnaB, dnaC, dnaG, dnaT, n, n"*, and the replication factor Y (DNA helicase). The initiation protein DnaA first complexes with ATP, binds to a specific site on single-stranded DNA (*primosome assembly site, PAS*; thought to be located on the →

lagging strand at the → replication fork), opens the PAS, which allows a replicative helicase (DnaB) to access. Onto this scaffold other proteins are assembled. Then the primosome moves along the DNA in 5' → 3' direction and occasionally synthesizes an → RNA primer. This polymerization reaction is catalyzed by the *dnaG*-encoded → RNA primase.

Primosome assembly site: See → primosome.

Primosome complex: See → primosome.

Principal component analysis (PCA): A classical statistical technique for a multivariate analysis to simplify (reduce) a complex highly dimensional dataset that involves a mathematical procedure transforming a number of (possibly) correlated variables into a smaller number of uncorrelated variables called *principal components*. The first principal component comprises as much of the variability in the data as possible, and each successive component accounts for as much of the remaining variability as possible. PCA is widely used in data analysis and data compression.

PRINS: See → primed *in* situ labeling.

Printed microarray: Any glass wafer (chip), onto which small volumes of oligonucleotides are uniformly spotted via a noncontact inkjet process. Alternatively, the term is also used for glass chips, on which oligonucleotides are synthesized base-by-base, using standard phosphoramidite chemistry.

Prion (*proteinaceous infectious particle*): A cellular protein (PrP^c) with as yet unknown function in synaptic transmission. The primary transcript for this protein encodes a pre-protein of 254 amino acids that is posttranslationally trimmed by 22 amino acids at the amino terminus (the released peptide is a → signal peptide) and 23 amino acids at the carboxy terminus. This trimmed protein contains two N-glycosylation sites and a disulfide bridge, and is anchored in the membrane of neuronal cells via a g*lyco*p*hosphoi*no*sitol* (GPI) anchor. The prion exists in two isoforms, PrP^c and PRP^{SC} (sc stands for *sc*rapie, a disease of sheep leading to loss of motoric control and degeneration of the brain). In contrast to PrP^c, the PrP^{SC} is extremely insoluble, resistant to proteolytic degradation and less sensitive to heat denaturation. PrP^{SC} is probably a derivative of PrP^c. Once PrP^c is converted to a PrP^{SC}, then an autocatalytic process progressively increases the concentration of PrP^{SC}, which is ultimately deposited as socalled amyloid plaques or rods in the brain, leading to the outbreak of the scrapie disease (*s*p*o*n*g*iform *e*ncephalopathy, SE). The human equivalent of the scrapie disease, the socalled *C*reutzfeld-*J*acob *d*isease (CJD), as well as the *b*ovine *s*pongiform *e*ncephalopathy (BSE) are also associated with prions that are able to "replicate" in the absence of → DNA or → RNA.

Privacy: In genetics, the right of people to restrict access to their genetic information.

Private polymorphism: Any genetic (or sequence) → polymorphism common in only specific populations that are usually reproductively isolated from other, larger groups. These private variations may be completely absent in other groups.

PRM: See → protein recognition molecule.

P-RNA: See → *p*yranosyl RNA.

pRNA: See → promoter RNA.

Probe:

a) A defined and radioactively or non-radioactively labeled nucleic acid sequence used in → molecular cloning to identify specific DNA molecules with complementary sequence(s) by → autoradiography or with non-radioactive → DNA-detection systems. The term is also used for proteins (e.g. a → monoclonal antibody reacting with its target protein).

b) A somewhat misleading, but widely accepted term for any defined nucleic acid sequence (e.g. an → oligonucleotide, a → cDNA) that is covalently bound to a carrier ("chip") made of glass, quartz, or plastic, and hybridized to a → target nucleic acid (mostly cDNAs). Thousands of such probes are assembled on socalled → cDNA expression arrays, → DNA chips, → microarrays, → sequencing arrays and serve for the simultaneous detection of multiple hybridization events ("massively parallel"). Note that the conventional term "probe" (definition a) has been converted to another meaning in chip technology. Here, probes are also termed "reporters".

Probe-based genotyping: The detection of individual-specific genomic profiles by binding complementary sequences ("probes") to target regions, and monitoring the binding event by various techniques. The probes can either be immobilized on a solid phase (e.g. as in → microarrays), or the hybridization of the probe and target occurs in solution (as e.g. with hybridization probes).

Probe complexity: A measure for the number of different nucleic acid sequences in a → probe.

Probe excess: The presence of very high → probe concentrations on the surface of a → microarray such that the array-bound target is saturated. Probe excess is undesirable, because it obscures quantitative differences in hybridisation signals.

Prober sequencing: An outdated laboratory slang term for a → DNA sequencing technique with fluorescent chain-terminating → dideoxynucleotides (ddNTPs), named after J.M. Prober and colleagues. As a variant of the → Sanger sequencing method, it starts with a → polymerase chain reaction driven by a → primer oligonucleotide and using ddNTPs labeled with slightly different succinyl *f*luoresceins (SFs). The different fluoresceins vary in their absorption maxima (e.g. SF-505: 486 nm; SF-512: 493 nm) and emission spectra (e.g. maximum of SF-505: 505 nm; SF-512: 512 nm), and their incorporation leads to chain termination, leaving each molecule labeled at its 3′-end with an SF-ddNTP. During the electrophoretic separation, the various chains are excited with an argon laser at 488 nm and the emitted → fluorescence light monitored by parallel filter/ photomultiplier systems.

Probe set: Any collection of → probes that altogether represent a target sequence (e.g. a gene). For example, a set of about 20 oligonucleotides of 50 nucleotides each, derived from different regions of a known gene, and thus representing the gene, can be spotted onto a → microarray. Then labeled → cDNAs can be hybridized to the microarray and the expression of the gene of interest be detected. The presence of probes complementary to various regions of one distinct gene allows to discriminate between various gene → homologues or → splicing variants.

Procapsid: The → capsid precursor formed during the assembly of viral capsomers.

Procaryotes: See → prokaryotes.

Processed gene: See → processed pseudogene.

Processed pseudogene (retropseudogene, retrosequence, retrogene, processed gene): An intron-less → pseudogene that contains a poly(A) tract at its 3′ end and is flanked by short → direct repeats of 10–20 nucleotides, which are potential → insertion sequences. Processed pseudogenes most probably arise through reverse transcription of → messenger RNA, after it has been processed (e.g. spliced and polyadenylated), and the integration of the product cDNA into an arbitrary site of the genome. Two classes of processed pseudogenes can be discriminated. The complete retropseudogene contains all the → exons of the gene from which it is transcribed, whereas the truncated retropseudogene harbors only a fraction of the exons.

Processing (editing):

a) The → post-transcriptional modification of → primary transcripts.
b) The → post-translational modification of proteins.

Processing body (P body, also DCP body or GW 182): A discrete dynamic cytoplasmic granule of *Saccharomyces cerevisiae* and mammalian cells that contains proteins of the → messenger RNA (mRNA) → decapping pathway and is the place of mRNA decapping and decay (i.e. 5′→3′ exonucleolytic degradation). P bodies, present in only few copies (2–3 per cell), harbor subunits Dcp1p and Dcp2p of the decapping enzyme, the 5′ → 3′mRNA exonuclease Xrn1p, the Lsm proteins (binding to mRNA after → deadenylation and acting as decapping activators) and Dhh1p (an RNA helicase) and Pat1p (enhancing decapping) that partly interact with each other. Additionally, P bodies contain Argonaute 1 and 2 proteins, and are therefore part of the → RNA interference pathway. The P body compartment may also store maternal mRNA for later → translation activation.

Processive DNA polymerase: Any → DNA-dependent DNA polymerase that remains associated with its → template during successive steps of nucleotide incorporations. See → distributive DNA polymerase, → processive enzyme, → processivity.

Processive enzyme: Any enzyme that does not dissociate from its substrate between repetitions of the catalytic event. For example, DNA-dependent DNA polymerase is such a processive enzyme, since it continues to polymerize nucleotides after adding the first nucleotide to e.g. a primer. See → processive DNA polymerase, → processivity.

Processive transcription: Any gene → transcription, where → initiation and elongation are highly efficient, so that high levels of polyadenylated RNAs accumulate. See → nonprocessive transcription.

Processivity (processivity index, processivity value): The extent to which DNA-dependent → DNA polymerases use their → template strand before they dissociate from it (expressed as number of nucleotides incorporated per binding event). The processivity of different, especially purified DNA polymerases *in vitro* is different. Thus some enzymes allow the synthesis of short DNA strands only, though the template strand is not yet fully copied. See → processive DNA polymerase, → processive enzyme.

Processivity clamp (sliding clamp): A ring-shaped protein (dimer or trimer) that encircles double-stranded DNA and binds to DNA-dependent DNA polymerase, thereby increasing its → processivity.

Producer gene: Synonym for → structural gene (→ Britten-Davidson model).

Productive base pairing: The pairing between two bases in DNA that are perfectly complementary to each other. For example, A-T, T-A, C-G and G-C are such productive base pairs. See → non-productive basepairing.

Productive infection: See → lytic infection.

Progeroid gene: Any gene, whose expression is reduced or abolished during aging, and therefore most likely involved in the aging process. For example, the progeroid gene *WRN* (*Wer*ner syndrome), encoding an RecQ → helicase, is either already mutated in the germline (causing the socalled Werner syndrome, a disorder with the clinical symptoms of premature aging as e.g. cataracts, type 2 diabetes, osteoporosis, arteriosclerosis and hypogonadism at a very young age), or the → CpG island of the *WRN* → promoter is hypermethylated, which in turn accompanies malignant transformation. Both symptoms are a consequence of defects in DNA replication and reapir. Or, the progeroid gene *LMNA* (nuclear *lamin* A/C), encoding nuclear filaments A and C (two → isoforms arising from → alternative RNA splicing) and involved in tumor suppression, is either mutated in the germline (which induces the socalled Hutchinson-Gilford syndrome, a disorder characterized by rapid premature aging), or its promoter CpG island is hypermethylated (associated with human cancer). The symptoms are a consequence of nuclear disintegration.

Prognostic reporter: A → transcript derived from a gene that is up-regulated at the onset of tumorigenesis or during the course of tumor establishment, and serves as a → molecular marker ("expression marker") for the clinical outcome (e.g. overall and relapse-free survival). Such prognostic reporters are identified from transcript profiles of a large number of tumor samples, generated by e.g. → expression microarrays, and may outperform currently used clinical parameters. See → reporter gene.

Programmable *autonomously-controlled* electrodes (PACE) gel electrophoresis: An improved version of the conventional → pulsed-field gel electrophoresis which utilizes a hexagonal array of 24 computer-controlled electrodes around an → agarose gel, allowing the generation of defined, homogeneous electric fields. The field direction may be alternated during a run in a preprogrammed way, since each electrode or set of electrodes can be individually controlled by a high-voltage operation amplifier driven by a power supply. With this specific arrangement a nearly linear separation of DNA fragments in the range from 500 to more than 10 million bp can be achieved. See → gel electrophoresis.

Programmable chip: Any → DNA chip produced in a completely automated process that is designed on a computer and customized to given and requested experimental conditions. After hybridisation of target RNA, the binding events are monitored in digital form.

Programmable *mel*ting display *m*icroplate-*a*rray *d*iagonal *g*el *e*lectrophoresis (meltMADGE): A variant of the → microplate-array diagonal gel electrophoresis (MADGE) technique that is based on the separation of amplified PCR products (see → polymerase chain reaction) by temporally changing the running temperature during → polyacrylamide gel electrophoresis, which readily distinguishes between the non-mutated (one single band) and the mutated sequences (four bands: two → homo-duplexes, two → heteroduplexes). This type of separation can also be achieved by → denaturing gradient gel electrophoresis. meltMADGE is therefore used for the *de novo* mutation scanning of target DNA, and requires only one hour for separation and little starting material and gel, and additionally is adapted to a → microplate (i.e. 96-well) format.

Programmable restriction endonuclease (programmed restriction endonuclease): Any → restriction endonuclease whose specificity is dramatically enhanced by its coupling to a specific → DNA-binding domain or a → triple-helix-*f*orming oligonucleotide (TFO), which anchors the enzyme adjacent to a socalled triple-helix forming site (TFS) on the target DNA. The target site then is a composite of a specific recognition site and a nearby TFS. For example, a single-chain variant of the endonuclease *Pvu*II (sc*Pvu*II) is covalently coupled to a 16-mer TFO (either 5'-NH$_2$-[CH$_2$]$_6$ or $_{12}$-MPMPMPMPMPPPPPPT-3', where M = 5-methyl-2'-deoxycytidine, and P = 5-[1-propynyl]-2'-deoxyuridine) with the aid of a bifunctional chemical cross linker specific for amino and sulfhydryl groups (e.g. N-[γ-maleimidobutyryloxy] succinimide ester, GMBS). The succinimide group of GMBS forms an amide bond with the 5'-NH$_2$-group of the TFO, which in turn is connected to the TFO by 6 or 12 methylene groups (linker). The TFO forms a triple helix by binding to the → major groove of the target duplex DNA via hydrogen bond contacts with the Hoogsteen faces of the → purine bases. → Polypyrimidine TFOs bind in a parallel, → polypurine TFOs in an antiparallel orientation with respect to the purine strand of the → Watson-Crick base pairs. The sc*Pvu*II-C$_6$/C$_{12}$-TFO conjugate binds and cuts an addressable *Pvu*II recognition site with 1000-fold more specificity, leaving the unaddressed sites (where no TFS sequence is available) uncleaved. Programmable restriction enzymes are used to map → chromosomal DNA and to clone very large DNA fragments.

Programmed DNA deletion: The programmed destruction of both single-copy and moderately repetitive DNA sequences ("deletion elements") from several hundred base pairs to more than 20 kb in size and specific for the micronucleus in *Tetrahymena thermophila*. In short, this ciliated protozoon contains one germinal nucleus (micronucleus) and one somatic nucleus (macronucleus) per cell. During sexual conjugation, the micronucleus goes through a series of events to produce a zygotic nucleus that divides and differentiates into the new macro- and micronucleus of the progeny cell. The old macronucleus is destroyed. Formation of the new macronucleus involves extensive genome-wide DNA rearrangements. Thousands of specific DNA segments (about 15% of the genome) are deleted, and the remaining DNA is fragmented and endoduplicated about 23-fold to form the somatic genome responsible for all transcriptional activities during growth. The programmed DNA deletion is triggered and guided by double-stranded RNA transcribed from germline sequences during conjugation.

Programmed restriction endonuclease: See → programmable restriction endonuclease.

Programmed ribosomal frameshifting: See → ribosomal frameshifting.

Prohibitin gene: An evolutionary conserved mitochondrial gene that encodes the protein prohibitin functioning as negative regulator of cell proliferation and life span. Prohibitin is associated with senescence and cell death in yeast and mammalian cells.

Prokaryotes (procaryotes): Members of the superkingdom that contains archaebacteria, eubacteria, and cyanobacteria (formerly, blue-green algae). Most of the prokaryotes did not evolve a membrane-bound nucleus with chromosomes, but instead possess a circular DNA genome anchored at the membrane. However, exceptions to this general rule exist: some species of the Planctomycetes as e.g. Gemmata obscuriglobus and Pirellula marina contain membrane-bound compartments with DNA, RNA and DNA- and RNA-processing proteins. These compartments may represent precursors of the eukaryotic nucleus. The prokaryotes also do not contain mitochondria, plastids or microtubules. See → eukaryotes.

Proline switch: The transition between *cis* and *trans* isomers of the amino acid proline in specific proteins that changes ("switches") the protein conformation from a closed (inactive) to an open (active) structure. For example, the filamentous → bacteriophage fd that infects *E. coli*, carries a gene encoding a socalled gene-3-protein (g3p). Three to five copies of this protein form the tip of the fd phage and are responsible for host infection. The C-terminus of each gene-3-protein is anchored within the phage coat, whereas domains N1 and N2 protrude and are firmly associated with each other (closed conformation: phage not infectious). During the infection process domain N2 first reacts with the tip of the bacterial F pilus, thereby exposing the N1 domain for binding of the phage to the TolA phage receptor on the host cell, activating the g3p. The exposed open conformation is temporarily stabilized by *trans*-isomerization of proline 213 of the g3p that functions therefore as proline switch. In the inactive phage, Pro213 is locked in the *cis* configuration, holding domains N1 and N2 of g3p tightly bound to each other. Interaction of domain N2 with the *E.coli* TolA receptor transforms *cis* proline into the more stable *trans* proline: the phage is active.

Promiscuous DNA: See → promiscuous gene.

Promiscuous gene (promiscuous DNA): Any → gene or DNA fragment which has been moved or is still being moved from one organelle to another in the eukaryotic cell (e.g. nuclear genes encoding mitochondrial or plastid functions and believed to originate from the respective organellar genomes).

Promiscuous plasmid: See → plasmid promiscuity.

Promoter (promotor): A → *cis*-acting DNA sequence, 80–120 bp long and located 5′

upstream of the initiation site of a gene to which → RNA polymerase may bind and initiate correct → transcription (see also → 5' flanking region). Prokaryotic promoters contain the sequences 5'-TATAATG-3' (→ Pribnow box) approximately at position –10, and 5'-TTGACA-3' at position –35. Eukaryotic promoters differ for the different DNA-dependent RNA polymerases (see also → core promoter). RNA polymerase I recognizes one single promoter for → rDNA transcription, RNA polymerase II transcribes a multitude of genes from very different promoters, which have specific sequences in common (e.g. the → TATA box at about position –25 and the → CAAT box at about position –90. The so-called → house-keeping genes contain promoters with multiple GC-rich stretches with a consensus core sequence, 5'-GGGCGG-3'. RNA polymerase III recognizes either single elements (e.g. in 5S RNA genes) or two blocks of elements (e.g. in all → transfer tRNA genes) within the gene (→ internal control regions). All these consensus sequences function as address sites for DNA-affine proteins (→ transcription factors) that promote or reduce transcription. Specific promoter sequences can be identified in → genomic DNA using → promoter trap vectors, in cloned DNA fragments using → promoter probe vectors. The promoter regions of various → inducible genes are useful tools for the construction of → cloning vectors with specific requirements; see for example → heat-shock promoter, → heavy metal resistance promoter, → light-inducible promoter, → tac, trc and → trp, and → tissue-specific promoters. See also → alternative promoter, → bidirectional promoter, → bifunctional promoter, → cell-specific promoter, → chemically inducible promoter, → chimeric promoter, → constitutive promoter, → core promoter, → cryptic promoter, → decoy promoter, → divergent promoter, → downstream promoter, → dual promoter, → Emu promoter, → hybrid promoter, → inducible promoter, → internal promoter, → low level promoter, → minimal promoter, → mutated promoter, → portable promoter, → promoter strength, → promoter-up mutant, → pseudo-promoter, → regulated promoter, → reverse promoter, → shared promoter, → single promoter, → split promoter, → strong promoter, → synthetic promoter, → *tac* promoter, → tandem promoter, → 35S promoter, → 3'-promoter, → *trc* promoter, → *trp* promoter, → twin promoter, → upstream promoter, → weak promoter, → *wun* promoter. Do not confuse with → tumor promoter.

Promoter array: Any solid support (e.g. a glass slide), onto which a whole series of → oligonucleotides complementary to a → promoter region are immobilized (usually in triplicate). Most frequently many promoter regions are interrogated with one single array. The array is then incubated with fluorescently labeled proteins to detect interactions between them and the promoter sequences. Promoter arrays are used for the identification of → transcription factors and for the mapping of their binding sites. For example, such promoter arrays allow to detect the binding of cell cycle regulator E2F4 to at least 2–3% of a total of 13,000 tested human promoters. However, other transcription factors associate with many more promoters, as e.g. c-Myc that binds to some 10–15% of human promoters, as does its dimerization partner Max. c-Myc in fact binds to a total of 25,000 sites in the human genome. Promoter arrays allow to establish transcription factor-binding maps.

Promoter-associated long RNA (PALR): Any one of a series of cytoplasmic and nuclear → poly-adenylated RNAs longer than 200 nucleotides (spanning from several hundreds of bases to more than 1 kb) that overlap the 5′-end of protein-encoding genes (i.e. usually the → core promoter and the first → exon and → intron region). PALRs map to the same position as → promoter-associated small RNAs. See → promoter RNA, →short RNA, → termini-associated small RNA.

Promoter-associated small RNA (PASR, promoter-associated transcript): Any one of a series of cytoplasmic and nuclear → poly-adenylated RNAs shorter than 200 nucleotides (spanning from 20–200 nucleotides) that are synthesized by → DNA-dependent RNA polymerase II, and map to the 5′-end of standard protein-encoding genes. PASRs are expressed at levels similar to transcripts from the protein-coding genes they overlap, but do not extend far enough to function as → messenger RNAs (mRNAs). Instead, at least some of these PASRs regulate (i.e. repress) the → transcription of downstream mRNAs. For example, *SRG1* RNA is such a non-coding promoter-associated RNA, whose expression regulates the adjacent *SER3* gene in *Saccharomyces cerevisiae*. Some of the PASRs are syntenic between human and mouse, and the boundaries of PASRs are conserved between human cell lines. Almost 50% of human protein-coding genes are bracketed by PASRs (and → TASRs). See → promoter-associated long RNA, → promoter RNA, → short RNA, → termini-associated sRNA.

Promoter-associated transcript: See → promoter-associated small RNA.

Promoter bashing: A laboratory slang term for a technique that allows to determine regulatory DNA sequence motifs in → promoter regions by → mutation, either by → deletions of internal sequences, successive → exonuclease digestion of the promoter 5′-end, or by → restriction endonuclease(s)-catalyzed removal of segments of the promoter. Subsequent residual promoter functions are tested with a variety of methods (e.g. → electrophoretic mobility shift assays, → reporter gene expression assays).

Promoter clearance: The ATP-dependent escape of a transcription elongation complex stalled after synthesis of the first 10–17 nucleotides of → messenger RNA, catalyzed by the ERCC3 subunit of transcription factor IIH. See → initiation.

Promoter core: See → core promoter.

Promoter hypermethylation: The methylation of most, if not all, → cytosine residues in a → promoter sequence (see → localized hypermethylation) that leads to the inactivation of the promoter and the adjacent gene. Loss-of-function by promoter hypermethylation is common in several cancer-related genes involved in → DNA repair, cell cycle control, → apoptosis, angiogenesis, cellular differentiation, metatstatic invasion, → transcription and signal transduction.

Promoter insertion: The integration of a → promoter or promoter-containing DNA segment in front of a promoter-less or otherwise inactivated gene with the result of activation of the gene. See also → insertional activation.

Promoter melting: The ATP-dependent and transient breakage of hydrogen bonds ("melting") of about one turn of DNA encompassing the → transcription start site to form the socalled "transcription bubble", catalysed by transcription factor

IIH (in humans) or analogous proteins in other organisms. See → promoter opening.

Promoter-methylated gene: A laboratory slang term for any gene, whose adjacent → promoter region is methylated at CpG residues. For example, about 5% of *Arabidopsis thaliana* expressed genes represent such promoter-methylated genes. Promoter-methylated genes are tissue-specifically expressed at comparatively low levels. See → body-methylated gene.

Promoter module: A structural and functional unit of eukaryotic → promoters that is composed of two (or more) → transcription factor binding sites (TF sites) in a defined distance from each other. This arrangement allows for synergistic or antagonistic influences of the different transcription factors – that bind to the different TF sites – on each other, which in turn result in stimulation or inhibition of the → transcription of the adjacent gene.

Promoter mutation: Any → mutation that occurs within the → promoter sequence of a gene. For example, socalled *aphakia* (*ak*) mouse mutants (aphak: without lens) that do not form any lens or pupil in their otherwise normal embryonic development, the underlying gene *Pitx3* on chromosome 19 is absolutely identical to the wild-type gene. However, two deletions of 652 and 1423 bp, respectively, in the promoter and in the transcription initiation region lead to an almost complete silencing of the *Pitx3* gene. As a consequence, the encoded → homeobox → transcription factor is not functional, and the eye development does not occur. See → mutated promoter.

Promoter occupancy: The extent to which all of the → transcription factor-recognition sites of a → promoter are occupied at a given time. Promoter occupancy can be tested with e.g. → chromatin cross linking with immune precipitation.

Promoter opening: The localized formation of a short stretch of melted DNA ("DNA bubble") near the 3' end of a → promoter. The opened structure is accommodated in a channel of the bacterial RNA polymerase, and serves to assemble the preinitiation complex.

Promoter plasmid: A → plasmid cloning vector that contains an → RNA polymerase → promoter that can drive genes cloned into a cloning site 3' downstream of it.

Promoter polymorphism: An imprecise term for any → sequence polymorphism that occurs in a → promoter of a gene. Usually such polymorphisms are caused by small → deletions, → insertions, or, most frequently, → single nucleotide polymorphisms. Promoter polymorphisms may be neutral (i.e. without effect on the → transcription of the adjacent gene) or inhibit the binding of specific proteins necessary for accurate transcription (e.g. → transcription factors), and account for differences in the expression of a distinct gene between two (or more) individuals. See → promoter SNP.

Promoter prediction program (PPP): Any software package for *in silico* discovery of → promoter sequences in → genomic DNA without support from → expressed sequence *t*ag (EST), → messenger RNA or → cDNA sequences. The underlying principle of all the different programs is the detection of sequence elements characteristic for various types of promoters that cannot be found in the bulk of genomic DNA. Such promoter-specific elements are either → CpG islands, → transcription start sites (TSSs), specific → transcription factor binding sites (TFBS), a combination of TFBSs and their mutual distances, the → CCAAT box, the → GC box and the → initiator region (Inr), homology with orthologous promoters, and/or a → TATA-box motif in a GC-rich domain, but also first → exon properties. For example, programs DragonPF, DragonGSF, McPromoter, NNPP2.2 and Promoter2.0 use artificial neural networks (ANNs) as part of their design, DragonPF, DragonGSF, Eponine, FirstEF are based on GC contents, Eponine, NNPP2.2 and Promoter 2.0 use the TATA motif, and CpGProD, DragonGSF andf FirstEF exploit different versions of CpG islands. Non-CpG-island promoters own specific sequence features, and can be detected by FirstEF with only low efficiency. Some of the programs predict potential promoter sequences close to, or worse than, random guess. The power of PPPs can be increased by masking repeat sequences (e.g. by Repeat Masker at http://repeat-masker.org/ or http://ftp.genome.washington.edu/RM/RepeatMasker.html), or by combining predictions from two, or better more different pograms. If only a single PPP is employed, then DragonGSF or Eponine provide good coverage of promoters. Additionally, Dragon GSF has a preference for CpG-island-related promoters.

Promoter primer: Any synthetic → oligodeoxyribonucleotide that is complementary to a conserved sequence in either → T7, → T3, or → SP6 RNA polymerase promoters. Such primers are used to sequence inserts in plasmids containing these promoters.

Promoter probe vector: A → cloning vector that contains appropriate cloning site(s) located just 5′ (upstream) of a promoter-less → reporter gene (e.g. a bacterial → β-glucuronidase gene). Any foreign DNA cloned into such a vector and possessing promoter elements will drive the expression of the reporter sequence. For example, the vector plasmid pPL 603 designed to clone promoter sequences from *Bacillus subtilis*, contains a → kanamycin resistance gene and a promoter less *Bacillus* → structural gene for → chloramphenicol acetyl transferase (CAT). Immediately upstream of the CAT coding sequence a unique *Eco* RI site is located into which foreign DNA can be inserted. If the foreign DNA contains a promoter element, then the CAT gene will be express ed. Compare → promoter trap vector, which allows the identification of promoter sequences in → genomic DNA after being itself inserted into chromosomal sites.

Promoter proximal pausing: The transient blockage of the → elongation of the nascent → messenger RNA by → DNA-dependent RNA polymerase II about 20–60 bases downstream of the → transcription start site. The release of this blockage relaxes the limiting step in the transcription of e.g. the human *c-myc* gene and the *Drosophila hsp70* gene. The pausing of RNA polymerase II can also be suppressed by elongation factors such as SII, TFIIF, ELL, or elongin.

Promoter region (PR): Any region in a sequenced → genome that may represent a → promoter on the basis of similarity (or identity) of specific sequence motifs to entries in the databases.

Promoter RNA (pRNA): An RNA → transcript derived from the → intergenic spacer (IGS) between two → ribosomal RNA gene units, whose synthesis is catalyzed by →

DNA-dependent RNA polymerase II driven by an IGS → promoter 2 kb upstream of the gene. This pRNA is complementary to the promoter of the rRNA gene and necessary for rDNA promoter methylation and silencing. The conserved pRNA secondary structure is recognized and bound by the socalled → *n*ucleolar *r*emodelling *c*omplex (NoRC), a prerequisite for rDNA silencing and transcription stop. Since pRNA is also present as → anti-sense RNA, it may form → double-stranded RNA, which then is cleaved by either Drosha or → Dicer to form → *s*mall *i*nterfereing RNA (siRNA). See → promoter-associated long RNA, → promoter-associated small RNA.

Promoter scanning: Any procedure to screen → promoter sequences for specific functions. For example, the → methylation interference analysis uses the *in vitro* methylation at → purine residues of the target sequence to prove that this sequence can bind proteins. If so, then this sequence is considered to be an address site for → DNA-binding proteins.

Promoter shuffling: The recombination of individual regulatory elements of different → promoters to create a new promoter with new function(s). Shuffling involves the gain or loss of such elements ("boxes") during evolution. Usually the length, copy number and relative location of the boxes vary in different → orthologous gene promoters, a results of gross changes such as → amplification, contraction, → deletion, duplication, elongation, → fusion, → inversion, → transposition, and the steady accumulation of → single nucleotide polymorphisms. Compare → exon shuffling, → intron shuffling.

Promoter single *n*ucleotide *p*olymorphism (promoter SNP, pSNP): Any → single nucleotide polymorphism that occurs in the → promoter sequence of a gene. If a pSNP prevents the binding of a → transcription factor to its recognition sequence in the promoter, the promoter becomes partly disfunctional. See → anonymous SNP, → candidate SNP, → coding SNP, → exonic SNP, → gene-based SNP, → human SNP, → intronic SNP, → non-synonymous SNP, → promoter polymorphism, → reference SNP, → regulatory SNP, → synonymous SNP.

Promoter strength: The frequency with which an → RNA polymerase molecule can bind to specific → consensus sequences within a → promoter and express the linked gene. It depends on specific sequences (e.g. → TATA box, → CAAT-box) and their exact → spacing within the promoter region. Compare → promoter-up mutant. See → strong promoter, → weak promoter.

Promoter tiling array: Any → microarray, onto which → oligonucleotides (from 40–85 nucleotides in length) are immobilized that represent → promoter regions at an average 100 bp spacing from –4000 to +1000 of a set of known genes of an organism (e.g. human, mouse, *Arabidopsis thaliana*). All annotated → messenger RNAs (mRNAs), including splice variants (see → alternative splicing) and mRNAs with alternative start and → alternative polyadenylation sites are represented. Preferentially, oligonucleotides are designed to be isothermal at a specific → T_m (e.g. T_m = 76 °C). Promoter tiling arrays are used for the identification of active promoters in a given genome by e.g. the determination of the sequences where binding of → TFIID complexes occurs.

Promoter transcript: Any RNA encoded by a → promoter and transcribed at a low level in the cell (see → promoter-associated long RNA, → promoter-associated small RNA). Therefore, synthetic small RNAs of approximately 21–28 bp in length can modulate transcriptional → gene silencing, when designed to target complementary regions of → RNA polymerase II promoters. These small RNAs can be called → small interfering RNAs, though their target is a promoter transcript.

Promoter trapping: The use of → promoter trap vectors for the isolation of → promoters.

Promoter trap vector: A → transformation vector that carries a transcriptionally and translationally incompetent (i.e. promoter-less) → reporter gene (e.g. → neomycin phosphotransferase, or → β-glucuronidase gene) and used to select → promoter sequences from a → genomic library. If such a vector integrates at a position in the host genome where promoter sequences are located, the reporter gene will be transcribed, allowing the identification of promoter-containing genomic fragments. See for example → T-DNA mediated gene fusion. Compare → promoter probe vector.

Promoter-up mutant (up-promoter mutant, up mutant): Any → mutant with a → mutation in the → promoter of one of its genes that leads to a higher rate of expression of this gene. In gene technology such up-promoters are used for the → overexpression of genes encoding useful proteins.

Promotor: See → promoter.

Pronase: A mixture of serine proteases and acid proteases from *Streptomyces griseus* that catalyze the cleavage of → peptide bonds in proteins. Pronase is used to degrade proteins in RNA and DNA isolation procedures, and is especially effective for the isolation of intact RNA molecules through its deleterious effect on → RNases. Compare → proteinase K.

Pronuclear injection (pronuclear microinjection): A variant of the → microinjection technique for the transfer of foreign genes into target cells, which works with a fine glass needle to inject about 200–300 linearized copies of a gene of interest (fused to a promoter) into previously fertilized egg cells. After injection, the eggs are briefly cultured and then implanted into surrogate mothers. Only 1–5% of the newborne animals are transgenic (because integration of the → transgene into the host cell chromosome is extremely rare and additionally a totally random process), and only a fraction of these transgenics express the foreign gene strongly. Pronuclear injection suffers from mosaic expression of the transgene (i.e. its expression in some, but not all cells of the → transgenic animal).

Proof-reading ("editing"): The correction of errors in DNA → replication or DNA → transcription. Proof-reading of replicative errors is catalyzed by DNA polymerase, which recognizes and removes incorrectly inserted bases by its 3'-5' exonucleolytic function.

Propeller twist: The relative rotation of bases with respect to the long axis of a base pair.

Propensity: The likelihood, with which a certain amino acid residue is contained in a specific secondary structure (e.g. a → domain) of a protein (e.g. glycine in an → α–helix).

Prophage: A → bacteriophage DNA integrated covalently into the chromosomal DNA of a bacterial host, and replicated as part of the host's genome. Compare → lysogenic bacteriophage, also → temperate phage. See also → phage exclusion.

Prophage interference: See → phage exclusion.

Prophage-mediated conversion: See → phage conversion.

Propidium *i*odide (PI): A → fluorochrome that binds to DNA. The propidium-iodide-DNA complex can be excited at a wave length of 488 nm, and emits red fluorescent light. PI is the most commonly used dye for → flow cytometry. See → ethidium bromide.

Proportion of *e*ssential genes (P_E): The fraction of essential genes in the sum of the total number of genes in a genome.

Propyne pyrimidine (C-5 propyne pyrimidine): Any → deoxycytidine or → deoxyuridine, in which a propyne group is introduced at the C-5 position of the base. This modification results in a stronger binding of the base to target RNA sequences.

Prosite: A database for proteins, protein families and protein domains maintained at the *E*uropean *M*olecular *B*iology *L*aboratory (EMBL). Prosite orders the huge protein diversity on the basis of similarities in their sequences into a limited number of families.

Protamine: An arginine-rich, highly basic protein that replaces → histones in sperm heads of various animals to package the DNA into an extremely compact form. The protamines have molecular weights of 3000–5000, contain up to 50% of arginine residues that are mostly clustered, and can be classified into monoprotamines with one type of basic amino acid (arginine; e.g. clupeine, salmine, iridine, truttine, esocine), diprotamines (arginine and lysine; e.g. barbine, cyprinine, pereine) or triprotamines (arginine, lysine, histidine; e.g. sturgeone).

Protease (proteinase): An enzyme catalyzing the hydrolysis of → peptide bonds in proteins or oligopeptides. Proteinases may be classified according to the chemical nature of the amino acids located in their reactive center (e.g. serine proteases, acid proteases). See for example → pronase, → proteinase K.

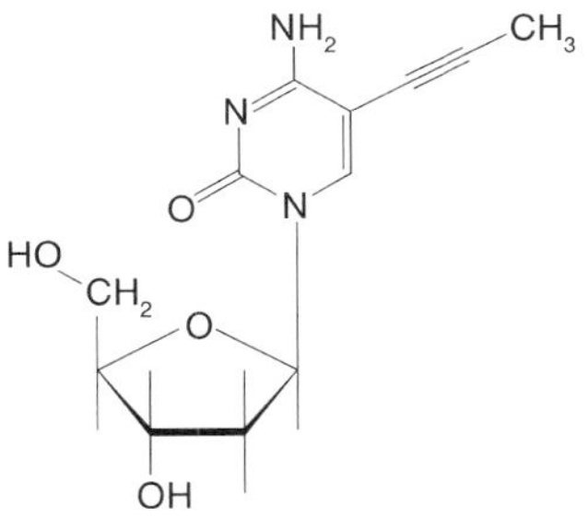

Propyne-dC

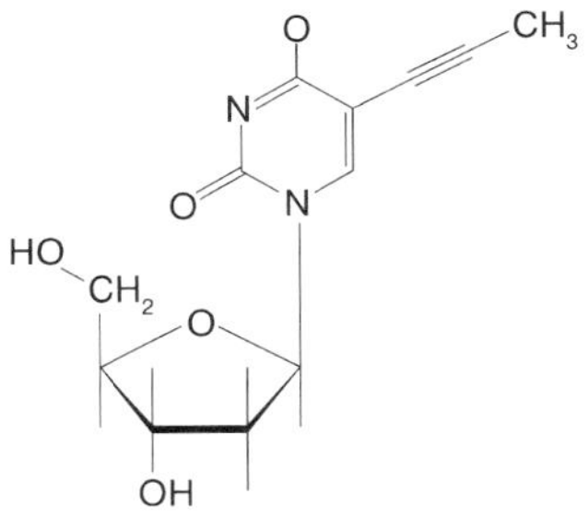

Propyne-dU

Proteasome (26S proteasome): A nuclear and cytoplasmic 2,000 kDa multi-protein complex consisting of 45–50 individual polypeptides organized in several sub-complexes that altogether function in the proteolytic degradation of abnormal proteins and the bulk turnover of all other proteins. Moreover, sub-sets of proteasomal proteins are involved in → antigen processing, cell cycle control, cell differentiation, stress response, pre-protein cleavage, and programmed cell death.

The degradation of proteins starts with their tagging by multiple enzymatic → ubiquitin (Ubq) binding to the ε-amino group of specific internal reactive lysine residues, catalyzed by a multi-enzyme complex consisting of Ubq-activating, Ubq-conjugating and Ubq-ligating enzymes. The conjugated ubiquitin moieties function as "secondary" → degradation signals (degrons) for pro teasomal proteases. The proteasome itself consists of a core complex with multiple peptidase activities (the 20S proteasome with chymotrypsin-like sites [cleavage after hydrophobic residues], trypsin-like sites [cleavage after basic residues], and caspases [cleavage after acidic residues]), a barrel-like structure with central catalytic centers. Each degron-tagged protein has to be completely unfolded to be channeled to these centers, which is achieved by an 11S particle (PA 700) consisting of 15–20 different polypeptides, associated with each end of the 20S complex. The whole degradation process requires ATP (at least six AAA-ATPases are part of the 11S particle that act as reverse → chaperones and unfold the substrate proteins). An average human cell contains about 30,000 such proteasomes. In addition, proteasomes also mediate amide bond formation. Defects in the proteasome pathway lead to diseases in humans. For example, a mutated sodium channel that is no longer recognized by ubiquitination enzymes, causes the socalled Liddle syndrome, a specific form of hypertension. Or, the von Hippel-Lindau syndrome, which predisposes for certain types of cancer as well as the Angelman syndrome, a mental retardation disorder, are each caused by defective E3 ligases. Moreover, the mutant huntingtin (causing Huntington disease) and mutant α-synuclein (accumulating in patients with Parkinson disorder) both inhibit proteasome function. Compare → degradosome. See → ubiquitin-proteasome system.

Figure see page 1257

Proteasome inhibitor: Any one of a series of peptides or peptide conjugates that inhibits the activity of one of the six proteolytic sites on the → rings of the → proteasome. For example, synthetic peptide aldehydes such as Z-leu-leu-leu-al, Z-leu-leu-nval-al, Ac-leu-leu-nle-al, or Z-ile-glu-ala-leu-al reversibly inhibit the proteasomal chymotrypsin-like activity (but also other proteases, e.g. calpains, serine proteases of digestive vacuoles, and lysosomal cysteine proteases). Other proteasome inhibitors are more specific, as e.g. peptide boronates (e.g. Z-leu-leu-leu-boronate), synthetic peptide vinyl sulfones, epoxyketones such as expoxomicin, or the non-peptide lactacystin (that converts to the active form β-lactone in aqueous solution). Proteasome inhibitors are used as antiviral drugs and allow to study the functions of proteasomes.

Protectifer: The direct transfer of nucleosomally organized or otherwise protected genes into eukaryotic recipient cells (see → direct gene transfer). Protection is usually achieved by complexing DNA with proteins (e.g. → histones, → protamine from salmon sperm, protein VII from

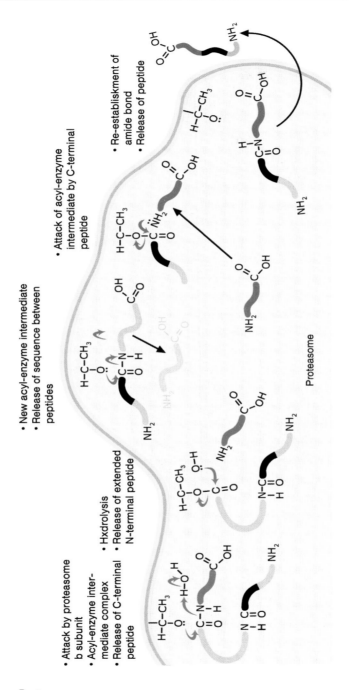

Proteasome

adenovirus type 2, HMG-1), but low molecular weight compounds such as spermidine have also been used successfully. Protectifer enhances the → transformation rate, the integration of *intact* copies of the transferred gene into the target genome, and generally stimulates expression of the transgene.

Protecting group: Any chemiacal group that minimizes or prevents undesirable side reactions (i.e. protects functional groups during chemical reactions as e.g. *in vitro* peptide or oligonucleotide synthesis). It must be resistant to the conditions of synthesis and be easily removed after completion of polymer synthesis. For protein synthesis, protecting groups fall into two categories: the amino protecting groups (e.g. urethanes, *benzoyloxy*carbonyl [Cbo] groups ["Z groups"] and derivatives, and tert-*butoxy*carbonyl [Boc] groups) and the carboxy protecting groups (e.g. methyl, benzoyl or tert-butyl groups that are esterified in the presence of hydrochloric acid, acetanhydride or thionyl chloride). For an oligonucleotide synthesis, acid-labile, substituted trityl groups protect the primary 5'-OH groups, and sylyl or acyl groups protect the 3'-OH groups of deoxyribonucleotides. In nucleosides, the 2'- and 3'-OH groups can be protected through the introduction of an acid-labile isopropylidene bridge:

Protective allele: Any → allele of a gene, whose → expression protects a carrier from developing a phenotype (e.g. a disease phenotype). See → contributing allele, → neutral allele.

Protein A: A cell wall protein from *Staphylococcus aureus* that recognizes the Fc portion of most → antibodies only when the antibody is bound to its → antigen. Protein A, either labeled with a fluorescence marker or a radioactive atom (e.g. ^{125}I) is used to detect antibodies in a variety of antibody screening procedures (e.g. → sandwich techniques, → direct immune assays), or functions as ligand in affinity chromatography of immunoglobulins and antigen-antibody complexes. Compare → protein G.

Protein affinity tag: Any synthetic, usually short amino acid sequence that is part of a → fusion protein encoded by a → fused gene. The tag allows to isolate and purify the target protein in a single → affinity chromatographic step. In principle, a highly tag-affine compound is covalently bound to a (preferentially) inert matrix, the fused protein as part of a more complex protein mixture then exposed to the matrix, the fused protein bound via tag-tag-affine component interactions, the non-bound proteins washed off, and the tagged protein eluted in almost pure form. For example, the socalled arg-tag, composed of 5-6 arginine residues (RRRRR, 0.8 kDa), can be fused to the → C-terminus of the target protein and serves to bind the construct onto → cation exchanger matrices, from which it can be eluted by a sodium chloride gradient. Another example is the family of his-tags (or poly-his tags), six or more histidine residues (HHHHHH, 0.84 kDa) that bind to metal ions as e.g. Co^{2+}, Ni^{2+}, Cu^{2+} or Zn^{2+} immobilized on a matrix via their imidazole ring. The longer the his-tag, the stronger the interaction with the metal ions. The tagged protein can then be eluted from the → metal affinity column by imidazol. Since his-tags may influence the tertiary structure of the target protein and its activity, an improved variant, the hat-tag, was designed that contains six histidines separated by other amino acids (KDHLIHNVHKEF HAHAHNK, 2.73 kDa). The 26 kDa glutathione-S-transferase tag (GST-tag) can be used for protein purification under physiological conditions (eluent: reduced glutathione). The biggest tag is the maltose-binding protein tag (MBP-tag) with 40 kDa that binds to cross-linked amylose. The tagged protein is eluted with maltose. The → streptavidin (strep)-tag family exploits streptavidin or modified streptavidins as chromatographic matrix. The original strepI tag (AWRHPQFGG, 1.01 kDa) could only be fused to the C-terminus of the target protein, the advanced strepII tag (WSHPQFEK, 1.06 kDa) to both termini, but both tags belong to the class of low-affinity binders. Higher-affinity streptavidin-binding tags are the streptavidin-binding peptide (SBP-tag; MDEKTT GWRGGHVVEGLAGELEQLRARLEHHP QGQREP; 4.30 kDa) and the so called nanotag (either MDVEAWLGAR or MDVEAWLGARVPLVET; 1.15 or 1.78 kDa, respectively). SBP- and nanotags bind to streptavidin in the nanomolar range and are optimal for interaction studies. The flag tag DYKDDDDK (1.01 kDa) and the advanced tandem flag tag DYKDHDG-DYKDHDIDYKDDDDK (2.73 kDa) bind to immobilized → antibodies M1, M2, and M5, and can eluted by chelators such as EDTA. Additionally, two short peptides, the c-myc tag (EQQKLISEEDL, 1.20 kDa) and the S-tag (KETAAAKFERQH-MDS, 1.75 kDa) bind to antibody 9E10 and the S-fragment of → RNase A, respectively. The calmodulin-binding peptide KRRWK-KNFIAVSAANRFKKISSSGAL (2.96 kDa) binds to calmodulin in the presence of potassium dichloride and in the nanomolar range. Other, less known tags are the cellulose-binding domain (CBD, various sequences, 3.00–20.00 kDa) and the chitin-binding domain tag TNPGVSAWQVN-TAYTAGQLVTYNGKTYKCLQPHTSLAGWE PSNVPALWQLQ (5.59 kDa) that binds to the respective polyglycans with high affinity. Altogether these protein affinity tags are valuable tools for the isolation of fusion proteins by affinity chromatography.

Protein array: See → protein chip.

Proteinase K (EC 3.4.21.14): An enzyme from the fungus *Tritirachium album* that catalyzes the cleavage of peptide bonds in proteins with a slight preference for aliphatic, aromatic or other hydro-phobic amino acids. The enzyme belongs to the subtilisin type serine proteases, and is effectively used to degrade proteins (including → RNases and → DNases) in RNA or DNA isolation procedures. The activity of proteinase K can be stimulated by denaturants (e.g. SDS, urea) and elevated temperature. Autolysis of the enzyme is minimal, and may be prevented by Ca^{2+}. Compare → pronase.

Protein association cloning (PAC): A term encompassing several → cloning techniques that use → antibodies to screen → cDNA → expression libraries. Usually the antibody is radioactively labeled and interacts with its target protein expressed in a → bacteriophage or → plasmid expression library. This interaction can be detected by → autoradiography and the corresponding clone carrying the target cDNA be easily isolated.

Protein atlas: A laboratory slang term for an inventory of all peptides and proteins of a cell at a given time, their interactions, and localization. See → protein interaction mapping, → protein linkage map, → protein profiling, → protein-protein interaction map, → protein signature, → proteome, → proteome mapping, → proteomic fingerprint.

Protein autoprocessing (autoprocessing): The autocatalytic modification of a peptide or protein. For example, *cis*- or *trans*-splicing, transpeptidation or self-cleavage are autoprocessive reactions. See → protein splicing.

***P*rotein-*b*inding *m*icroarray (PBM):** A high-throughput *in vitro* technique for the detection and analysis of DNA-protein interaction(s) and DNA sites for protein binding. In short, a purified → epitope-tagged DNA-binding protein is first incubated with a → microarray, onto which DNA in the form of short synthetic double-stranded → oligonucleotides or also longer → PCR products are spotted. The microarray is then washed gently and stained with a fluorophore-conjugated anti-tag → antibody (in case of a GST-tagged protein an anti-GST antibody is used, for example), washed, dried and scanned. Fluorescence indicates DNA-protein interaction(s). The proteins can be dissociated from the DNA, and the DNA sequenced to identify the DNA-binding motif. See → DNA immunoprecipitation followed by microarray analysis.

Protein biochip: See → protein chip.

Protein biomarker: Any peptide or protein that is associated with and characteristic for a specific state of a cell, a tissue, an organ, or an organism. Such markers are identified with high-throughput screening techniques such as → protein chips, e.g. using normal and diseased samples.

Protein biosensor: Any protein tagged with a fluorophore that responds to any environmentally induced change in structure, ligand binding, or catalytic activity of the protein with an increase or decrease of emitted fluorescence. Usually the biosensors are transferred to a target cell by → microinjection. The construction of a protein biosensor may require → protein engineering. For example, to engineer a myosin light chain kinase biosensor, a cysteine residue has to be introduced into the myosin light chain next to the phosphorylated amino acid. This cystein is then specifically labeled with the fluorochrome acrylodan that reacts upon phosphorylation/dephosphorylation events of the light chain with a change in fluorescence.

Protein blot: See → Western blotting.

Protein blotting: The transfer of electrophoretically separated proteins from a gel (usually a → polyacrylamide gel) to a membrane by diffusion (diffusion blotting), by liquid flow (using capillary forces, or vacuum, → vacuum blotting) or by electrophoretic transfer (→ electroblotting). Most commonly used are → nitrocellulose (binding capacity: $80\,\mu g/cm^2$), but also nylon based (binding capacity: $480\,\mu g/cm^2$) and polyvinylidene difluoride membranes (binding capacity: about $500\,\mu g/cm^2$). After transfer, the filters may be used for → Western blotting or → South-Western blotting experiments.

Compare → cDNA expression array, → DNA chip, → expression array, → gene array, → microarray, → sequencing array, → sequencing by hybridization.

Protein booster: Any protein that is augmenting an immune response, if injected into an immunresponsive organism after, or simultaneously with the → antigen. For human applications, usually a protein or protein mixture from the envelope of a virus is used as protein booster.

Protein chip (protein array, protein biochip, protein microarray, protein microchip, protein-protein interaction chip): A glass slide or other solid support, onto which thousands of proteins are spotted in an ordered array to allow the visualization of protein-protein (e.g. protein-antibody) or protein-ligand interaction(s). In short, the glass slides are first treated with an aldehyde-containing silane, because aldehydes react with primary amines of proteins to establish a Schiff's base linkage (covalent coupling). Alternatively, the glass surfaces can be coated with poly-L-lysine (non-covalent coupling), or reactive epoxide (covalent coupling). Then purified proteins, each in a nanodroplet (containing billions of protein copies; the concentration should be 100 µg/ml or less) are spotted onto the glass slide that is additionally coated with → bovine serum albumin (BSA) to prevent the denaturation of the spotted protein, which is then covalently bound to the glass surface. For peptides or very small proteins, a molecular layer of BSA is first attached to the glass, which is then activated with N,N'-disuccinimidyl carbonate to yield BSA-N-hydroxysuccinimide, and quenched with glycine. The peptides are then displayed on top of the BSA monolayer, which readily exposes them for interaction(s) with target molecules. Then a ligand is fluorescently labeled and exposed to the protein array, and binding of the ligand to a protein (or to many proteins) detected by fluorescence. Using different fluorophores with non-overlapping excitation and emission spectra expands the number of simultaneously detectable interactions. Such protein chips can be used to study the → proteome of a cell. Alternatively, → monoclonal antibodies can be bound onto the surface of microarray chips, to which peptides or proteins from a cell extract are bound. Compare → antibody array, → cDNA expression array, → DNA chip, → expression array, → functional protein array, → microarray, → protein domain array, → recombinant protein array, → sequencing array, → sequencing by hybridization.

Reactive epoxide surface chemistry binds proteins covalently in several different ways and, therefore, a high number of protein molecules can be captured on the surface of various binding reactions. Proteins are not oriented.

Figure see page 1262

Protein-coding sequence (COD): A more general term for any sequence in genomic DNA that encodes a peptide or protein.

Protein complementation assay (CPA): An *in vivo* selection system for peptides or proteins with novel properties that is based on the functional complementation of the bacterial dihydrofolate reductase (DHFR) with its murine equivalent (mDHFR). In short, the mDHFR gene is genetically dissected into two fragments, each encoding a part of the enzyme necessary for its function. Each of these fragments is then separately fused to a library of peptides or proteins. *E. coli* cells are then co-transformed with both fusion libraries and plated on selective medium (containing the antibiotic → trimethoprim, TMP). TMP in turn specifically inhibits bacterial DHFR, thereby prevents the synthesis of essential purines, also pantothenate and methionine and therefore blocks cell division. The complemented mDHFR is insensitive to low TMP concentrations and

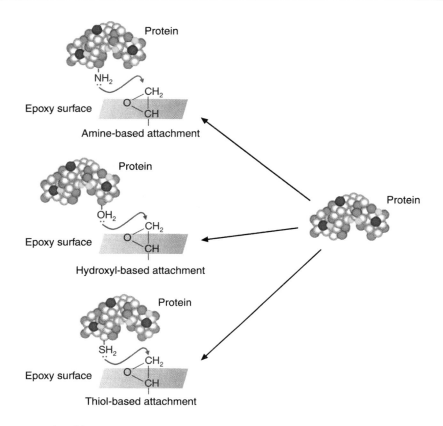

Protein chip

allows *E. coli* cells to grow on minimal medium containing TMP. Therefore, the complementation of DHFR as a result of the interaction of members of the different libraries (heterodimerization) restores DHFR enzymatic activity and guarantees survival (growth of colonies). The surviving bacteria are directly accessible for the sequencing of the interacting library partners. See → directed molecular evolution, → ribosome display.

Protein complex: An aggregate of various proteins, whose coordinated activities exert a novel function that the individual proteins do not possess. For example, the → primosomes, → proteasomes, → repairosomes, → ribosomes, → spliceosomes, and → transcriptosomes are such protein complexes, to name only few.

Protein computer: See → peptide computer.

Protein conformation: The tertiary structure of proteins. See → α-helix, → arginine fork, → β-sheet, → helix-loop-helix, → helix-turn-helix, → random coil; → zinc cluster, → zinc finger and → zinc twist proteins.

Protein connectivity: The total number of connections that a particular protein establishes with other proteins under a certain experimental condition. Protein connectivity in this narrow sense disregards the identity of interacting protein partners as well as the type of interaction.

Protein correlation profiling: A technique for the identification and characterization of proteins from a cellular organelle (e.g. the centrosome) that is based on the traditional fractionation of cell organelles in a → sucrose gradient, and the subsequent analysis of all fractions (not only the fractions enriched for the target organelle, as e.g. the centrosome) by → mass spectrometry and sequencing of the peptides. Correlation with already known proteins of the target organelle and their known *in vivo* localization identifies true organellar (here: centrosomal) proteins. Then the peptides of all fractions are quantified and socalled abundance profiles generated. Finally, common profiles are mathematically calculated and considered to be indicative for proteins from the same organelle. Protein correlation profiling is used to analyze multiprotein complexes that can be enriched by e.g. sucrose density centrifugation, but not purified to homogeneity.

Protein cross-linking: The formation of covalent bonds between free amino groups of two individual proteins by the action of bifunctional cross-linking chemicals (as e.g. *N-h*ydroxysuccinimide [NHS] esters such as disuccinimidyl tartarate, disulfosuccinimidyl tartarate, dithio bis[succinimidyl propionate], or ethylene glycolbis [succinimidylsuccinate]) that react primarily with the ε-amino groups of lysine side-chains. Protein cross-linking *in vivo* is possible through the use of water-soluble (or even water-insoluble) agents that permeate cellular membranes and e.g. preserve intracellular structures (as e.g. actin filaments, actin-binding proteins such as gelsolin, and tubulins) for subsequent analysis.

Protein crowding: An undesirable accumulation of overexpressed proteins in host cells that leads to unspecific protein-protein interactions via hydrophobic or electrostatic interactions, and/or reduced solubility and mobility of the protein(s). All these consequences of macromolecular crowding eventually lead to the precipitation of the proteins. Protein crowding also causes the formation of plaques or inclusion bodies that are indicative for some neurodegenerative diseases (as e.g. Alzheimer disease).

Protein crystallization: The formation of crystals from a purified, solubilized protein that starts from its transfer into a state of supersaturation. First, the reaction conditions (e.g. the ionic strength of the medium and its pH and temperature) have to be optimized to avoid any precipitation of the target protein. Then several techniques can be used. The socalled vapor diffusion comes in two variants, the hanging drop or sitting drop version. In both variants, a crystallization droplet, formed by combining a protein solution with a socalled reagent solution is incubated together with a larger volume of the same reagent solution in a closed chamber. The composition of the reagent solution varies from laboratory to laboratory, and may contain a wide range of salts, buffers, and precipitants. After mixing both solutions, the concentration of reagents within the droplet becomes lower than the concentration of the reagent solution itself, which in turn causes water to evaporate from the droplet unless equilibrium is attained. During vapor diffusion, the concentrations of protein and chemicals in the droplet are continuously rising, which ultimately leads to protein crystal formation. In another technique, the socalled microbatch crystallization, the protein-containing droplet is covered with oil (usually paraffin wax), which prevents any water diffusion from the droplet. Therefore, all ingredients including the protein remain more or less constant. Then the paraffin wax is mixed with silicone oil, which allows water to diffuse from the droplet: the protein concentration is increasing, and crystals form under optimized conditions. Both techniques allow high-throughput (96 or 384 well microtiter plates). A free interface diffusion technique requires microfluidics chips. Imaging of the crystals in various stages combines lenses with CCD cameras and bright Xenon lamps or LCD illumination. Crystals can also be detected by crossed polarizers. Under the exposure to polarized light, crystals become light in color (non-crytalline material remains dark). Subsequent polarimetry identifies even minute crystals.

Protein data bank (PDB): An international repository for NMR, X-ray crystallography or homology data of peptides, proteins, RNA, DNA, viruses and polysaccharides.

Protein design: The development of synthetic (i.e. not naturally occurring) proteins with predictable structural and functional properties, supported by special computer programs. The amino acid sequence of such designed proteins may then be translated into a nucleotide sequence, see → gene design. Compare → protein engineering.

Protein domain array ("protein binder array"): A glass slide or other solid support, onto which native or synthetic peptides representing specific → domains of a protein are spotted at high density. Domain arrays serve to map protein-protein interactions, to detect interactions between ligands and domains and to profile the specificity of enzymes (e.g. kinases or proteases) for their protein substrates. See → functional protein array, → protein chip, → recombinant protein array.

Protein dynamics: The multitude of conformational changes in a protein as a result of its interaction(s) with intrinsic or environmental factors (e.g. ligands).

Protein electroblotting: See → protein blotting, and → electroblotting.

Protein engineering: The modification of the physico-chemical or biological properties (i.e. the reaction kinetics, substrate affinity, substrate specificity, effector sensitivity, thermostability or -lability, intracellular location) of a naturally occurring protein with the ultimate aim of improving the protein's quality for biotechnological processes. One of the techniques used in protein engineering is → *in vitro* mutagenesis that allows to change the coding capacity of a gene at defined locations (e.g. within the region encoding the active center of the protein). In consequence, the engineered protein adopts different properties that are useful for biotechnology. Compare also → protein design.

Protein epitope signature tag (PrEST): Any unique subfragment of a protein that forms a conformational → epitope correctly representing the folded part of the native protein and triggers the formation of a highly selective → antibody. PrEST design is based on the identification of subfragments with no (or minimal) similarity to any other protein (or protein →

domain) to avoid cross-reactivity by the generated antibodies. Further, transmembrane regions are excluded (difficult to express in *E. coli*), signalpeptides avoided (unsuitable as epitopes, since they are cleaved off during translocation), → restriction sites not allowed, and the PrEST optimally sized (~100–150 amino acids long). After the basic design the → oligonucleotide encoding the PrEST is synthesized, cloned into an appropriate → expression vector and expressed in *E.coli*.

Protein equalizer technology: Any one of several techniques for the reduction of the abundancies of the different proteins in biological samples such that also the socalled → low abundance proteome (a → subproteome that contains preferentially low-abundance proteins, i.e. proteins present in low copy numbers or even traces) becomes manageable. For example, socalled combinatorial solid-phase ligand libraries or → ligand library beads can be employed. Each bead in auch a library carries a different hexapeptide ligand with affinity to a specific protein in a sample. A complex protein sample is incubated with this bead library, and proteins find their capture peptides and bind to them. Excess high-abundance proteins do not find hexapeptide ligand partners and are depleted after washing the beads. Low-abundance proteins, however, will be concentrated on their specific affinity ligands and enriched in contrast to high-abundance proteins.

Protein expression: An infelicitous and misleading term for → gene expression. The terms "protein expression" and "protein expression profiling" should be avoided, because only genes can be expressed (into → messenger RNAs, which in turn are translated into proteins, the products of gene expression).

Protein expression array: See → antibody array.

Protein expression map: A graphical depiction of the subcellular distribution, abundance and relative concentration of (preferably) all peptides and proteins, also their post-translational modifications, in a cell at a given time. A protein expression map is the result of → protein expression mapping.

Protein expression mapping (PEM): The description of the abundance and subcellular distribution of preferably all peptides and proteins of a cell, a tissue, an organ or an organism under defined physiological conditions at a given time. PEM exploits all → proteomics techniques. See → protein interaction mapping.

Protein family expansion: The evolutionary increase in the number of proteins of a protein family, reflecting an increase in the number of corresponding genes. For example, a comparison of protein families in *Anopheles gambiae* versus *Drosophila melanogaster* reveals an expansion of various protein families in the former: many more genes encode proteins for cell adhesion, anabolic and catabolic enzymes involved in protein and lipid metabolism, lysosomal enzymes and salivary gland proteins, reflecting blood feeding and oviposition.

Protein farnesylation: The covalent transfer of a 15-C isoprene *f*arnesyl *d*iphosphate (FPP) residue onto target proteins, catalyzed by the cytoplasmic enzyme protein *f*arnesyltransferase (PFT). Protein farnesylation is conserved in fungi, plants and animals, including humans. The transferase is composed of two subunits α and β, of which the β-subunit harbors the substrate- and lipid-binding pocket and determines substrate-protein specificity through sequence-specific recognition of a carboxy-terminal CaaX motif (C = cystein, a = aliphatic amino acid, X=cysteine, methionine, serine, alanine, or glutamine). After recognition of the substrate, PFT catalyzes the covalent attachment of the 15-carbon farnesyl chain via a thioester bond to the cysteine of the CaaX-box on the target protein. After farnesylation, the modified protein is transported to the endoplasmic reticulum (ER), where the carboxyterminal three amino acids aaX in the CaaX-box sequence are cleaved by an endopeptidase (e.g. sterility 24, STE24). The exposed COOH-group of the farnesylated cysteine is then methylated by a prenyl-cysteine methyl transferase (PCM). The farnesylated protein is finally targeted to the plasma membrane, or interacts with other proteins. Prenylated proteins function in cellular signaling and membrane trafficking. Members of the Ras family of GTPases, several γ-subunits of G proteins, nuclear lamins, cyclic nucleotide phosphodiesterase subunits, protein kinases, and type-I inositol-1,4,5-triphosphate-5-phosphatase are such farnesylated proteins. Since several oncogenic forms of farnesylated proteins (e.g. Ras) depend on prenyl modification for their transforming activity, the inhibition of PFT has potential for medical intervention.

Protein fingerprinting (peptide fingerprinting): A technique to characterize a protein by partial proteolytic cleavage that generates a pattern of → peptide fragments characteristic for this protein. In short, the purified protein is cleaved separately with endoproteinases (e.g. Glu-C that cleaves at the carboxylic side of glutamic acids, Lys-C that cleaves at the carboxylic side of lysine residues or Arg-C [clostripain] that cleaves at the carboxylic side of arginine residues), and alkaline protease (that cleaves preferentially at aromatic residues). The proteolysis products are then separated according to size by electrophoresis or chromatography and visualized by staining. Protein fingerprints can be used to confirm the identity between the product of a cloned gene and its natural counterpart, or for the analysis of multi-subunit proteins.

Protein fold: The specific three-dimensional structure of a protein adopted in solution that can be analysed by e.g. X-ray crystallography. See → RNA fold.

Protein fold disease (protein folding-related disease): Any human disease that is caused by a misfolding of a protein (see → protein fold). For example, so called amyloidoses such as the Alzheimer or Creutzfeld-Jakob diseases (involving deposits of aggregated proteins ["amyloid plaques"] in many tissues of a patient), lung diseases such as cystic fibrosis or hereditary emphysema (involving → mutations leading to the degradation of proteins with vital functions for the respiratory tract), diabetes (misfolded proteins disrupt carbohydrate metabolism or accumulate in the endoplasmic reticulum with toxic consequences), cancer (misfolding of tumor-suppressor proteins abolish their tumor-suppressor function. For example, mutation(s) in the gene encoding → p53, such a tumor-suppressor protein, causes a misfold of the corresponding protein that in turn fails to recognize its target. About half of all tumors are probably caused by such a p53 misfolding), or infection diseases (pathogens interfere with the host ER-associated degradation (ERAD) system that is responsible for the removal of terminally misfolded proteins) are such folding-related diseases, to name only few.

Protein folding: The spontaneous and autonomous self-organization of a nascent or unfolded → polypeptide chain into a unique three-dimensional structure (the → protein fold) that is usually necessary for the function of the protein. Accurate folding is a prerequisite for the interaction of a protein with other proteins and low molecular weight compounds (e.g. substrates, effectors). Any misfolding, or failure to remain correctly folded, may lead to protein aggregates and serious cellular misfunction of the protein. See → protein fold disease.

Proteinformatics (protein informatics): The whole repertoire of software packages for the evaluation of terabytes of data generated by high-throughput → proteomics. Proteinformatics employs software for the identification of peptides and proteins and the prediction of their structure (e.g. Piums), the computation of → peptide fingerprints (e.g. BioAnalyst, biotools, Ettan MALDI-TOF software, Knexus, Mascot, ProtoCall, Radars), → mass spectrometer (MS) peak extraction, including peak picking and reporting (e.g. Pepex, ProID, SNAP), cross-correlation of MS/MS mass spectra of peptides with protein (or nucleic acid) databases (e.g. Mascot, Turbo-SEQUEST), → post-translational modification analysis and *de novo* sequencing (ProteinLynx Global SERVER). Frequently appropriate software is developed for a specific instrument (e.g. → MALDI) and sold in combination.

Protein fragment complementation assay (PCA): A technique for the detection of protein-protein interactions in living cells (*in vivo*) and *in vitro*, based on the dissection of a protein (e.g. an enzyme or → fluorescent protein, called a reporter) into two fragments that are fused to two proteins thought to interact with each other. The interaction of these proteins, to which the fragments are fused, restores the unique three-dimensional structure ("fold") of the reporter protein. If the reporter is an enzyme, then the PCA restores its catalytic activity that can be measured. Should the reporter be a protein that binds a fluorescent → probe, then its reconstitution is monitored by → fluorescence. Do not confuse with *p*erchloric *a*cid (PCA).

Protein-free medium: Any medium for the cultivation of cells that does not contain proteins.

Protein function array (protein interaction array): Any → microarray, onto which peptides or proteins (or protein domains) are immobilized in a native state (i.e. in correct three-dimensional folding) such that specific function(s) can be tested in high-throughput (as e.g. the binding requirements and strength of interaction of the protein with other proteins, DNA, RNA, or a low molecular weight ligand, or a specific activity). The correctly folded state of a target protein on a protein function array can e.g. be achieved with the *b*iotin *c*arboxyl *c*arrier *p*rotein (BCCP), which is biotinylated *in vivo* only if it is correctly folded. Now BCCP is covalently attached to the target protein and serves as indicator for the native state of this protein, binds very strongly to a → streptavidin-coated microscope slide and at the same time presents the target protein at a distance of 50Å above the underlying glass slide (thereby exposing functional sites to the environment). See → antibody array.

Protein fusion and purification technique (PFP): A method for the expression and purification of proteins from cloned genes by fusing them to the *m*altose-*b*inding *p*rotein (MBP). The fusion protein is extracted from the producer cell and purified in a one-step affinity column chromatography. The affinity column retains any MBP (together with the fused protein). The bound proteins can be released in almost pure form. PFP is a variant of → protein tagging.

Protein G: A cell wall protein of certain strains of *Streptococcus* that binds to a wide variety of IgG → antibodies by a non-immune mechanism (i.e. does not involve the antigen-binding site of IgG). Protein G is used to purify antibodies, and to detect antibodies in a variety of antibody screening procedures (e.g. → sandwich techniques, → direct immune assays) in which it is either labeled with a fluorescence marker or a radioactive atom (e.g. ^{125}I). Compare → protein A.

Protein homeostasis: The balance between the denaturation, renaturation, degradation, *de novo* synthesis and intracellular distribution of proteins.

Protein insert: See → intein.

Protein *in situ* array (PISA): A → protein chip, on which tagged functional proteins are immobilized that are synthesized by → cell-free expression directly from PCR-amplified genes and simultaneously bound to the chip surface. PISAs can therefore be made in a single step from DNA, avoiding cloning, cell-based expression and protein purification.

Protein interaction array: See → protein function array.

Protein interaction domain: Any protein → domain (an independently folding module of 30–150 amino acids, whose N- and C-termini are in close proximity to each other, while the ligand-binding site is on the opposite face of the domain) that mediates interaction with another peptide or protein, or other proteins. Protein interaction domains can be characterized by their common sequences or ligands. For example, many cytoplasmic proteins contain one or two socalled SH2 (Src-homology 2) domains recognizing phosphotyrosine moieties and 3–6 residues C-terminal to the phosphotyrosine of e.g. activated receptors for growth factors, antigens or cytokines. Other proteins harbor socalled PTB domains, recognizing phosphotyrosine in the context of Asn–Pro–X–Tyr (NPXY) motifs (which represent β-turns). These molecular interactions lead to protein networks mediating e.g. signaling from cell surface receptors to intracellular metabolic pathways.

Protein interaction map: See → protein-protein interaction map.

Protein interaction mapping (PIM): The process of establishing a → protein-protein interaction map. See → protein expression mapping.

Protein interactome: See → interactome.

Protein interference (Pi): The modulation of the function of a particular protein by an → intrabody, an immunoglobulin expressed intracellularly and directed to defined subcellular compartments. Such intrabodies can be directed towards specific → domains of the target protein, and prevents its interaction with other proteins or ligands. Compare → RNA interference.

Protein intron: See → intein.

Protein knock-out: The specific inactivation of a particular protein or the inhibition of transcription of its gene, so that the function of the protein is missing and (preferably) leads to a visible or measureable phenotype. Protein knock-outs can be generated by e.g. → chromophore-assisted laser inactivation. See → gene knock-down, → gene knock-in, → gene knock-out.

Protein L: A cell wall-bound receptor protein of 719 amino acids in some strains of the anaerobic bacterium *Peptostreptococcus magnus* that binds to all → immunoglobulin (Ig) classes (e.g. IgA, IgD, IgE, IgG, IgM), to kappa light chains (without interfering with the binding of the → antigen), and to single chain variable fragments (ScFv). Binding to target proteins is mediated through 5 socalled B domains (each 72–76 amino acids long, sharing 70–80% homology at both the nucleotide and amino acid level). Protein L can therefore be used to detect and purify ScFv and Fab fragments containing kappa light chains, and IgA, IgD, IgE, IgG, IgM and IgY. See → protein A, → protein LA, → protein L-agarose.

Protein LA: A recombinant protein that combines the immunoglobulin-binding domains of → protein A with those of → protein L, so that it binds strongly to diverse immunoglobulins from many different species with high affinity. Therefore LA is coupled to → agarose, and the resulting LA-agarose used to purify antibodies in a single affinity chromatography step.

Protein L-agarose: A beaded → agarose coupled to → protein L that is used for the separation of immunoglobulins (e.g. IgA, IgD, IgE, IgG and IgM containing kappa light chains) from a variety of sources, and the purification of → monoclonal antibodies. See → protein A, → protein LA.

Protein library ("proteotheque"): Any collection of artificially created mutant proteins, each differing by a few amino acid residues that change the three-dimensional structure in an interesting → domain. Usually such libraries are established by systematic variation of the corresponding gene sequence, transformation of *E. coli* host cells, selection of the clones with the desirable mutant proteins by → phage display or colony screening techniques for an interaction with an immobilized target protein ("ligand"). Interacting proteins can then be isolated, sequenced and analyzed.

Protein ligand affinity column: Any glass or plastic column filled with a matrix (e.g. → agarose) that is covalently bound to a specific → ligand (e.g. an → antibody, a peptide, or a protein), and allows to separate specific proteins from complex protein mixtures. A variety of different protein ligand affinity matrices are available. For example, amine-reactive agarose (in which a reactive glyoxal group is coupled to agarose, whose aldehyde group reacts with primary amines to form intermediate Schiff base complexes) binds to exposed primary amines of peptides or proteins. The Schiff's bases are stabilized by sodium cyanoborohydride that only reduces the Schiff base intermediates and not the aldehyde groups. Therefore the resulting ligands are more stable than ligands coupled with cyanogen bromide. Or, sulfhydryl-reactive agarose binds proteins via their reduced sulfhydryl groups. The agarose beads carry a reactive aminoethyl group terminating in a primary amine ("activated agarose") that is cross-linked with a ligand using e.g. *sulfo*succinimidyl

4-(N-maleimidomethyl)cyclohexane-1-carboxylate (sulfoSMCC). The N-hydroxysuccinimide (NHS) ester of sulfoSMCC reacts with the primary amine of the aminoethyl group to form covalent amide bonds. Then the sulfhydryl groups of the ligand react with the maleimide group to stable thioether bonds.

Protein linkage map: A misleading term for all proteins physically interacting with each other in a cell, and identified by chemical crosslinking or → two hybrid or → three-hybrid system techniques.

Protein lipidation: The covalent conjugation of phosphatidylethanolamine to a protein via an amide bond between a C-terminal glycine of the protein and the amino group of phosphatidylethanolamine, which is catalyzed by a ubiquitination system. Protein lipidation plays an essential role in membrane dynamics.

Protein machine: A generic name for any intracellular complex of many proteins that interact physically and cooperate synergistically. For example, → nucleosomes, → primosomes, → repairosomes, → ribosomes, → spliceosomes and → transcriptosomes are such multi-protein machines.

Protein maturation: See → post-translational modification.

Protein microarray: See → protein chip.

Protein microsequencing: See → microsequencing.

Protein mimic: Any peptide or protein possessing special domains ("mimotopes") that mimic the immunogenic properties of antigens. For example, a peptide mimic of the pneumococcal serotype 4 polysaccharide (pep4) induces much stronger immunoglobulin G (IgG) responses in mice than the polysaccharide itself. Therefore, the mimotopes can structurally mimic carbohydrate antigens, induce carbohydrate-reactive B- and T-cell responses after immunization, and enhance boosting responses after priming with carbohydrate. See → DNA mimic, → RNA mimic.

Protein minimization: A somewhat imprecise term for the reduction of the level of cellular proteins during evolution to the minimum necessary for their function. Protein minimization avoids molecular crowding that would affect functionality of the various proteins.

Protein misfolding cyclic amplification: A technique for the amplification of infectious → prions (PrP^{Sc}) in the test tube. In short, a brain of a Scrapie-infected animal is first homogenized and mixed with the brain homogenate from a non-infected animal. The PrP^{Sc} from the first animal's brain start transforming any PrP^C to PrP^{Sc}. The homogenates are periodically sonicated to break growing PrP^{Sc} fibers. The mixture is then diluted into more non-infected brain homogenate, and the process repeated.

Protein misrouting: The aberrant transport of newly synthesized proteins into cellular compartments (e.g. mitochondria, plastids, endoplasmic reticulum, vacuoles, lysosomes, nuclei, to name few), into which they are never incorporated under normal conditions.

Protein nanoarray: Any solid support (e.g. a gold-coated glass chip), onto which peptides or proteins are spotted via e.g. →

dip-pen nanolithography in arrays of 100 nm (or less) diameter and 100 nm (or less) distance between spots (see → ultra-high density microarray). In short, a protein nanoarray is fabricated by first patterning 16-*m*ercapto*h*exadecanoic *a*cid (MHA) as dots on a gold film substrate. The areas surrounding the spots are passivated with 11-mercaptoundecyl-tri (ethylene glycol), and the proteins absorbed on the preformed MHA patterns by immersing the substrate with the target proteins. Interactions between the proteins on the nanoarray and target molecules are detected by atomic force microspcopy.

Protein network: The entirety of → protein-protein interactions within a cell. See → protein expression map, → protein expression mapping, → protein interaction map, → protein interaction mapping.

Protein palmitoylation: The reversible transfer of a palmitate group onto a cysteine residue of an acceptor protein, catalyzed by *p*almitoyl *t*ransferase (PAT). The enzyme uses the activated form palmitoylCoA and attaches the palmitate group to the cystein via a thioester linkage. The reaction can be reversed by a *p*rotein *t*hio*e*sterase (PTE). Protein palmitoylation is necessary for the fusion of cellular membranes. See → SNARE.

Protein PCR: See → protein polymerase chain reaction.

Protein pharmaceutical: Any one of a series of pharmaceutically active proteins. See → recombinant protein pharmaceutical.

Protein *p*olymerase *c*hain *r*eaction (protein PCR): A misleading term for a specific → coupled transcription/translation system, in which an → *o*pen *r*eading *f*rame (ORF) is first ligated into the → multiple cloning site of a specific vector and efficiently transcribed into an RNA template that is immediately used in a linked translation system to produce the corresponding protein. The ORF is first amplified in a conventional → polymerase chain reaction, using → *Taq* DNA polymerase. The extendase activity of this enzyme adds an adenosyl residue to the 3' end of the amplicon, which allows easy annealing to a T-overhang of the vector. Subsequent ligation of the ORF to the vector obviates the need for cloning. Then a new round of amplification with a vector-specific and an insert-specific → primer (which carries a → T7 RNA polymerase → promoter) generates the template for an effective T7 RNA polymerase-driven transcription. Subsequent → *in vitro* translation in a → rabbit reticulocyte lysate produces the protein of interest that can then be analyzed by → denaturing polyacrylamide gel electrophoresis.

Protein polymorphism: Any difference in the amino acid sequence between two (or more) homologous proteins from different cells, tissues, organs, or organisms. These polymorphisms are altogether a consequence of mutations in the → exons of the corresponding genes.

Protein processing: See → post-translational modification.

Protein profile (protein signature): The complete inventory of peptides and proteins in a specific sample (e.g. a patient's blood sample, buccal sweep, needle biopsy, or tumor specimen) by → protein profiling.

Protein profiling: The establishment of a – preferably complete – inventory of all proteins (and peptides) of a cell at a given time. Protein profiling encircles the isolation of the various types of proteins (e.g. membrane proteins, "soluble" proteins, glyco- and phospho-proteins), their characterization (e.g. electrophoretic behaviour, estimation of molecular weight), their sequencing (see → microsequencing, → protein finger-printing, → protein sequencing) and the correlation of the sequence data with entries in the protein data banks (e.g. SWISSPROT) in order to functionally categorize the proteins and peptides. In a wider sense, protein profiling may also comprise the identification of specific functions of the proteins (e.g. enzymatic activity, affinity towards ligands, → protein-protein-, protein-DNA-, protein-RNA- and protein-ligand interactions, posttranslational modifications, and others). See → protein profile.

Protein-protein interaction (PPI): The noncovalent and usually transient *in vitro* or *in vivo* interaction(s) between two (or more) proteins necessary for the execution of a function or functions that each protein by itself cannot perform. Protein-protein interactions are the basis of cellular life, and lead e.g. to the aggregation of → protein machines, enable the transfer of signals through signal transduction chains and guarantee high efficiency in metabolic pathways, to name very few. The multitude of PPIs form the basis of the (presumed) dynamic proteome network, are detected by e.g. → yeast two-hybrid systems, and can be represented by a → protein-protein interaction map.

Protein-protein interaction map (protein interaction map, protein linkage map): The establishment of the complex network of interactions between (preferably all) proteins of a given cell at a given time point.

Usually → two-hybrid screening procedures are employed to systematically screen for such interactions. See → protein-protein interaction, → proteome.

Figure see page 1273

Protein quantitative locus (PQL): Any one of a series of genes contributing to the complexity and composition of the → proteome of a cell, a tissue, an organ or an organism. Compare → quantitative trait locus.

Protein recognition molecule (PRM): Any synthetic or naturally occurring molecule that recognizes a target peptide or protein and bind it with high affinity. For example, → antibodies, → aptamers or → molecularly imprinted polymers (MIPs) or molecularly imprinted electrosynthesized polymers (MIEPs) are such PRMs.

Protein sequencing: The determination of the sequence of amino acids in purified oligo- and poly-peptides. Protein sequencing starts with the identification of both terminal amino acids. The amino-terminus reacts with → dansyl chloride (dansylation) or, alternatively, with dimethyl aminoazobenzoyl-isothiocyanate (DABITC), phenyl-isothiocyanate (PITC), or fluorodinitrobenzol (FDNB), and can be isolated after acid hydrolysis as the corresponding derivative (e.g. dansyl derivative) e.g. by → high-pressure liquid chromatography. The carboxy-terminus is released by carboxypeptidase, using specific reaction conditions. The bulk of the protein is then cleaved into a series of peptide fragments, either by chemical means (e.g. with cyanogen bromide that cleaves specifically at the carboxy group of a methionine) or enzymatically (e.g. by endopeptidases). The sequencing of these fragments again starts with the modifica-

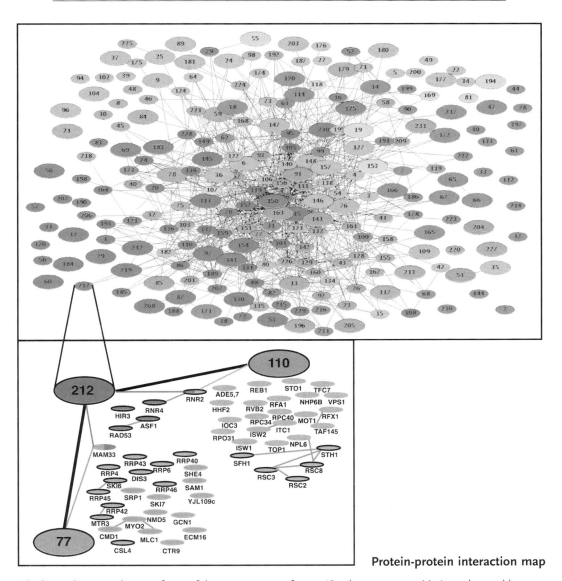

Protein-protein interaction map

A higher order network map of part of the → proteome of yeast (*Saccharomyces cerevisiae*), as detected by → tandem affinity purification (TAP) tagging and → matrix-assisted laser desorption-ionization time-of-flight mass spectrometry (MALDI-TOF-MS).

Upper panel: The functions of the individual protein complexes in cell metabolism are encoded in colors (*dark blue*: → transcription, DNA maintenance, → chromatin structure; *light blue*: membrane synthesis and membrane trafficking; *red*: cell cycle; *dark green*: signalling; *pink*: protein and RNA transport; *orange*: RNA metabolism; *light green*: protein synthesis and protein turnover; *brown*: cell polarity and structure; *violet*: intermediate and energy metabolism).

Lower panel: Details of the connection between one yeast protein complex (212) and two other complexes (77 and 110). Red lines: physical interactions.
(Kind permission of Dr. Monika Blank. Cellzome, Heidelberg, Germany; see Acknowledgements).

See → protein atlas, → protein complex, → protein interaction map, → protein interactome. → protein network, → protein-protein interactions, → proteome map.

tion of their amino terminal amino acids by phenyl-isothiocyanate, the subsequent cleavage of the neighboring peptide bond, the separation of the modified amino acid from the residual peptide, and its identification. Then the peptide is again modified at its new amino terminus, and the whole cycle (Edman cycle) reinitiated until the peptide fragment is sequenced. The whole process is called Edman degradation. See also → microsequencing.

Protein signature: The complex pattern of all proteins and peptides (see → proteome) of a cell at a given time, as revealed by e.g. → two-dimensional or → three dimensional polyacrylamide gel electrophoresis and subsequent characterization of the individual spots by e.g. → electrospray ionization or → time-of-flight mass spectrometry.

Protein splicing: The posttranslational precise excision of → inteins from a precursor protein, the first product of → messenger RNA → translation. The → excision (definition c) of the intein(s) is catalyzed by intein-extein border-specific endoproteinases, and leads to free exteins, which are subsequently linked to each other by peptide bond formation. Highly conserved amino acid residues mark both splice junctions. A hydroxyl or thiol containing residue (e.g. serine, threonine, or cysteine) is always present at the position that immediately follows both junctions. All inteins own an invariant C-terminal asparagine and, in most cases, a histidine residue in the penultimate position. Protein splicing is a multi-step process involving N-S shift, transesterification, succinimide formation and an S-N shift, is very efficient (the precursor protein rarely accumulates) and leads to a → mature protein. Compare → DNA splicing, → protein autoprocessing, → RNA splicing, → splicing.

Protein superfamily: Any cluster of proteins related by sequence homology or structurally and functionally similar or identical conformations. This sequence homology is reflected by the sequence homology of the encoding genes. See → gene superfamily.

Protein tagging: A technique for the detection of a specific protein by fusing it to a second protein that can be easily monitored (by e.g. a specific antibody or by an enzyme assay). Protein tagging is used to screen for → translational fusions. Such fusion proteins are often produced using the α-peptide of → β-galactosidase. See also → protein fusion and purification technique, → expression vector (→ fusion vector).

Protein target site (PTS): Any one of a multitude of (usually short) DNA sequence motifs in a genome that are recognized by specific proteins (e.g. → transcription factors) and serve as address sites for these protein (i.e. the protein strongly binds to the PTS). PTSs are usually components of → promoters, but also located in → introns, at the 3' end of genes, or in the → intragenic space. Such PTSs can be detected by e.g. → two-dimensional electrophoretic mobility shift assays.

Protein targeting: The translocation of a newly synthesized protein from the cytoplasm into a cellular compartment such as the chloroplast, endoplasmic reticulum, lysosome, nucleus, mitochondrium, peroxysome, or vacuole. Protein targeting requires specific → targeting sequences ("targeting signals") within the protein that may or may not be cleaved by specific proteases ("cleavases"). Many proteins are either channeled into the secretory pathway or onto the lysosomal route. Other proteins are endocytotically taken up from the extracellular medium and either targeted to endosomes or lysosomes. Still other proteins are initially targeted to the endoplasmic reticulum (some remain in the lumen of the ER), then become translocated to the *cis*- and *trans*-Gologi network, from which they may be transferred to endosomes, lysosomes or secretory granules. Proteins in the granules are transported to the cell surface and are secreted into the extracellular space.

Protein therapy: The replacement of nonfunctional peptides or proteins in a cell by the transfer of correctly folded and functional peptides or proteins via e.g. → microinjection, resulting in → complementation. See → gene therapy.

Protein thermometer: A laboratory slang term for any protein that senses temperature changes in the environment and reacts with a conformational change of one (or more) of its → domains. For example, the *Salmonella typhimurium* DNA-binding protein TlpA, encoded on a → virulence plasmid, is a transcriptional → repressor of its own synthesis. If the pathogen enters the host organism, then the increased temperature, which is sensed by long coiled-coil helices of TlpA, leads to a switch from the folded coiled-coil oligomeric states to an unfolded monomeric state. Since only the folded form of TlpA acts a s a repressor, the host temperature favors a → derepression and subsequent expression of the *TlpA* genes (and probably other genes as well). Likewise, the *Drosophila* HSF protein activates the expression of target → heat shock genes in response to elevated temperature in the environment. This → transcription factor is normally present in a monomeric form that does not bind its cognate sequence in DNA. Initial activation of HSF results in the conversion of monomers to homotrimers that in turn bind DNA with high affinity. The HSF trimerization domain is composed of several hydrophobic heptad repeats that assume a three-stranded coiled-coil structure in the HSF trimer, but is precluded from forming in a monomer. See → RNA thermometer.

Protein transduction: The direct transfer of peptides and proteins (e.g. → antibodies) into target cells without the interference of a receptor. In nature, transducing proteins (e.g. the HIV-1 TAT protein, or the *Drosophila* Antennapedia transcription factor) contain short peptide sequences rich in basic amino acids (see → protein transduction domain) that catalyze the internalization of the cargo protein. See → non-covalent protein delivery. But see also → cell penetrating peptide.

Protein transduction domain (PTD, Trojan horse): A short peptide sequence motif of about 10–16 amino acids (with a high number of positively charged lysine and arginine residues) in viral or cellular proteins that transduces across the plasma membrane without the necessity of a receptor. PTDs are present in transducing proteins such as e.g. HIV-1 TAT (consensus sequence: YGRKKRRQRRR), HSV-1 VP22, and the *Drosophila* Antennapedia transcription factor, to name few. The PTD is always covalently bound to the cargo protein, which is rapidly internalised, but excluded from the endocytotic pathway. PTDs can be used to import target proteins almost independently of their size (e.g. proteins in excess of 700 kDa), even → antisense oligonucleotides and → liposomes (with a diameter of more than 200 nm) *in vivo* and *in vitro*. For this purpose, artificial PTDs are synthesized (e.g. with the sequence H_2N-RRRRRRRR-COOH, or H_2N-KKKKKKKK-COOH, or H_2N-KKKKKKKKK-COOH). A TAT derivative (H_2N-YARAAARQARA-COOH) is 33 times more efficient than the Tat peptide itself. The bonds between a cargo protein and a PTD can also be made reversible, if e.g. disulfide or ester linkages are used that are reduced or cleaved within the target cell. See → cell-penetrating peptide, → non-covalent protein delivery.

Protein transfection: A technique for the delivery of biologically active peptides or proteins (e.g. antibodies) into living cells. For example, the target protein can be transferred into a cell by → lipofection, i.e. noncovalently complexing cationic lipids with the negatively charged protein, fusing the complex with the cell membrane (if it is not endocytotically taken up), and releasing the protein into the cytoplasm. Ideally, the protein should be noncovalently complexed with a transfective carrier, which protects and stabilizes the protein and preserves its structure and function during the transfection process. Once internalized, the complex should dissociate and the protein be liberated in an active conformation. Usually proteins with a high positive overall charge or minimal hydrophobicity are less efficiently transfected as compared to negatively charged or hydrophilic proteins. Compare → transduction. See → transfection.

Protein transformation: The → transformation of a cell by a protein rather than a nucleic acid. For example, yeast cells contain a → prion-like factor called PSI^+ that is inherited by classical Mendelian rules (transmitted from parents to their progeny both in mitoses and meioses). A cross between PSI^+ and psi^- therefore produces a PSI^+ progeny. The yeast prion PSI^+ results from self-propagating aggregation of Sup35p, a protein required for efficient → termination of translation. In psi^- state, Sup35p is not aggregated and active (sup35[u]), whereas in the PSI^+ state it is aggregated and inactive (Sup35[ag]). The PSI^+ aggregates of Sup35p enhance the → read-through of → nonsense mutations by → ribosomes. The PSI^+ prion propagates when a misfolded version of the Sup35 protein templates the aggregation of the properly folded Sup35 protein, thereby converting the psi^- to a PSI^+ cell. The trait PSI^+ newly acquired by protein transformation is then transmitted from generation to generation.

Protein translocation: The vectorial movement of a protein from the cytoplasm (where it is synthesized) across a membrane (e.g. into a chloroplast, a mitochondrion, the nucleus, or the intraluminal space of the endoplasmic reticulum). See → protein translocon.

Protein translocon: Any membrane-bound multiprotein complex that catalyzes the import, but also export of proteins through membranes (e.g. from cytoplasm into the nucleus or mitochondria, or plastids in higher plants, from cytoplasm to the cell wall, the apoplastic space in plants). For example, the 400 kDa multiprotein translocase of the *o*uter *m*embrane (TOM) complex is such a translocon. It consists of receptor protein TOM20 (that recognizes and binds the helical N-terminal signal sequence of the import protein), the TOM70 protein (that binds the chaperones associated with the precursor protein), and the socalled *g*eneral *i*mport *p*ore, GIP) with its central protein TOM40 (forming a ~22 Å pore in the outer mitochondrial membrane and translocating the precursor protein across it). Associated with TOM40 are at least three smaller TOM proteins and the central receptor protein TOM22 (altogether called the GIP complex). The transport of the import protein through the inter-membrane gap and across the inner membrane requires the socalled *t*ranslocase of the *i*nner *m*embrane, TIM), another translocon. The core of TIM contains TIM17 and TIM23, two intergral membrane proteins. TIM23 is the inner membrane receptor protein, recognizing the N-terminal signal sequence of the precursor protein ("pre-sequence") and forming a 13 Å channel, through which the pre-sequence is guided. TIM50 as additional 40 kDa component protrudes into the inter-membrane gap, binds to TIM23, and directs the pre-sequence together with the complete protein into the inner translocation pore. The membrane potential $\Delta\psi$ across the inner membrane activates TIM23 that pulls the positively charged pre-sequence electrophoretically through the inner membrane. The final ATP-dependent import into the mitochondrial matrix is achieved by a complex consisting of TIM44, the *mit*ochondrial Hsp70 (mtHsp70) and its co-chaperones. After (or already during) the translocation, the pre-sequence is removed by a *m*atrix *p*rocessing *p*eptidase (MPP). The proteins without a pre-sequence, but internal signal sequences, are imported by another translocon (consisting of the TIM9-TIM10 complex, the 300 kDa TIM22 complex with the integral membrane proteins TIM12, TIM18, TIM22, and TIM54).

Protein truncation assay: See → protein truncation test.

Protein *t*runcation *t*est (PTT; *protein truncation assay*, PTA): A technique for the detection of one or more → translation termination → mutations occuring in a → gene that lead to early → termination of → transcription and the synthesis of a protein with shorter amino acid chain length (truncated protein) as compared to the wild type protein. This truncated protein is usually inactive ("loss of function"). In short, → genomic DNA or → messenger RNA is first isolated and the target sequence amplified in a conventional → polymerase chain reaction. To that end, the mRNA is reverse transcribed into → cDNA. The amplification uses specially designed primers carrying either a T7, or → SP6 or T3 phage promoter sequence as adapter at its 5′ end. The resulting PCR products serve as → templates for an *in vitro* RNA synthesis, employing the corresponding phage RNA polymerases. The RNAs in turn are translated into the corresponding proteins, which are then subjected to denaturing → SDS polyacrylamide gel electrophoresis. If a specific mutation in the target gene sequence leads to a premature stop of the translation of the encoded protein, it will be shorter ("truncated") and migrate further in the gel than the wild-type encoded protein. PTTs discover frequent mutations especially in comparatively long coding sequences (e.g. → megagenes; for example the *D*uchenne *M*uscular *D*ystrophy, DMD gene) and additionally allow the detection of → splicing variants.

Figure see page 1278

Proteolytic processing: The highly specific and limited *in vivo* cleavage of substrate peptides and proteins by a variety of proteases. Proteolytic processing controls a series of cellular pathways, as e.g. angiogenesis, → apoptosis, cell cycle progression, cell proliferation, → DNA replication, immunity, or wound healing).

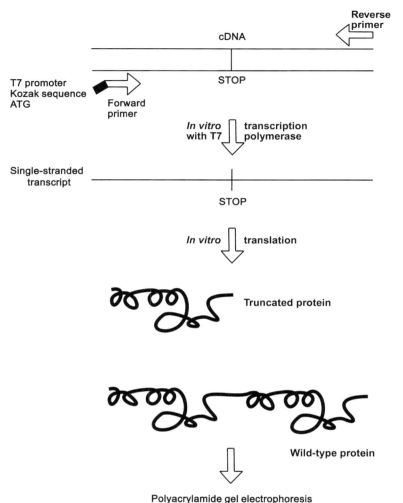

Protein truncation test

Proteome (*protein-genome*): The complete set of proteins in a cell, an organelle, or any sub-cellular or sub-organellar compartment at a given time. The proteome of a particular cell changes dynamically with its developmental stage and changing environment, and comprises many more proteins than → genes are present in the → genome, because (1) → alternative splicing might give rise to → mRNAs that are different from the original mRNA, and (2) proteins are frequently modified by co- or post-translational modification (e.g. acetylation, ADP-ribosylation, C-terminal truncation, → glycosylation, methylation, N-terminal truncation, phosphorylation, sulphation). See → proteome mapping, → proteomics. Compare → proteosome.

Proteome array: A misleading term for a → microarray, onto which full-length → cDNAs or → open reading frame clones, preferably covering the whole → genic space of a genome, are spotted. Compare → proteome chip.

Proteome chip (proteome microarray): Any → microarray, onto which preferably all highly purified proteins of an organism are spotted. The genes encoding these proteins are first cloned, then overexpressed in a bacterium, the overexpressed proteins isolated and purified, and then covalently bound onto the chip's surface. For example, a total of 5,800 *Saccharomyces cerevisiae* genes can be expressed as fusions to sequences encoding glutathione-S-transferase, the corresponding fusion proteins isolated and purified and arrayed on glass slides. Such a chip covers about 94% of all yeast genes (or proteins), and therefore represents a proteome chip. Proteome chips are used for studies of protein-protein interaction(s). Compare → proteome array.

Proteome map: The graphical description of the abundance, quantity and subcellular distribution of (preferably) all peptides and proteins in a cell at a given time. Mostly identical to → protein expression map. See → protein expression mapping, → protein interaction mapping.

Proteome mapping: A misleading term for the categorization of all proteins of a given cell at a given time, using the whole technology of → proteomics, and reflecting the number of → active genes only, without any clue as to where in the → genome these genes map. But see → gene mapping.

Proteome microarray: See → proteome chip.

Proteome signature: The specific patterns of the various → sub-proteomes of a cell at a specific time and a particular developmental stage, as detected by e.g. → two-dimensional polyacrylamide gel electrophoresis of the individual sub-proteomes and subsequent staining of the protein spots. Proteome signatures (or more precisely, characteristic indicator proteins) are characteristic for a specific situation of a cell (e.g. after stress, during mitotic change, after a knock-out of a gene).

Proteomic diversity (proteome diversity): The variations of protein number, protein complexity and protein functions in two (or more) → proteomes. Proteomic diversity is partly a result of → alternative exon usage, → alternative splicing, or → exon shuffling, but is also influenced by various → post-translational modifications.

Proteomic fingerprint: A pattern of proteins that is diagnostic for a specific state of a cell, a tissue, an organ, or organism. First, a global pattern of proteins (comprising potentially all proteins in a given cell type at a given time, usually some 10–30,000 different molecules, but not simply their presence or absence, yet their relative amounts in relation to each other) is established, then compared to other cells of the same type, and characteristic proteins selected. These are then taken as a reference proteomic fingerprint, with which other fingerprints from other cellular states are compared to detect differentially expressed proteins. Compare → peptide fingerprint.

Proteomics (*prote*in-gen*omics*, proteinomics): The whole repertoire of techniques to analyze and characterize the → proteome of an organelle or a cell, including protein isolation, fractionation, separation, post-separation analysis (e.g. identification of different proteins by → microsequencing), analysis of post-translational modification by → MALDI mass spectometry (e.g. acetylation, ADP-ribosylation, C-terminal truncation, glycosylation, methylation, N-terminal truncation, phosphorylation, sulphation), bioinformatics (storage and analysis of the resulting informations) and robotics (automation for high-throughput analysis). See → functional genomics, → functional proteomics, → genomics, → industrial proteomics, → *in silico* proteomics, → one cell proteomics, → phosphoproteomics, → recognomics, → subcellar proteomics, → 3D proteomics, → tissue proteomics, → topological proteomics.

Proteosome: The complete set of proteins expressed by a particular → genome in a cell. See → proteome.

Proteozyme: Any protein that possesses an active center and consequently catalytic (enzymatic) properties. Compare → ribozyme.

Protochromosome: An ancestral → chromosome, from which homologous chromosomes of one species or homoeologous chromosomes of related species are derived by various → deletions, → insertions, → translocations, generally genomic → rearrangements.

Protoclone: Any progeny derived from a single → protoplast.

Proto-exon shuffling: The hypothetical combination of precursors of → exons (proto-exons) to functional genes.

Proto-microsatellite: A sequence element of eukaryotes (e.g. diptera), from which → microsatellites are generated by as yet not fully understood processes. For example, the socalled → mini-me elements of *Drosophila melanogaster* contain a $(TA)_n$ repeat 5', and a $(GTCY)_n$ repeat 3' of the highly conserved 33 bp core motif. Frequently the 3' repeat is absent, and instead a cryptically simple sequence (consisting of only C, G and T) can be found (see → cryptic simplicity). Both loci are considered as proto-microsatellites, since they can expand to new $(TA)_n$ and $(GTCY)_n$ repeats, respectively, once inserted at new genomic locations by a possible retroposition of the mini-me element (catalyzed by genes *in trans*).

Proto-oncogene: See → cellular oncogene.

Protoplast: A bacterial, yeast or plant cell from which the cell wall has been removed experimentally (e.g. bacterial walls can be digested with → lysozyme, plant cell walls with a mixture of cellulase, pectinase and polygalacturonase). See → protoplast fusion, → protoplast transformation.

Protoplast fusion: The combination of two related or unrelated → protoplasts, or of an enucleated protoplast and a → karyoplast to form a hybrid cell by either chemical (with → polyethylene glycol or Ca^{2+}) or electrical treatment (electrofusion). Compare → cell fusion.

Protoplast transformation: The integration of foreign DNA into plant DNA using plant cells without cell walls (→ protoplasts). Originally such protoplasts were cocultivated with virulent cells of → *Agrobacterium tumefaciens*. During cocultivation the → T-region from the *Agrobacterium* → Ti-plasmid was transferred to the protoplast genome. After the procedure, the bacteria were killed by an → antibiotic (e.g. carbenicillin, Claforan), and the transformed protoplasts selected by their ability to grow without added growth hormones. Today, the term protoplast transformation is also used when protoplasts are incubated with purified Ti-plasmids or any other DNA. The transformation frequency of protoplasts from especially competent plants such as tobacco is in the range of maximally 10%.

Prototrophy: The ability of a wild-type organism to grow on a → minimal medium.

Protruding end: See → protruding terminus.

Protruding terminus (protruding end, overhanging end, overhang, extension): The end(s) of a DNA duplex molecule where one strand is longer ("protruding") than the other which is referred to as "recessed". See also → cohesive end, → recessed 5'-terminus, → recessed 3'-terminus.

Proviral insertion site (PIS): Any site in a host genome, into which a viral (also retroviral) DNA is inserted.

Provirus: Any viral DNA that is an integral part of the host cell chromosome and as such is transmitted from one cell generation to another without → lysis of the host, for example an integrated → retrovirus. See also → prophage.

Proximal: Located close to any fixed point. See → distal.

Proximal promoter: The DNA region in a → promoter that extends 250 bp → upstream of the → transcription start site, i.e. from to +1 to −250. This proximal promoter sequence engulfes → core promoter, and carries most of the → transcription factor binding motifs. See → distal promoter.

Proximal promoter array: Any → microarray, onto which → oligodeoxynucleotides spanning the → proximal promoter region (usually from −800 to +200) of many different genes are spotted. Such microarrays are used to identify → transcription factor binding sites.

Proximal sequence element (PSE): A sequence motif at position −40 and −79 of the 5'-flanking region of the U6 → small nuclear RNA (U6snRNA) gene that is necessary and sufficient to direct precise U6 transcription *in vitro* and *in vivo* (e.g. in *Xenopus laevis* oocytes). See → distal sequence element.

Proximity ligation assay (PLA): A technique for the detection of proteins in the low femtomole range in complex biological samples that is based on socalled proximity probes (e.g. → antibodies) containing → oligonucleotide extensions, designed to bind pairwise to a target protein and to form amplifiable tag sequences when in close proximity to each other. In short, each of a pair of such proximity probes (e.g. → polyclonal antibodies, or matched pairs of → monoclonal antibodies, also called affinity probes) is first coupled to a specific → oligonucleotide sequence using e.g. the bifunctional crosslinker succinimidyl 4-(p-maleimidophenyl)butyrate. Then the → probes are bound to the target protein such that the oligonucleotide extensions come close to each other ("into close proximity") and can be ligated together with the support of a templating oligonucleotide ("connector oligonucleotide") that hybridizes to the ends of nearby strands and guides the ligation process. As a result of → ligation, a new DNA sequence is generated that can be amplified by conventional → polymerase chain reaction and analyzed by → sequencing. The sequence of this diagnostic oligonucleotide identifies the antibodies, and by inference, the protein. PLA is also used for the detection of protein-DNA interactions. See → triple-binder proximity ligation assay.

PR protein: See → pathogenesis-related protein.

Ps: See → pseudouridine.

pSC 101: A small non-conjugative → plasmid carrying a → tetracycline resistance gene that was used for the construction of the universal → cloning vector → pBR 322. Used as a → low copy number plasmid vector.

Pseudocomplementary peptide nucleic acid (pseudocomplementary PNA): Any → peptide nucleic acid that contains modified nucleobases, as e.g. 2,6-diaminopurine·2-thiothymine or thiouracil, where the former substitutes for adenine, the latter for thymine. Such pseudocomplementary bases recognize their natural A·T and G·C counterparts, but cannot recognize each other. Therefore, pseudocomplementary PNAs are used for → double duplex invasion techniques, where any binding of the two PNA strands to each other would prevent invasion of the DNA target.

Pseudoexon: Any exon-like sequence in eukaryotic → split genes, flanked by → pseudosplice sites that is ignored by the → spliceosome. Pseudoexons are probably not used, because no functional → splicing enhancer sequences are present in the host gene.

Pseudogene ("silent gene", truncated gene, "dead gene", ψ): A non-functional derivative of a functional eukaryotic gene that suffered → rearrangements and → mutations preventing normal expression (e.g. lacks → introns and → promoter regions or contains one or more → stop codons). Since pseudogenes are not under selective pressure, they frequently degenerate more rapidly than their functional counterparts. Pseudogenes are thought to represent the DNA copies of mRNA, because they usually carry a poly(dA) sequence at the 3' end. Some pseudogenes may also have arisen from gene duplication and concomitant → deletion of the promoter region or parts of it. Such truncated genes may be present in a particular genome in appreciable numbers (e.g. the human → high mobility group HMG-17 protein → multigene family contains about 30, the actin gene family about 20, and the glyceraldehyde-3-phosphate dehydrogenase gene family about 25 retropseudogenes). Pseudogenes are often located in → introns and → intergenic regions. At least some pseudogenes are also transcribed (in some cases up to 20% of all pseudogenes in a genome), and some pseudogenes possess important function(s). See → gene-pseudogene chimeric transcript, → processed pseudogene.

Pseudogene messenger RNA (pseudogene mRNA, pseudogene transcript): Any → messenger RNA (mRNA) that is transcribed from a → pseudogene under the control of a nearby or adjacent → promoter. Usually such mRNAs represent read-throughs from promoters of adjacent genes.

Pseudogene transcript: See → pseudogene messenger RNA.

Pseudogenization: The process of transformation of a gene into a → pseudogene.

Pseudogenome: An infelicitous and misleading term for the complete set of → pseudogenes in a genome.

Pseudo-intron: Any sequence in a → transcript that is normally spliced out (i.e. is treated as → intron), but sometimes remains in the mature → messenger RNA (i.e. is not spliced).

Pseudoknot: A helical complex formed between a single-stranded loop and another single-stranded region of the same or another DNA molecule.

Pseudomolecule: An at best infelicitous term for a → bacterial artificial chromosome → tiling path stretching from the → centromere of a chromosome (e.g. human chromosome) to the → telomere, thus comprising one chromosome arm only.

Pseudopromoter: A DNA sequence element that allows → *in vitro* transcription of linked genes, but does not function *in vivo*.

Pseudo resistance gene analogue (pseudo RGA): Any non-functional (e.g. promoter-less) → resistance gene analogue. See → pseudogene.

Pseudosplice site (pseudosplice junction, pseudosite): Any → splice junction that matches the → consensus sequence of a real splice junction, but is efficiently ignored by the → spliceosome. Pseudosplice sites are abundant in eukaryotic genes. For example, the 42kb human *hprt* gene contains eight real 5′-splice sites, but over 100 5′- and 683 3′-pseudosplice sites.

Pseudouridine **(Ps, 5-β-D-ribofuranosyl uracil, ψ):** One of the so-called → rare bases, unusual nucleotides found in some → transfer RNAs where the glycosidic bond is associated with position 5 of uracil. See for example → TcC loop.

p73: A tetrameric → transcription factor, structurally similar to its homolog → p53 (see → guardian-of-the-genome) or → p63, and encoded by the *p73* gene that binds to DNA recognition motifs of p53 and transcativates target genes of p53, thereby inducing cell cycle arrest and → apoptosis. The C-terminus of p73 is highly variable, and at least 9 different isoforms arise from → alternative splicing. The N-terminus of the protein contains a socalled *s*terile *a*lpha *m*otif (SAM) domain that is missing in p53. In addition to the full-length anti-cancer protein (TAp73), N-terminally truncated variants exist that lack a complete transactivation domain (ΔNp73). These variants arise by read-through from a → second promoter in the p73 gene or via alternative splicing. The ΔNp73variant forms complexes with Tap73 and p53, thereby preventing the → transcription of their target genes and blocking their protective effect. ΔNp73 also competes with Tap73 and p53 for binding motifs on the DNA and thereby inhibit the effect of these proteins. The truncated variant protein also accumulates in tumor cells as a consequence of a higher → expression rate. The p73 status of a tumor owns diagnostic and prognostic value.

ψ:

a) See → pseudogene.
b) See → pseudouridine.

PSI-BLAST: See → Position-Specific Iterative BLAST.

P-site (peptidyl-tRNA binding site): The site on the → ribosome to which the growing peptide chain is attached during protein synthesis.

p63: A homolog of p53 (see → guardian-of-the-genome) encoded by the *p63* gene that is involved in normal human development and whose mutation(s) inevitably lead to deformities, especially outer extremities (knock-out mice possess only a single-layer skin, and do not develop teeth, eyelids, mammary glands, salivary glands, and lachrymal glands). p63 consists of an N-terminal → transactivation domain, a central → DNA-*b*inding *d*omain (DBD), and a C-terminal oligomerization domain. About 65% of the DBD sequence is identical to p53. The C-terminal end of p63 occurs in three different variations and therefore three different proteins exist: p63γ with a 50 amino acid long terminus, p63α with a 245 amino acid end, which is structured in three different domains, of which one is the socalled SAM (*s*terile *a*lpha *m*otif) domain. The N-terminus of the protein also exists in variants: a complete, and a truncated transactivation domain. The combination of the different sequence variants produces six different forms of p63, of which those with a missing N-terminal transactivation domain naturally do not activate the cognate → promoter. Mutations in the human p63 gene, most of them in the DBD domain, lead to the socalled EEC syndrome (*e*ctrodactyly, *e*ctodermal dysplasia and *c*left lip with or without cleft palate).

PSMEA: See → *p*rimer-*s*pecific and *m*ispair *e*xtension *a*nalysis.

Psoralen: A photoreactive 6-hydroxy-5-benzofuranacrylic acid-δ-lactone that intercalates into the two → strands of → double-stranded nucleic acids (→ dsRNA and → dsDNA). Upon UV irradiation, psoralen reacts with the 5-6 position of the → pyrimidine bases (particularly → thymidine) and thereby forms interstrand cross-links. Photoinduced cross-linking can be targeted to a specific genomic sequence by attaching a psoralen to an oligonucleotide complementary to this sequence. Psoralen can be linked to the 5′ or/and 3′ terminus of an oligonucleotide via a hexamethylene arm. Psoralen-modified oligonucleotides are used in → antisense and triple-helix studies, and in → denaturing gradient gel electrophoresis, → temperature gradient gel electrophoresis, and → temporal temperature gradient gel electrophoresis as an alternative to a → GC clamp. See → triple helix-directed DNA cross-linking.

Psoralen

Psoralen-biotin labeling: A technique for the labeling of nucleic acids that is based on the covalent attachment of a → biotin moiety to → psoralen, the → intercalation of the psoralen adduct into the double-stranded nucleic acid, and the irradiation with light of a wavelength of 320–400 nm. Under these conditions, psoralen reacts via cycloaddition with → thymidine (in DNA) or → uridine (in RNA), and thereby labels the nucleic acid with biotin.

Psoralen footprinting: See → photo-footprinting, definition a.

Psoralen labeling: The introduction of a psoralen molecule (covalently bound to either → biotin or → fluorescein) into

DNA. Psoralen binds to pyrimidine bases (preferentially thymine or uracil, less to cytosine) via hydrophobic interaction(s), which is stabilized by UV crosslinking (i.e. simple irradiation with long wavelength [365 nm] UV light). After removal of unincorporated psoralen-biotin by n-butanol extraction, the probe can be used for DNA-DNA- or DNA-RNA-hybridization. Since psoralen is conjugated to e.g. biotin, the hybridization can be detected by → streptavidin-alkaline phosphatase conjugate (or, in case of fluorescein, with anti-fluorescein-alkaline phosphatase antibody conjugate). Psoralen labeling avoids the problems inherent in → random priming and → nick translation. See → triple helix-directed DNA cross-linking.

pSP 64: A derivative of a → pUC plasmid that contains the → RNA polymerase promoter of phage SP6 (see → SP6 vector).

PSSA: See → phosphorylation site-specific antibody.

pSUPER: See → suppression of endogenous RNA.

PSV: See → paralogous sequence variant.

Psychiatric genetics: A branch of genetics that focusses on the genes underlying psychiatric disorders (e.g. mental retardations, neurodegenerative diseases as Alzheimer disease, also depressions and psychoses, and addiction to the various drugs, to name few), and their inheritance.

PTA: See → protein truncation test.

PTC: See → premature termination codon.

Pteridine nucleoside: Any → nucleoside, in which the conventional → base is sub-

Pteridine nucleoside

stituted by a fluorogenic pteridine system or its derivatives. This pteridine is a structural analog of → guanosine or → adenosine, but in contrast to the conventional bases absorbs light around 340 nm (3-MI: 348 nm; 6-MI: 340 nm; 6-MAP: 330 nm; DMAP: 330 nm) and emits → fluorescence light (3-MI: 431 nm; 6-MI: 430 nm; 6-MAP: 435 nm; DMAP: 437 nm) with very high fluorescence quantum yields. Like native nucleosides, pteridine nucleosides incorporate into → oligonucleotides via a deoxyribose linkage, can be site-specifically inserted into DNA during automated DNA synthesis and are stable through the various de-blocking and purification steps.

PTGS:

a) See → co-suppression.
b) See → posttranscriptional gene silencing.

pTi: See → Ti-plasmid.

PTI: See → putative transcript isoform.

PTM:

a) See → post-transcriptional modification.
b) See → post-translational modification.

PTIS: See → portable translation initiation site.

PTS:

a) See → peroxin.
b) See → peroxysomal targeting signal.
c) See → protein target site.

PTT: See → protein truncation test.

Pu: Abbreviation for → purine.

PU (palindromic unit): See → repetitive extragenic palindromic element.

Public sequence databases: A series of databases that are collected and maintained by public institutions and whose data are fully available to the public (e.g. GenBank, the EMBL data library, and DDBJ. In contrast, the private databases are not available to the public (without licensing fees).

PubMed: A retrieval system containing abstracts, citations, and indexing terms for journal articles in the biomedical sciences, including literature citations supplied directly to NCBI by publishers as well as URLs to full text articles on the publishers' web sites. PubMed includes the complete contents of the MEDLINE and PREMEDLINE databases.

pUC (pUC vector): Any one of a series of relatively small, versatile *E. coli* → plasmid cloning vectors that contain the *Pvu*II/*Eco*RI fragment of → pBR 322 with the β-lactamase gene (coding for → ampicillin resistance), an → origin of replication, and a sequence coding for the α-peptide of the *lac Z* (b-galactosidase) gene with an inserted → polylinker. Insertion of foreign DNA at the polylinker leads to the interruption of the α-peptide gene, so that no functional protein can be synthesized. Consequently, the host cells produce colorless colonies, if grown on media with ampicillin and → X-gal. Strains that are transformed with non-recombinant vectors, develop blue colonies on the same medium. Thus the recombinants can easily be selected. pUC vectors are present in 500–700 copies per host cell.

Figure see page 1287

pUC vector: See → pUC.

Puff (chromosome puff): A local unwinding of → polytene chromosomes where the → chromatin is less condensed and the

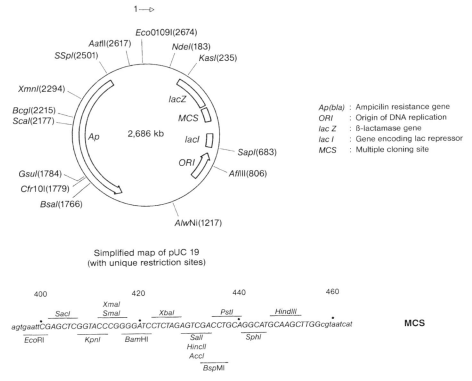

Simplified map of pUC 19
(with unique restriction sites)

pUC vector

genes are actively transcribed. Usually a single chromosome band is unwound, though two or more bands can be involved in puffing (Balbiani ring). The pattern of puffing in various animals, especially in the salivary gland cells of *Drosophila* larvae, is cell-specific, organ-specific, and developmentally regulated. It may also be influenced by a series of environmental factors (e.g. ions, or a heat shock).

Pull-down: A laboratory slang term for the isolation of specific compounds from complex mixtures by their high-affinity binding to immobilized capture molecules (that are bound to → magnetic beads or stationary phases of affinity columns). See → glutathione-S-transferase pull-down assay.

Pulse-chase analysis: An experiment designed to follow the course of degradation of a molecule within the living cell. First, cells (or cell extracts) are incubated with a radioactively labeled precursor compound *in vitro* for a short period of time ("pulse"), then a large excess of the same, unlabeled compound is added to dilute and to prevent further significant incorporation of radioactivity into potential metabolites. Samples are taken at various time intervals ("chase period") to estimate precursor metabolism.

Pulse proteolysis: A technique for the determination of protein stability. In short, a mixture of proteins, in which folded and unfolded proteins are in equilibrium, is exposed to a high dosage ("pulse") of a protease (e.g. thermolysin, 0.2 mg/ml final concentration) for a short time (e.g. one minute maximum) and rising concentrations of urea. The population of folded (i.e. more stable) proteins is then determined by measuring the amount of remaining protein after the pulse (f_{fold}). Pulse proteolysis is designed such that only the unfolded (i.e. partly or totally denatured) proteins are digested, and is used to characterize the stabilities of specific enzymes and their variants, to monitor the binding of ligands to proteins and to detect the effect of low molecular weight compounds on the stability of the cognate protein.

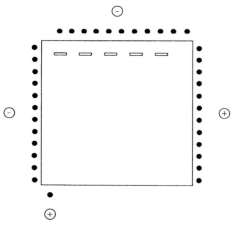

Electrode configuration of PFGE

Pulsed-field gel electrophoresis (PFGE): A technique for the electrophoretic separation of DNA molecules from the size of ordinary → restriction fragments (<10 kb) to intact chromosomal DNAs of up to 15 million base pairs (15 Mb). The DNA is subjected alternately to two electrical fields at different angles for specific time intervals called the pulse time (t). Under appropriate conditions reorientation of DNA segments with different size and topology is different, which – among other parameters – leads to the separation of these segments. For example, with each reorientation of the electric field relative to the gel, smaller DNA fragments move in the new direction much faster than the larger ones. Thus, the larger DNA molecules lag behind, resulting in their separation from the smaller DNAs. See → contour-clamped homogeneous electric field gel electrophoresis, → field inversion gel electrophoresis, → gel electrophoresis, → orthogonal-field alternation gel electrophoresis, → programmable autonomously controlled electrodes gel electrophoresis, → pulsed homogeneous orthogonal-field gel electrophoresis, → rotating gel electrophoresis, → secondary pulsed field gel electrophoresis, → transverse alternating field electrophoresis, → zero integrated field electrophoresis.

Pulsed homogeneous orthogonal-field gel electrophoresis (PHOGE): See → orthogonal-field alternation gel electrophoresis.

Pure culture: Any → cell culture that is made up of only one cell type, or one strain of cells.

Pure line: Any organism or population of organisms made homozygous through extended inbreeding.

pUR expression vector: Any one of a series of 5.2 kb → plasmid → expression vectors that is designed for the expression of *lac Z* → fusion genes in *E. coli*. Each pUR vector contains a → polylinker with → recognition sites for *Bam* HI, *Sal* I, *Pst* I, *Xba* I, *Hind* III and *Cla* I in all three → reading frames at the 3'-terminus of a *lac Z* gene. This gene is driven by the *lac* UV 5 promotor. Insertion of a cDNA sequence into the appropriate cloning site allows the expression of a → fusion protein consisting of → β-galactosidase and the peptide encoded by the cDNA.

Purifying selection: Any → selection against DNA sequence changes (→ mutations) that have a deleterious effect on the organism. Compare → adaptive evolution.

Purine (Pu): A heterocyclic molecule consisting of a pyrimidine and an imidazole ring. Purines are constituents of nucleic acids (→ DNA, → RNA). See → adenine, → guanosine; compare → pyrimidine.

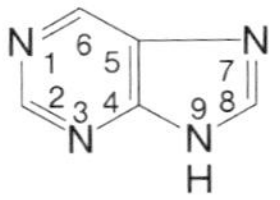

Puromycin (6-dimethyl-3'-deoxy-3'-p-methoxy-L-phenylalanylamino adenosine): A → nucleoside antibiotic from *Streptomyces alboniger* (may also be synthesized chemically) that is structurally similar to the 3'-terminal aminoacyl adenosine residue of a → tRNA, binds to the A site on the → ribosome, forms a → peptide bond with the growing peptide chain (see arrow), leaves the ribosome as peptidyl-puromycin and causes termination of elongation.

Puromycin-mediated fusion: A technique for the direct covalent coupling of a distinct → messenger RNA to its encoded peptide or protein that is based on the *in vitro* synthesis of a chimeric DNA linker-puromycin-oligomer, where → puromycin is covalently bound via its 5'-hydroxyl group to the 3'-end of the → linker. Then complex messenger RNA (mRNA) libraries are coupled to the DNA linkers of these chimeric oligomers by either enzymatic or photochemical techniques such that mRNA-DNA linker-puromycin complexes are formed that may serve as template for → translation ("mRNA display template"). If such construct libraries are used as templates for an → *in vitro* translation, the → ribosome pauses at the border between mRNA and DNA linker, so that

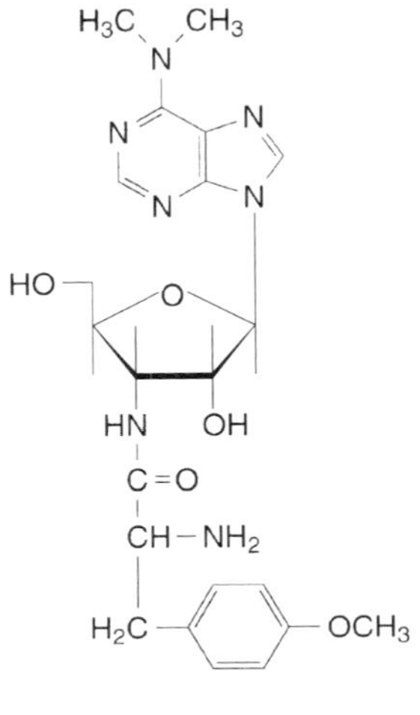

Puromycin

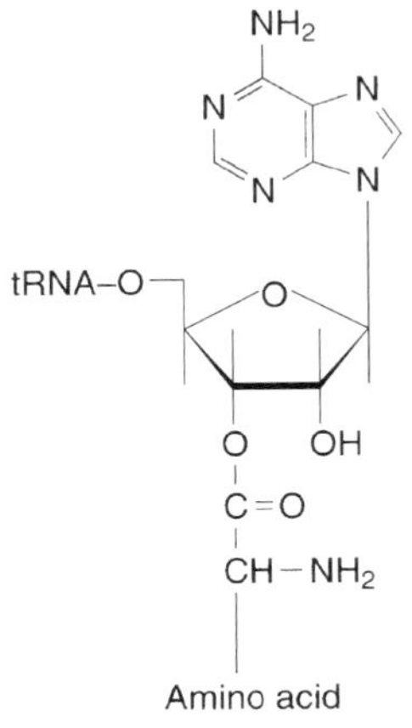

3'-Terminus of a tRNA

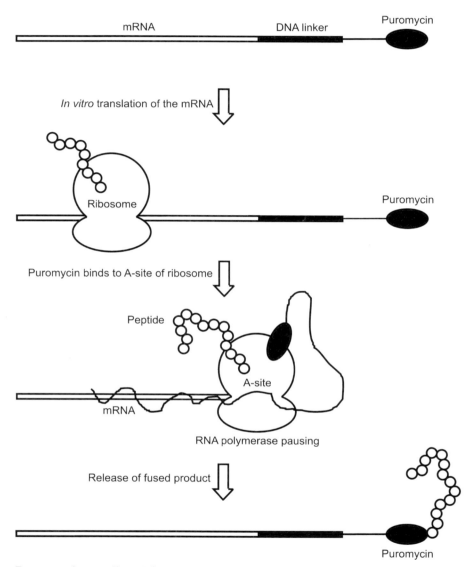

Puromycin-mediated fusion

the puromycin can occupy the A site of the ribosome, and is covalently bound to the peptide (or protein), catalyzed by the peptidyltransferase activity of the ribosome. As a result, a stable, covalently linked fusion complex between the mRNA, the DNA linker, puromycin, and the protein is produced. This complex represents a stable covalent junction between → genotype and → phenotype.

Push column: A device for the separation of radioactively labeled DNA or RNA probes from unincorporated nucleotides after → nick-translation or → random priming procedures. The separation is accomplished by the application of pressure to the upper chamber of the push column which in its simplest form is a modified syringe, and is complete after minutes. The high-molecular weight probes elute first from the push column.

Putative alien gene (pA gene): Any gene in a → genome whose → codon usage is different from an average gene in this genome. The difference should (1) exceed a high threshold, i.e. be pronounced, and (2) also be characteristic for genes encoding ribosomal proteins, → chaperones and proteins functional in the protein synthesis machinery. For example, the cagA domain of *Helicobacter pylori* strongly differs from the rest of the genome in its → codon bias.

Putative gene: Any → genic sequence that has homology to proven → genes, whose sequences are deposited in relevant databanks (e.g. GenBank). Compare → predicted gene. See → putative protein.

Putative protein ("probable protein"): Any protein, whose amino acid sequence has only limited similarity to already characterized proteins (as e.g. checked by sequence comparison via SWISS-PROT database).

Putative transcript isoform (PTI): Anyone of several possible combinations of alternative → exons of a particular gene (or → open reading frame, or → transcription unit). Such PTIs remain purely hypothetical, unless they are isolated from an organism.

P-value: The probability in a → BLAST search to obtain, by chance, a pair-wise sequence comparison of the observed similarity, given the length of the query sequence and the size of the searched database. Low P-values symbolize sequence similarities of high significance. Compare → E value.

Pwo DNA polymerase: See → *Pyrococcus woesii* DNA polymerase.

Py: See → *py*rimidine.

pYAC: See → *y*east *a*rtificial *c*hromosome.

pYC: See → *y*east *c*entromere *p*lasmid.

pYE: See → *y*east *e*pisomal *p*lasmid.

pYH: See → *y*east *h*ybrid *p*lasmid.

pYI: See → *y*east *i*ntegrative *p*lasmid.

Pyknon (Greek pyknos = dense, serried): Any non-random pattern of short repeated elements first detected in the human genome, where each element consists of at least 16 bases in length and occurs at least 40 times, with an average spacing of 18 to 22 nucleotides between copies. Nearly all pyknons overlap with repetitive elements, and more frequently reside in the → 3'-*u*ntranslated *r*egion (3'-UTR) of genes than in other regions of the human genome (as e.g. in genes or → 5'-UTRs). In non-genic regions of the human genome, clusters of about 40 such pyknons can be found, whose function is obscure. Many pyknons are species-specific. In humans, pyknons appear in more than 20,000 unrelated genes and in more than 30,000 transcripts. Pyknon-containing transcripts and → antisense transcripts form → double-stranded RNAs that induce → RNA interference.

pYL: See → *y*east *l*inear *p*lasmid.

pYP: See → *y*east *p*romoter *p*lasmid.

pYR: See → *y*east *r*eplicative *p*lasmid.

Pyranosyl-RNA (p-RNA; ribopyranosyl RNA): A synthetic ribopyranosyl isomer of RNA, a β-D-ribopyranosyl-(4'→2') oligonucleotide, in which the ribose units exist in the pyranose form, and neighbouring ribopyranosyl units are connected by phosphodiester linkages between the positions C (4') and C (2') instead of the 5'→3' phosphodiester bonds in RNA. Double strands of p-RNAs adopt a rigid linear structure and are held together by Watson-Crick purin-pyrimidine and purin-purine base pairing, where e.g. adenine-uracil pairing is stronger than in the corresponding RNA duplexes. Base pairing in such duplexes is also more selective than in RNA, due to the conformational rigidity of the p-RNA pyranosyl ring relative to that of the furanosyl ring of the RNA. p-RNA folds into → hair-pin structures and is able to replicate by non-enzymatic template-directed ligation of 2', 3'-cyclophosphates of short oligomers (e.g. tetramers).

Pyrimidine (Py): A heterocyclic 1,3-diazine ring. Pyrimidines are constituents of nucleic acids (→ DNA, → RNA). See → thymine, → cytosine, → uracil; compare → purine. See also → pyrimidine dimer.

Pyrimidine

Pyrimidine dimer: A structure formed by UV irradiation of DNA in which two → thymidine (or → cytidine) residues or one thymine and one cytosine residue at adjacent positions in the same DNA strand become covalently linked to each other. Such dimers block DNA → transcription and → replication. See → thymine dimer.

Pyrimidine-pyrimidone(6-4)photoproduct (TC pyrimidine-pyrimidone(6-4) photoproduct): A DNA photoproduct, resulting from the UV-induced opening of the double bond of the 5' pyrimidine and its reaction across the exocyclic group of the 3' pyrimidine in adjacent pyrimidines. In the (6-4) photoproduct, a rotation of the 3' base by 90° makes it resemble an → abasic site. Pyrimidine-pyrimidone (6-4) photoproducts cause mutations of the C→T or CC→TT type. If these mutations occur in e.g. one of the tumor suppressor genes, for example *Trp53*, they may lead to cancerous proliferation of the afflicted cells.

Pyrithiamine: A thiamine analogue that is effective against a series of *Aspergillus* species (e.g. *A. fumigatus*, *A. niger*, *A. oryzae*, *A. terris*) and serves as → selectable marker for transformed strains.

Pyrococcus abyssi (Pab or Isis) DNA polymerase: A thermostable → DNA polymerase from the hyperthermophilic archaebacterium *Pyrococcus abyssi* that operates at an optimal temperature of 70–80°C for → extension, produces errors at a rate of 2×10^{-6} to 6.6×10^{-7}, generates blunt ends (i.e. does not produce → overhangs) and has a half-life time of 5 hours at 100°C.

Pyrococcus furiosus DNA polymerase (Pfu DNA polymerase; EC 2.7.7.7): A highly thermostable and hyperactive monomeric → DNA polymerase with a 5' → 3' polymerase and a 3' → 5' → proof-reading exonuclease activity, isolated from the hyperthermophilic marine archaebacterium *Pyrococcus furiosus* (*Pfu*) that has originally been detected in geothermal vents off the coast of Vulcano (Italy). This enzyme is used for the → polymerase chain reaction and is considered superior to the convential → *Thermus aquaticus* DNA polymerase, since its 3'→5' proof-reading capacity will excise mismatched 3'-terminal nucleotides from primer-template complexes and incorporate the correct, complementary nucleotides, whereas *Taq* polymerase lacks this activity. Compare → *Pyrococcus* species DNA polymerase, → *Pyrococcus woesii* DNA polymerase.

Pyrococcus species DNA polymerase (Psp DNA polymerase; EC 2.7.7.7): A highly processive (see → processivity) and thermostable → DNA polymerase with a 3' → 5' exonuclease (proof-reading) activity, isolated from the thermophilic archaebacterium *Pyrococcus* species. The enzyme does not possess any 5' → 3'exonuclease activity and is used for → polymerase chain reaction experiments. Compare → *Pyrococcus furiosus* DNA polymerase, → *Pyrococcus woesii* DNA polymerase

Pyrococcus woesii DNA polymerase (Pwo DNA polymerase): A → DNA polymerase of the thermophilic bacterium *Pyrococcus woesii* (*Pwo*) with high thermostability and 3'-5'-exonuclease ("proofreading") activity, but no terminal transferase (→ extendase)

activity. Therefore *Pwo* DNA polymerase generates blunt-ended amplification fragments that facilitate their → cloning and → sequencing. In concert with → *Taq* DNA polymerase, the *Pwo* enzyme complements the high → processivity with proofreading, so that in this mixed system ("long template PCR system") fragments of up to 27 kb in size are synthesized on genomic DNA as → template. Compare → *Pyrococcus furiosus* DNA polymerase, → *Pyrococcus* species DNA polymerase.

Pyrogram: The graphical depiction of a → pyrosequencing procedure as a series of spikes, each of which represents a light pulse generated by the oxidative decarboxylation of → luciferin and monitored with a sensitive luminometer or a charge-coupled device (CCD) camera.

PyroMethA: See → pyrosequencing methylation analysis.

Pyrophosphat*ase* (PPase; inorganic pyrophosphatase): An enzyme that catalyzes the hydrolysis of pyrophosphate into two molecules of orthophosphate. The enzyme is used for → DNA sequencing (especially when selective band weakening occurs).

Pyrophosphorolysis-*activated* polymerization (PAP): A technique for the detection of large heterozygous chromosomal → deletions and gene → duplications that employs 30 nucleotides long → oligonucleotides blocked at their 3′-end by a → *di*deoxy*n*ucleotide (ddNTP, mostly ddCMP) not extendable by → DNA polymerase. When such blocked oligonucleotides specifically and completely anneal to a complementary target → template, pyrophosphorolysis removes the blocking ddNTP in the presence of pyrophosphate (PPi). After removal of the blocking ddNTP, DNA polymerase can now extend the activated oligonucleotide. The amplification products are then electrophoresed through → denaturing polyacrylamide gels. See → multiplex dosage pyrophosphorolysis-activated polymerization.

Pyrosequencing (real-time pyrophosphate detection; PP$_i$-based sequencing; minisequencing): A technique to determine the sequence of bases in DNA that avoids → sequencing gel electrophoresis and radioactivity or fluorescence, needed in other sequencing procedures (e.g. → Sanger sequencing). Instead, pyrosequencing quantitatively measures the pyrophosphate (PP$_i$) released during the DNA polymerase reaction by coupling it to the generation of light by firefly → luciferase. In short, the single-stranded → template DNA is first annealed to a short → sequencing primer. Then DNA polymerase together with an apyrase, ATP sulfurylase and firefly luciferase and only one dNTP (e.g. dGTP) are added. If dG does not form a base pair with the first free base on the template, it is rapidly removed by the apyrase (a mixture of nucleoside 5′-triphosphatase and nucleoside 5′-diphosphatase). Then the next base is added (e.g. dTTP). If a base pair T=A can be formed, the DNA polymerase extends the primer and releases PP$_i$ that is quantitatively converted to ATP by ATP sulfurylase. Now → luciferase utilizes this ATP to oxidatively decarboxylate → luciferin, and the light produced in this reaction is detected by a sensitive luminometer or a *c*harge-*c*oupled *d*evice (CCD) camera. The light pulse signaling incorporation of dT is shown in real time on a PC. Since apyrase removes the excess of dTTP, and luciferase utilizes the ATP, light emission is transient. Then e.g. dATP is added, which either forms a base pair on the template DNA (initiating a new reaction), or is

(1) Template DNA + primer $\rightleftharpoons$ [template-primer]

(2) [Template-primer] + dNTPs $\xrightarrow{\text{DNA polymerase}}$ template + extended primer + dNMP + PP$_i$

(3) PPi + APS $\xrightarrow{\text{ATP sulfurylase}}$ ATP

(4) ATP + luciferin + O$_2$ $\xrightarrow{\text{luciferase}}$ oxiluciferin + AMP + PP$_i$ + CO$_2$ + h·ν

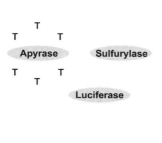
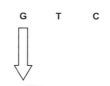
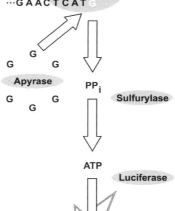
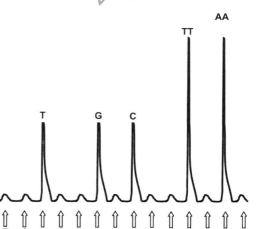

Pyrosequencing

rapidly removed by apyrase. By repeatedly adding the deoxynucleotides (here in the series G, T, A and C) and monitoring the light pulses, the sequence of the template DNA can be derived. One base is recorded every minute, and 96 samples can be read simultaneously in an automated pyrosequencer. Pyrosequencing is only suitable for the diagnostic sequencing of relatively short DNA fragments (up to 200 bases).

Pyrosequencing methylation analysis (PyroMethA): A technique for the quantification of methylated → cytosines at CpG sites in a distinct genomic region that combines the → combined bisulfite restriction analysis (COBRA) and → pyrosequencing. In short, the genomic target fragment is first amplified by conventional → polymerase chain reaction (PCR), using sequence-specific up-and down-stream flanking → primers. Preferentially PCR fragments of 100–150 bp are employed for PyroMethA. Then bisulfite treatment converts unmethylated cytosines to uracil, leaving methylated cytosines unaffected. In essence, this process results in a chemically induced methylation-dependent C→T transition site that is subsequently detected by pyrosequencing.

L-Pyrrolysine: Any lysine with its epsilon nitrogen in amide linkage with (4R, 5R)-4-substituted pyrroline-5-carboxylate. L-pyrrolysine is encoded by a UAG codon in some Archaeal genes (e.g. the methylamine methyltransferase [MtmB] gene[s] of *Methanosarcina barkeri*) and represents the 22nd genetically encoded amino acid in nature. See → selenocysteine.

pYX: See → yeast expression plasmid.

Q

Q: Abbreviation for base Q (queuosine), the nucleoside of queuine. See → rare bases.

Q-beta (Qβ): A small phage of *E. coli* (→ coliphage) with a single-stranded RNA genome (Qb plus-strand) of 4.2 kb, encoding a coat protein, a maturation protein, and an RNA-dependent → RNA polymerase (Qb replicase). The → plus strand is used directly as mRNA for the synthesis of these phage proteins. After infection of bacteria that contain an → F-factor, the Qb replicase begins to synthesize so-called → minus strands, using the plus strand of the phage as → template. The minus strands then serve as templates for the synthesis of Qb-RNA (the plus strand) which is packaged into phage heads, leading to the generation of fully infectious Qb phages.

Qβ: See → Q-beta.

Q-beta (Qβ) replicase amplification: The exponential multiplication of an RNA sequence inserted into the template RNA (Qb plus strand) for Q-beta replicase (Qb phage RNA-dependent RNA polymerase). Since the inserted RNA does not interfere with the function of Q-beta replicase, the whole recombinant molecule is amplified to about 10^9 copies. The extent of the amplification may be quantified by → ethidium bromide fluorescence. See → Q-beta.

Q-beta plus-strand: See → Q-beta.

QCM-D: See → quartz crystal microbalance with dissipation monitoring.

QCP: See → quantitative chromatin profiling.

QDFM: See → quantitative DNA fiber mapping.

Q-FISH: See → quantitative fluorescence *in situ* hybridisation.

Q gene: A gene (or genes) that confers the free-thrashing character of wheat (*Triticum vulgare* L.) and pleiotropically influences other → traits selected by domestication. The gene encodes a member of the APETALA2 (AP2) family of → transcription factors. Compare → QT gene.

Qiagen column: The trademark for a small disposable anion exchange column, used for the fast and simple isolation and purification of → nucleotides, → oligonucleotides and polynucleotides (especially → plasmids).

Q-PCR: See → quantitative polymerase chain reaction.

QRT-PCR: See → quantitative reverse transcriptase polymerase chain reaction.

QSAR: See → quantitative structure-activity relationship.

QT gene: Any gene that underlies a → quantitative trait locus (QTL). Usually, a

QTL is first genetically mapped with → molecular or → genic markers, the most tightly linked markers flanking the QTL then used to isolate the responsible gene (see → map-based cloning) and the gene functionally verified as responsible for the trait (or most of the trait) by e.g. → cDNA complementation, or → gene knockout techniques, or → small interfering RNA (siRNA) → gene knock-down. Compare → Q gene.

QTL: See → quantitative trait locus.

Quadrome: A hybrid → hybridoma that is generated by the fusion of two hybridomas.

Quadruplet codon: Any four-base → codon (instead of the canonical three-base codon, → triplet codon), whose information content can be incorporated into a protein, provided a → frameshift suppressor → transfer RNA (FS-tRNA), as e.g. a chemically acylated yeast tRNA[Phe] with an expanded → anticodon loop (one more nucleotide [eight] rather than the usual seven bases) can be synthesized that is no substrate for any endogenous synthetase and efficiently decodes the quadruplet codon (i.e. reads the four-base codon), and thereby serves to expand the genetic code in pro- and eukaryotes. Additionally, an aminoacyl-tRNA synthetase must be evolved that is highly selective for the new tRNA and aminoacylates it with the → unnatural amino acid (UAA) of interest, but cannot use endogenous amino acids. For example, an orthogonal synthetase/tRNA pair derived from archaeal tRNA[Lys] sequences efficiently and selectively incorporates an unnatural amino acid into protein in response to the quadruplet codon, AGGA. "Orthogonal" means that the tRNA cannot be charged by natural aminoacyl-tRNA synthetases, and the tRNA synthetase must not charge natural tRNAs.

For the tRNA to read the AGGA quadruplet codon, the sequence of the anticodon loop has to be changed to the complementary UCCU. In addition, the acceptor stem has to be mutated to allow the attachment of the unnatural amino acid homoglutamine. Ensuring that the ARS charged the mutated tRNA[UCCU] with homoglutamine required two changes. First, specific residues in the ARS active site were changed so it would bind both the tRNA[UCCU] acceptor stem and homoglutamine. Second, the part of the ARS that recognized the tRNA anticodon loop was deleted. This saved having to change to ARS to recognize a quadruplet codon on the tRNA. To expand the genetic code for specification of multiple non-natural amino acids, unique codons for these novel amino acids are needed.

Quantitative chromatin profiling (QCP): A technique for the identification and localization of cis-regulatory sequences (e.g. → promoters) and other functionally relevant sequence motifs (e.g. boundary elements, → enhancers, insulators, locus-control regions, or → silencers) that is based on the profiling of → DNAseI-hypersensitive sites across a genome or genomic regions in vivo. In short, intact nuclei are first isolated and divided into two samples. One sample is then DNAseI-treated (see → indirect end-label technique), the other one serves as untreated control. DNA is then purified from both samples separately, and tiles of contiguous (or minimally overlapping) ~250 bp amplicons spanning the → locus of interest are amplified by specially designed primers in a conventional → polymerase chain reaction (PCR). The relative number of intact copies of the genomic DNA corresponding to each amplicon is determined repeatedly by e.g. → real-time PCR in both samples. The resulting DNAseI-sensitivity ratios (copies in treated versus untreated samples) are then plotted on a genomic axis (e.g. against a region on a specific chromosome). Mean values for the replicates from each amplicon are computed to form a DNAseI-sensitivity ground level. Outliers exhibiting low variance from experiment to experiment are then identified, and represent potential hypersensitive amplicons that can be further analyzed.

Quantitative chromosome map (idiogram): The physical arrangement and copy number of → genes, generally DNA sequences, along a → chromosome, as measured by → in situ hybridization of fluorochrome-labeled → probes (representing e.g. genes) to chromosome spreads and quantification of the emitted fluorescence. The higher the copy number of the target sequence, the higher is the number of bound probe molecules, and the higher is the fluorescence light emission, which allows a semi-quantitative analysis of the gene content at the detected locus.

Quantitative DNA fiber mapping (QDFM): A technique for the construction of high resolution → physical maps at a resolution of few kilobases that relies on the hybridization of specific fluorescently labeled probes to target DNA immobilized on a specially prepared glass surface and stretched by hydrodynamic forces to produce linear templates ("fibers") of about 2–2.5 kb per µm. The hybridization events and the distribution of different probes along the stretched fibers can be detected by fluorescence and be imaged for further analysis. See → molecular combing.

Quantitative fluorescence in situ hybridisation (Q-FISH): A variant of the conventional → fluorescent in situ hybridisation technique for the quantitative estimation of → telomere lengths that employs → peptide nucleic acid (PNA) → probes labeled with → fluorochromes (e.g. → cyanin 3) rather than the DNA or RNA probes in the traditional methods. PNA possesses an uncharged backbone and can hybridise to the target DNA (of a chromosome) under extremely low ionic conditions, which inhibit target DNA → renaturation. Under these conditions, fluorescence intensity correlates linearly with the number of bound fluorophores, and therefore allows a quantitative estimate of telomeric repeat numbers.

Quantitative polymerase chain reaction (Q-PCR; kinetic PCR; real-time PCR; real-time detection PCR, RTD-PCR); TaqMan technique): The detection of the accumulation of amplification products during conventional → polymerase chain reactions and their quantification. Basically, the various techniques of Q-PCR fall into two broad categories. First, the intercalator-based methods include intercalating dyes (as e.g. → ethidium bromide) in each amplification reaction, irradiate the sample with UV-light in a specialized → thermocycler, and detect the resulting fluorescence light with a computer-controlled, cooled, charge-coupled device (CCD) camera. By plotting fluorescence increase versus cycle number, amplification plots are generated, allowing to quantify the products. However, this type of Q-PCR suffers from the disadvantage that both specific and non-specific products generate fluorescence signals, which makes quantitation obsolete. Second, the socalled → 5′ nuclease PCR and similar probe-based quantification protocols allow to

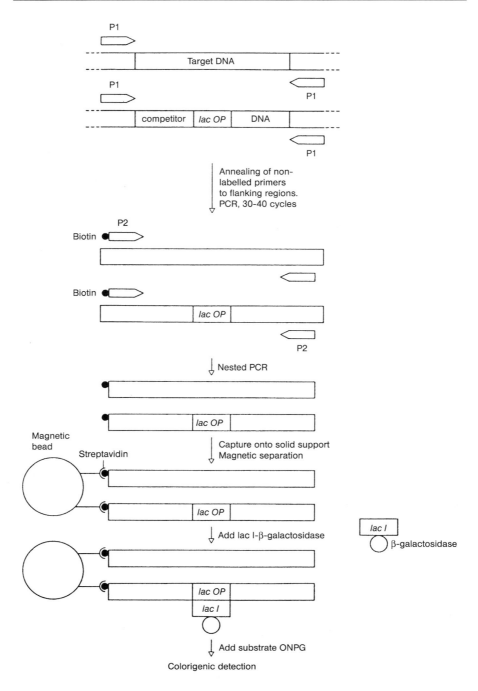

Quantitative polymerase chain reaction

detect only specific amplification products in real-time. The 5′ nuclease PCR assay exploits the 5′ nuclease activity of *Taq* DNA polymerase to cleave probe-target hybrids during amplification, when the enzyme extends from an → upstream primer into the region of the probe. This cleavage can be visualized by increased fluorescence, if the oligonucleotide probe contains both a reporter fluorochrome at its 5′ end and a quencher dye at its 3′ end. The close proximity of both fluorochromes (1.5–6.0 nm) results in a Förster type fluorescence energy transfer, leading to the suppression of the reporter ("quenching") which is relaxed when the probe is hydrolyzed.

Probe-based Q-PCR has been refined to be reproducible. First, the endpoint measurement of the amount of accumulated PCR products is skipped in favour of the more reliable threshold cycle (T_c), which is defined as the fractional cycle number at which the reporter fluorescence generated by cleavage of the probe passes a fixed threshold above base-line. T_c is inversely proportional to the number of target copies in the sample. The quantification is made by calculating the unknown target concentration relative to an absolute standard (e.g. a known copy number of plasmid DNAs, or a house-keeping gene as internal control). In contrast to the endpoint approach, T_c is measured when PCR amplification is still in the exponential phase (i.e. the amplicons accumulate at a constant rate, the amplification efficiency is not influenced by variations and limitations of the reaction components, and the enzymes and reactants are still stable). Also, primer-primer artifacts are low in number.

In another version of Q-PCR, the → competitive PCR, a synthetic DNA or RNA is used as internal standard (competitor amplicon) that contains the same primer binding sites and (optimally) has the same amplification efficiency as the target, but has a different size to discriminate it from the target. A known amount of this competitor is co-amplified with the target nucleic acid in the same tube. If the amplification efficiency of target and competitor is identical, then the ratio target / competitor will be constant throughout the PCR process. By determining the target / competitor ratio at the end of the process, and accounting for the starting amount of the spiked-in competitor, the initial amount of target can be calculated. As opposed to the superior real-time Q-PCR, competitive PCR is tedious, as it requires to find the most suitable ratio of target to competitor by dilution series, and moreover necessitates construction and characterization of a different competitor for every target to be quantified. Also, a series of experiments have to insure that the amplification efficiencies of target and competitor are in fact identical. See → anti-primer-based quantitative real-time polymerase chain reaction.

Quantitative proteomics: The whole repertoire of techniques to quantify protein content, masses of the various proteins, numbers of protein isoforms and the stochiometry of protein-protein interaction of a cell. See → proteomics.

Quantitative reverse transcriptase polymerase chain reaction (QRT-PCR): A variant of the → reverse transcription PCR (more precisely, the → relative quantitative reverse transcriptase polymerase chain reaction) that allows accurate quantitation of → messenger RNAs of a cell. This technique is based on the serial dilution of → competitor RNA or → mimic standards that are then added to a constant amount of sample cDNA. Quantification is achieved by comparison of the relative fluorescence intensities of target and competitor bands after their amplification in conventional → polymerase chain reaction and their electrophoretic separation in an → agarose or → polyacrylamide gel. See → polymerase chain reaction-aided transcript titration assay.

Quantitative structure-activity relationship (QSAR): A computer-assisted process for the prediction of biological activity of a compound or a class of compounds, which is based on physicochemical parameters of the compound such as geometry, energies, electronic and spectroscopic attributes and volume, and its biological or chemical activity. QSAR is able to predict the putative biological activity (or activities) of a new chemical compound, and is also competent to e.g. eliminate biologically incompetent candidates from a pool of potential drugs.

Quantitative trait locus (QTL): A genomic region with several genes, or two or more separate genetic → loci that contribute cooperatively to the establishment of a specific phenotype. See → cis QTL, → trans QTL.

Quantitative trait locus mapping (QTL mapping): A procedure to localize a → quantitative trait locus (or loci) on a → genetic or also a → physical map.

Quantitative trait modifying factor (QTMF): Any protein(s) encoded by gene(s) mapping ouside of a distinct → quantitative trait locus (QTL) that positively or negatively affect the → expression of the genes underlying the QTL through epistasis.

Quantum dot (QD, qdot): A 1–5 nm nonionic semiconductor nanocrystal (NC) with a cadmium-selenium (CdSe) or also cadmium-tellurium (CdTe) core and a zinc sulfide cover that can be excited by UV light and emits fluorescence light over long periods of time (depending on the crystal sizes, over several months) without the photobleaching characteristic for normal → fluorochromes. QDs are chemically synthesized at high temperatures, and normally possess a hydrophobic coat that can be replaced by polyacrylate (which allows e.g. coupling to antibodies) or modified by negatively charged dihydroliponic acid (DHLA). DHLA allows electrostatic binding of specific antibodies, or other proteins. Also, the pores of polystyrene microbeads can be loaded with QDs of different size and emission spectra, by coupling them to short oligonucleotide sequences. These can be hybridised to specific target sequences, and the hybridisation events detected by long-term fluorescence light of different wave-lengths. The surface of qdots can also be functionalized by primary and secondary antibodies, receptor ligands, recognition peptides ("peptide-qdot") and affinity pairs such as biotin-avidin. In QDs, electrons exist at discrete energy levels ("bands"). Any energy input (e.g. as photons) raises an electron from a lower ("valence") to a higher band ("conduction band"). Upon return to the lower band, the excess energy is released as a photon with an energy roughly equalling the gap between the bands (see → photoluminescence). This band gap increases with decreasing size of the dots. Therefore smaller dots release more energy (i.e. emit blue light). QDs fluoresce up to 100 times longer than conventional fluorochromes (increased photostability), are relatively biocompatible (although they contain a toxic cadmium core), and are used for the sensitive non-isotopic detection of biomolecules and ligand-receptor interactions, for time-resolved single molecule analysis, and visualization of molecular transportation in living cells. Single qdots can be tracked over a time period of several hours with confocal microscopy, basic wide-field epifluorescence microscopy, or → total internal reflection microscopy. See → colloidal quantum dot, → core shell quantum dot, → giant quantum dots.

Quantum dot nanocrystal (QD nanocrystal): A nanometer scale atom cluster with a few hundred to a few thousand atoms of cadmium-selenium or cadmium-tellurium mixed crystals that is coated with a semiconductor zinc sulfide shell, and absorbs light of a specific wavelength in near-UV, but emits fluorescence of different wavelengths depending on the size of the nanocrystal. The larger the diameter of the nanocrystal, the more red-shifted is the emitted light, and vice versa. Quantum dot nanocrystals are used for multicolor immunofluorescence detection of different proteins, employing primary antibodies raised against the target proteins, and secondary antibodies directed against the primary antibodies that are covalently linked to quantum dots. In contrast to natural or artificial → fluorochromes, quantum dot nanocrystals provide bright and photostable fluorescence.

Quantum efficiency: See → quantum yield.

Quantum yield (QY, quantum efficiency, f): The number of fluorescence photons emitted (N_{fl}) over the number of photons absorbed (N_{abs}) by a → fluorochrome. QY is a characteristic feature of fluorochromes and is used to describe their quality.

Quartz crystal microbalance with dissipation monitoring (QCM-D): A technique for the real-time detection of interactions of molecules with a surface or with each other, the determination of the mass of extremely thin surface-bound layers and their viscoelastic (structural) properties. A QCM-D sensor consists of a thin plate of crystalline quartz sandwiched between two electrodes. If an AC voltage is applied over these electrodes, an oscillation is induced in the sensor. This oscillation exponentially decays after the AC voltage is switched off. This decay is recorded and the resonance frequency (f) and the dissipation (D) calculated. Since the resonance frequency of the sensor crystal depends on the total oscillating mass, it is used to determine the mass of the molecules deposited on the sensor surface. Also, the thickness can be calculated by dividing the mass by the density of the layer. The dissipation energy (energy dissipated per oscillation over 2p times total energy stored in the system), which represents the frictional losses in the surface film, informs about the structure of this film attached to the sensor surface (which can be composed of metals, polymers such as proteins, but also living bacterial cells). For example, a compact globular protein attached to the surface results in a low dissipation, whereas an elongated protein with many coupled water molecules increases dissipation. An adapted software calculates correct thickness, viscosity and elasticity from the resulting data. QCM-D is used to measure protein adsorption to and their desorption from membranes, and protein-protein interactions.

Quaternary structure: The specific three-dimensional arrangement of two (or more) identical, similar or different molecules (e.g. the individual proteins in a multicomponent enzyme, the substrate and its cognate enzyme, protein-RNA complexes or RNA-RNA interactions with the formation of → helices). See → primary structure, → secondary structure, → tertiary structure.

Quelling: See → co-suppression.

Quencher: Any molecule that reduces or completely deactivates ("quenches") an excited state of another molecule (e.g. a → fluorochrome), either by energy transfer, electron transfer, or by a chemical reaction. For example, → DABCYL and → TAMRA, the socalled → black hole quenchers (BHQs) and the QSY series are such quenchers. See → quenching.

Quencher extension (QEXT): A single-step closed-tube real-time technique for the detection and quantification of specific → single nucleotide polymorphisms (SNPs) in pooled samples (i.e. containing many different → genomic DNAs). In short, the target DNA (containing the SNP or SNPs) is first amplified in a conventional → polymerase chain reaction (PCR), and then a → probe complementary to the DNA flanking the SNP(s) and coupled to a 5'-reporter fluorochrome (e.g. → FAM) is hybridized to it. The reporter FAM emits fluorescent light after excitation by e.g. laser light. Subsequently, the 3'-end of the probe is extended by → DNA polymerase, which incorporats a → dideoxynucleotide containing a quencher fluorophore (e.g. → TAMRA), if the target SNP allele is present. After sequence-specific incorporation (=extension) of the quencher nucleotide, the emitted fluorescence of the reporter is quenched (reduced).

Quencher

Quenching: The reduction or complete deactivation of an excited state of a distinct molecule (e.g. a → fluorochrome) by a → quencher.

Query sequence: Any amino acid or nucleotide sequence that is used in a database search (e.g. for sequence → homology). See → BLAST.

Questionable open reading frame (questionable ORF): Any → open reading frame (ORF), defined by the presence of a → coding sequence flanked by a → start and → stop codon that is detectable in a → genome sequence, but neither is transcribed (i.e. has no → SAGE or → transposon tag, and is not detected on → expression microarrays), nor is conserved in evolution. See → disabled open reading frame, → essential open reading frame, → homology-based open reading frame, → known open reading frame, → short open reading frame, → transposon identified open reading frame.

QEXT: See → quencher extension.

Quick blot: See → quick blotting

Quick blotting (quick blot, fast blot): A technique to immobilize RNA and DNA from cellular extracts without extensive purification on → nitrocellulose filters, in which sodium iodide (NaI) instead of baking is used to fix the nucleic acids onto the support. In short, cellular extracts are deproteinized enzymatically, detergents and NaI are added, and the mixture is filtered through nitrocellulose. NaI promotes dissolution of proteins, causes selective binding of mRNA and DNA (depending on the temperature used), and preserves the biological activity of e.g. → messenger RNA (which still can be reverse-transcribed into → cDNA or translated into protein). Filters are washed and used directly for molecular → hybridization.

Quick-stop mutant: A → mutant of *E. coli* that rapidly stops its DNA synthesis after a temperature increase to 42 °C.

Quinacrine: A → fluorochrome that intercalates into DNA duplex molecules and allows the staining of chromosomes so that a typical pattern of fluorescing bands (Q bands) is generated. Q banding is exploited for the specification of chromosomes and the detection of gross → rearrangements, → deletions, and other abnormalities.

Quorum sensing (QS): A mechanism, by which Gram-negative bacteria that live in close association with plants, recognize their population density. QS is mediated by small signal molecules, mostly derivatives of N-*a*cyl-*h*omoserine*l*actone (AHL) that differ by the length of the N-acyl side chain and modifications at position C-3. QS systems consist of an AHL synthase (member of the Lux1 protein family) and an AHL receptor (member of the LuxR protein family). At low population densities the AHL synthase produces low AHL concentrations. With increasing population density, more and more AHL accumulates in the environment. Since AHL freely moves through bacterial membranes, the intracellular AHL concentration increases, and finally AHL binds to its receptor. As a consequence, the AHL-receptor complex in turn binds to socalled *lux* boxes in the → promoters of more than 20 different responsive genes. These genes are then activated, and the coordinated action of the encoded proteins finally leads to a specific phenotype (e.g. → virulence, → plasmid transfer, antibiotic synthesis, motility, biofilm production). For example, → *Agrobacterium tumefaciens* produces 3-oxo-C8-homoserine lactone as signal molecule, and its accumulation results in the → conjugative transfer of the socalled → Ti-plasmid.

Q vector: A bicistronic → self-inactivating retroviral vector for the transfer of target genes into animal cells that provides high virus titers, reliable expression levels of the target gene, and reduces → promoter interference. For example, a specialized Q vector contains 3' → *l*ong *t*erminal *r*epeats (LTRs) of the self-inactivating class; a 5' LTR that is a hybrid of the *c*yto*m*egalo*v*irus (CMV) enhancer and an engineered *m*ouse *s*arcoma *v*irus (MSV) → promoter; an internal CMV immediate early region promoter to drive the expression of the transduced gene and the → neomycin phosphotransferase (or any other) → selectable marker; an expanded → *m*ultiple *c*loning *s*ite (MCS) and a eukaryotic → *i*nternal *r*ibosome *e*ntry *s*ite (IRES). The vector backbone contains an SV40 origin to promote high copy number replication in packaging cell lines that express the SV40 large T antigen. The target gene is expressed together with the selectable marker, and after integration of the vector into the host cell genome the promoter in the 5' LTR is inactivated. Self-inactivation is a function of a deletion in the U3 region of the 3' LTR. After → reverse transcription of the plus strand of the retroviral genome, it is integrated. During integration, a circular intermediate is formed that causes the duplication of the deletion in the U3 region of the 3' LTR. This inactivates the CMV-MSV hybrid promoter in the 5' LTR. Therefore, transcription is only driven by the internal CMV promoter immediately upstream of the target gene.